CAMBRIDGE TRACTS IN MATHEMATICS

General Editors

B. BOLLOBÁS, W. FULTON, A. KATOK, F. KIRWAN,
P. SARNAK, B. SIMON, B. TOTARO

---

200   Singularities of the Minimal Model Program

CAMBRIDGE TRACTS IN MATHEMATICS

GENERAL EDITORS

B. BOLLOBÁS, W. FULTON, A. KATOK, F. KIRWAN, P. SARNAK,
B. SIMON, B. TOTARO

A complete list of books in the series can be found at www.cambridge.org/mathematics.
Recent titles include the following:

Singularities of the Minimal Model Program

JÁNOS KOLLÁR
Princeton University

with the collaboration of
SÁNDOR KOVÁCS
University of Washington

CAMBRIDGE
UNIVERSITY PRESS

CAMBRIDGE
UNIVERSITY PRESS

University Printing House, Cambridge CB2 8BS, United Kingdom

Published in the United States of America by Cambridge University Press, New York

Cambridge University Press is part of the University of Cambridge.

It furthers the University's mission by disseminating knowledge in the pursuit of education, learning and research at the highest international levels of excellence.

www.cambridge.org
Information on this title: www.cambridge.org/9781107035348

First published 2013

A catalogue record for this publication is available from the British Library

Library of Congress Cataloguing in Publication data
Kollár, János.
Singularities of the minimal model program / János Kollár, Princeton University ; with the collaboration of Sándor Kovács, University of Washington.
pages cm. – (Cambridge tracts in mathematics ; 200)
Includes bibliographical references and index.
ISBN 978-1-107-03534-8
1. Singularities (Mathematics)   2. Algebraic spaces.   I. Kovács, Sándor J. (Sándor József)
II. Title.
QA614.58.K685   2013
516.3'5 – dc23   2012043204

ISBN 978-1-107-03534-8 Hardback

Contents

Preface

In 1982 Shigefumi Mori outlined a plan – now called *Mori's program* or the *minimal model program* – whose aim is to investigate geometric and cohomological questions on algebraic varieties by constructing a birational model especially suited to the study of the particular question at hand.

The theory of minimal models of surfaces, developed by Castelnuovo and Enriques around 1900, is a special case of the 2-dimensional version of this plan. One reason that the higher dimensional theory took so long in coming is that, while the minimal model of a smooth surface is another smooth surface, a minimal model of a smooth higher dimensional variety is usually a *singular* variety. It took about a decade for algebraic geometers to understand the singularities that appear and their basic properties. Rather complete descriptions were developed in dimension 3 by Mori and Reid and some fundamental questions were solved in all dimensions.

While studying the compactification of the moduli space of smooth surfaces, Kollár and Shepherd-Barron were also led to the same classes of singularities.

At the same time, Demailly and Siu were exploring the role of singular metrics in complex differential geometry, and identified essentially the same types of singularities as the optimal setting.

The aim of this book is to give a detailed treatment of the singularities that appear in these theories.

We started writing this book in 1993, during the 3rd Salt Lake City summer school on Higher Dimensional Birational Geometry. The school was devoted to moduli problems, but it soon became clear that the existing literature did not adequately cover many properties of these singularities that are necessary for a good theory of moduli for varieties of general type. A few sections were written and have been in limited circulation, but the project ended up in limbo.

The main results on terminal, canonical and log terminal singularities were treated in Kollár and Mori (1998) and for many purposes of Mori's original program these are the important ones.

There have been attempts to revive the project, most notably an AIM conference in 2004, but real progress did not restart until 2008. At that time several long-standing problems were solved and it also became evident that for many problems, including the abundance conjecture, a detailed understanding of log canonical and semi-log canonical singularities and pairs is necessary. In retrospect we see that many of the necessary techniques have not been developed until recently, so the earlier efforts were rather premature.

Although the study of these singularities started only 30 years ago, the theory has already outgrown the confines of a single monograph. Thus many of the important developments could not be covered in detail. Our aim is to focus on the topics that are important for moduli theory. Many other areas are developing rapidly and deserve a treatment of their own.

Sections 6.1, 8.4, 8.5 and 10.6 were written by SK. Sections 2.5, 6.2 and the final editing were done collaboratively.

Acknowledgments Throughout the years many of our colleagues and students listened to our lectures or read early versions of the manuscript; we received especially useful comments from A. Chiecchio, L. Erickson, O. Fujino, K. Fujita, S. Grushevsky, C. Hacon, A.-S. Kaloghiros, D. Kim, M. Lieblich, W. Liu, Y.-H. Liu, S. Rollenske, B. Totaro, C. Xu, R. Zong and M. Zowislok.

Much of the basic theory we learned from Y. Kawamata, Y. Miyaoka, S. Mori, M. Reid and V. Shokurov.

Conversations with our colleagues D. Abramovich, V. Alexeev, F. Ambro, F. Bogomolov, S. Casalaina-Martin, H. Clemens, A. Corti, J.-P. Demailly, T. de Fernex, O. Fujino, Y. Gongyo, D. Greb, R. Guralnick, L. Ein, P. Hacking, C. Hacon, B. Hassett, S. Ishii, M. Kapovich, N. Katz, M. Kawakita, J. McKernan, R. Lazarsfeld, J. Lipman, S. Mukai, M. Mustaţă, M. Olsson, Zs. Patakfalvi, M. Popa, C. Raicu, K. Schwede, S. Sierra, Y.-T. Siu, C. Skinner, K. Smith, M. Temkin, Y. Tschinkel, K. Tucker, C. Voisin, J. Wahl and J. Włodarczyk helped to clarify many of the issues.

Partial financial support to JK was provided by the NSF under grant number DMS-07-58275. Partial financial support for SK was provided by the NSF under grant number DMS-0856185 and the Craig McKibben and Sarah Merner Endowed Professorship in Mathematics at the University of Washington. Portions of this book were written while the authors enjoyed the hospitality of RIMS, Kyoto University.

Introduction

In the last three decades Mori's program, the moduli theory of varieties and complex differential geometry have identified five large and important classes of singularities. These are the basic objects of this book.

Terminal. This is the smallest class needed for running Mori's program starting with smooth varieties. For surfaces, terminal equals smooth. These singularities have been fully classified in dimension 3 but they are less understood in dimensions ≥ 4.

Canonical. These are the singularities that appear on the canonical models of varieties of general type. The classification of canonical surface singularities by Du Val in 1934 is the first appearance of any of these classes in the literature. These singularities are reasonably well studied in dimension 3, less so in dimensions ≥ 4.

For many problems a modified version of Mori's program is more appropriate. Here one starts not with a variety but with a pair (X, D) consisting of a smooth variety and a simple normal crossing divisor on it. These lead to the "log" versions of the above notions.

Log terminal. This is the smallest class needed for running the minimal model program starting with a simple normal crossing pair (X, D). There are, unfortunately, many different flavors of log terminal; the above definition describes "divisorial log terminal" singularities. From the point of view of complex differential geometry, log terminal is characterized by finiteness of the volume of the smooth locus $X \setminus \text{Sing } X$; that is, for any top-degree holomorphic form ω, the integral $\int_X \omega \wedge \bar{\omega}$ is finite.

Log canonical. These are the singularities that appear on the log canonical models of pairs of log general type. Original interest in these singularities came from the study of affine varieties since the log canonical model of a pair (X, D) depends only on the open variety $X \setminus D$. One can frequently view log canonical singularities as a limiting case of the log terminal ones, but they are technically much more complicated. They naturally appear in any attempt to use induction on the dimension.

The relationship of these four classes to each other seems to undergo a transition as we go from dimension 3 to higher dimensions. In dimension 3 we understand terminal singularities completely and each successive class is understood less. In dimensions ≥ 4, our knowledge about the first three classes has been about the same for a long time while very little was known about the log canonical case until recently.

Semi-log-canonical. These are the singularities that appear on the stable degenerations of smooth varieties of general type. The same way as stable degenerations of smooth curves are non-normal nodal curves, stable degenerations of higher dimensional smooth varieties also need not be normal. In essence "semi-log canonical" is the straightforward non-normal version of "log canonical," but technically they seem substantially more complicated. The main reason is that the minimal model program fails for varieties with normal crossing singularities, hence many of the basic techniques are not available.

The relationship between the study of these singularities and the development of Mori's program was rather symbiotic. Early work on the minimal models of 3-folds relied very heavily on a detailed study of 3-dimensional terminal and canonical singularities. Later developments went in the reverse direction. Several basic results, for instance adjunction theory, were first derived as consequences of the (then conjectural) minimal model program. When they were later proved independently, they provided a powerful inductive tool for the minimal model program.

Now we have relatively short direct proofs of the finite generation of the canonical rings, but several of the applications to singularity theory depend on more delicate properties of minimal models in the non-general-type case. Conversely, recent work on the abundance conjecture relies on subtle properties of semi-log canonical singularities. In writing the book, substantial effort went into untangling these interwoven threads.

The basic definitions and key results of the minimal model program are recalled in Chapter 1.

Canonical, terminal, log canonical and log terminal singularities are defined and studied in Chapter 2. As much as possible, we develop the basic theory for arbitrary schemes, rather than just for varieties over $\mathbb{C}$.

Chapter 3 contains a series of examples and classification theorems that show how complicated the various classes of singularities can be.

The technical core of the book is Chapter 4. We develop a theory of higher-codimension Poincaré residue maps and apply it to a uniform treatment of log canonical centers of arbitrary codimension. Key new innovations are the sources and springs of log canonical centers, defined in Section 4.5.

These results are applied to semi-log canonical singularities in Chapter 5. The traditional methods deal successfully with the normalization of a semi-log canonical singularity. Here we show how to descend information from the normalization of the singularity to the singularity itself.

In Chapter 6 we show that semi-log canonical singularities are Du Bois; an important property in many applications. The log canonical case was settled earlier in Kollár and Kovács (2010). With the basic properties of semi-log canonical singularities established, the induction actually runs better in the general setting.

Two properties of semi-log canonical singularities that are especially useful in moduli questions are treated in Chapter 7.

Chapter 8 is a survey of the many results about canonical, terminal, log canonical and log terminal singularities that we could not treat adequately.

Chapter 9 contains results on finite equivalence relations that were needed in previous Chapters. Some of these are technical but they should be useful in different contexts as well.

A series of auxiliary results are collected in Chapter 10.

1

Preliminaries

We usually follow the definitions and notation of Hartshorne (1977) and Kollár and Mori (1998). Those that may be less familiar or are used inconsistently in the literature are recalled in Section 1.1.

The rest of the chapter is more advanced. We suggest skipping it at first reading and then returning to these topics when they are used later.

The classical theory of minimal models is summarized in Section 1.2. Minimal and canonical models of pairs are treated in greater detail in Section 1.3. Our basic reference is Kollár and Mori (1998), but several of the results that we discuss were not yet available when Kollár and Mori (1998) appeared. In Section 1.4 we collect various theorems that can be used to improve the singularities of a variety while changing the global structure only mildly. Random facts about some singularities are collected in Section 1.5.

Assumptions Throughout this book, all schemes are assumed noetherian and separated. Further restrictions are noted at the beginning of every chapter.

All the concepts discussed were originally developed for projective varieties over $\mathbb{C}$. We made a serious effort to develop everything for rather general schemes. This has been fairly successful for the basic results in Chapter 2, but most of the later theorems are known only in characteristic 0.

1.1 Notation and conventions

Notation 1.1 The *singular locus* of a scheme X is denoted by $\mathrm{Sing}\, X$. It is a closed, reduced subscheme if X is excellent. The open subscheme of nonsingular points is usually denoted by X^{ns}. For regular points we use X^{reg}.

The *reduced scheme* associated to X is denoted by red X.

Divisors and $\mathbb{Q}$-divisors

Notation 1.2 Let X be a normal scheme. A *Weil divisor*, or simply *divisor*, on X is a finite, formal, $\mathbb{Z}$-linear combination $D = \sum_i m_i D_i$ of irreducible and reduced subschemes of codimension 1. The group of Weil divisors is denoted by $\text{Weil}(X)$ or by $\text{Div}(X)$.

Given D and an irreducible divisor D_i, let $\text{coeff}_{D_i} D$ denote the *coefficient* of D_i in D. That is, one can write $D = (\text{coeff}_{D_i} D) \cdot D_i + D'$ where D_i is not a summand in D'. The *support* of D is the subscheme $\cup_i D_i \subset X$ where the union is over all those D_i such that $\text{coeff}_{D_i} D \neq 0$.

A divisor D is called *reduced* if $\text{coeff}_{D_i} D \in \{0, 1\}$ for every D_i. We sometimes identify a reduced divisor with its support. If $D = \sum_i a_i D_i$ (where the D_i are distinct, irreducible divisors) then $\text{red } D := \sum_{i:a_i \neq 0} D_i$ denotes the reduced divisor with the same support. One can usually identify $\text{red } D$ and $\text{Supp } D$.

Linear equivalence of divisors is denoted by $D_1 \sim D_2$.

For a Weil divisor D, $\mathcal{O}_X(D)$ is a rank 1 reflexive sheaf and D is a Cartier divisor if and only if $\mathcal{O}_X(D)$ is locally free. The correspondence $D \mapsto \mathcal{O}_X(D)$ is an isomorphism from the group $\text{Cl}(X)$ of Weil divisors modulo linear equivalence to the group of rank 1 reflexive sheaves. (This group does not seem to have a standard name but it can be identified with $\text{Pic}(X \setminus \text{Sing } X)$.) In this group the product of two reflexive sheaves L_1, L_2 is given by $L_1 \hat{\otimes} L_2 := (L_1 \otimes L_2)^{**}$, the double dual or reflexive hull of the usual tensor product. For powers we use the notation $L^{[m]} := (L^{\otimes m})^{**}$.

One can think of the Picard group $\text{Pic}(X)$ as a subgroup of $\text{Cl}(X)$.

A Weil divisor D is $\mathbb{Q}$-*Cartier* if and only if mD is Cartier for some $m \neq 0$. Equivalently, if and only if $\mathcal{O}_X(mD) = (\mathcal{O}_X(D))^{[m]}$ is locally free for some $m \neq 0$.

A normal scheme is *factorial* if every Weil divisor is Cartier and $\mathbb{Q}$-*factorial* if every Weil divisor is $\mathbb{Q}$-Cartier. See Boissière *et al.* (2011) for some foundational results.

Note further that if L is a reflexive sheaf and $D = \sum a_i D_i$ a Weil divisor then $L(D)$ denotes the sheaf of rational sections of L with poles of multiplicity at most a_i along D_i. It is thus the double dual of $L \otimes \mathcal{O}_X(D)$.

More generally, let X be a reduced, pure dimensional scheme that satisfies Serre's condition S_2. Let $\text{Cl}^*(X)$ denote the abelian group generated by the irreducible Weil divisors not contained in $\text{Sing } X$, modulo linear equivalence. (Thus, if X is normal, then $\text{Cl}^*(X) = \text{Cl}(X)$.) As before, $D \mapsto \mathcal{O}_X(D)$ is an isomorphism from $\text{Cl}^*(X)$ to the group of rank 1 reflexive sheaves that are locally free at all codimension 1 points of X. For more details, see (5.6).

Aside If X is not S_2, then one should work with the group of rank 1 sheaves that are S_2. Thus $\mathscr{O}_X(\sum a_i D_i)$ should denote the sheaf of rational sections of $\mathscr{O}_X$ with poles of multiplicity at most a_i along D_i. Unfortunately, this is not consistent with the usual notation $\mathscr{O}_X(D)$ for a Cartier divisor D since on a non-S_2 scheme a locally free sheaf is not S_2, hence we will avoid using it.

Definition 1.3 (Q-Divisors) If in the definition of a Weil divisor $\sum_i m_i D_i$ we allow $m_i \in \mathbb{Q}$ (resp. $m_i \in \mathbb{R}$), we get the notion of a Q-*divisor* (resp. $\mathbb{R}$-*divisor*). We mostly work with Q-divisors. For singularity theory, (2.21) reduces every question treated in this book from $\mathbb{R}$-divisors to Q-divisors.

We say that a Q-divisor D is a *boundary* if $0 \leq \mathrm{coeff}_{D_i} D \leq 1$ for every D_i and a *subboundary* if $\mathrm{coeff}_{D_i} D \leq 1$ for every D_i.

A Q-divisor D is Q-*Cartier* if mD is a Cartier divisor for some $m \neq 0$.

Note the difference between a Q-Cartier divisor and a Q-Cartier Q-divisor.

Since the use of Q-divisors is rather pervasive in some parts of the book, we sometimes call a divisor a $\mathbb{Z}$-*divisor* to emphasize that its coefficients are integers.

Two Q-divisors D_1, D_2 on X are Q-*linearly equivalent* if mD_1 and mD_2 are linearly equivalent $\mathbb{Z}$-divisors for some $m \neq 0$. This is denoted by $D_1 \sim_{\mathbb{Q}} D_2$.

Let $f\colon X \to Y$ be a morphism. Two Q-divisors D_1, D_2 on X are *relatively* Q-*linearly equivalent* if there is a Q-Cartier Q-divisor B on Y such that $D_1 \sim_{\mathbb{Q}} D_2 + f^*B$. This is denoted by $D_1 \sim_{\mathbb{Q},f} D_2$.

For a Q-divisor $D = \sum_i a_i D_i$ (where the D_i are distinct irreducible divisors) its *round down* is $\lfloor D \rfloor := \sum_i \lfloor a_i \rfloor D_i$ where $\lfloor a \rfloor$ denotes the largest integer $\leq a$. We will also use the notation $D_{>1} =: \sum_{i:a_i>1} a_i D_i$ and similarly for $D_{<0}$, $D_{\leq 1}$ and so on.

Definition 1.4 Let $f\colon X \to S$ be a proper morphism and D a Q-Cartier Q-divisor on X. Let $C \subset X$ be a closed 1-dimensional subscheme of a closed fiber of f. Choose $m > 0$ such that mD is Cartier. Then

$$(D \cdot C) := \tfrac{1}{m} \deg_C(\mathscr{O}_X(mD)|_C)$$

is called the *intersection number* or *degree* of D on C.

We say that D is f-*nef* if $(D \cdot C) \geq 0$ for every such curve C. If S is the spectrum of a field, we just say that D is *nef*.

We say that D is f-*semiample* if there are proper morphisms $\pi\colon X \to Y$ and $g\colon Y \to S$ and a g-ample Q-divisor H on Y such that $D \sim_{\mathbb{Q}} \pi^*H$. Thus f-semiample implies f-nef.

If S is a point, the difference between semiample and nef is usually minor, but for $\dim S > 0$ the distinction is frequently important; see Section 10.3.

Pairs

Mori's program was originally conceived to deal with smooth projective varieties. Later it became clear that one needs to handle certain singular varieties, schemes, algebraic or analytic spaces, and also add a divisor to the basic object.

Our main interest is in pairs (X, Δ) where

- X is either a normal variety over a field k or a normal scheme of finite type over a regular, excellent base scheme B. (In practice, the most important cases are when B is the spectrum of a field or a Dedekind ring.)
- Δ is a $\mathbb{Q}$-divisor such that $0 \leq \operatorname{coeff}_D \Delta \leq 1$ for every prime divisor D.

However, in some applications we need to work with non-normal varieties, with schemes that are not of finite type and with noneffective divisors Δ. Thus we consider the following general setting.

Definition 1.5 (Pairs) We consider pairs (X, Δ) over a base scheme B satisfying the following conditions.

(1) B is regular, excellent and pure dimensional.
(2) X is a reduced, pure dimensional, S_2, excellent scheme that has a canonical sheaf $\omega_{X/B}$ (1.6). (We will frequently simply write ω_X instead.)
(3) The canonical sheaf $\omega_{X/B}$ is locally free outside a codimension 2 subset. (This is automatic if X is normal.)
(4) $\Delta = \sum a_i D_i$ is a $\mathbb{Q}$-linear combination of distinct prime divisors none of which is contained in Sing X. We allow the a_i to be arbitrary rational numbers. (See (2.20) for some comments on real coefficients.)

Although we will always work with schemes, the results of Chapters 1–2 all apply to algebraic spaces and to complex analytic spaces satisfying the above properties.

Comments Assumption (1) is a very mild restriction since most base schemes can be embedded into a regular scheme. However, changing the base scheme B changes $\omega_{X/B}$.

If B itself is of finite type over a field k, then we are primarily interested in the "absolute" canonical sheaf ω_X of X and not in the relative canonical sheaf $\omega_{X/B}$ for $p\colon X \to B$. There is, however, no "absolute" canonical sheaf on a scheme; the above "absolute" canonical sheaf on a k-variety is in fact $\omega_{X/\operatorname{Spec} k}$. If B is a smooth k-variety then

$$\omega_{X/B} \simeq \omega_{X/\operatorname{Spec} k} \otimes p^*\omega_B^{-1}$$

and $p^*\omega_B$ is a line bundle which is trivial along the fibers of p. In defining the singularities of Mori's program for k-varieties, we use various natural maps between various "absolute" canonical sheaves. If we work over a smooth base B, these maps just get tensored with the pull-back of ω_B^{-1}; the definitions and theorems remain unchanged. This is the reason why one can work over a regular base scheme in general.

It should be possible to define everything over a Gorenstein base scheme, but we do not see any advantage to it.

Definition 1.6 (Canonical class and canonical sheaf)   For most applications, the usual definition of the canonical class and canonical sheaf given in Hartshorne (1977) and Kollár and Mori (1998) is sufficient. More generally, if X has a dualizing complex (Hartshorne (1966); Conrad (2000)), then its lowest (that is, in degree $-\dim X/S$) cohomology sheaf is the dualizing sheaf $\omega_{X/S}$. Unfortunately, the existence of a dualizing complex is a thorny problem for general schemes. Instead of getting entangled in it, we discuss a simple special case that is sufficient for most purposes. The uninitiated reader may also find the discussion in Kovács (2012b, section 5) useful.

Let B be a regular base scheme and $X \to B$ be a pure dimensional scheme of finite type over B that satisfies the following:

Condition 1.6.1   There is an open subscheme $j: X^0 \hookrightarrow X$ and a (locally closed) embedding $\iota: X^0 \hookrightarrow \mathbb{P}_B^N$ such that

(a) $Z := X \setminus X^0$ has codimension ≥ 2 in X, and
(b) $\iota(X^0)$ is a local complete intersection in $\mathbb{P}_B^N$.

Let I denote the ideal sheaf of the closure of $\iota(X^0)$. Then I/I^2 is a locally free sheaf on $\iota(X^0)$ and, as in Hartshorne (1977, II.8.20), we set

$$\omega_{X^0/B} := \iota^*(\omega_{\mathbb{P}^N/B} \otimes \det^{-1}(I/I^2)). \qquad (1.6.2)$$

Finally define the *canonical sheaf* of X over B as

$$\omega_{X/B} := j_*\omega_{X^0/B}. \qquad (1.6.3)$$

If $\omega_{X/B}$ is locally free, it is frequently called the *canonical bundle*. We frequently drop B from the notation.

If X^0 is smooth over B, then one can define ω_X using differentials, but in general, differential forms give a different sheaf.

(We are mainly interested in three special cases and one extension of this construction. First, if X is normal and quasi-projective then $Z = \operatorname{Sing} X$ works. Second, in dealing with stable varieties, we consider schemes X that have ordinary nodes at some codimension 1 points. Third, we occasionally use the dualizing sheaf for nonreduced divisors on a regular scheme. Finally, we sometimes use that if $p \in X$ is a point and $\hat{X}_p$ the completion of X at p then $\hat{\omega}_{X/B}$ is the canonical sheaf of $\hat{X}_p$.)

If X is reduced, the corresponding linear equivalence class of Weil divisors is denoted by K_X.

Note that while Hartshorne (1977, II.8.20) is a theorem, for us (1.6.2–3) are definitions. Therefore we need to establish that $\omega_{X/B}$ does not depend on the projective embedding chosen. This is easy to do by comparing two different embeddings ι_1, ι_2 with the diagonal embedding

$$(\iota_1, \iota_2) : X^0 \hookrightarrow \mathbb{P}_B^{N_1} \times_B \mathbb{P}_B^{N_2} \hookrightarrow \mathbb{P}_B^{N_1 N_2 + N_1 + N_2}.$$

We also need that $\omega_{X/B}$ is the relative *dualizing sheaf*. If X itself is projective, a proof is in Kollár and Mori (1998, section 5.5). For the general case see Hartshorne (1966) and Conrad (2000).

Normal crossing conditions

Normal crossing means that something "looks like" the coordinate hyperplanes. Depending on how one interprets "looking like," one gets different notions. In many instances, the difference between them is minor or merely a technical annoyance. This is reflected by inconsistent usage in the literature. However, in some applications, especially when the base field is not algebraically closed, the differences are crucial. We have tried to adhere to the following conventions.

Definition 1.7 (Simple normal crossing for pairs) Let X be a scheme. Let $p \in X$ be a regular (not necessarily closed) point with ideal sheaf m_p and residue field $k(p)$. Then $x_1, \ldots, x_n \in m_p$ are called *local coordinates* if their residue classes $\bar{x}_1, \ldots, \bar{x}_n$ form a $k(p)$-basis of m_p/m_p^2.

Let $D = \sum a_i D_i$ be a Weil divisor on X. We say that (X, D) has *simple normal crossing* or *snc* at a (not necessarily closed) point $p \in X$ if X is regular at p and there is an open neighborhood $p \in X_p \subset X$ with local coordinates $x_1, \ldots, x_n \in m_p$ such that $X_p \cap \operatorname{Supp} D \subset (x_1 \cdots x_n = 0)$. Alternatively, if for each D_i there is a $c(i)$ such that $D_i = (x_{c(i)} = 0)$ near p. We say that (X, D) is *simple normal crossing* or *snc* if it is snc at every point. It is important to note that being simple normal crossing is local in the Zariski topology, but not in the étale topology.

This concept is frequently called *strict normal crossing* or, if X is defined over an algebraically closed field, *global normal crossing*.

A *stratum* of an snc pair $(X, \sum_{i \in I} a_i D_i)$ is any irreducible component of an intersection $\cap_{i \in J} D_i$ for some $J \subset I$. Sometimes X itself is allowed as a stratum corresponding to $J = \emptyset$. All the strata of an snc pair are regular.

We say that (X, D) has *normal crossing* or *nc* at a point $p \in X$ if there is an étale neighborhood $\pi : (p' \in X') \to (p \in X)$ such that $(X', \pi^{-1} D)$ is snc at p'. Equivalently, if $(\hat{X}_K, D|_{\hat{X}_K})$ is snc at p where $\hat{X}_K$ denotes the completion at p

and K is a separable closure of $k(p)$. We say that (X, D) is *normal crossing* or *nc* if it is nc at every point. Being normal crossing is local in the étale topology.

If (X, D) is defined over a perfect field, this concept is also called *log smooth*.

Examples Let $p \in D$ be a nc point of multiplicity 2. If the characteristic is different from 2, then, in suitable local coordinates, D can be given by an equation $x_1^2 - ux_2^2 = 0$ where $u \in \mathscr{O}_{p,X}$ is a unit. D is snc at p if and only if u is a square in $\mathscr{O}_{p,X}$.

Thus $(y^2 - (1 + x)x^2) \subset \mathbb{A}^2$ is nc but it is not snc at the origin. Similarly, $(x^2 + y^2 = 0) \subset \mathbb{A}^2$ is nc but it is snc only if $\sqrt{-1}$ is in the base field k.

Given (X, D), the largest open set $U \subset X$ such that $(U, D|_U)$ is snc is called the *snc locus* of (X, D). It is denoted by $\mathrm{snc}(X, D)$. Its complement is called the *non-snc locus* of (X, D) and denoted by $\mathrm{non\text{-}snc}(X, D)$.

We also use the analogously defined *nc locus*, denoted by $\mathrm{nc}(X, D)$, or its complement $\mathrm{non\text{-}nc}(X, D)$.

Finally, $p \in D$ is called a *double, triple*, etc. snc (or nc) point if D has multiplicity 2, 3, etc. at p. The *double-snc locus* (resp. the *double-nc locus*) of (X, D) is the largest open set $U \subset X$ such that $(U, D|_U)$ is snc (resp. nc) and each point of D has multiplicity ≤ 2.

Definition 1.8 (Simple normal crossing schemes) Let Y be a scheme. We say that Y has *simple normal crossing* or *snc* at a point $p \in Y$ if there is an open neighborhood $p \in Y_p \subset Y$ and a closed embedding $Y_p \hookrightarrow X_p$ of Y_p into a regular scheme X_p such that (X_p, Y_p) has simple normal crossing (1.7). We say that Y is *snc* or has *simple normal crossings* if it is snc at every point. As before, being simple normal crossing is local in the Zariski topology, but not in the étale topology.

Let X be an snc scheme with irreducible components $X = \cup_{i \in I} X_i$. A *stratum* of X is any irreducible component of an intersection $\cap_{i \in J} X_i$ for some $J \subset I$.

We say that Y has *normal crossing* or *nc* at a point $p \in Y$ if there is an étale neighborhood $\pi \colon (p' \in Y_p') \to (p \in Y)$ and a closed embedding $Y_p' \hookrightarrow X_p'$ such that (X_p', Y_p') has simple normal crossing (1.7). Equivalently, if $\hat{Y}_K$ is snc at p where $\hat{Y}_K$ denotes the completion at p and K is a separable closure of $k(p)$. We say that Y is *normal crossing* or *nc* if it is nc at every point. Being normal crossing is local in the étale topology.

Note that, étale locally, snc schemes and nc schemes look the same. An nc scheme is snc if and only if its irreducible components are regular.

If X is defined over a perfect field, this concept is also called *log smooth*.

Given Y, the largest open set $U \subset Y$ such that U is snc is called the *snc locus* of Y. It is denoted by $\mathrm{snc}(Y)$. Its complement is called the *non-snc locus* of Y and denoted by $\mathrm{non\text{-}snc}(Y)$. We sometimes use the analogously defined *nc locus*, denoted by $\mathrm{nc}(Y)$, or its complement $\mathrm{non\text{-}nc}(Y)$.

Remark 1.9 Note that an snc scheme X of dimension n can locally everywhere be realized as an snc divisor on a regular scheme Y^{n+1} but usually X itself cannot be an snc divisor on a regular scheme.

A simple obstruction arises from the following observation. Let Y be a nonsingular variety and $D_1 + D_2$ an snc divisor on Y. Set $Z := D_1 \cap D_2$. Then $N_{Z,D_2} \simeq N_{D_1,Y}|_Z$ where $N_{X,Y}$ denotes the normal bundle of $X \subset Y$.

Thus if $X = X_1 \cup X_2$ is an snc variety with $Z := X_1 \cap X_2$ such that N_{Z,X_2} is not the restriction of any line bundle from X_1 then X is not an snc divisor on a nonsingular variety.

We do not know which snc varieties can be realized as an snc divisor on a nonsingular variety. This explains some of the complications in Section 3.4.

A common generalization of (1.7) and of (1.8) is the following.

Definition 1.10 (Semi-snc pairs) Let W be a regular scheme and $\sum_{i \in I} E_i$ an snc divisor on W. Write $I = I_V \cup I_D$ as a disjoint union. Set $Y := \sum_{i \in I_V} E_i$ as a subscheme and $D_Y := \sum_{i \in I_D} a_i E_i|_Y$ as a divisor on Y for some $a_i \in \mathbb{Q}$. We call (Y, D_Y) an *embedded semi-snc pair*. A pair (X, D) is called *semi-snc* if it is Zariski locally isomorphic to an embedded semi-snc pair. (We try to avoid the long version "semi-simple normal crossing pair" since "semi-simple" is too well established with other meanings.) Note that it follows from the definition that $D_Y \cap E_i$ is a divisor for every $i \in I_V$.

A *stratum* of a semi-snc pair (X, D) is a closure of a stratum of any of the local charts $(W, \sum_{i \in I} E_i)$ that is contained in X. The global strata are in fact unions of local strata, hence they are regular subschemes of X.

A pair (X, D) is called *semi-nc* if it is étale locally isomorphic to an embedded semi-snc pair.

There are three étale local models of semi-nc surface pairs (S, D) over $\mathbb{C}$.

(1) $S = (z = 0) \subset \mathbb{A}^3$ and $D = a_x(x|_S = 0) + a_y(y|_S = 0)$. This is the usual normal case.
(2) $S = (yz = 0) \subset \mathbb{A}^3$ and $D = a_x(x|_S = 0)$. Note that as a Weil divisor, D has two irreducible components, namely $D_1 := (x = y = 0)$ and $D_2 := (x = z = 0)$. The support of the $\mathbb{Q}$-divisor $a_1 D_1 + a_2 D_2$ is always snc, but the pair $(S, a_1 D_1 + a_2 D_2)$ is semi-snc only if $a_1 = a_2$. It is easy to see that $a_1 D_1 + a_2 D_2$ is $\mathbb{Q}$-Cartier only if $a_1 = a_2$.
(3) $S = (xyz = 0) \subset \mathbb{A}^3$ and $D = 0$.

Note that, by our definition, neither of the following is semi-snc:

$$((xy = 0), (x = y = 0)) \subset \mathbb{A}^3 \quad \text{or} \quad ((xy = 0), (x = z = 0)) \subset \mathbb{A}^3.$$

Resolution of singularities

Definition 1.11 (Birational maps)   A map of reduced schemes $f: X \dashrightarrow Y$ is called *birational* if there are dense open subschemes $U_X \subset X$ and $U_Y \subset Y$ such that $f|_{U_X}: U_X \to U_Y$ is an isomorphism. Equivalently, if one can index the irreducible components $X = \cup_i X_i$ and $Y = \cup_i Y_i$ such that each $f|_{X_i}: X_i \dashrightarrow$ Y_i is birational. Birational equivalence is denoted by $X \overset{\text{bir}}{\sim} Y$.

Among all such pairs $\{U_X, U_Y\}$ there is a unique maximal one $\{U_X^m, U_Y^m\}$. The complement $\mathrm{Ex}(f) := X \setminus U_X^m$ is called the *exceptional set* or *locus* of f.

Let $Z \subset X$ be a closed subscheme such that $Z \cap U_X^m$ is dense in Z. Then the closure of $f(Z \cap U_X^m) \subset Y$ is called the *birational transform* of Z. It is denoted by $f_* Z$.

Warning   If f is not a morphism then f_* need not preserve linear or algebraic equivalence. Furthermore, if $D := Z$ is a divisor, then $\mathcal{O}_Y(f_* D)$ and $f_* \mathcal{O}_X(D)$ agree on U_Y but can be quite different elsewhere. (The latter need not even be coherent in general.)

Applying this to $W \subset Y$ and f^{-1} we get the slightly unusual looking notation $f_*^{-1} W := (f^{-1})_* W$.

All birational maps of X to itself form a group, denoted by $\mathrm{Bir}\, X$.

Let X, Y be reduced schemes, proper over a base scheme S. A birational map $f: X \dashrightarrow Y$ is called a *birational contraction* if $\mathrm{Ex}(f^{-1})$ has codimension ≥ 2. Note that if Y is normal and f is a morphism then it is a birational contraction.

If f and f^{-1} are both birational contractions, we say that X and Y are *isomorphic in codimension 1*.

There does not seem to be a good notion of birational equivalence for pairs (X_i, D_i). See (2.23–2.24) for some special cases and examples.

Definition 1.12 (Resolution and log resolution)   Let X be a reduced scheme. A *resolution* of X is a proper birational morphism $f: X' \to X$ such that X' is regular.

Let X be a reduced scheme and D a Weil divisor on X. A *log resolution* of (X, D) is a proper birational morphism $f: X' \to X$ such that X' is regular, $\mathrm{Ex}(f)$ has pure codimension 1 and $(X', D' := \mathrm{Supp}(f^{-1}(D) + \mathrm{Ex}(f)))$ is snc. We also say that $f: (X', D') \to (X, D)$ is a log resolution.

The existence of log resolutions is proved in Section 10.4 over a field of characteristic 0.

1.12.1 It is sometimes useful to know if there is an effective, f-exceptional divisor F such that $-F$ is f-ample. This is always the case if X is $\mathbb{Q}$-factorial (Kollár and Mori, 1998, 2.62) or if f is the composite of blow-ups $X_{i+1} := B_{Z_i} X_i \to X_i$ such that the image of Z_i in $X = X_0$ has codimension ≥ 2. Thus, in characteristic 0, every log resolution of a normal variety is dominated by another log resolution that carries such an exceptional divisor.

Though the above definition makes sense for any scheme, we usually use it for normal schemes. It is actually not clear what is the best notion of resolution for non-normal schemes. The following variant is quite useful; see Section 10.4 for other versions and conjectures.

1.13 (Semi-resolutions) Let X be a reduced scheme over a field of characteristic $\neq 2$ and $X^0 \subset X$ an open subset that has only nonsingular points $(x_1 = 0)$, double nc points $(x_1^2 - ux_2^2 = 0)$ and pinch points $(x_1^2 = x_2^2 x_3)$ such that $X \setminus X^0$ has codimension ≥ 2. A *semi-resolution* of X is a projective birational morphism $f \colon X' \to X$ such that

(1) X' has only nonsingular points, double nc points and pinch points,
(2) f is an isomorphism over X^0 and
(3) f maps $\operatorname{Sing} X'$ birationally onto the double normal crossing locus $D \subset \operatorname{Sing} X$.

Note that by (3), X' is nonsingular at the generic point of any f-exceptional divisor.

We show in (10.54) that semi-resolutions exist in characteristic 0. See Section 10.4, especially (10.53) for more details. For semi-resolutions of pairs, see (10.56).

1.2 Minimal and canonical models

This section is intended as a quick introduction to the theory of minimal and canonical models. For more details see Kollár and Mori (1998, chapters 2–3).

1.14 (Canonical models) Let X be a nonsingular projective variety over a field k. Its *canonical ring* is the graded ring

$$R(X) = R(X, K_X) := \sum_{m \geq 0} H^0(X, \mathscr{O}_X(mK_X)).$$

The canonical ring depends only on the birational equivalence class of X. Conversely, if X is of *general type*, that is, if $|mK_X|$ gives a birational map for $m \gg 1$, then the canonical ring determines the birational equivalence class of X.

If the canonical ring is finitely generated and X is of general type, then

$$X^{\mathrm{can}} := \mathrm{Proj}_k R(X, K_X)$$

is called the *canonical model* of X (or of its birational equivalence class).

For various reasons, we are also interested in *minimal models* of X. These are varieties X^{min} that are birational to X, have terminal singularities and nef canonical class. More generally, X^w is a *weak canonical model* of X if X^w has canonical singularities, nef canonical class and is birational to X. Thus canonical and minimal models are also weak canonical models.

Theorem 1.15 (Reid, 1980) *Let X be a nonsingular, proper variety of general type such that its canonical ring $R(X, K_X)$ is finitely generated. Then*

(1) *its canonical model X^{can} is normal, projective, birational to X,*
(2) *the canonical class $K_{X^{\mathrm{can}}}$ is $\mathbb{Q}$-Cartier and ample,*
(3) *X^{can} has canonical singularities,*
(4) *$H^0(X, \mathscr{O}_X(mK_X)) \simeq H^0(X^{\mathrm{can}}, \mathscr{O}_{X^{\mathrm{can}}}(mK_{X^{\mathrm{can}}}))$ for every $m \geq 0$ and*
(5) *if K_X is nef then $X \dashrightarrow X^{\mathrm{can}}$ is a morphism.*

Proof Since $\sum_{m \geq 0} H^0(X, \mathscr{O}_X(mK_X))$ is finitely generated, there is an $r > 0$ such that $\sum_{m \geq 0} H^0(X, \mathscr{O}_X(mrK_X))$ is generated by $H^0(X, \mathscr{O}_X(rK_X))$. Thus $D := rK_X$ satisfies the assumptions of (1.16). Therefore, there is a birational map $\phi \colon X \dashrightarrow Z$ to a normal variety and a very ample Cartier divisor H on Z such that $H \sim \phi_*(D)$ and $\phi^* \colon H^0(Z, \mathscr{O}_Z(mH)) \to H^0(X, \mathscr{O}_X(mD))$ is an isomorphism for every $m > 0$. Thus $Z = X^{\mathrm{can}}$.

Using (1.16.3) we see that the push-forward of the canonical class of X is the canonical class of Z, thus

$$rK_Z \sim \phi_*(rK_X) \sim \phi_*(D) \sim H.$$

Thus K_Z is $\mathbb{Q}$-Cartier and ample. Property (4) is exactly what led to the definition of canonical singularities; see (2.17) where we also prove (3). ☐

Proposition 1.16 *Let X be a proper, normal variety and D a Cartier divisor on X. Assume that $\sum_{m \geq 0} H^0(X, \mathscr{O}_X(mD))$ is generated by $H^0(X, \mathscr{O}_X(D))$ and that $h^0(X, \mathscr{O}_X(mD)) > cm^{\dim X}$ for some $c > 0$ and $m \gg 1$. Let $\phi \colon X \dashrightarrow \mathbb{P}^N$ denote the map given by $|D|$. Let Z be the closure of $\phi(X)$ and $|H|$ the hyperplane class on Z. Then*

(1) ϕ *is birational,*
(2) Z *is normal,*
(3) $Z \setminus \phi(X \setminus \mathrm{Bs}\,|D|)$ *has codimension* ≥ 2 *in* Z,
(4) *every divisor in* $\mathrm{Bs}\,|D|$ *is contracted by* ϕ,
(5) $\phi_*|D| = |H|$ *and*
(6) *if* D *is nef then* $|D|$ *is base point free.*

Proof Let $X \leftarrow X' \rightarrow Z$ be the normalization of the closure of the graph of ϕ with projections π and ϕ'. Set $D' := \pi^*D$. Then X' and D' satisfy our assumptions. Moreover, $\mathrm{Bs}\,|D'| = \pi^{-1}\,\mathrm{Bs}\,|D|$ implies that it is enough to show the conclusions for X' and D'. Replacing X by X' we may assume from now on that ϕ is a morphism, $|D| = \phi^*|H| + F$ and F is the base locus of $|D|$. Since $\sum_{m \geq 0} H^0(X, \mathcal{O}_X(mD))$ is generated by $H^0(X, \mathcal{O}_X(D))$, we conclude that

$$H^0(Z, \mathcal{O}_Z(mH)) = H^0(X, \mathcal{O}_X(mD)) > cm^{\dim X} \quad \text{for } m \gg 1.$$

In particular, $\dim Z = \dim X$ and ϕ is generically finite.

Let $p: Z' \rightarrow Z$ be the normalization of Z in X. For large m, $m(p^*H)$ is very ample on Z' and

$$H^0(Z, \mathcal{O}_Z(mH)) \subset H^0(Z', \mathcal{O}_{Z'}(m(p^*H)))$$
$$\subset H^0(X, \mathcal{O}_X(mD)) = H^0(Z, \mathcal{O}_Z(mH)).$$

Thus $Z = Z'$ is normal, proving (1) and (2).

Let B be any irreducible divisor on X such that $\phi(B)$ is a divisor. We can view $\mathcal{O}_Z(\phi(B))$ as a subsheaf of the sheaf of rational functions. It has thus an inverse image $\mathcal{O}_Z(\phi(B)) \cdot \mathcal{O}_X$ which is a coherent subsheaf of the sheaf of rational functions on X. Thus there is a ϕ-exceptional divisor E such that

$$\mathcal{O}_Z(\phi(B)) \cdot \mathcal{O}_X \subset \mathcal{O}_X(B + E).$$

Let H_0 be a Cartier divisor on Z that vanishes along $\phi(E)$ but not along $\phi(B)$. Then, for some $m_0 > 0$,

$$\mathcal{O}_Z(\phi(B) - m_0 H_0) \cdot \mathcal{O}_X \subset \mathcal{O}_X(B + E - m_0\phi^*H_0) \subset \mathcal{O}_X(B)$$

and these three sheaves agree generically along B.

Take now $m \gg 1$ such that $\mathcal{O}_Z(mH + \phi(B) - m_0 H_0)$ is generated by global sections. Then $\mathcal{O}_X(m\phi^*H + B)$ has a global section that does not vanish along B. In particular, such a B is not in the base locus of $|m\phi^*H + B|$. Since $|mD| = |m\phi^*H| + mF$, we see that B is not in $F = \mathrm{Bs}\,|D|$.

Thus $\phi(\mathrm{Bs}\,|D|)$ has codimension ≥ 2 in Z, which implies (3) and (4). Therefore $\phi_*|D| = \phi_*\phi^*|H| + \phi_*F = |H|$, proving (5).

Finally, if D is nef then F is effective, ϕ-exceptional and ϕ-nef. Thus $F = 0$ by (1.17), proving (6). $\qquad\square$

The next lemma is useful in many situations. (It is stated for varieties in Kollár and Mori (1998, 3.39), but the proof applies for schemes once we use that the Hodge index theorem holds for schemes (10.1).)

Lemma 1.17 (Kollár and Mori, 1998, 3.39)   *Let $h: Z \to Y$ be a proper birational morphism between normal schemes. Let $-B$ be an h-nef $\mathbb{Q}$-Cartier $\mathbb{Q}$-divisor on Z. Then*

(1) *B is effective if and only if h_*B is.*
(2) *Assume that B is effective. Then for every $y \in Y$, either $h^{-1}(y) \subset \operatorname{Supp} B$ or $h^{-1}(y) \cap \operatorname{Supp} B = \emptyset$.*

Thus if B is also h-exceptional then B is numerically h-trivial if and only if $B = 0$.   □

1.3 Canonical models of pairs

We generalize the notions of canonical and minimal models to pairs (X, Δ) and to the relative setting. We are mostly interested in the case when Δ is a boundary, that is, the coefficient of every divisor in Δ is in the interval $[0, 1]$, but most of the basic results hold for any Δ.

We use the basic notions of singularities of pairs that are developed in Section 2.1. Thus it makes sense to postpone reading various parts of this section until they are actually needed.

The guiding principle is that the *(log) canonical ring*

$$R(X, K_X + \Delta) := \sum_{m \geq 0} H^0(X, \mathscr{O}_X(mK_X + \lfloor m\Delta \rfloor))$$

should play the role of the canonical ring.

Aside   Note that $\lfloor A + B \rfloor \geq \lfloor A \rfloor + \lfloor B \rfloor$ for any $\mathbb{Q}$-divisors A, B, thus, for any $\mathbb{Q}$-divisor D, $R(X, D) := \sum_{m \geq 0} H^0(X, \mathscr{O}_X(\lfloor mD \rfloor))$ is a ring. In particular, the canonical ring is indeed a ring. On the other hand, using round-ups instead of round-downs leads to $R^u(X, D) := \sum_{m \geq 0} H^0(X, \mathscr{O}_X(\lceil mD \rceil))$, which is, in general, not a ring. However, $\lceil A + B \rceil \geq \lfloor A \rfloor + \lceil B \rceil$, thus $R^u(X, D)$ is an $R(X, D)$-module.

A new twist is that while it is straightforward to define when (X^c, Δ^c) is a canonical model, it is harder to pin down when is (X^c, Δ^c) a canonical model of another pair (X, Δ). The main reason is that (X^c, Δ^c) carries no information on those irreducible components of Δ that are exceptional for $X \dashrightarrow X^c$.

Given a pair (X, Δ), what should its canonical model be?

First of all, it is a pair (X^c, Δ^c) which is a canonical model (1.18). Second, X^c should be birational to X. More precisely, there should be a birational map

$\phi\colon X \dashrightarrow X^c$ which is a birational contraction (1.11). Then the only sensible choice is to set $\Delta^c := \phi_*\Delta$.

However, these conditions are not yet sufficient, as shown by the next example.

Let $(Y, 0)$ be log canonical with ample K_Y and $f : X \to Y$ a resolution with exceptional divisors E_i. Write

$$f^*K_Y \sim_{\mathbb{Q}} K_X + \textstyle\sum_i b_i E_i$$

where $-b_i = a(E_i, Y, 0)$ is the discrepancy as in (2.4). For some $0 \le c_i \le 1$, set $\Delta_X := \sum c_i E_i$. Then $f_*\Delta_X = 0$. As in (2.17) and (2.18), we see that the two rings

$$\textstyle\sum_{m\ge 0} H^0(Y, \mathcal{O}_Y(mK_Y)) \quad \text{and} \quad \textstyle\sum_{m\ge 0} H^0(X, \mathcal{O}_X(mK_X + \lfloor m\Delta_X \rfloor))$$

agree only if $\Delta_X \ge \sum_i b_i E_i$, that is, if $c_i \ge b_i$ for every i.

Note that this problem did not occur in the canonical case. Indeed, if $(Y, 0)$ is canonical then $b_i = -a(E_i, Y, 0) \le 0$ and on X we take $\Delta_X = 0$. Thus $\Delta_X \ge \sum_i b_i E_i$ is automatic.

By contrast, in the log canonical case, only $b_i = -a(E_i, X, 0) \le 1$ is assumed, hence the condition $c_i \ge b_i$ is nontrivial.

Keeping this example in mind, we see that we have to compare the discrepancies of divisors with respect to the canonical model and the discrepancies with respect to the original pair.

The discrepancy inequality should hold for all divisors, but it turns out (see (1.22)) that it is enough to require it for ϕ-exceptional divisors (1.19.5).

Eventually we also need these concepts in the relative setting, that is, over rather general base schemes.

Definition 1.18   Let (X, Δ) be a pair as in (1.5), Δ a boundary (1.3) and $f\colon X \to S$ a proper morphism. We say that (X, Δ) is an

f-canonical					
f-minimal	$\Big\}$ model if (X, Δ) is	$\left\{\begin{array}{c}\text{lc}\\ \text{dlt}\\ \text{lc}\end{array}\right.$	and $K_X + \Delta$ is	$\left\{\begin{array}{c}f\text{-ample.}\\ f\text{-nef.}\\ f\text{-nef.}\end{array}\right.$	
f-weak canonical					

Minimal models are quite close to canonical models but they have milder singularities. Weak canonical models form a convenient class that includes canonical models, minimal models and everything in between them.

(Recall that, as in (1.6), we work over a regular base scheme B which we suppress in the notation. In the classical theory of Section 1.2, $S = B = \operatorname{Spec} k$ where k is a field. However, in many applications, S is quite different from B and S is allowed to be singular. We will be careful to distinguish K_X and $K_{X/S}$, especially since the latter is frequently not defined.)

Warning   Note that a canonical model (X, Δ) has *log* canonical singularities, not necessarily canonical singularities. This, by now entrenched, unfortunate terminology is a result of an incomplete shift. Originally everything was defined only for $\Delta = 0$. When Δ was introduced, its presence was indicated by putting "log" in front of adjectives. Later, when the use of Δ became pervasive, people started dropping the prefix "log." This is usually not a problem. For instance, the canonical ring $R(X, K_X)$ is just the $\Delta = 0$ special case of the log canonical ring $R(X, K_X + \Delta)$.

However, canonical singularities are not the $\Delta = 0$ special cases of log canonical singularities.

Definition 1.19   Let (X, Δ) be a pair as in (1.5) such that X is normal and $K_X + \Delta$ is $\mathbb{Q}$-Cartier. Let $f: X \to S$ be a proper morphism. A pair $(X^{\mathrm{w}}, \Delta^{\mathrm{w}})$ sitting in a diagram

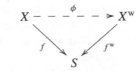

is called a *weak canonical model* of (X, Δ) over S if

(1) f^{w} is proper,
(2) ϕ is a birational contraction (1.11),
(3) $\Delta^{\mathrm{w}} = \phi_* \Delta$,
(4) $K_{X^{\mathrm{w}}} + \Delta^{\mathrm{w}}$ is $\mathbb{Q}$-Cartier and f^{w}-nef and
(5) $a(E, X, \Delta) \le a(E, X^{\mathrm{w}}, \Delta^{\mathrm{w}})$ for every ϕ-exceptional divisor $E \subset X$.

(We will see in (1.22) that if Δ is a boundary and (X, Δ) is lc then $(X^{\mathrm{w}}, \Delta^{\mathrm{w}})$ is an f^{w}-weak canonical model.) As in (2.6), write

$$K_X + \phi_*^{-1}\Delta^{\mathrm{w}} \sim_{\mathbb{Q}} \phi^*(K_{X^{\mathrm{w}}} + \Delta^{\mathrm{w}}) + \textstyle\sum_i a(E_i, X^{\mathrm{w}}, \Delta^{\mathrm{w}})E_i, \qquad (1.19.6\mathrm{a})$$

where the E_i are ϕ-exceptional. Note that by (3), $\Delta = \phi_*^{-1}\Delta^{\mathrm{w}} + \Delta^{\mathrm{ex}}$ where Δ^{ex} is the ϕ-exceptional part of Δ. We can thus rewrite the above equation as

$$K_X + \Delta - \phi^*(K_{X^{\mathrm{w}}} + \Delta^{\mathrm{w}}) \sim_{\mathbb{Q}} \Delta^{\mathrm{ex}} + \textstyle\sum_i a(E_i, X^{\mathrm{w}}, \Delta^{\mathrm{w}})E_i. \qquad (1.19.6\mathrm{b})$$

We can thus restate (5) as

(5′) $\Delta^{\mathrm{ex}} + \sum_i a(E_i, X^{\mathrm{w}}, \Delta^{\mathrm{w}})E_i$ is effective.

A weak canonical model $(X^{\mathrm{m}}, \Delta^{\mathrm{m}}) = (X^{\mathrm{w}}, \Delta^{\mathrm{w}})$ is called a *minimal model* of (X, Δ) over S if in addition to (1–4), we have

(5ᵐ) $a(E, X, \Delta) < a(E, X^{\mathrm{m}}, \Delta^{\mathrm{m}})$ for every ϕ-exceptional divisor $E \subset X$.
      Equivalently, if $\Delta^{\mathrm{ex}} + \sum_i a(E_i, X^{\mathrm{w}}, \Delta^{\mathrm{w}})E_i$ is effective and its support contains all ϕ-exceptional divisors.

(We will see in (1.22) that if Δ is a boundary and (X, Δ) is klt then (X^m, Δ^m) is an f^m-minimal model. The dlt case is more delicate, see (1.23).)

A weak canonical model $(X^c, \Delta^c) = (X^w, \Delta^w)$ is called a *canonical model* of (X, Δ) over S if, in addition to (1–3) and (5) we have

(4^c) $K_{X^c} + \Delta^c$ is $\mathbb{Q}$-Cartier and f^c-ample.

(We will see in (1.22) that if Δ is a boundary and (X, Δ) is lc then (X^c, Δ^c) is an f^c-canonical model.)

Comments For any (X, Δ) as above, its canonical model is unique (1.26) (but it need not exist). Despite their names, minimal models are the largest among the weak canonical models, see (1.21) and (1.36). Weak canonical models should be thought of as intermediate objects between minimal models and canonical models.

Warning 1.20 (Pull-back by rational maps) If $g: X \dashrightarrow Y$ is a dominant rational map then one can define the pull-back maps $g^*: \mathrm{CDiv}(Y) \to \mathrm{WDiv}(X)$ and $g^*: \mathrm{Pic}(Y) \to \mathrm{Cl}(X)$. Note, however, that if $h: Y \dashrightarrow Z$ is another dominant rational map then usually $(h \circ g)^* \neq g^* \circ h^*$.

A simple example is the following. Let $Y = \mathbb{P}^2$, $h: Y \dashrightarrow \mathbb{P}^1$ the projection from a point $y \in Y$ and $g: X := B_y Y \to Y$ the blow-up with exceptional curve E.

Then $h \circ g: B_y Y \to \mathbb{P}^1$ is a morphism.

Let $D \subset \mathbb{P}^1$ be any effective divisor. Then $(h \circ g)^* D$ consists of the corresponding fibers of $B_y Y \to \mathbb{P}^1$ and it never contains E. By contrast, $h^* D$ consists of a bunch of lines through y and so $g^*(h^* D)$ contains E with multiplicity $\deg D$.

(1.20.1) If h is a morphism, then $(h \circ g)^* = g^* \circ h^*$.

The next result shows how similar the various minimal (resp. weak canonical) models are to each other.

Proposition 1.21 *Let (X, Δ) be a pair as in (1.19) and $\phi_w: (X, \Delta) \dashrightarrow (X^w, \Delta^w)$ a weak canonical model. Then*

(1) $a(E, X^w, \Delta^w)$ *is independent of (X^w, Δ^w) for any divisor E on any birational model of X.*

(2) *If $\phi_m: (X, \Delta) \dashrightarrow (X^m, \Delta^m)$ is a minimal model then $\phi_w \circ \phi_m^{-1}: X^m \dashrightarrow X^w$ is a birational contraction. In particular, any two minimal models of (X, Δ) are isomorphic in codimension one.*

Proof Let $\phi_i: (X, \Delta) \dashrightarrow (X_i^w, \Delta_i^w)$ for $i = 1, 2$ be two weak canonical models of (X, Δ). Choose $g: Y \to X$ such that $h_i: Y \to X_i^w$ are both morphisms.

Write

$$g^*(K_X + \Delta) \sim_{\mathbb{Q}} h_i^*\big(K_{X_i^w} + \Delta_i^w\big) + Z_i$$

where the Z_i are both effective by (1.19.5). Subtracting the two formulas we get

$$h_1^*\big(K_{X_1^w} + \Delta_1^w\big) - h_2^*\big(K_{X_2^w} + \Delta_2^w\big) \sim_{\mathbb{Q}} Z_2 - Z_1.$$

Applying (1.17) to $h_1 \colon Y \to X_1^w$ (resp. to $h_2 \colon Y \to X_2^w$) we obtain that $Z_2 - Z_1$ (resp. $Z_1 - Z_2$) is effective. Thus $Z_2 = Z_1$. This implies (1) since the discrepancies are exactly the coefficients in $Z_i - \Delta_Y$.

If (X_1^w, Δ_1^w) is a minimal model, then $\operatorname{Supp} Z_1$ contains all exceptional divisors of ϕ_1 by (1.19.5^{m}). Thus every ϕ_1-exceptional divisor also appears in Z_2 with positive coefficient, hence is also ϕ_2-exceptional. Therefore $\phi_2 \circ \phi_1^{-1} \colon X_1^w \dashrightarrow X_2^w$ is a birational contraction.

Finally, if (X_2^w, Δ_2^w) is also a minimal model then $\phi_1 \circ \phi_2^{-1}$ is also a birational contraction, giving (2).   □

Next we consider results that connect properties of (X, Δ) and of its weak canonical models.

Proposition 1.22 *Let $\phi \colon (X, \Delta) \dashrightarrow (X^w, \Delta^w)$ be a weak canonical model. Then*

(1) $a(E, X^w, \Delta^w) \geq a(E, X, \Delta)$ *for every divisor E,*
(2) $a(E, X^w, \Delta^w) > a(E, X, \Delta)$ *if (X^w, Δ^w) is the canonical model and ϕ is not a local isomorphism at the generic point of $\operatorname{center}_X E$,*
(3) $\operatorname{totaldiscrep}(X^w, \Delta^w) \geq \operatorname{totaldiscrep}(X, \Delta)$,
(4) $\operatorname{discrep}(X^w, \Delta^w) \geq \min\{\operatorname{discrep}(X, \Delta), a(E_i, X, \Delta) : i \in I\}$ *where the $\{E_i \colon i \in I\}$ are the ϕ-exceptional divisors and*
(5) *if (X, Δ) is lc (resp. klt) then so is (X^w, Δ^w).*

Proof Fix E and consider any diagram

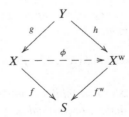

where (X^w, Δ^w) is a weak canonical model and $\operatorname{center}_Y E$ is a divisor. Write

$$K_Y \sim_{\mathbb{Q}} g^*(K_X + \Delta) + E_1, \quad \text{and}$$
$$K_Y \sim_{\mathbb{Q}} h^*(K_{X^w} + \Delta^w) + E_2.$$

Notice that $a(E, X^w, \Delta^w) - a(E, X, \Delta)$ is the coefficient of E in $E_2 - E_1$. Set

$$B := g^*(K_X + \Delta) - h^*(K_{X^w} + \Delta^w) \sim_{\mathbb{Q}} E_2 - E_1.$$

Then $-B$ is g-nef and $g_*B = g_*(E_2 - E_1)$ is effective by (1.19.5). Thus $E_2 - E_1$ is effective by (1.17).

If (X^w, Δ^w) is the canonical model and ϕ is not a local isomorphism at the generic point of center$_X$ E then center$_X$ $E \subset \mathrm{Supp}\, g_*(E_2 - E_1)$, thus $E \subset \mathrm{Supp}(E_2 - E_1)$ by (1.17.2).

Since discrep(Y, D) is the minimum of all discrepancies of divisors which are exceptional over Y, we see that the ϕ-exceptional E_i are taken into account when computing discrep(X^w, Δ^w) but not in the computation of discrep(X, Δ).   □

A given pair (X, Δ) usually has many different minimal models, and so far we have viewed them as essentially equivalent objects. In some applications it is important to understand exactly when the inequality in (1.22.1) is strict. There is a very clear answer for some minimal models.

Corollary 1.23 (Minimal models arising from MMP) *Assume that a minimal model (X^m, Δ^m) is obtained from (X, Δ) by running an MMP as in Kollár and Mori (1998, section 3.7), that is, ϕ is a composite of maps ϕ_i that sit in diagrams*

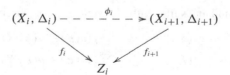

where $(X, \Delta) = (X_0, \Delta_0)$, $(X^m, \Delta^m) = (X_{j+1}, \Delta_{j+1})$ for some $j \geq 0$ and each (X_{i+1}, Δ_{i+1}) is the canonical model of (X_i, Δ_i) over Z_i. Then

(1) $a(E, X^m, \Delta^m) > a(E, X, \Delta)$ *iff $\phi : X \dashrightarrow X^m$ is not a local isomorphism at the generic point of center$_X$ E.*

(2) *If (X, Δ) is dlt then so is (X^m, Δ^m).*

Proof   (1.22.2) applies to each ϕ_i and hence to their composite ϕ, proving (1). Assume next that $a(E, X^m, \Delta^m) = -1$. Then

$$-1 = a(E, X^m, \Delta^m) \geq a(E, X, \Delta) \geq -1,$$

hence equality holds in both places and ϕ is a local isomorphism at the generic point of center$_X$ E by (1). Since (X, Δ) is dlt, it is snc at the generic point of center$_X$ E and so is (X^m, Δ^m).   □

Next we consider a result that makes it easy to decide when two pairs have the same canonical rings.

Theorem 1.24 *Let $f_1: X_1 \to S$ and $f_2: X_2 \to S$ be proper morphisms of normal schemes and $\phi: X_1 \dashrightarrow X_2$ a birational map such that $f_1 = f_2 \circ \phi$. Let Δ_1 and Δ_2 be $\mathbb{Q}$-divisors such that $K_{X_1} + \Delta_1$ and $K_{X_2} + \Delta_2$ are $\mathbb{Q}$-Cartier. Then*

$$f_{1*}\mathcal{O}_{X_1}(mK_{X_1} + \lfloor m\Delta_1 \rfloor) = f_{2*}\mathcal{O}_{X_2}(mK_{X_2} + \lfloor m\Delta_2 \rfloor) \quad \text{for } m \geq 0$$

if the following conditions hold:

(1) $a(E, X_1, \Delta_1) = a(E, X_2, \Delta_2)$ *if ϕ is a local isomorphism at the generic point of E,*
(2) $a(E, X_1, \Delta_1) \leq a(E, X_2, \Delta_2)$ *if $E \subset X_1$ is ϕ-exceptional and*
(3) $a(E, X_1, \Delta_1) \geq a(E, X_2, \Delta_2)$ *if $E \subset X_2$ is ϕ^{-1}-exceptional.*

Proof Let Y be the normalization of the main component of $X_1 \times_S X_2$ and $g_i: Y \to X_i$ the projections. We can write

$$K_Y \sim_{\mathbb{Q}} g_1^*(K_{X_1} + \Delta_1) + \sum_E a(E, X_1, \Delta_1)E \quad \text{and}$$
$$K_Y \sim_{\mathbb{Q}} g_1^*(K_{X_2} + \Delta_2) + \sum_E a(E, X_2, \Delta_2)E.$$

Set $b(E) := \max\{-a(E, X_1, \Delta_1), -a(E, X_2, \Delta_2)\}$. By (2.19), it is sufficient to prove that

$$f_{i*}\mathcal{O}_{X_i}(mK_{X_i} + \lfloor m\Delta_i \rfloor) = (f_i \circ g_i)_*\mathcal{O}_Y(mK_Y + \sum_E \lfloor mb(E) \rfloor E) \quad (1.24.4)$$

holds for $i = 1, 2$ and any sufficiently divisible $m > 0$. Observe that

$$K_Y + \sum_E b(E)E \sim_{\mathbb{Q}} g_1^*(K_{X_1} + \Delta_1) + \sum_E(b(E) + a(E, X_1, \Delta_1))E, \quad \text{and}$$
$$K_Y + \sum_E b(E)E \sim_{\mathbb{Q}} g_1^*(K_{X_2} + \Delta_2) + \sum_E(b(E) + a(E, X_2, \Delta_2))E.$$

Note that $\sum_E(b(E) + a(E, X_i, \Delta_i))E$ is effective by the definition of $b(E)$. Furthermore, if E is not g_1-exceptional then either ϕ is a local isomorphism at the generic point of E (and thus $b(E) = -a(E, X_1, \Delta_1) = -a(E, X_2, \Delta_2)$) or $E \subset X_1$ is ϕ-exceptional (and thus $b(E) = -a(E, X_1, \Delta_1) \geq -a(E, X_2, \Delta_2)$). A similar argument applies to g_2. Therefore $\sum_E(b(E) + a(E, X_i, \Delta_i))E$ is effective and g_i-exceptional for $i = 1, 2$, Thus, for sufficiently divisible $m > 0$,

$$(f_i \circ g_i)_*\mathcal{O}_Y(mK_Y + \sum_E mb(E)E)$$
$$= f_{i*}g_{i*}\mathcal{O}_Y(g_1^*(mK_{X_1} + m\Delta_1) + \sum_E(mb(E) + ma(E, X_1, \Delta_1))E)$$
$$= f_{i*}\mathcal{O}_{X_i}(mK_{X_i} + m\Delta_i),$$

giving (1.24.4). □

Applying (1.19) to $X_1 = X$ and $X_2 = X^w$ gives the following.

Corollary 1.25 *Let $f: (X, \Delta) \to S$ be a pair and $f^w: (X^w, \Delta^w) \to S$ a weak canonical model as in (1.19). Then*

$$f_*\mathcal{O}_X(mK_X + \lfloor m\Delta \rfloor) = f_*^w\mathcal{O}_{X^w}(mK_{X^w} + \lfloor m\Delta^w \rfloor) \quad \text{for every } m \geq 0.$$

Thus if (X, Δ) has a canonical model $f^c \colon (X^c, \Delta^c) \to S$ then

$$\sum_{m \geq 0} f_* \mathscr{O}_X(m K_X + \lfloor m \Delta \rfloor) = \sum_{m \geq 0} f_*^c \mathscr{O}_{X^c}(m K_{X^c} + \lfloor m \Delta^c \rfloor)$$

is a finitely generated sheaf of $\mathscr{O}_S$-algebras.

Proof The only thing that remains to be established is that the last sheaf of $\mathscr{O}_S$-algebras is finitely generated. We claim that, more generally, if $g \colon W \to S$ is a proper morphism and D a g-ample $\mathbb{Q}$-Cartier $\mathbb{Q}$-divisor then $\sum_{m \geq 0} g_* \mathscr{O}_W(\lfloor m D \rfloor)$ is a finitely generated sheaf of $\mathscr{O}_S$-algebras.

Pick m_0 such that $H := m_0 D$ is Cartier. For $0 \leq i < m_0$ set $D_i := \lfloor i D \rfloor$ and note that $r H + D_i = \lfloor (r m_0 + i) D \rfloor$. Then $\sum_{r \geq 0} g_* \mathscr{O}_W(r H)$ is a finitely generated sheaf of $\mathscr{O}_S$-algebras and each $\sum_{r \geq 0} g_* \mathscr{O}_W(r H + D_i)$ is a finitely generated $\sum_{r \geq 0} g_* \mathscr{O}_W(r H)$-module (cf. Lazarsfeld (2004, section 1.8)). Thus

$$\sum_{m \geq 0} g_* \mathscr{O}_W(\lfloor m D \rfloor) = \sum_{0 \leq i < m_0} \sum_{r \geq 0} g_* \mathscr{O}_W(r H + D_i)$$

is a finitely generated sheaf of $\mathscr{O}_S$-algebras. $\qquad\square$

The following converse is a straightforward generalization of (1.15).

Theorem 1.26 *Let X be an irreducible, regular scheme, $f \colon X \to S$ a proper morphism and Δ an snc boundary on X. Assume that the (relative, log) canonical sheaf of rings*

$$R(X/S, K_X + \Delta) := \sum_{m \geq 0} f_* \mathscr{O}_X(m K_X + \lfloor m \Delta \rfloor)$$

is a finitely generated sheaf of $\mathscr{O}_S$-algebras and $K_X + \Delta$ is big on the generic fiber of f. Then

(1) *$X^{\mathrm{can}} := \mathrm{Proj}_S R(X/S, K_X + \Delta)$ is normal, projective over S and there is a natural birational map $\phi \colon X \dashrightarrow X^{\mathrm{can}}$.*
(2) *The class $K_{X^{\mathrm{can}}} + \Delta^{\mathrm{can}}$ is $\mathbb{Q}$-Cartier and ample over S where $\Delta^{\mathrm{can}} := \phi_* \Delta$.*
(3) *$(X^{\mathrm{can}}, \Delta^{\mathrm{can}})$ is the unique canonical model of (X, Δ).*
(4) *Push-forward by ϕ gives an isomorphism*

$$\sum_{m \geq 0} f_* \mathscr{O}_X(m K_X + \lfloor m \Delta \rfloor) \simeq \sum_{m \geq 0} f_*^{\mathrm{can}} \mathscr{O}_{X^{\mathrm{can}}}(m K_{X^{\mathrm{can}}} + \lfloor m \Delta^{\mathrm{can}} \rfloor).$$

(5) *If (X, Δ) is log canonical then so is $(X^{\mathrm{can}}, \Delta^{\mathrm{can}})$.*
(6) *If $K_X + \Delta$ is f-nef then $\phi \colon X \dashrightarrow X^{\mathrm{can}}$ is a morphism.* $\qquad\square$

Next we investigate when different choices of Δ lead to the same models.

Proposition 1.27 *Let $\pi \colon X' \to X$ be a proper birational morphism. Let Δ and Δ' be $\mathbb{Q}$-divisors on X and X' such that $\pi_* \Delta' = \Delta$. Assume that $a(E, X', \Delta') \leq a(E, X, \Delta)$ for every π-exceptional divisor $E \subset X'$. (This holds if (X, Δ) is lc and $\mathrm{coeff}_E \Delta' = 1$ for every π-exceptional divisor E or if (X, Δ) is klt and $\mathrm{coeff}_E \Delta' \geq 1 - \varepsilon$ for $0 < \varepsilon \ll 1$.) Then*

(1) *Every weak canonical (resp. canonical) model of (X, Δ) is also a weak canonical (resp. canonical) model of (X', Δ').*
(2) *A canonical model of (X', Δ') is also a canonical model of (X, Δ).*
(3) *If $a(E, X', \Delta') < a(E, X, \Delta)$ for every π-exceptional divisor E then every weak canonical (resp. minimal) model of (X', Δ') is also a weak canonical (resp. minimal) model of (X, Δ).*

Proof Let (X^w, Δ^w) be a weak canonical model of (X, Δ). If E is any divisor on X', then $a(E, X', \Delta') \le a(E, X, \Delta)$ (and equality holds if E is not π-exceptional). Thus, by (1.22.1), $a(E, X', \Delta') \le a(E, X, \Delta) \le a(E, X^w, \Delta^w)$. The other assumptions in (1.19) hold automatically, hence (X^w, Δ^w) is also a weak canonical model of (X', Δ').

Conversely, let (X'^w, Δ'^w) be a weak canonical model of (X', Δ') and $\phi': X' \dashrightarrow X'^w$ the corresponding birational map. Set $\phi := \phi' \circ \pi^{-1}: X \dashrightarrow X'^w$. Note that ϕ is a contraction if and only if the following assumption is satisfied.

(4) Every π-exceptional divisor $E \subset X'$ is contracted by ϕ'.

If this holds, then $\Delta'^w = \phi'_* \Delta' = \phi_* \Delta$, thus the assumptions (1.19.1–4) are satisfied. Furthermore, if $F \subset X$ is ϕ-exceptional, then $a(E, X, \Delta) = a(E, X', \Delta') \le a(E, X'^w, \Delta'^w)$, thus (1.19.5) also holds and so (X'^w, Δ'^w) is a weak canonical model of (X, Δ).

We still need to show that either of the assumptions in (2) or (3) implies (1.19.5). As in (1.19.6.b) we have

$$K_{X'} + \Delta' \sim_{\mathbb{Q}} \phi'^*(K_{X'^w} + \Delta'^w) + F_1,$$

where F_1 is effective and supported on the ϕ'-exceptional locus. By our assumptions

$$K_{X'} + \Delta' \sim_{\mathbb{Q}} \pi^*(K_X + \Delta) + F_2,$$

where F_2 is effective and supported on the π-exceptional locus. Subtracting these two from each other, we obtain that

$$F_1 - F_2 \sim_{\mathbb{Q}} \pi^*(K_X + \Delta) - \phi'^*(K_{X'^w} + \Delta'^w).$$

Note that $\pi_*(F_1 - F_2) = \pi_* F_1$ is effective and $-(F_1 - F_2)$ is π-nef since $K_{X'^w} + \Delta'^w$ is f'^w-nef. Thus, by (1.17), $F_1 - F_2$ is effective.

Assume to the contrary that (1.19.5) fails for E. Then E appears in F_1 with coefficient 0 and in F_2 with coefficient $a(E, X, \Delta) - a(E, X', \Delta')$. Since $F_1 - F_2$ is effective these imply that $a(E, X, \Delta) = a(E, X', \Delta')$, which is impossible if (3) holds. Furthermore, (1.17.2), shows that $\phi'^*(K_{X'^w} + \Delta'^w)|_E$ is numerically π-trivial. This is impossible if (2) holds. $\square$

Lemma 1.28 *Consider a diagram as in* (1.19).

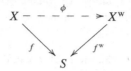

Let Δ_1, Δ_2 be $\mathbb{Q}$-divisors on X such that $\Delta_1 \sim_{\mathbb{Q}} \Delta_2$. Assume that $(X^{\mathrm{w}}, \phi_* \Delta_1)$ is an f-weak canonical (resp. f-canonical, f-minimal) model of (X, Δ_1).

Then $(X^{\mathrm{w}}, \phi_* \Delta_2)$ is also an f-weak canonical (resp. f-canonical, f-minimal) model of (X, Δ_2).

Proof By assumption there is an $m \geq 1$ and a rational function h on X such that $\Delta_1 - \Delta_2 = \frac{1}{m}(h)_X$ where $(h)_X$ denotes the divisor of h on X. Thus

$$\phi_* \Delta_1 - \phi_* \Delta_2 = \tfrac{1}{m}(h)_{X^{\mathrm{w}}}.$$

In particular, $K_{X^{\mathrm{w}}} + \phi_* \Delta_2$ is $\mathbb{Q}$-Cartier (resp. f-nef or f-ample) if and only if $K_{X^{\mathrm{w}}} + \phi_* \Delta_2$ is. As in (1.19.6.b), write

$$K_X + \Delta_i - \phi^*(K_{X^{\mathrm{w}}} + \phi_* \Delta_i) \sim_{\mathbb{Q}} \Delta_i^{\mathrm{ex}} + \textstyle\sum_j a_{ij} E_j. \qquad (1.28.1.i)$$

By definition, $(X^{\mathrm{w}}, \phi_* \Delta_i)$ is an f-weak canonical (resp. f-minimal) model of (X, Δ_i) if and only if the right hand side of $(1.28.1.i)$ is effective (resp. effective and with support $\mathrm{Ex}(\phi)$).

Subtracting the equations $(1.28.1.i)$ from each other we obtain that

$$\Delta_1^{\mathrm{ex}} + \textstyle\sum_j a_{1j} E_j \sim_{\mathbb{Q}} \Delta_2^{\mathrm{ex}} + \textstyle\sum_j a_{2j} E_j.$$

Thus, a multiple of their difference is a function divisor $(g)_X$. Since these divisors are ϕ-exceptional, we see that $(g)_{X^{\mathrm{w}}} = 0$. Thus g is a regular function on X^{w}, hence on S and also on X. Therefore

$$\Delta_1^{\mathrm{ex}} + \textstyle\sum_j a_{1j} E_j = \Delta_2^{\mathrm{ex}} + \textstyle\sum_j a_{2j} E_j. \qquad (1.28.2)$$

This completes the proof. □

Existence of minimal and canonical models

Let us start with a very general form of the minimal model conjecture.

Conjecture 1.29 (Minimal model conjecture) *Let $f \colon X \to S$ be a proper, dominant morphism between normal, irreducible schemes (or algebraic spaces) with generic fiber X_{gen}. Let Δ be an effective $\mathbb{R}$-divisor on X such that (X, Δ) is lc. Then*

(1) *(X, Δ) has a minimal model $(X^{\mathrm{m}}, \Delta^{\mathrm{m}})$ if and only if the restriction of $K_X + \Delta$ to the generic fiber X_{gen} is pseudo-effective (that is, it is numerically*

equivalent to a limit of effective $\mathbb{Q}$-divisors). Furthermore, if X is dlt (resp. $\mathbb{Q}$-factorial) then one can choose X^m to be dlt (resp. $\mathbb{Q}$-factorial).

(2) *(X, Δ) has a canonical model if and only if the restriction of $K_X + \Delta$ to X_{gen} is big.*

1.30 (Known special cases)   The minimal model conjecture (1.29) is known in the following instances.

1.30.1 (Surfaces)   If X is a smooth surface over a field k, this is the classical theory of minimal models, see Beauville (1983, chapter II) or Barth *et al.* (1984, section III.4). The general case can be established along the same lines using resolution of singularities (Shafarevich, 1966) and general contractibility criteria for curves on surfaces (Lipman, 1969).

Assumption   For (1.30.2–8), we always assume that S is essentially of finite type over a field of characteristic 0.

1.30.2 (3-folds) $\dim X \leq 3$.
    The case when X is terminal and $\Delta = 0$ is due to Mori (1988). The first part in the klt case was proved by Shokurov (1992). A different proof and a seminar-style work out of both parts in the lc case is in Kollár (1992).

1.30.3 (4-folds) $\dim X \leq 4$ and $K_{X_{\text{gen}}} + \Delta_{\text{gen}}$ is effective.
    The first part is due to Shokurov (2006) and the second part to Fujino (2000). See Corti (2007) for a detailed treatment.

1.30.4 (5-folds) $\dim X \leq 5$ and $K_{X_{\text{gen}}} + \Delta_{\text{gen}}$ is effective.
    The first part of the conjecture is proved in Birkar (2010).

1.30.5 (n-folds) (X, Δ) is klt and $K_{X_{\text{gen}}} + \Delta_{\text{gen}}$ is big or Δ_{gen} is big.
    This is the main result of Birkar *et al.* (2010). It easily implies the case $(X, \Delta + H)$ where (X, Δ) is lc and H is an f-ample $\mathbb{Q}$-divisor.
    An analytic approach to the existence of the canonical model is outlined in Siu (2008); see Păun (2010) and the references there for a more complete treatment. A direct algebraic approach is developed in Cascini and Lazić (2010, 2012).

The following three results seem rather special, but they are tailor-made for the applications in this book.

1.30.6 (n-folds) (Fujino, 2010)   Assume that $f \colon X \to S$ is birational, (X, Δ) is $\mathbb{Q}$-factorial, dlt and $K_X + \Delta \sim_{f,\mathbb{Q}} E$ where E is effective and f-exceptional.
    Then (X, Δ) has a dlt, $\mathbb{Q}$-factorial, minimal model over S.

Warning   The proofs of the next two cases use some of the results that we develop in Chapters 4–5. Thus we will avoid using them until Chapter 7.

1.30.7 (n-folds) (Hacon and Xu, 2011a)   Assume that $f: X \to S$ is projective, (X, Δ) is dlt, there is a dense open $S^0 \subset S$ such that every log canonical center (4.15) of (X, Δ) intersects $f^{-1}(S^0)$ and $(f^{-1}(S^0), \Delta|_{f^{-1}(S^0)}) \to S^0$ has a canonical model.

Then (X, Δ) has a dlt, $\mathbb{Q}$-factorial, minimal model over S and also a canonical model over S.

1.30.8 (n-folds) (Birkar, 2011; Hacon and Xu, 2011a)   Assume that $f: X \to S$ is projective and there is an effective divisor Δ' such that $(X, \Delta + \Delta')$ is dlt and $K_X + \Delta + \Delta' \sim_{\mathbb{Q},f} 0$.

Then (X, Δ) has a dlt, $\mathbb{Q}$-factorial, minimal model over S.

1.30.9 *Positive and mixed characteristic*   Very little is known in general. The first steps of the MMP for 3-folds in positive characteristic are treated in Kollár (1991a). For semistable families of surfaces over Dedekind rings the existence of minimal models is discussed in Kawamata (1994, 1999b).

1.4 Canonical models as partial resolutions

So far we have used the MMP to improve the global structure of a smooth variety Y or of an snc pair (Y, Δ_Y) but have paid a price by introducing singularities. Next we consider a quite different point of view. We start with a – possibly very singular – variety X (or a pair (X, Δ)) and our aim is to find a proper birational morphism $g: X' \to X$ whose global structure is as simple as possible yet the singularities of X' (or of (X', Δ')) are better than before.

It would be interesting to have such a program as part of a resolution process, but this seems completely out of reach for now. The proofs of the theorems below work by taking first a resolution of X (or of (X, Δ)) and then running a suitable MMP. Nonetheless, the end results are very useful.

Assumption   The results in this section hold if X is proper over a scheme S that is essentially of finite type over a field of characteristic 0.

For arbitrary S, the results would follow from the minimal model conjecture (1.29) and the existence of log resolutions.

Canonical and log canonical modifications

Let X be a normal variety and Δ a boundary on X. Let $f: X' \to X$ be a log resolution of (X, Δ) such that the irreducible components of $\Delta' := f_*^{-1}\Delta$ are disjoint from each other. Then (X', Δ') is canonical hence it has a canonical model over X. If $\lfloor \Delta \rfloor = 0$ this follows from (1.30.5). The general case is implied by (1.30.7). This gives the following.

Theorem 1.31 (Canonical modification) *Let X be a normal variety and Δ a boundary (1.3) on X. Then there is a unique, projective, birational morphism $g^{\text{can}}\colon X^{\text{can}} \to X$ such that*

(1) $(X^{\text{can}}, (g^{\text{can}})_*^{-1}\Delta)$ *is canonical and*
(2) $K_{X^{\text{can}}} + (g^{\text{can}})_*^{-1}\Delta$ *is g^{can}-ample.* □

The above argument also has a log version. Let E' be the sum of all f-exceptional divisors with coefficient 1 and take the canonical model of $(X', \Delta' + E')$ over X. Its existence would follow directly from the still unknown conjecture (1.29). If $K_X + \Delta$ is $\mathbb{Q}$-Cartier then Odaka and Xu (2012) showed that one can use Hacon and Xu (2011a) instead of (1.29).

Theorem 1.32 (Log canonical modification) *Let X be a normal variety and Δ a boundary on X. Assume that $K_X + \Delta$ is $\mathbb{Q}$-Cartier. Then there is a unique projective, birational morphism $g^{\text{lc}}\colon X^{\text{lc}} \to X$ such that*

(1) $(X^{\text{lc}}, (g^{\text{lc}})_*^{-1}\Delta + E^{\text{lc}})$ *is log canonical, where E^{lc} is the sum of all g^{lc}-exceptional divisors with coefficient 1 and*
(2) $K_{X^{\text{lc}}} + (g^{\text{lc}})_*^{-1}\Delta + E^{\text{lc}}$ *is g^{lc}-ample.* □

A great advantage of the above models is their uniqueness, but their singularities can still be rather complicated, especially in the lc case. We can improve the singularities if in the above proofs we stop at a minimal model over X. The price we pay is that we have to give up relative ampleness, hence uniqueness. (Therefore, it is not clear that the existence of such models is a local question on X, so we better assume that X is quasi-projective.) These results follow from Birkar *et al.* (2010). The dlt case needs additional arguments as in Odaka and Xu (2012).

Theorem 1.33 (Terminal modification) *Let X be a normal, quasi-projective variety and Δ a boundary on X such that $\lfloor \Delta \rfloor = 0$. Then there is a non-unique, projective, birational morphism $g^{\text{min}}\colon X^{\text{min}} \to X$ such that*

(1) $(X^{\text{min}}, (g^{\text{min}})_*^{-1}\Delta)$ *is terminal and*
(2) $K_{X^{\text{min}}} + (g^{\text{min}})_*^{-1}\Delta$ *is g^{min}-nef.* □

Theorem 1.34 (Dlt modification) *Let X be a normal, quasi-projective variety and Δ a boundary on X. Assume that $K_X + \Delta$ is $\mathbb{Q}$-Cartier. Then there is a non-unique, projective, birational morphism $g^{\text{dlt}}\colon X^{\text{dlt}} \to X$ such that*

(1) $(X^{\text{dlt}}, (g^{\text{dlt}})_*^{-1}\Delta + E^{\text{dlt}})$ *is dlt, where E^{dlt} is the sum of all g^{dlt}-exceptional divisors with coefficient 1, and*
(2) $K_{X^{\text{dlt}}} + (g^{\text{dlt}})_*^{-1}\Delta + E^{\text{dlt}}$ *is g^{dlt}-nef.* □

Next we turn the lack of uniqueness in (1.33)–(1.34) to our advantage.

Cornucopia of dlt models

The following result gives us great flexibility in choosing a birational model.

1.35 (Picking birational models) Let (X, Δ) be lc, Δ a boundary and $f: Y \to (X, \Delta)$ a log resolution. Let $\{E_i: i \in I\}$ be the f-exceptional divisors and pick rational numbers c_i such that

$$\max\{0, -a(E_i, X, \Delta)\} \le c_i \le 1 \quad \text{for every } i \in I.$$

Set $\Delta' := \sum_{i \in I} c_i E_i$ and run the $(Y, f_*^{-1}\Delta + \Delta')$-MMP. Note that

$$K_Y + f_*^{-1}\Delta + \Delta' \sim_{f,\mathbb{Q}} \sum_{i \in I} (c_i + a(E_i, X, \Delta))E_i,$$

and the right hand side, call it F, is effective and f-exceptional. Furthermore Supp F is the union of those E_i for which $c_i > -a(E_i, X, \Delta)$. By (1.30.6), a suitable MMP terminates with a $\mathbb{Q}$-factorial, dlt, minimal model

$$f^{\mathrm{m}}: (Y^{\mathrm{m}}, (f^{\mathrm{m}})_*^{-1}\Delta + \Delta'_{\mathrm{m}}) \to X.$$

Note that F has ≥ 0 intersection with any curve not contained in Supp F and it has negative intersection with some curve contained in F and contracted by f by (1.17). This continues to hold after any number of steps of the MMP, thus the MMP contracts all the divisors in Supp F and it does not contract any other divisor.

In applications the key question is the choice of the coefficients c_i. Varying the c_i gives different models Y^{m}. Let us see how this works in some examples.

Corollary 1.36 (Thrifty dlt modification) *Let (X, Δ) be lc and Δ a boundary. Then there is a proper birational morphism $g: X^{\mathrm{dlt}} \to X$ with exceptional divisors E_i such that*

(1) $(X^{\mathrm{dlt}}, g_*^{-1}\Delta + \sum E_i)$ *is dlt,*
(2) X^{dlt} *is $\mathbb{Q}$-factorial,*
(3) $K_{X^{\mathrm{dlt}}} + g_*^{-1}\Delta + \sum E_i \sim_{\mathbb{Q}} g^*(K_X + \Delta)$ *and*
(4) g *is an isomorphism over* $\mathrm{snc}(X, \Delta)$ (1.7).

Proof Start with a log resolution that is an isomorphism over $\mathrm{snc}(X, \Delta)$ as in (10.45.2), choose $c_i = 1$ for every i and apply (1.35). The only exceptional divisors that survive in X^{dlt} are those divisors $E_i \subset Y$ for which $a(E_i, X, \Delta) = -1$. $\qquad\square$

Corollary 1.37 (Small $\mathbb{Q}$-factorial modification) *Let (X, Δ) be dlt and Δ a boundary. Then there is a proper birational morphism $g: X^{\mathrm{qf}} \to X$ such that*

(1) X^{qf} *is $\mathbb{Q}$-factorial,*
(2) g *is small, that is, without g-exceptional divisors,*

(3) $(X^{\mathrm{qf}}, g_*^{-1}\Delta)$ *is dlt and*

(4) $\mathrm{discrep}(X^{\mathrm{qf}}, g_*^{-1}\Delta) = \mathrm{discrep}(X, \Delta)$. *Thus if (X, Δ) is terminal (resp. canonical, resp. klt) then so is $(X^{\mathrm{qf}}, g_*^{-1}\Delta)$.*

Proof By (10.45.2) we can take a log resolution $f\colon Y \to (X, \Delta)$ such that every f-exceptional divisor has discrepancy > -1. Pick all $c_i = 1$ and apply (1.35) to get $g\colon X^{\mathrm{qf}} \to X$. All the f-exceptional divisors get killed, hence $g\colon X^{\mathrm{qf}} \to X$ is small. The last claim follows from (2.23). $\qquad\square$

Cones over elliptic curves show that (1.37) fails in general if (X, Δ) is only lc.

Even for 3-folds, X^{qf} depends subtly on the coefficients of the equations defining X. For instance, if X_h is the analytic germ $(0 \in (x_1 x_2 + h(x_3, x_4) = 0)) \subset \mathbb{C}^4$ then $X_h^{\mathrm{qf}} \simeq X_h$ iff $h(x_3, x_4)$ is irreducible (as a power series); see Kollár (1991b, 2.2.7).

Corollary 1.38 (Extracting specific divisors) *Let (X, Δ) be lc, Δ a boundary and $\{E_i : i \in I\}$ finitely many exceptional divisors over X such that $-1 \le a(E_i, X, \Delta) \le 0$. Then there is a $\mathbb{Q}$-factorial, dlt model $g\colon (X^E, \Delta^E) \to (X, \Delta)$ such that*

(1) *the $\{E_i : i \in I\}$ are among the exceptional divisors of g and*

(2) *every other exceptional divisor F of g has discrepancy -1.*

If (X, Δ) is dlt and $-1 < a(E_i, X, \Delta) < 0$ then we can arrange that the $\{E_i : i \in I\}$ are all the exceptional divisors of g.

Proof Pick any log resolution $Y \to X$ such that every E_i is a divisor on Y. Choose all $c_j = 1$ except $c_i = -a(E_i, X, \Delta)$ for $i \in I$ and apply (1.35). The only divisors that survive are $\{E_i : i \in I\}$ and those with discrepancy -1.

If (X, Δ) is dlt then by (1.37) we may assume that X is $\mathbb{Q}$-factorial. Then $(X, (1 - \varepsilon)\Delta)$ is klt and $-1 < a(E_i, X, (1 - \varepsilon)\Delta) < 0$ still holds for $0 < \varepsilon \ll 1$. Thus we get $g\colon (X^E, \Delta^E) \to (X, \Delta)$ such that the $\{E_i : i \in I\}$ are all the exceptional divisors of g.

We still need to prove that (X^E, Δ^E) is dlt. Let F be a divisor over (X^E, Δ^E) with discrepancy -1. Then X is smooth and Δ is Cartier at the generic point η of $\mathrm{center}_X F$. Thus there are no exceptional divisors with non-integral discrepancy over η and $\mathrm{Ex}(g)$ has pure codimension 1 over η. Thus g is an isomorphism over η. $\qquad\square$

If we allow (X^E, Δ^E) to be lc, most likely one can arrange that the $\{E_i : i \in I\}$ are exactly the exceptional divisors of g, but in the dlt case both of the inequalities $-1 < a(E_i, X, \Delta) < 0$ are necessary. As a simple example, take $(X, \Delta) = (\mathbb{A}^2, (xy = 0))$. Let E_0 be obtained by blowing up the origin and E_1 by blowing up a point on E_0. Depending on where the point is, $a(E_1, X, \Delta)$

is either 0 or -1. The only model where E_1 is the sole exceptional divisor is obtained by contracting the birational transform of E_0. We get a non-dlt point this way.

If $a(E, X, \Delta) > 0$ then (1.38) may fail, even if X is smooth and we allow X^E to have arbitrary singularities. Such examples are given in Cossart *et al.* (2000) and Küronya (2003).

Corollary 1.39 (Extracting one divisor)   *Let (X, Δ) be dlt and E an exceptional divisor over X such that $-1 < a(E, X, \Delta) < 0$. Then there is a model $g: (X^E, \Delta^E) \to (X, \Delta)$ such that*

(1) (X^E, Δ^E) *is dlt,*
(2) X^E *is $\mathbb{Q}$-factorial,*
(3) $-E$ *is g-nef,*
(4) $E = \operatorname{Supp} g^{-1}(g(E))$,
(5) E *is the sole exceptional divisor of g and*
(6) $E = \operatorname{Ex}(g)$ *iff $X \setminus g(E)$ is $\mathbb{Q}$-factorial.*

The main advantage of this result is that if $Z \subset X$ is a subvariety that intersects $g(E)$ then E cannot be disjoint from $g_*^{-1}Z$.

Proof   Let $h: (X', \Delta') \to (X, \Delta)$ be a $\mathbb{Q}$-factorial, dlt model whose sole exceptional divisor is E (1.38).

If (X, Δ) is klt, then so is $(X', h_*^{-1}\Delta)$, thus the latter has a $\mathbb{Q}$-factorial minimal model $g: (X^E, g_*^{-1}\Delta) \to X$ by (1.30.5). Note that

$$K_{X'} + h_*^{-1}\Delta \sim_{\mathbb{Q}} h^*(K_X + \Delta) - cE \sim_{h,\mathbb{Q}} -cE, \qquad (1.39.7)$$

where $c = -a(E, X, \Delta) > 0$.

We claim that $X' \dashrightarrow X^E$ does not contract E. Indeed, if $\phi_i: X_i' \to X_{i+1}'$ contracts E then E has negative intersection with some contracted curve by (1.17), but this contradicts (1.39.7). Thus the MMP ends with $-E$ becoming g-nef. Since g has connected fibers, this implies (4–6).

If (X, Δ) is only dlt, then, by Kollár and Mori (1998, 2.43) or (3.71), it can be approximated by klt pairs (X, Δ_ε). We claim that the model constructed for (X, Δ_ε) also works for (X, Δ). The only question is to show that (X^E, Δ^E) is dlt. To see this note that every divisor whose center intersects $\operatorname{snc}(X, \Delta)$ has integer discrepancy. Thus $\operatorname{center}_X E \subset \operatorname{non-snc}(X, \Delta)$ and so every divisor whose center is contained in $\operatorname{center}_X E$ has discrepancy > -1. Thus every divisor F such that $\operatorname{center}_{X^E} F \subset E$ has discrepancy > -1 by (2.23) and therefore (X^E, Δ^E) is dlt. $\square$

One can extend (1.39) to the lc case using (1.30.8). An even more general result will be proved in (7.3).

Non-normal cases

The minimal model program does not seem to work for non-normal varieties. Kollár (2011e) gives examples of surfaces S with only normal crossing singularities whose canonical ring $\sum_{m \geq 0} H^0(S, \omega_S^m)$ is not finitely generated. Here we construct similar examples for reducible 3-folds.

Example 1.40 (Nonexistence of slc modification) In the first example, X will be a reducible 3-fold with two irreducible components X_1, X_2. There are surfaces $D_i \subset X_i$ and X is obtained by gluing X_1 and X_2 using an isomorphism $\sigma \colon D_1 \simeq D_2$.

As we see in Chapter 5, the only way to get a (semi-)log canonical modification of X is to construct the lc modification $(X_i^{\mathrm{lc}}, D_i^{\mathrm{lc}})$ and glue them together by an isomorphism $\sigma^{\mathrm{lc}} \colon D_1^{\mathrm{lc}} \simeq D_2^{\mathrm{lc}}$.

In the example below, $D_1^{\mathrm{lc}} \not\simeq D_2^{\mathrm{lc}}$, thus the two components $(X_i^{\mathrm{lc}}, D_i^{\mathrm{lc}})$ cannot be glued together to a (semi-)log canonical modification of X.

Take $(X_1, D_1) := (\mathbb{A}^3_{uvw}/\frac{1}{3}(1, 1, 1), (w = 0)/\frac{1}{3}(1, 1) \simeq \mathbb{A}^2_{uv}/\frac{1}{3}(1, 1))$ (for the notation used here see (3.19.2)). Then (X_1, D_1) is lc, even plt, hence its log canonical modification is itself $\pi_1 \colon (X_1^{\mathrm{lc}}, D_1^{\mathrm{lc}}) \simeq (X_1, D_1)$.

Let D_2 be the image of the embedding $\mathbb{A}^2_{uv}/\frac{1}{3}(1, 1) \hookrightarrow \mathbb{A}^3_{xyzt}$ given by $\sigma \colon (u, v) \mapsto (u^3, u^2v, uv^2, v^3)$ and set $(X_2, D_2) := ((xt - yz = 0), D_2) \subset \mathbb{A}^3_{xyzt}$. Use $\sigma \colon D_1 \simeq D_2$ to glue (X_1, D_1) and (X_2, D_2) to obtain X.

To compute the log canonical modification of (X_2, D_2), note that D_2 also satisfies the equation $xz = y^2$ and $X_2 \cap (xz = y^2)$ is the union of D_2 and of a residual plane $P := (x = y = 0)$.

Let $\pi_2 \colon X_2^{\mathrm{lc}} \to X_2$ be the blow-up of the plane P and $C \subset X_2^{\mathrm{lc}}$ the exceptional curve. Let D_2^{lc} (resp. P^{lc}) denote the birational transforms of D_2 (resp. P). Then $\pi_2^*(D_2 + P) = D_2^{\mathrm{lc}} + P^{\mathrm{lc}}$ and $(C \cdot P^{\mathrm{lc}}) = -1$. Thus $(C \cdot D_2^{\mathrm{lc}}) = 1$ hence $K_{X_2^{\mathrm{lc}}} + D_2^{\mathrm{lc}}$ is π_2-ample. By explicit computation, $X_2^{\mathrm{lc}}, D_2^{\mathrm{lc}}$ are both smooth, thus $\pi_2 \colon (X_2^{\mathrm{lc}}, D_2^{\mathrm{lc}}) \to (X_2, D_2)$ is indeed the log canonical modification. Furthermore, $\pi_2 \colon D_2^{\mathrm{lc}} \to D_2$ is the blow-up of the origin, hence it is not an isomorphism.

Thus the isomorphism $\sigma \colon D_1 \simeq D_2$ gives a birational map $\sigma^{\mathrm{lc}} \colon D_1^{\mathrm{lc}} \dashrightarrow D_2^{\mathrm{lc}}$ that is not an isomorphism. Therefore $(X_1^{\mathrm{lc}}, D_1^{\mathrm{lc}})$ and $(X_2^{\mathrm{lc}}, D_2^{\mathrm{lc}})$ cannot be glued together.

The following, more singular example applies in all dimensions.

Let $Z \subset \mathbb{P}^{n-1}$ be a smooth hypersurface of degree n and $D_1 \subset X_1 := \mathbb{A}^n$ the cone over Z. Then (X_1, D_1) is an lc pair by (3.1).

Fix a point $p \in \mathbb{P}^1$, embed $Z \times \mathbb{P}^1$ into $\mathbb{P}^{2n-1}$ by the global sections of $\mathscr{O}_{Z \times \mathbb{P}^1}(1, 1)$ and let (X_2, D_2) be the cone over $((Z \times \mathbb{P}^1), (Z \times \{p\}))$.

One can glue (X_1, D_1) and (X_2, D_2) using the natural isomorphism $D_1 \simeq D_2$ to obtain a demi-normal variety X. We claim that it has no slc modification.

First note that $(X_1^{lc}, D_1^{lc}) \simeq (X_1, D_1)$.

One can describe (X_2^{lc}, D_2^{lc}) as follows. Resolve the cone X_2 by blowing up its vertex. The exceptional divisor is then $Z \times \mathbb{P}^1$ and $\pi_2 \colon X_2^{lc} \to X_2$ is obtained by contracting the $\mathbb{P}^1$-factor of the exceptional divisor. Thus $(\pi_2)_*^{-1} D_2$ is the blow-up of the cone over Z at its vertex.

Thus we see that (X_1^{lc}, D_1^{lc}) and (X_2^{lc}, D_2^{lc}) cannot be glued together to an slc modification of X.

It is not hard to construct many similar examples, but in all of them K_X is not $\mathbb{Q}$-Cartier. It is proved in Odaka and Xu (2012) that the natural non-normal version of (1.32) holds as long as $K_X + \Delta$ is $\mathbb{Q}$-Cartier.

1.5 Some special singularities

1.41 (Nodes)   We say that a scheme S has a *node* at a point $s \in S$ if its local ring $\mathcal{O}_{s,s}$ can be written as $R/(f)$ where (R, m) is a regular local ring of dimension 2, $f \in m^2$ and f is not a square in m^2/m^3. If char $k(s) \neq 2$ then a node is a double nc point.

Claim 1.41.1   Let (A, n) be a 1-dimensional local ring with residue field k and normalization $\bar{A}$. Let $\bar{n}$ be the intersection of the maximal ideals of $\bar{A}$. Assume that (A, n) is a quotient of a regular local ring. Then (A, n) is nodal if and only if $\dim_k(\bar{A}/\bar{n}) = 2$ and $\bar{n} \subset A$.

Proof   If (A, n) is nodal then blowing up n gives the normalization and these properties are clear.

Conversely, if $\dim_k(\bar{A}/\bar{n}) = 2$ then $\dim_k(\bar{n}^r/\bar{n}^{r+1}) = 2$ for every r. Let $x, y \in \bar{n}$ be a k-basis of $\bar{n}/\bar{n}^2$. Then x^2, xy, y^2 are k-linearly dependent in $\bar{n}^2/\bar{n}^3$. The relation is not a perfect square since

$$(\bar{n}/\bar{n}^2) \otimes_{\bar{A}/\bar{n}} (\bar{n}/\bar{n}^2) \to (\bar{n}^2/\bar{n}^3) \quad \text{is an isomorphism.}$$

Thus x, y generate n and and the resulting relation is not a square modulo n^3.   □

1.41.2 *Deformation of nodes*   Let (B', n') be a complete DVR with maximal ideal n' and (R, m) a complete, local B'-algebra such that R is flat over B'.

If B'/n' is perfect, or, more generally, if R/m is separably generated over B'/n', then there is a complete DVR $(B', n') \subset (B, n) \subset (R, m)$ such that R is a quotient of a formal power series ring $B[[x_1, \ldots, x_r]]$ (cf. Matsumura, 1986, section 29).

If R/nR has a node at m, then R has dimension 2 and embedding dimension ≤ 3, hence we can write

$$R \simeq B[[x, y]]/(G(x, y))$$

for some $G \in B[[x, y]]$ that defines a node over B/n. That is, there is a quadratic form $q(x, y) := ax^2 + bxy + cy^2$ with $a, b, c \in B$ such that

$$(\tfrac{\partial q}{\partial x}, \tfrac{\partial q}{\partial y}) = (x, y) \quad \text{and} \quad G - q(x, y) \in nB[[x, y]] + (x, y)^3.$$

Assume that there are coordinates (x_r, y_r) such that $G - q(x_r, y_r) \in m^r(x, y) + n$. For suitable $h_x, h_y \in m^r$, we can define new coordinates $x_{r+1} := x_r + h_x, y_{r+1} := y_r + h_y$ such that $G - q(x_{r+1}, y_{r+1}) \in m^{r+1}(x, y) + n$. We repeat this until, at the end, in suitable coordinates (x_∞, y_∞),

$$R \simeq B[[x_\infty, y_\infty]]/(ax_\infty^2 + bx_\infty y_\infty + cy_\infty^2 + d) \quad \text{where } a, b, c \in B \text{ and } d \in n.$$

If char $B/n \neq 2$ then this can be further simplified to $B[[x, y]]/(x^2 + cy^2 + d)$ where $c \in B \setminus n$ and $d \in n$. If, in addition, B/n is algebraically closed, then it can be written as $B[[x, y]]/(xy + d)$.

Finally, if B is a power series ring $\mathbb{C}[[z]]$ then we obtain the simplest forms

$$R \simeq \mathbb{C}[[x, y, z]]/(xy + z^r) \quad \text{for } r \geq 1 \text{ or } \quad R \simeq \mathbb{C}[[x, y, z]]/(xy).$$

1.42 (*cA*-type singularities) Let Y be a smooth variety and $X \subset Y$ a divisor.

At a singular point $p \in X$ choose local coordinates y_i and let $f(y_1, \ldots, y_{n+1}) = 0$ be a local equation of X. Since p is a singular point, $f \in (y_1, \ldots, y_{n+1})^2$ and we have a well-defined quadratic form $\bar{f} \in (y_1, \ldots, y_{n+1})^2/(y_1, \ldots, y_{n+1})^3$.

We say that X has *cA-type singularities* if rank $\bar{f} \geq 2$ for every singular point $p \in X$. Equivalently, if there are at least two linearly independent partials $\partial f/\partial y_i$ modulo $(y_1, \ldots, y_{n+1})^2$. Thus cA-type singularities form an open subset of Sing X.

Over an algebraically closed field, up-to formal or analytic coordinate change, one can define a cA-type singularity by an equation $y_1 y_2 - g(y_3, \ldots, y_{n+1}) = 0$. These are exactly the singularities of the total spaces we get by deforming a node over a smooth base space.

By an easy computation, if X has cA-type singularities and $Z \subset$ Sing X is a smooth subvariety then the blow-up $B_Z X$ again has cA-type singularities. (By contrast, for $C := (x = y = z = 0) \subset (xy + zt^n = 0) =: X$ the blow-up $B_C X$ has a singular point with equation $xyz + t^n = 0$.)

An induction as in (2.10) shows that a normal cA-type singularity is canonical and if codim$_X$ Sing $X \geq 3$ then X is terminal.

These singularities also appear in the following Bertini-type theorem of Kollár (1997, 4.4): Let X be a smooth variety over a field of characteristic 0 and $|H|$ a linear system. Assume that for every $x \in X$ there is an $H_x \in |H|$

such that either $x \notin H_x$ or H_x is smooth at x. Then the general $H \in |H|$ has only cA-type singularities.

Definition 1.43 (Pinch points) Let X be a nonsingular variety over a field of characteristic $\neq 2$ and $D \subset X$ a divisor. We say that D has a *pinch point* at $p \in D$ if, in suitable local coordinates, D can be defined by the equation $x_1^2 - x_2^2 x_3 = 0$.

Note that this notion is invariant under field extensions and even completion. Indeed, if the singular set of D is a codimension 2 nonsingular subvariety, then D can be locally given by an equation $ax_1^2 + bx_1x_2 + cx_2^2 = 0$ where a, b, c are regular functions. If the quadratic part of the equation is a square times a unit, then, after a coordinate change, we can write the equation as $x_1^2 + cx_2^2 = 0$. This gives a pinch point after a field extension and completion if and only if the linear term of c is independent of x_1, x_2. Thus we can take $x_3 = -c$ to get the equation $x_1^2 - x_2^2 x_3 = 0$.

Let us blow-up $Z := (x_1 = x_2 = 0)$. The normalization of D is contained in the affine chart with coordinates $x_1' := x_1/x_2, x_2, \ldots, x_n$. If we introduce $x_3' := x_3 - x_1'^2$ then the normalization of D is given by $(x_3' = 0)$. The preimage of Z is the nonsingular divisor $(x_2 = 0)$ and Z is the quotient of it by the involution $\tau \colon (x_1', 0, 0, x_4, \ldots, x_n) \mapsto (-x_1', 0, 0, x_4, \ldots, x_n)$.

A function f defines a τ-invariant divisor if and only if

$$f(x_1', x_2, x_4, \ldots, x_n) = \begin{cases} g\big(x_1'^2, x_4, \ldots, x_n\big) + x_2 h(x_1', x_2, x_4, \ldots, x_n), & \text{or} \\ x_1' g\big(x_1'^2, x_4, \ldots, x_n\big) + x_2 h(x_1', x_2, x_4, \ldots, x_n). \end{cases}$$

In the first case f is τ-invariant and descends to a regular function on D. In the second case f is not τ-invariant, but f^2 descends to a regular function on D.

In particular, $(x_1 = x_3 = 0) \subset (x_1^2 = x_2^2 x_3)$ is not a Cartier divisor but it is $\mathbb{Q}$-Cartier since $2(x_1 = x_3 = 0) = (x_3 = 0)$ is Cartier.

Conversely, let Y be a nonsingular variety, $B \subset Y$ a nonsingular divisor and τ an involution on B whose fixed point set $F \subset B$ has pure codimension 1 in B. Let $Z := B/\tau$ and X the universal push-out of $Z \leftarrow B \hookrightarrow Y$ (9.30). Then X has only nc and pinch points.

To see this, pick a point $p \in F$ and local coordinates $y_1, \ldots, y_n$ such that $B = (y_1 = 0)$, $\tau^* y_2|_B = -y_2|_B$ and $\tau^* y_i|_B = y_i|_B$ for $i > 2$. Then

$$x_1 := y_1 y_2, x_2 := y_1, x_3 := y_2^2 \quad \text{and} \quad x_i := y_{i-1} \quad \text{for } i > 3$$

give local coordinates on X with the obvious equation $x_1^2 - x_2^2 x_3 = 0$.

Definition 1.44 An excellent scheme Y is called *unibranch* or *topologically normal* if the following equivalent conditions hold.

(1) For every $y \in Y$ the completion $\hat{Y}_y$ is irreducible.

(2) For every $y \in Y$, every étale neighborhood $(y' \in Y') \to (y \in Y)$ is irreducible.

(3) The normalization $\pi \colon \bar{Y} \to Y$ is a universal homeomorphism.

(4) The normalization $\pi \colon \bar{Y} \to Y$ is universally open.

(5) Let $f \colon X \to Y$ be a quasi-finite morphism such that every irreducible component of X dominates an irreducible component of Y. Then $f \colon X \to Y$ is universally open. (That is, for every $Z \to Y$, the induced morphism $X \times_Y Z \to Z$ is open.)

(6) For every $y \in Y$, the field extension $k(\operatorname{red} \pi^{-1}(y)) \supset k(y)$ is purely inseparable.

The equivalence, especially (5), is called *Chevalley's criterion;* see Grothendieck (1960, IV.14.4.4).

By (3) normal schemes are unibranch. More precisely, Y is normal if and only if it is unibranch and seminormal.

If $f \colon X \to Y$ is surjective, universally open and X is unibranch then so is Y.

Note that if Y is any irreducible curve then the normalization $\pi \colon \bar{Y} \to Y$ is open. Thus in (4) it is not enough to assume that $\pi \colon \bar{Y} \to Y$ open.

2

Canonical and log canonical singularities

In this chapter we introduce the singularities – *terminal, canonical, log terminal* and *log canonical* – that appeared in connection with the minimal model program.

Section 2.1 contains the basic definitions and their fundamental properties. Whenever possible, we work with schemes, rather than just varieties over a field of characteristic 0.

Section 2.2 starts a detailed study of log canonical surface singularities. Here we focus on their general properties; a complete classification is postponed to Section 3.3.

Ramified covers are studied in Section 2.3. This method gives a good theoretical framework to study log canonical singularities and their deformations in all dimensions, as long as the degree of the cover is not divisible by the characteristic.

Ramified covers have been especially useful in dimensions 2 and 3, leading to a classification of 3-dimensional terminal, canonical and log terminal singularities in characteristic 0. A short summary of this approach is given in Section 2.4. These topics are excellently treated in Reid (1980, 1987).

Divisorial log terminal singularities and their rationality properties are investigated in Section 2.5. Roughly speaking, these form the largest well-behaved subclass of log canonical singularities. We prove that they are rational and many important sheaves on them are Cohen-Macaulay (2.88), at least in characteristic 0.

This section also shows the two, rather pervasive, problems that arise when one tries to generalize results of birational geometry to positive and mixed characteristics. The first is the lack of resolution of singularities and the second is the failure of Kodaira-type vanishing theorems. Resolution of singularities is expected to hold in general and in many cases one can get around it. For now, the failure of vanishing theorems seems to present a bigger challenge.

Assumptions   In this chapter we work with arbitrary schemes as in (1.5–1.6).

2.1 (Log) canonical and (log) terminal singularities

While our eventual main interest is in varieties over a field of characteristic zero, in this section we define and prove everything for general schemes as in (1.5). The definitions and results are all local in the étale or analytic topology, hence they carry over to algebraic and complex analytic spaces.

As motivation, we start with varieties over a perfect field.

Definition 2.1 (Discrepancy I)   Let X be a normal variety over a perfect field such that mK_X is Cartier for some $m > 0$. Suppose $f: Y \to X$ is a birational morphism from a normal variety Y. Let $E \subset Y$ be an irreducible divisor and $e \in E$ a general point of E. Let $(y_1, \dots, y_n)$ be a local coordinate system at $e \in Y$ such that $E = (y_1 = 0)$. Then, locally near e,

$$f^*(\text{local generator of } \mathcal{O}_X(mK_X) \text{ at } f(e))$$
$$= y_1^{c(E,X)} \cdot (\text{unit}) \cdot (dy_1 \wedge \cdots \wedge dy_n)^{\otimes m}$$

for some integer $c(E, X)$. The rational number $a(E, X) := \frac{1}{m} c(E, X)$ is called the *discrepancy* of E with respect to X. It is independent of the choice of m.

We refer to any such E as a *divisor over* X. The closure of $f(E) \subset X$ is called the *center* of E on X. It is denoted by $\text{center}_X E$.

If $f: Y \to X$ is a birational morphism then mK_Y is linearly equivalent to

$$f^*(mK_X) + \sum_i (m \cdot a(E_i, X))E_i,$$

where the E_i are the f-exceptional divisors. We can formally divide by m and write

$$K_Y \sim_{\mathbb{Q}} f^*K_X + \sum_i a(E_i, X)E_i.$$

Assume that $f': Y' \to X$ is another birational morphism and $E' \subset Y'$ an irreducible divisor such that the rational map $f^{-1} \circ f': Y' \to X \dashrightarrow Y$ is a local isomorphism mapping a general point $e \in E$ to a general point $e' \in E'$. Then we see from the definition that $a(E, X) = a(E', X)$ and $\text{center}_X E = \text{center}_X E'$. Because of this, in discrepancy considerations, we frequently do not distinguish E from E'.

Warning   While the discrepancy $a(E, X)$ is independent of the choice of the birational model Y that contains E, it is very sensitive to any changes of the base variety X. It is, in fact, frequently useful to fix $E \subset Y$ and study how $a(E, X)$ changes as we vary X in its birational equivalence class.

A basic property of discrepancy is that it is positive when X is smooth.

Proposition 2.2 *Let X be a smooth variety defined over a perfect field k. Then $a(E, X) \geq 1$ for every exceptional divisor E over X.*

Proof Let $f: Y \to X$ be a birational morphism, Y normal, $E \subset Y$ an exceptional divisor and $e \in E$ a general point. Choose local coordinates $(y_1, \ldots, y_n)$ near $e \in Y$ and $(x_1, \ldots, x_n)$ near $f(e) \in X$. Then

$$f^*(dx_1 \wedge \cdots \wedge dx_n) = \text{Jac}\left(\frac{x_1, \ldots, x_n}{y_1, \ldots, y_n}\right) \cdot dy_1 \wedge \cdots \wedge dy_n, \qquad (2.2.1)$$

thus $a(E, X)$ is the order of vanishing of the Jacobian along E. Hence $a(E, X) \geq 0$ and $a(E, X) = 0$ if and only if f is étale over an open subset of E. (Over $\mathbb{C}$ this is the inverse function theorem.) Thus $a(E, X) \geq 1$ for every exceptional divisor E. $\qquad \square$

2.3 (Discrepancies and sections of ω_X) Let Y be a smooth variety over a field of characteristic 0 and $g: Y' \to Y$ a proper, birational morphism. The above computation gives a proof of the equality $g_*\omega_{Y'} = \omega_Y$. More generally, let X be a normal variety and $f: Y \to X$ a resolution of singularities. By comparing different resolutions using (2.2.1), we conclude that the image of

$$f_*\omega_Y \hookrightarrow \omega_X \quad \text{is independent of } Y. \qquad (2.3.1)$$

How does this image compare with the discrepancies? The answer is very simple if ω_X is locally free but more complicated otherwise.

Claim 2.3.1 Assume that ω_X is locally free. Then $f_*\omega_Y = \omega_X$ if and only if $a(E, X) \geq 0$ for every exceptional divisor E.

Proof We can write $\omega_Y = f^*\omega_X(F)$ where $F = \sum_i a(E_i, X)E_i$. Thus

$$f_*\omega_Y = f_*(f^*\omega_X(F)) = \omega_X \otimes f_*\mathcal{O}_Y(F).$$

If $a(E, X) \geq 0$ for every divisor E then F is effective and exceptional, hence $f_*\mathcal{O}_Y(F) = \mathcal{O}_X$ and so $f_*\omega_Y = \omega_X$. Conversely, if $f_*\omega_Y = \omega_X$ then $f_*\mathcal{O}_Y(F) = \mathcal{O}_X$ hence the constant section $1_X \in \mathcal{O}_X$ lifts to a section f^*1_X whose zero divisor is F. Thus F is effective. Every exceptional divisor appears on some resolution, hence $a(E, X) \geq 0$ for every divisor E. $\qquad \square$

Claim 2.3.2 Assume that K_X is $\mathbb{Q}$-Cartier. If $a(E, X) > -1$ for every exceptional divisor E then $f_*\omega_Y = \omega_X$. The converse usually does not hold.

Proof If mK_X is Cartier, we have an equality $\mathcal{O}_Y(mK_Y) = f^*\mathcal{O}_X(mK_X)(mF)$ where $F = \sum_i a(E_i, X)E_i$. Set $G := \sum_i E_i$. Since $ma(E_i, X) > -m$ is an integer, we know that $(m - 1)G + mF$ is effective. Thus the previous argument yields that

$$f_*\mathcal{O}_Y(mK_Y + mG)(-G) = \mathcal{O}_X(mK_X). \qquad (2.3.4)$$

Take any local section ϕ of $\mathcal{O}_X(K_X)$. Then $f^*\phi$ is a rational section of $\mathcal{O}_Y(K_Y)$, so $(f^*\phi)^m$ is a rational section of $\mathcal{O}_Y(mK_Y)$ and hence also of $\mathcal{O}_Y(mK_Y + mG)$. On the other hand, ϕ^m is a section of $\mathcal{O}_X(mK_X)$, thus (2.3.4) shows that $(f^*\phi)^m$ is a section of $\mathcal{O}_Y(mK_Y + mG)$ that vanishes along G. If an mth power vanishes along G, it vanishes with multiplicity $\geq m$. Thus $(f^*\phi)^m$ is a section of $\mathcal{O}_Y(mK_Y + mG)(-mG) = \mathcal{O}_Y(mK_Y)$, hence $f^*\phi$ is a section of $\mathcal{O}_Y(K_Y)$. $\square$

Now we switch to the general case as in (1.5) and extend the definition (2.1) to pairs (X, Δ) such that $K_X + \Delta$ is $\mathbb{Q}$-Cartier.

Definition 2.4 (Discrepancy II) Let $(X, \Delta = \sum a_i D_i)$ be a pair as in (1.5). (As usual, we suppress the base scheme B in the notation.) Assume that $m(K_X + \Delta)$ is Cartier for some $m > 0$. Equivalently, $m\Delta$ has integral coefficients and $\omega_X^{[m]}(m\Delta)$ is locally free.

Suppose $f: Y \to X$ is a (not necessarily proper) birational morphism from a reduced scheme Y. Let $E \subset Y$ denote the exceptional locus of f and $E_i \subset E$ the irreducible exceptional divisors. We assume that Y is regular at the generic point of each E_i. (This is automatic if Y is normal.) Let

$$f_*^{-1}\Delta := \sum a_i f_*^{-1} D_i$$

denote the birational transform of Δ. (The latter is well defined since the D_i are not contained in Sing X.) There is a natural isomorphism of invertible sheaves

$$\iota_{Y \setminus E}: \omega_Y^{[m]}(mf_*^{-1}\Delta)|_{Y \setminus E} \simeq f^*\left(\omega_X^{[m]}(m\Delta)\right)\big|_{Y \setminus E}. \qquad (2.4.1)$$

Thus there are rational numbers $a(E_i, X, \Delta)$ such that $m \cdot a(E_i, X, \Delta)$ are integers, and $\iota_{Y \setminus E}$ extends to an isomorphism

$$\iota_Y: \omega_Y^{[m]}(mf_*^{-1}\Delta) \simeq f^*(\omega_X^{[m]}(m\Delta))(\textstyle\sum_i m \cdot a(E_i, X, \Delta)E_i). \qquad (2.4.2)$$

This defines $a(E, X, \Delta)$ for exceptional divisors. Set $a(D, X, \Delta) := -\text{coeff}_D \Delta$ for nonexceptional divisors $D \subset X$.

The rational number $a(E, X, \Delta)$ is called the *discrepancy* of E with respect to (X, Δ). As in the $\Delta = 0$ case, $a(E_i, X, \Delta)$ depends only on E_i but not on f.

As we see in (2.7), for most cases of interest to us, $a(E, X, \Delta) \geq -1$. For this reason, some authors use *log discrepancies*, defined as $1 + a(E, X, \Delta)$.

We want to stress that this definition works for non-normal schemes as well. For instance, let C be a curve with a single node $p \in C$. Let $f: C' \to C$ denote the normalization and p', p'' the preimages of the node. Then p', p'' are exceptional divisors of f and the definition gives that $a(p', C, 0) = a(p'', C, 0) = -1$.

The discrepancies measure the singularities of a pair (X, Δ) together. Positive discrepancies indicate that (X, Δ) is mildly singular and negative values indicate that (X, Δ) is more singular.

When $\Delta = 0$, we measure the singularities of X and when X is regular, we measure the singularities of Δ, but the interplay between X an Δ is quite subtle. In general, the discrepancies have the following obvious linearity property.

Lemma 2.5 *With the above notation, assume that Δ' is $\mathbb{Q}$-Cartier and let $E \subset Y$ be a divisor. Then*

$$a(E, X, \Delta) = a(E, X, \Delta + \Delta') + \operatorname{coeff}_E f^*\Delta'.$$

In particular, if Δ' is effective then $a(E, X, \Delta) \geq a(E, X, \Delta + \Delta')$ for every divisor E over X and strict inequality holds if and only if $\operatorname{center}_X E \subset \operatorname{Supp} \Delta'$. □

Notation 2.6 The pull-back and the discrepancies can be conveniently packaged into any of the equivalent forms:

$$K_Y + f_*^{-1}\Delta \sim_{\mathbb{Q}} f^*(K_X + \Delta) + \sum_{E_i:\text{exceptional}} a(E_i, X, \Delta) E_i, \quad \text{or}$$
$$K_Y \qquad \sim_{\mathbb{Q}} f^*(K_X + \Delta) + \sum_{E_i:\text{arbitrary}} a(E_i, X, \Delta) E_i, \quad \text{or}$$
$$K_Y + \Delta_Y \sim_{\mathbb{Q}} f^*(K_X + \Delta) \quad \text{where } f_*\Delta_Y = \Delta \text{ if } X \text{ is normal, or}$$
$$K_Y + \Delta_Y \sim_{\mathbb{Q}} f^*(K_X + \Delta) \quad \text{where } f_*(K_Y + \Delta_Y) = K_X + \Delta \text{ in general.}$$

Note, however, that these formulas do not show that the isomorphisms in (2.4.1–2) are canonical.

We frequently refer to these formulas by saying, for instance: "write $K_Y \sim_{\mathbb{Q}} f^*(K_X + \Delta) + A$." In this case it is understood that A is chosen as above. That is, we have to make sure that $A = \sum_i a(E_i, X, \Delta) E_i$.

It is easy to see that $a(E, Y, \Delta_Y) = a(E, X, \Delta)$ for every divisor E (Kollár and Mori, 1998, 2.30). A birational morphism satisfying these properties will be called *crepant*; see (2.23).

The basic example is the following version of (2.2).

Proposition 2.7 *Let X be a regular scheme and $\Delta = \sum_{i \in I} a_i D_i$ an snc subboundary (1.3). For a divisor E over X set $I(E) := \{i \in I : \operatorname{center}_X E \subset D_i\}$. Then*

$$a(E, X, \Delta) \geq -1 + \sum_{i \in I(E)} (1 - a_i).$$

In particular, $a(E, X, \Delta) \geq -1$ and $a(E, X, 0) \geq 0$.

Proof Assume first that we are over a perfect field and $a_i = 1$ for every i. Let $f: Y \to X$ be a birational morphism from a normal variety Y such that E is an f-exceptional divisor. Let $e \in E$ be a general point and choose local coordinates $(y_1, \ldots, y_n)$ such that $E = (y_1 = 0)$.

Let $(x_1, \ldots, x_n)$ be local coordinates in a neighborhood $U \ni f(e)$ such that $\Delta \cap U = (x_1 \cdots x_r = 0)$. A local generator of $\mathscr{O}_U(K_U + \Delta)$ is given by

$$\sigma_U := \frac{dx_1}{x_1} \wedge \cdots \wedge \frac{dx_r}{x_r} \wedge dx_{r+1} \wedge \cdots \wedge dx_n.$$

Near e, we can write $f^* x_i = y_1^{a_i} \cdot u_i$ where u_i is a unit. Thus

$$f^* \frac{dx_i}{x_i} = \frac{d\left(y_1^{a_i} \cdot u_i\right)}{y_1^{a_i} \cdot u_i} = a_i \frac{dy_1}{y_1} + \frac{du_i}{u_i}.$$

Since $dy_1 \wedge dy_1 = 0$, we obtain that

$$f^* \sigma_U = \frac{dy_1}{y_1} \wedge \omega_{n-1} + \omega_n,$$

where ω_{n-1} is a regular $(n-1)$-form and ω_n is a regular n-form. Thus $f^* \sigma_U$ is a section of $\mathscr{O}_Y(K_Y + E)$ near E and so $a(E, X, \Delta) \geq -1$, as required.

The case when $a_i \leq 1$ follows from this and the equality (2.5). For regular schemes we work out a more general formula in (2.10). $\qquad \square$

Next we define the six classes of singularities that are most important for the minimal model program.

Definition 2.8 Let (X, Δ) be a pair as in (1.5) where X is a normal scheme of dimension ≥ 2 and $\Delta = \sum a_i D_i$ is a subboundary, that is, a formal sum of distinct prime divisors and a_i are rational numbers, all ≤ 1. Assume that $m(K_X + \Delta)$ is Cartier for some $m > 0$. We say that (X, Δ) is

terminal		> 0	for every exceptional E,
canonical		≥ 0	for every exceptional E,
klt	if $a(E, X, \Delta)$ is	> -1	for every E,
plt		> -1	for every exceptional E,
dlt		> -1	if center$_X E \subset$ non-snc(X, Δ),
lc		≥ -1	for every E.

Here klt is short for "*Kawamata log terminal,*" plt for "*purely log terminal,*" dlt for "*divisorial log terminal*" and lc for "*log canonical.*" As in (1.7), non-snc(X, Δ) denotes the set of points where (X, Δ) is not snc.

We usually say that X is terminal (canonical, ...) or that it has terminal (canonical, ...) singularities if and only if $(X, 0)$ is terminal (canonical, ...).

(The frequently used phrase "(X, Δ) has terminal, etc. singularities" may be confusing since it could refer to the singularities of X instead.)

Each class contains the previous one, except canonical does not imply klt if Δ contains a divisor with coefficient 1. The key point is to show that if $a(E, X, \Delta) \geq -1$ for every exceptional E then $a(E, X, \Delta) \geq -1$ for every E (Kollár and Mori, 1998, 2.31). This last claim fails if dim $X = 1$ since there are no exceptional divisors at all. If dim $X = 1$ then $(X, \sum a_i D_i)$ is terminal/klt if and only if $a_i < 1$ for every i; the other four concepts all coincide and they hold if and only if $a_i \le 1$ for every i.

Warning on effectivity   In final applications, the above concepts are useful only if Δ is an effective divisor, and frequently in the literature the definitions assume that $\Delta \ge 0$. However, in some inductive proofs, it is very convenient to allow Δ to contain some divisors with negative coefficients. (See (2.23) for a typical example.) The usage is inconsistent in the literature, probably even in this book.

Each of these six notions has an important place in the theory of minimal models:

(1) Terminal: Assuming $\Delta = 0$, this is the smallest class that is necessary to run the minimal model program for nonsingular varieties. The $\Delta \ne 0$ case appears less frequently.

(2) Canonical: Assuming $\Delta = 0$, these are precisely the singularities that appear on the canonical models of varieties of general type. This class is especially important for moduli problems.

   If $K_X + \Delta$ is Cartier then the discrepancies are all integers, and so plt implies canonical.

(3) Kawamata log terminal: The proofs of the vanishing theorems seem to run naturally in this class but it is not suitable for inductive proofs.

   If $\Delta = 0$ then the notions klt, plt and dlt coincide and in this case we say that X has *log terminal* (abbreviated as *lt*) singularities.

   This class is also easy to connect to analysis. If M is a smooth complex manifold and f is a meromorphic function and $\Delta := (f = 0) - (f = \infty)$, then $(M, c\Delta)$ is klt if and only if $|f|^{-c}$ is locally L^2, see (8.3) or Kollár (1997, 3.2).

(4) Purely log terminal: This class was invented for inductive purposes. We will see that (X, Δ) is plt if and only if it is dlt and $\lfloor \Delta \rfloor$ is normal (4.16).

(5) Divisorial log terminal: These are the singularities we obtain if we start with an snc pair (X, Δ) and run the MMP. These are much better behaved than log canonical pairs.

(6) Log canonical: This is the largest class where discrepancy still makes sense. It contains many cases that are rather complicated from the cohomological point of view, therefore it is quite hard to work with. However, these singularities appear naturally on the stable varieties at the boundary of the moduli spaces of varieties of general type and in many applications.

For basic examples illustrating the nature of these singularities see Chapter 3.

Given (X, Δ), the most important value for us is the minimum of $a(E, X, \Delta)$ as E runs through various sets of divisors. We use several versions:

Definition 2.9   The *discrepancy* of (X, Δ) is given by

$$\mathrm{discrep}(X, \Delta) := \inf_{E}\{a(E, X, \Delta) : E \text{ is an exceptional divisor over } X\}.$$

(That is, E runs through all the irreducible exceptional divisors of all birational morphisms $f: Y \to X$.)

The *total discrepancy* of (X, Δ) is defined as

$$\mathrm{totaldiscrep}(X, \Delta) := \inf_{E}\{a(E, X, \Delta) : E \text{ is a divisor over } X\}.$$

(That is, $E \subset Y$ runs through all the irreducible exceptional divisors for all birational morphisms $f: Y \to X$ and through all the irreducible divisors of X.)

If $W \subset X$ is a subscheme, the *minimal log discrepancy* of (or over) W is defined as

$$\mathrm{mld}(W, X, \Delta) := \inf_{E}\{1 + a(E, X, \Delta) : \mathrm{center}_X E = W\}.$$

(Using *log* discrepancies seems the usual variant here.) Let $w \in W$ be the generic point. Localizing at w we obtain $w \in X_w$ and $\Delta_w := \Delta|_{X_w}$. Then $\mathrm{mld}(W, X, \Delta) = \mathrm{mld}(w, X_w, \Delta_w)$, thus any question about minimal log discrepancies can be usually reduced to the case when W is a closed point.

For a reducible subscheme $W \subset X$ with irreducible components W_i we set $\mathrm{mld}(W, X, \Delta) := \max_i\{\mathrm{mld}(W_i, X, \Delta)\}$.

The minimal log discrepancy is a refinement of the discrepancy since

$$\mathrm{totaldiscrep}(X, \Delta) = \inf_W\{\mathrm{mld}(W, X, \Delta) : W \subset X\} - 1 \quad \text{and}$$

$$\mathrm{discrep}(X, \Delta) = \inf_W\{\mathrm{mld}(W, X, \Delta) : W \subset X, \mathrm{codim}_X W \geq 2\} - 1.$$

One problem with the above definitions is that one needs to check discrepancies on all possible birational maps to X. We start by computing $\mathrm{mld}(W, X, \Delta)$ for an snc pair (X, Δ). Then we use resolution of singularities to reduce the general case to the snc case (2.23).

In particular, these imply that over a field of characteristic 0 all the infima in (2.9) are minima; this is not known in general.

2.10 (Computing minimal log discrepancies)   Let X be a regular scheme, $W \subset X$ a regular subscheme and $\pi: B_W X \to X$ the blow-up with exceptional divisor $E \subset B_W X$. Then $K_{B_W X} = \pi^* K_X + (\mathrm{codim}_X W - 1)E$. (This formula is given in Hartshorne (1977, exercise II.8.5). The set-up there tacitly assumes that X is over a perfect field, but the suggested proof works in general.) Thus

$$\pi^*(K_X + \Delta) \sim K_{B_W X} + \pi_*^{-1}\Delta + (1 - \mathrm{codim}_X W + \mathrm{mult}_W \Delta)E.$$

If $\Delta = \sum a_i D_i$ is an snc divisor then $\text{mult}_W \Delta = \sum_{i:D_i \supset W} \text{coeff}_{D_i} \Delta$. As in Kollár and Mori (1998, 2.31), we can use this inductively to compute discrepancies in towers of blow-ups of regular subschemes.

Let now $W \subset X$ be an arbitrary reduced subscheme. Since $\text{mld}(W, X, \Delta)$ concerns only those divisors that dominate W, we can harmlessly replace X with any open subscheme that contains the generic point of W. We can thus replace X by $X \setminus \text{Sing } W$ and assume that we blow up a regular subscheme at each step.

Finally we use (2.22) to see that such towers of blow-ups account for all exceptional divisors over X. Thus we conclude that if Δ is a subboundary then

$$\text{mld}(W, X, \Delta) = \text{codim}_X W - \sum_{i:D_i \supset W} \text{coeff}_{D_i} \Delta. \qquad (2.10.1)$$

In particular, if $\text{mld}(W, X, \Delta) < 1$, then W is a stratum of $\Delta_{>0}$ (1.3). For such W, we can rewrite the formula as

$$\text{mld}(W, X, \Delta) = \sum_{i:D_i \supset W}(1 - \text{coeff}_{D_i} \Delta) = \sum_{i:D_i \supset W} \text{mld}(D_i, X, \Delta). \qquad (2.10.2)$$

This method also gives a formula for all possible discrepancies that occur over an snc pair $(X, \sum a_i D_i)$.

Claim 2.10.3   Let E be a divisor over X. Then one can write

$$1 + a(E, X, \Delta) = m_0 + \sum_i m_i(1 - a_i)$$

where $m_0 \geq 0, m_i \geq 1$ are integers and the summation is over those i such that $\text{center}_X E \subset D_i$.

For a non-snc divisor Δ it is much harder to understand what happens after several blow-ups and there is only one case with a simple answer. If $\text{mult}_p \Delta \leq 1$ for every $p \in X$ then the coefficient $(1 - \text{codim}_X W + \text{mult}_W \Delta)$ of the exceptional divisor E of the first blow-up is ≤ 0. Thus, by (2.23) and (2.5) (X, Δ) is canonical (resp. terminal) if $(B_W X, \pi_*^{-1}\Delta)$ is canonical (resp. terminal).

We have to be a little careful with induction, since in general $\pi_*^{-1}\Delta$ can have points of multiplicity > 1. However, as above, we can shrink X before each blow-up and assume that Δ is equimultiple along W. Then $\pi_*^{-1}\Delta$ has only points of multiplicity $\leq \text{mult}_W \Delta$ over W. Using (2.22) we get the following.

Claim 2.10.4   Let X be a regular scheme and Δ an effective $\mathbb{Q}$-divisor. If $\text{mult}_p \Delta \leq 1$ (resp. < 1) for every $p \in X$ then (X, Δ) is canonical (resp. terminal). □

These imply the following global version (cf. Kollár and Mori, 1998, 2.31).

Corollary 2.11 *Let X be a regular scheme and $\Delta = \sum a_i D_i$ an snc sub-boundary. Then*

$$\operatorname{discrep}(X, \Delta) = \min\left\{1, \min_i\{1 - a_i\}, \min_{D_i \cap D_j \neq \emptyset}\{1 - a_i - a_j\}\right\}.$$

In particular, $-1 \leq \operatorname{discrep}(X, \Delta) \leq 1$ *and* $\operatorname{discrep}(X, \Delta) = \operatorname{discrep}(X, \Delta_{>0})$. $\quad\square$

We can put these results together to get the following.

Corollary 2.12 *Let X be a normal scheme such that K_X is $\mathbb{Q}$-Cartier and $f: Y \to X$ a resolution of singularities. Write $K_Y \sim_{\mathbb{Q}} f^*K_X + \sum a_i E_i$ where the sum runs over all f-exceptional divisors E_i. Then*

(1) *X is canonical if and only if $a_i \geq 0$ for every i.*
(2) *X is terminal if and only if $a_i > 0$ for every i.*
(3) *If X is canonical then $\operatorname{discrep}(X) = \min\{1, \min_i\{a_i\}\}$.*

Proof Since $a(E_i, X) = a_i$, the conditions are necessary. Conversely, assume that $a_i \geq 0$. If F is an exceptional divisor of f then $a(F, X)$ equals the coefficient of F in $\sum a_i E_i$, hence it is positive by assumption. If F is any divisor that is exceptional over Y, then, by (2.23), $a(F, X) = a(F, Y, -\sum a_i E_i)$ and, using (2.5) and (2.7), we obtain that $a(F, Y, -\sum a_i E_i) \geq a(F, Y) \geq 1$. $\quad\square$

The proof of the version with boundary is the same:

Corollary 2.13 *Let X be a normal scheme and $\Delta = \sum d_j D_j$ a $\mathbb{Q}$-divisor such that $K_X + \Delta$ is $\mathbb{Q}$-Cartier. Let $f: Y \to X$ be a log resolution. Write $K_Y \sim_{\mathbb{Q}} f^*(K_X + \Delta) + \sum a_i E_i$. Then*

(1) *(X, Δ) is log canonical if and only if $a_i \geq -1$ for every i.*
(2) *(X, Δ) is klt if and only if $a_i > -1$ for every i.* $\quad\square$

(Note that by our conventions the birational transforms $f_*^{-1}(D_j)$ are among the E_i with coefficient $a_i = -d_j$. Thus the restrictions on the a_i imply that $d_j \leq 1$ (resp. $d_j < 1$) for every j. A formula for $\operatorname{discrep}(X, \Delta)$ is in (2.11).)

The following basic result is easy if (X, Δ) has a log resolution, but takes more work in general.

2.14 (Smooth base change and discrepancies) Let (X, Δ) be a pair as in (1.5). Let $p: Y \to X$ be a smooth surjection and set $\Delta_Y := p^*\Delta$. Our aim is to compare the discrepancies occuring over (X, Δ) with the discrepancies occuring over (Y, Δ_Y).

Let $g: X' \to X$ be a birational morphism and $F_X \subset X'$ a g-exceptional divisor. Set $Y' := Y \times_X X'$ with projections $p': Y' \to X'$, $g': Y' \to Y$ and $F_Y := p'^{-1}F_X$. Then F_Y is a union of g'-exceptional divisors on Y' and

$a(F_Y, Y, \Delta_Y) = a(F_X, X, \Delta)$ since $K_{Y'/X'} = g'^* K_{Y/X}$. We have thus proved the following.

(1) The correspondence p^*: $F_X \mapsto$ {irreducible components of F_Y} is a multi-valued map p^*: (divisors over X) $\to$ (divisors over Y) which preserves discrepancies.
(2) If p is étale then p^* is surjective by (2.22). In particular discrep$(Y, \Delta_Y) =$ discrep(X, Δ).

If $\dim Y > \dim X$ then the image of p^* misses two types of divisors E_Y.

First, it can happen that E_Y dominates X. In this case center$_Y E_Y$ is a regular point of Y hence $a(E_Y, Y, \Delta_Y) \geq 1$ by (2.7).

Second, it is possible that there is a divisor F_X over X such that center$_{Y'} E_Y \subset F_Y$ dominates F_X but is not one of the irreducible components of F_Y. In this case again center$_{Y'} E_Y$ is a regular point of Y' and using (2.7) we see that $a(E_Y, Y, \Delta_Y) \geq 1 + a(F_X, X, \Delta)$. In particular

(3) discrep$(Y, \Delta_Y) =$ discrep(X, Δ).

To get a sharper result, let $S \subset Y$ be a divisor such that $p|_S$: $S \to X$ is smooth. Then the above considerations also show that

(4) discrep$(Y, S + \Delta_Y) =$ totaldiscrep(X, Δ).

The most important special cases of these equalities are the following.

Proposition 2.15 *Let (X, Δ) be a pair with Δ effective and p: $Y \to X$ a smooth surjection. Then (X, Δ) is terminal (resp. canonical, klt, plt, lc) if and only if $(Y, p^*\Delta)$ is terminal (resp. canonical, klt, plt, lc).* $\square$

Warning By contrast, dlt is *not* an étale local notion since its definition involves snc which is not étale local. If (X, Δ) is dlt then so is $(Y, p^*\Delta)$ but the converse does not hold. For instance, the pair $(\mathbb{A}^2, (x^2 + y^2 = 0))$ is not dlt over $\mathbb{R}$ but it is dlt over $\mathbb{C}$.

2.16 (Local nature of discrepancy) Essentially by definition, the discrepancy of a pair (X, Δ) is a local invariant. That is, if X_p denotes the spectrum of the local ring of a point $p \in X$ and Δ_p is the restriction of Δ to X_p then

$$\text{discrep}(X, \Delta) = \inf\{\text{discrep}(X_p, \Delta_p) : p \in X\}.$$

By (2.14.2) discrepancies can be computed after an étale base change, hence

$$\text{discrep}(X, \Delta) = \inf\left\{ \text{discrep}\left(X_p^h, \Delta_p^h\right) : p \in X\right\}$$

where X_p^h denotes the henselisation of X_p (4.38). By contrast, we do not know how to prove the similar formula for the completions $\hat{X}_p$ in general.

If X_p has a log resolution $Y_p \to X_p$, then the completion gives a log resolution of $\hat{X}_p$ and (2.11) shows that discrep$(\hat{X}_p, \hat{\Delta}_p) = $ discrep(X_p, Δ_p). Thus, in characteristic 0, discrep$(X, \Delta) = \inf\{$discrep$(\hat{X}_p, \hat{\Delta}_p) : p \in X\}$. Furthermore, every point $p \in X$ has an open neighborhood $p \in U_p \subset X$ such that

$$\text{discrep}(\hat{X}_p, \hat{\Delta}_p) = \text{discrep}(U_p, \Delta|_{U_p}).$$

In particular, in characteristic 0, each pair (X, Δ) has a smallest closed subscheme $Z \subset X$ such that $(X \setminus Z, \Delta|_{X \setminus Z})$ is lc. This Z is called the *non-lc locus* of (X, Δ). It is denoted by non-lc(X, Δ).

Similarly, there is a smallest closed subscheme $W \subset X$ called the *non-klt locus* such that $(X \setminus W, \Delta|_{X \setminus W})$ is klt. It is denoted by non-klt(X, Δ) or nklt(X, Δ). (Some authors call this the "log canonical locus," but this violates standard usage.)

In (3.76) we give examples of singular rational surfaces whose canonical class is ample. Thus, for singular varieties, the plurigenera are not birational invariants. The following result shows that canonical singularities form the largest class where the plurigenera are birational invariants. This was the question that led to the classification of canonical singularities of surfaces in Du Val (1934).

Proposition 2.17 *Let X be a normal projective variety such that K_X is $\mathbb{Q}$-Cartier and ample. Let $f: Y \dashrightarrow X$ be a birational map from a nonsingular proper variety to X. Then X has canonical singularities if and only if*

$$H^0(X, \mathscr{O}_X(mK_X)) = H^0(Y, \mathscr{O}_Y(mK_Y)) \quad \text{for every } m \geq 0.$$

Proof Let Y' be a normal, proper variety and $Y' \to Y$ a birational morphism such that the composite $g: Y' \to X$ is a morphism.

Pick m such that mK_X is Cartier and write $mK_{Y'} \sim g^*(mK_X) + A$ where A is g-exceptional. Then

$$\begin{aligned} H^0(Y, \mathscr{O}_Y(mK_Y)) &= H^0(Y', \mathscr{O}_{Y'}(mK_{Y'})) \\ &= H^0(X, g_*(\mathscr{O}_{Y'}(mK_{Y'}))) \\ &= H^0(X, \mathscr{O}_X(mK_X) \otimes g_*\mathscr{O}_{Y'}(A)). \end{aligned}$$

If $A \geq 0$ then $g_*\mathscr{O}_{Y'}(A) = \mathscr{O}_X$ hence the last term equals $H^0(X, \mathscr{O}_X(mK_X))$. If A is not effective then $g_*\mathscr{O}_{Y'}(A) \subsetneq \mathscr{O}_X$, thus, for $m \gg 1$,

$$H^0(Y, \mathscr{O}_Y(mK_Y)) = H^0(Y', \mathscr{O}_{Y'}(mK_{Y'})) \subsetneq H^0(X, \mathscr{O}_X(mK_X)).$$

This takes care of all sufficiently divisible values of m. The rest follows from (2.19). $\qquad\square$

An essentially identical proof shows the following.

Proposition 2.18 *Let* (X, Δ) *be a normal pair and* $p_X \colon X \to S$ *a proper morphism. Assume that* Δ *is a boundary and* $K_X + \Delta$ *is* $\mathbb{Q}$-*Cartier and* p_X-*ample.*

Let (Y, Δ_Y) *be an snc pair and* $p_Y \colon Y \to S$ *a proper morphism. Assume that* Δ_Y *is a boundary. Finally let* $f \colon Y \dashrightarrow X$ *be a birational contraction such that* $p_Y = p_X \circ f$, $f_*(\Delta_Y) = \Delta$ *and every irreducible component of* $\mathrm{Ex}(f)$ *appears in* Δ_Y *with coefficient 1. Then* (X, Δ) *is log-canonical if and only if*

$$(p_X)_* \mathscr{O}_X(mK_X + \lfloor m\Delta \rfloor)$$
$$= (p_Y)_* \mathscr{O}_Y(mK_Y + \lfloor m\Delta_Y \rfloor) \quad \textit{for every } m \geq 0. \qquad \square$$

Lemma 2.19 *Let* X *be a normal scheme,* D *a* $\mathbb{Q}$-*divisor and* m *a positive integer such that* mD *is an integral divisor. Let* s *be a rational section of* $\mathscr{O}_X(\lfloor D \rfloor)$. *Then* s *is a regular section of* $\mathscr{O}_X(\lfloor D \rfloor)$ *if and only if* s^m *is a regular section of* $\mathscr{O}_X(mD)$.

Proof Since X is normal, it is enough to check this at the generic point of every divisor. So pick a prime divisor $E \subset X$ and assume that E has coefficient r in D. Let s have a pole of order n along E. Then s is a regular section of $\mathscr{O}_X(\lfloor D \rfloor)$ along E if and only if $n \leq \lfloor r \rfloor$ and s^m is a regular section of $\mathscr{O}_X(mD)$ along E if and only if $mn \leq mr$. Since n is an integer, these are equivalent. $\qquad \square$

2.20 (Real coefficients) One can also define discrepancies for pairs (X, Δ) where X is a normal scheme and $\Delta = \sum a_i D_i$ is a linear combination of distinct prime divisors with real coefficients, as long as the pull-back $f^*(K_X + \Delta)$ can be defined. The latter holds if $K_X + \Delta$ is a linear combination of Cartier divisors with real coefficients. (Unlike in the rational case, this is weaker than assuming that a real multiple of $K_X + \Delta$ be Cartier.) All the results of this section work for real divisors. See (2.21), Birkar *et al.* (2010) or Kollár (2010b, section 4) for some foundational issues.

The following result can be used to reduce many problems from $\mathbb{R}$-divisors to $\mathbb{Q}$-divisors directly.

Proposition 2.21 *Let* X *be a normal scheme and* Δ *an* $\mathbb{R}$-*divisor such that* $K_X + \Delta$ *is* $\mathbb{R}$-*Cartier. Let* $\{E_i : i \in I\}$ *be finitely many divisors over* X. *Fix rational numbers* $a_i \leq a(E_i, X, \Delta) \leq b_i$ *and* $r \leq \mathrm{discrep}(X, \Delta)$. *Then there is a* $\mathbb{Q}$-*divisor* Δ' *(depending on the* E_i, a_i, b_i *and* r*) such that*

(1) $K_X + \Delta'$ *is* $\mathbb{Q}$-*Cartier,*
(2) *Supp* $\Delta' = $ *Supp* Δ *and* Δ' *is effective if* Δ *is,*
(3) $a_i \leq a(E_i, X, \Delta') \leq b_i$ *for all* $i \in I$ *and*
(4) $a(E, X, \Delta') = a(E, X, \Delta)$ *whenever the latter is rational.*

Moreover, if (X, Δ) has a log resolution (for instance, if we are over a field of characteristic 0) then we can also guarantee that

(5) $r \leq \mathrm{discrep}(X, \Delta')$.

Proof Fix a canonical divisor $K'_X \in |K_X|$. By assumption there are Cartier divisors D_j and real numbers d_j such that $K'_X + \Delta = \sum d_j D_j$. (Actual equality, not just linear equivalence.) Set $\Delta_\mathbf{x} := -K'_X + \sum x_j D_j$. Let $f: Y \to X$ be any birational morphism, Y normal. There is a unique canonical divisor K'_Y such that $f_* K'_Y = K'_X$. Then, by the definition of discrepancies,

$$K'_Y - \sum_j x_j f^* D_j = \sum_F a(F, X, \Delta_\mathbf{x}) F,$$

where F runs through all divisors on Y. This shows that each $a(F, X, \Delta_\mathbf{x})$ is a linear function of the x_j with integral coefficients.

The conditions (2–4) represent a collection of linear rational equalities and a finite collection of linear rational inequalities. The common solution set is therefore a rational polyhedron P. By assumption P has a real point (corresponding to Δ), hence P also has a rational point, giving Δ'.

If (X, Δ) has a log resolution $g: X' \to X$, we can add all exceptional divisors of g and all irreducible components of Supp Δ to the E_i. We can further add all divisors obtained by blowing up an irreducible component of the intersection of 2 such divisors. Then, by Kollár and Mori (1998, 2.31),

$$\mathrm{discrep}(X, \Delta') = \min_i \{ a(E_i, X, \Delta') : E_i \text{ is exceptional} \}.$$

We can also assume that $r \leq a_i$ whenever E_i is exceptional. With these choices, $r \leq \min_i \{ a_i : E_i \text{ is exceptional} \} \leq \mathrm{discrep}(X, \Delta')$. $\square$

We have used the following result of Zariski and Abhyankar.

Lemma 2.22 (Blowing up centers of divisors) *Let $f: Y \to X$ be a dominant morphism of finite type of schemes. Let $y \in Y$ be a codimension 1 regular point that does not dominate any of the irreducible components of X.*

Starting with $X_0 := X$, $f_0 := f$, we define a sequence of schemes and rational maps as follows:

If we already have $f_i: Y \dashrightarrow X_i$ and f_i is defined at y, then let $Z_i \subset X_i$ be the closure of $x_i := f_i(y)$. Set $X_{i+1} = B_{Z_i} X_i$ and $f_{i+1}: Y \dashrightarrow X_{i+1}$ the induced map. It is again defined at y.

Then eventually we get $f_n: Y \dashrightarrow X_n$ such that $x_n := f_n(y)$ is a codimension 1 regular point of X_n.

Proof If f is birational, the proof is given in Artin (1986, section 5) or Kollár and Mori (1998, 2.45). (The latter assumes that X is a variety, but it is not needed in the proof.) The proof in the general case is essentially the same, with one small change that we explain below.

Note that y defines a valuation ring $R \subset k(X)$ with maximal ideal m_R whose center is $x := x_0 \in X$. In the birational case the above mentioned proofs use only that $\operatorname{trdeg}(R/m_R : k(x)) \geq \dim_x X - 1$; we do not need equality.

The general case of relative dimension $m \geq 0$ follows from these by noting that

$$\operatorname{trdeg}(R/m_R : k(x)) \geq \operatorname{trdeg}(k(y) : k(x)) - m \geq \dim_x X - 1. \quad \square$$

Crepant morphisms and maps

Definition 2.23 A proper birational morphism $f : (Y, \Delta_Y) \to (X, \Delta_X)$ satisfying the assumptions of (2.6) is called *crepant*. (The word "crepant" was coined by Reid as a back-formation of "discrepant.") If f is crepant then $a(F, Y, \Delta_Y) = a(F, X, \Delta_X)$ for any divisor F over X by (2.6). In particular, if $\{E_i : i \in I\}$ are the f-exceptional divisors then

$$
\begin{aligned}
&\operatorname{totaldiscrep}(X, \Delta_X) = \operatorname{totaldiscrep}(Y, \Delta_Y) \quad \text{and} \\
&\operatorname{discrep}(X, \Delta_X) = \min\{\operatorname{discrep}(Y, \Delta_Y), a(E_i, X, \Delta_X) : i \in I\}.
\end{aligned}
\tag{2.23.1}
$$

Thus, if (X, Δ_X) has a log resolution (Y, Δ_Y) then, in principle, any discrepancy computation reduces to the snc case which we already treated in (2.10).

Note that even if we are interested in the $\Delta_X = 0$ case, Δ_Y is almost always nonzero and contains divisors both with positive and negative coefficients.

2.23.2 Let (X_i, Δ_i) be normal pairs proper over a base scheme S and $\phi : X_1 \dashrightarrow X_2$ a birational map. Then $\phi : (X_1, \Delta_1) \dashrightarrow (X_2, \Delta_2)$ is called *crepant* if the following equivalent conditions hold.

(a) $a(E, X_1, \Delta_1) = a(E, X_2, \Delta_2)$ for every divisor E or
(b) there is a third pair (Y, Δ_Y) and crepant morphisms $h_i : (Y, \Delta_Y) \to (X_i, \Delta_i)$ such that the following diagram is commutative

$$
\begin{array}{ccc}
& (Y, \Delta_Y) & \\
{\scriptstyle h_1}\swarrow & & \searrow{\scriptstyle h_2} \\
(X_1, \Delta_1) & \overset{\phi}{\dashrightarrow} & (X_2, \Delta_2).
\end{array}
$$

Note that even if the Δ_i are effective, usually one can not choose (Y, Δ_Y) such that Δ_Y is effective. In the literature this concept is also called being *K-equivalent*, see Kawamata (2002). We denote *crepant birational equivalence* by

$$(X_1, \Delta_1) \overset{\mathrm{cbir}}{\sim} (X_2, \Delta_2).$$

Example 2.23.3 It is a somewhat subtle condition for a birational map to be crepant. As an example, for a_0, a_1, a_2 consider the pair

$$(\mathbb{P}^2, \Delta(a_0, a_1, a_2, b) := b(x_0 + x_1 + x_2 = 0) + \textstyle\sum_i a_i(x_i = 0)).$$

Then $(\mathbb{P}^2, \Delta(a_0, a_1, a_2, b))$ is lc if and only if $0 \le a_i, b \le 1$ and $K_{\mathbb{P}^2} + \Delta(a_0, a_1, a_2, b) \sim_{\mathbb{Q}} 0$ if and only if $a_0 + a_1 + a_2 + b = 3$. Let $\phi \colon \mathbb{P}^2 \dashrightarrow \mathbb{P}^2$ be the standard quadratic transformation $(x_0 \colon x_1 \colon x_2) \mapsto (x_0^{-1} \colon x_1^{-1} \colon x_2^{-1})$. There is a unique $\mathbb{Q}$-divisor $\Delta^*(a_0, a_1, a_2, b)$ such that

$$\phi \colon (\mathbb{P}^2, \Delta(a_0, a_1, a_2, b)) \dashrightarrow (\mathbb{P}^2, \Delta^*(a_0, a_1, a_2, b))$$

is crepant. The latter is lc if and only if, in addition, $a_i + a_j \ge 1$ for every $i \ne j$. Let us now fix b, change the a_i to a_i' and consider the birational map

$$\phi \colon (\mathbb{P}^2, \Delta(a_0, a_1, a_2, b)) \dashrightarrow (\mathbb{P}^2, \Delta^*(a_0', a_1', a_2', b)).$$

The only boundary divisor whose center on both sides is a divisor is $(x_0 + x_1 + x_2 = 0)$ and it appears on both sides with the same coefficient. Despite this, ϕ is crepant birational only if $a_i = a_i'$ for every i.

The first characterization (2.23.2a) shows that all crepant, birational maps of a pair (X, Δ) to itself form a group. It is denoted by $\mathrm{Bir}^c(X, \Delta)$.

The following special cases of (2.23.2) are especially useful.

Claim 2.23.4   Let $\phi \colon (X_1, \Delta_1) \dashrightarrow (X_2, \Delta_2)$ be a crepant, birational map. If (X_1, Δ_1) is klt (resp. lc) then so is (X_2, Δ_2). If (X_1, Δ_1) is plt and none of the irreducible components of $\lfloor \Delta_1 \rfloor$ are contracted by ϕ then (X_2, Δ_2) is also plt. $\square$

Aside 2.24   There does not seem to be a good notion of when two proper pairs (X_i, Δ_i) should be birational. In the general type case one could declare (X_1, Δ_1) and (X_2, Δ_2) to be "birational" if they have isomorphic canonical models. This is quite sensible. Following (1.24) this implies that a morphism $g \colon (X_1, \Delta_1) \to (X_2, \Delta_2)$ is "birational" if $g_*(\Delta_1) = \Delta_2$ and $\mathrm{coeff}_E \Delta_1 \ge -a(E, X_2, \Delta_2)$ for every g-exceptional divisor E.

This, however, has some counterintuitive consequences for pairs that are not of general type. As an example, consider a pair $(\mathbb{P}^2, (1-c)\sum_i L_i)$ where $L_i = (x_i = 0)$. Blow up the three vertices to get $g \colon S \to \mathbb{P}^2$ with exceptional divisors E_i. Then

$$g \colon (S, (1-c)\sum_i L_i' + (1-2c)\sum_i E_i) \to (\mathbb{P}^2, (1-c)\sum_i L_i)$$

is crepant where L_i' denotes the birational transform of L_i. We can now contract the L_i' to get

$$h \colon (S, (1-c)\sum_i L_i' + (1-2c)\sum_i E_i) \to (\mathbb{P}^2, (1-2c)\sum_i L_i)$$

where we can identify $L_i = h(E_i)$. The map h is not crepant but it is "birational" under the above rules. By iterating this, we get that every pair $(\mathbb{P}^2, \sum_i a_i(x_i = 0))$ where $0 \le a_i < 1$ for every i is "birational" to $(\mathbb{P}^2, \emptyset)$.

2.2 Log canonical surface singularities

In this section we study the surface singularities that appear in the minimal model program. The arguments we present work in arbitrary characteristic and even for excellent surfaces. Foundational theorems needed for this generality are proved in Section 10.1.

Over $\mathbb{C}$, the method of index 1 covers gives a quicker way to many of the theorems. This is outlined in (2.53).

The results give a rather complete description of log canonical surface singularities in terms of the combinatorial structure of the exceptional curves on the minimal resolution (2.25). In the theory of minimal models this seems to be the most useful. Also, as it turns out, this form of the answer is independent of the characteristic.

One can then go further and determine the completed local rings of the singularities. Most of these have been worked out in characteristic 0 but subtle differences appear in positive characteristic Lipman (1969). We will not discuss this aspect here.

The starting point is the following.

Theorem 2.25 (Log minimal resolution for surfaces) *Let Y be a 2-dimensional, normal, excellent scheme, $B_i \subset Y$ distinct, irreducible Weil divisors and $B := \sum b_i B_i$ a linear combination of them with $0 \leq b_i \leq 1$ for every i.*

Then there is a proper birational morphism $f : X \to Y$ such that

(1) *X is regular,*
(2) *$K_X + f_*^{-1}B$ is f-nef,*
(3) *$\mathrm{mult}_x f_*^{-1}B \leq 1$ for every $x \in X$,*
(4) *$f_*^{-1}\lfloor B \rfloor = \sum_{i:b_i=1} f_*^{-1}B_i$ is regular and*
(5) *we can choose either of the following conditions to hold:*
   (a) *the support of $\mathrm{Ex}(f) + f_*^{-1}\lfloor B \rfloor$ has a node (1.41) at every point of $\mathrm{Ex}(f) \cap f_*^{-1}\lfloor B \rfloor$, or*
   (b) *$K_X + (1 - \varepsilon)f_*^{-1}B$ is f-nef for $0 \leq \varepsilon \ll 1$.*

Proof Resolution is known for excellent surfaces (cf. Shafarevich, 1966), hence there is a proper birational morphism $f_1 : X_1 \to Y$ such that X_1 is regular, the support of $(f_1)_*^{-1}B$ is regular and the support of $\mathrm{Ex}(f_1) + (f_1)_*^{-1}B$ has only nodes.

If $K_{X_1} + (f_1)_*^{-1}B$ is not f_1-nef then there is an irreducible and reduced exceptional curve $C \subset X_1$ such that $C \cdot (K_{X_1} + (f_1)_*^{-1}B) < 0$. Since $C \cdot (f_1)_*^{-1}B \geq 0$, this implies that $C \cdot K_{X_1} < 0$. By the Hodge Index theorem (10.1), $(C \cdot C) < 0$, thus ω_C is anti-ample.

If we are in characteristic 0, then C is geometrically reduced and every geometric irreducible component of C is a smooth rational curve with self-intersection -1. Thus we can contract C by Castelnuovo's theorem. Furthermore, $(C \cdot (f_1)^{-1}_* B) < -(C \cdot K_{X_1}) \le 1$ which shows that we do not contract any curve that meets $(f_1)^{-1}_* \lfloor B \rfloor$ and the multiplicity of the birational transform of B is still ≤ 1 at every point after contraction. Thus, after contracting C, we get $f_2 \colon X_2 \to Y$ and the conditions (3–4) still hold. We can continue until we get $f \colon X \to Y$ as required.

There is only one case that forces us to choose between the two alternatives in (5). If

$$C \cdot (K_{X_i} + (f_i)^{-1}_* B) = 0 \quad \text{but} \quad C \cdot (K_{X_i} + (1 - \varepsilon)(f_i)^{-1}_* B) < 0$$

then we can still contract C but the image of $\lfloor B \rfloor$ may pass through a singular point of $\mathrm{Ex}(f_{i+1})$. So we have to decide which alternative we want. Both have some advantages.

Without assuming characteristic 0, we check in (10.5) that Castelnuovo's contraction theorem holds even if C is not geometrically reduced. The rest of the argument then goes as before. $\qquad\qquad\square$

2.26 (Dual graphs) Let $(y \in Y)$ be a surface singularity and $f \colon X \to Y$ a resolution with exceptional curve $C = \cup C_i$. It is frequently very convenient to represent the curve C by a graph Γ. The convention is to use the irreducible components of C as vertices and to connect two vertices by an edge if and only if the corresponding curves intersect.

Over an algebraically closed field, we use the negative of the self-intersection number $-(C_i \cdot C_i)$ to represent a vertex and employ multiple edges or put the intersection number $(C_i \cdot C_j)$ over an edge. We add $\deg \omega_{C_i}$ as an extra datum.

In general, the C_i are $k(y)$-schemes and we compute degrees over $k(y)$. Thus $\dim_{k(y)} \mathcal{O}_{C_i} \cap \mathcal{O}_{C_j}$ is the intersection number and $\deg_{k(y)} \mathcal{O}_Y(-C_i)|_{C_i}$ represents a vertex. We will also keep track of the number $r_i := \dim_{k(y)} H^0(C_i, \mathcal{O}_{C_i})$.

As an example, let $y^3 - az^3$ be an irreducible cubic in $k[y, z]$ and consider the surface

$$Y := (x^2 = y^3 - az^3 + y^4 + z^4) \subset \mathbb{A}^3_k.$$

(The 4th powers are there to make things work if char $k = 3$.) The dual graph of the minimal resolution of the singularity at the origin is

$$C_1 \equiv C_2 \quad \text{or} \quad (2) \equiv (3 \cdot 2)$$

This shows that $(C_1 \cdot C_2) = 3$, $(C_1^2) = -2$ and $(C_2^2) = -6$. The curve C_2 is irreducible over k. Let K be the splitting field of $y^3 - a$. If char $k \ne 3$ then, $(C_2)_K$ splits into three disjoint copies of $\mathbb{P}^1_K$. If char $k = 3$ then $(C_2)_K \simeq \mathbb{P}^1_K \times \mathrm{Spec}_K K[t]/(t^3)$.

These data satisfy the following.

Condition 2.26.1 $(C_i \cdot C_j)$ is divisible by $r_i = \dim_{k(y)} H^0(C_i, \mathcal{O}_{C_i})$ for every i, j and $\deg \omega_{C_i} = r_i(2p_i - 2)$ for some $p_i \in \mathbb{N}$. We write $-(C_i \cdot C_i) = r_i \cdot c_i$.

This graph, with the extra data added, is called the *dual graph* of the reducible curve $C = \cup C_i$. Note that $c_i \geq 2$ for every i if $f \colon X \to Y$ is a minimal resolution.

For log canonical singularities, most of the exceptional curves C_i are conics (10.6) and the dual graphs have few edges, so the picture is rather transparent.

Let $\det(\Gamma)$ denote the determinant of the negative of the intersection matrix of the dual graph. This matrix is positive definite for exceptional curves. For instance, if $\Gamma = \{2 \text{——} 2 \text{——} 2\}$ then

$$\det(\Gamma) = \det \begin{pmatrix} 2 & -1 & 0 \\ -1 & 2 & -1 \\ 0 & -1 & 2 \end{pmatrix} = 4.$$

Let B be another $\mathbb{Q}$-divisor on X which does not contain any of the C_i. The *extended dual graph* (Γ, B) has an additional vertex connected to C_i if $(B \cdot C_i) \neq 0$. We write $(B \cdot C_i)$ on the edge if this value is different from r_i. The new vertices are usually represented by $\bullet$ if the corresponding curve has coefficient 1 and by $\circledast$ in general.

One can also think of (extended) dual graphs as combinatorial objects, without reference to any particular collection of curves. To preserve the geometric origin, we always assume that the conditions (2.26.1) are satisfied.

(One can see that every (extended) dual graph satisfying this condition arises from a collection of curves on a smooth projective surface over $\mathbb{Q}$.)

Let $f \colon X \to Y$ be a resolution, $y \in Y$ a point and $\Gamma = \Gamma(y \in Y)$ the dual graph of the curves red $f^{-1}(y) = \cup_i C_i$. Since the intersection matrix $(C_i \cdot C_j)$ is invertible (10.1), there is a unique $\Delta = \Delta(y, Y, B) := \sum d_j C_j$ such that

$$(\Delta \cdot C_i) = -((K_X + f_*^{-1}B) \cdot C_i) \qquad \forall i. \tag{2.26.2}$$

Using the adjunction formula, we can rewrite these equations as

$$(\Delta \cdot C_i) = (C_i \cdot C_i) - \deg \omega_{C_i} - (f_*^{-1}B \cdot C_i) \qquad \forall i. \tag{2.26.3}$$

Note that these equations make sense for any extended dual graph (Γ, B).

By (2.25.2) the right hand side of (2.26.2) is always ≤ 0 for a log minimal resolution. Therefore, we call an extended dual graph (Γ, B) *log minimal* if the right hand sides of (2.26.3) are all ≤ 0.

Following the terminology of curves on surfaces, a curve C_i in an extended dual graph (Γ, B) is called a (-1)-*curve* if $-(C_i \cdot C_i) = r_i$ (that is, $c_i = 1$) and $\deg \omega_{C_i} = -2r_i$. We see that (Γ, B) is log minimal if and only if $(C_i \cdot B) \geq r_i$ for every (-1)-curve C_i.

A simple linear algebra lemma (10.3.5) gives the following.

Claim 2.26.4   Let (Γ, B) be a connected, negative definite, log minimal extended dual graph. Then the system (2.26.3) has a unique solution $\Delta = \sum d_j C_j$. Furthermore

(1) either $d_j = 0$ for every j,
(2) or $d_j > 0$ for every j.   □

If $K_Y + B$ is $\mathbb{Q}$-Cartier (which will always be the case for us) then d_j is the negative of the discrepancy $a(C_j, Y, B)$ defined in (2.4). Therefore, an extended dual graph (Γ, B) is called *numerically canonical* (resp. *numerically log terminal* or *numerically log canonical*) if it is connected, negative definite, log minimal and $d_j \leq 0$ for every j (resp. $d_j < 1$ for every j or $d_j \leq 1$ for every j).

2.27 (Numerically log canonical)   Let Y be a normal, excellent surface and B an effective $\mathbb{Q}$-divisor. We say that (Y, B) is *numerically log canonical* at a point $y \in Y$ if there is a log minimal resolution $f: X \to Y$ as in (2.25) such that the extended dual graph of $f^{-1}(y) = \cup_i C_i$ is numerically log canonical.

This is slightly more general than the terminology in Kollár and Mori (1998, 4.1). However, the notion of a numerically log canonical pair is only a temporary convenience, so this should not cause any problems.

It is more general than the traditional notion of log canonical in two aspects.

We assume the discrepancy condition $a(E_i, Y, B) \geq -1$ only for those curves that appear on a given resolution. Even over $\mathbb{C}$, this leads to a few extra cases which are, however, easy to enumerate.

By working directly on X, we do not worry at the beginning whether $K_Y + B$ is $\mathbb{Q}$-Cartier or not. This turns out to be automatic in the surface case.

First we show that most lc surface singularities are rational. This will be generalized to higher dimensions in Section 2.5.

Proposition 2.28   *Let $(y \in Y, B)$ be a numerically log canonical surface germ and $f: X \to Y$ a log minimal resolution with exceptional curves $C = \cup C_i$. Let $\Delta := \Delta(y \in Y, B)$ be the discrepancy divisor as in (2.26). Then*

(1) *either $y \notin \operatorname{Supp} B$ and $\Delta = \sum_i C_i$ (see (3.27) for a list),*
(2) *or $(y \in Y)$ is a rational singularity (10.7).*

Proof   Note that $K_X + f_*^{-1} B + \Delta$ has zero intersection number with every exceptional curve. Thus (10.4) applies to $L = \mathcal{O}_X$ and gives (2), unless $f_*^{-1} B \cdot C_i = 0$ for every i (and hence $y \notin \operatorname{Supp} B$) and every exceptional curve appears in Δ with coefficient 1.   □

For the rest of the section we discuss those topics concerning surface singularities that are most important for the general theory. A detailed classification of lc surface pairs is postponed to Section 3.3.

Terminal and canonical pairs

Theorem 2.29 *Let S be a normal, excellent surface, $s \in S$ a closed point and B a boundary. Then*

(1) $(s \in S, B)$ *is numerically terminal* $\Leftrightarrow$ *terminal* $\Leftrightarrow$ $s \in S$ *is regular and* $\mathrm{mult}_s B < 1$.

(2) $(s \in S, B)$ *is numerically canonical* $\Leftrightarrow$ *canonical* $\Leftrightarrow$
   (a) *either $s \in S$ is regular and* $\mathrm{mult}_s B \leq 1$,
   (b) *or $s \notin \mathrm{Supp}\, B$, K_S is Cartier and there is a resolution $f \colon T \to S$ such that $K_T \sim f^* K_S$. (See (3.26) for a complete list.)*

Proof Let $(s \in S, B)$ be numerically canonical with minimal resolution $f \colon T \to S$ satisfying the alternative (2.25.5.b). Then $\Delta := \Delta(s \in S, B)$ defined by the equations (2.26.3) contains every exceptional curve with coefficient ≤ 0. On the other hand, by (2.26.4), Δ contains every curve with coefficient ≥ 0, so $\Delta = 0$.

Thus $K_T + f_*^{-1} B$ is numerically f-trivial and $K_T + (1 - \varepsilon) f_*^{-1} B$ is f-nef. Thus $-f_*^{-1} B$ is f-nef. This can happen only in two degenerate ways: either f is an isomorphism or $B = 0$. In the latter case, the isomorphism $\mathcal{O}_T(K_{T/S}) \simeq f^* \mathcal{O}_S$ follows from (2.28) and (10.9.2).

Conversely, assume that $s \in S$ is regular and $\mathrm{mult}_s B \leq 1$. Then $a(E_1, S, B) = 1 - \mathrm{mult}_s B \geq 0$ as above. If E is any other exceptional curve over S then by (2.23) and (2.5)

$$a(E, S, B) = a(E, S_1, \pi_*^{-1} B + (\mathrm{mult}_s B - 1) E_1) \geq a(E, S_1, \pi_*^{-1} B).$$

Since $\mathrm{mult}_p \pi_*^{-1} B \leq \mathrm{mult}_{\pi(p)} B \leq 1$, we are done by induction on the number of blow-ups necessary to reach E. $\square$

This immediately implies the following higher dimensional version.

Corollary 2.30 *Let X be a normal scheme as in (1.5) and Δ a boundary. Then there is a closed subset $W \subset X$ of codimension ≥ 3 such that the following hold.*

(1) *If (X, Δ) is terminal then $X \setminus W$ is regular and $\mathrm{mult}_x \Delta < 1$ for every $x \in X \setminus W$.*

(2) *If (X, Δ) is canonical then $X \setminus W$ has rational singularities, $K_{X \setminus W}$ is Cartier and there is a resolution $f \colon Y \to (X \setminus W)$ such that $K_Y \sim f^* K_{X \setminus W}$.* $\square$

The reduced boundary

For a log canonical pair $(S, B = \sum b_i B_i)$, the singularities of Supp B can be arbitrary if the coefficients b_i are small. By contrast, if the b_i are bounded from below, we end up with a restricted class of singularities. In the extreme case, when all the b_i equal 1, we show that B has only ordinary nodes. The following theorem is a more precise version of this assertion.

Theorem 2.31   *Let $(S, B + B')$ be numerically lc and $B = \sum B_i$ a sum of curves, all with coefficients 1. Then, for any $s \in S$,*

(1) *either B is regular at s,*
(2) *or B has a node at s, Supp B' does not contain s and every exceptional curve of the minimal log resolution has discrepancy -1.*

This immediately implies the following higher dimensional version.

Corollary 2.32   *Let (X, Δ) be an lc pair, Δ effective and $x \in X$ a codimension 2 point. Then*

(1) *either $\lfloor \Delta \rfloor$ is regular at x,*
(2) *or $\lfloor \Delta \rfloor$ has a node at x and no other component of Δ contains x.*   □

Proof   The question is local on S. Fix a point $s \in S$ and let $f: T \to S$ be a log minimal resolution as in (2.25.5a).

Write $K_T + B_T + \Delta_T + f_*^{-1} B' \sim_{\mathbb{Q}} f^*(K_S + B + B')$ where $B_T := f_*^{-1} B$ is regular by (2.25.4). Note that $\Delta_T = \sum d_i C_i$ is effective by (2.26.4), f-exceptional and thus $0 \le d_i \le 1$ for every i. We can rewrite the linear equivalence as

$$-B_T \sim_{\mathbb{Q}} K_T + \Delta_T + f_*^{-1} B' - f^*(K_S + B + B'). \qquad (2.32.3)$$

Pushing forward the exact sequence

$$0 \to \mathscr{O}_T(-B_T) \to \mathscr{O}_T \to \mathscr{O}_{B_T} \to 0,$$

we get the exact sequence

$$\mathscr{O}_S \simeq f_* \mathscr{O}_T \to f_* \mathscr{O}_{B_T} \to R^1 f_* \mathscr{O}_T(-B_T).$$

By (10.4), either $R^1 f_* \mathscr{O}_T(-B_T) = 0$ or $\Delta_T = C_1 + \cdots + C_n$ where the C_i are all the exceptional curves and $s \notin$ Supp B'. We consider these two possibilities separately.

Case 1   If $R^1 f_* \mathscr{O}_T(-B_T) = 0$ then the composite

$$\mathscr{O}_S \to \mathscr{O}_B \hookrightarrow f_* \mathscr{O}_{B_T} \quad \text{is surjective.}$$

Thus $B \simeq B_T$ and so B is regular.

Case 2 If $\Delta_T = C_1 + \cdots + C_n$ then we prove that B has a node at s. We can drop B' and rewrite (2.32.3) as

$$-B_T - \Delta_T \sim_{\mathbb{Q}} K_T - f^*(K_S + B). \tag{2.32.4}$$

Then $R^1 f_* \mathscr{O}_T(-B_T - \Delta_T) = 0$ by (10.4) and, as above, we get that $I_{s,s} = f_* \mathscr{O}_T(-\Delta_T) \to f_* \mathscr{O}_{B_T}(-\Delta_T|_{B_T})$ is surjective. By (2.25.5a), $\mathscr{O}_{B_T}(-\Delta_T|_{B_T})$ is the ideal of the closed points of B_T. In particular,

$$\mathscr{O}_B(-s) = f_* \mathscr{O}_{B_T}(-\operatorname{red} f^{-1}(s)). \tag{2.32.5}$$

Since $(C_i \cdot (K_T + B_T + \sum C_i)) = 0$ for every i, the adjunction formula gives that

$$\deg \omega_{C_i} = -\sum_{j \neq i}(C_i \cdot C_j) - (C_i \cdot B_T) \quad \forall \, i. \tag{2.32.6}$$

Assume first that we are over an algebraically closed field. Then each $C_i \simeq \mathbb{P}^1$ and the dual graph is

$$B_1 \relbar\joinrel\relbar C_1 \relbar\joinrel\relbar \cdots \relbar\joinrel\relbar C_n \relbar\joinrel\relbar B_2$$

Thus B has two irreducible components and, by (2.32.5), the two branches intersect transversally.

Essentially the same argument works in general, but one has to be more careful, especially over imperfect fields.

Let $k(s)$ be the residue field of $s \in S$. We have already dealt with the second condition of (1.41.1), thus it remains to show that

$$(B_T \cdot \Delta_T) = \dim_{k(s)}(\mathscr{O}_{B_T}/\mathscr{O}_{B_T}(-\Delta_T|_{B_T})) \leq 2.$$

By (10.6), for each i, the right hand side of (2.32.6) has either 1 nonzero term or two nonzero terms which are equal. This gives two possibilities for the dual graph:

$$B_1 \overset{r}{\relbar\joinrel\relbar} C_1 \overset{r}{\relbar\joinrel\relbar} \cdots \overset{r}{\relbar\joinrel\relbar} C_n \overset{r}{\relbar\joinrel\relbar} B_2 \tag{2.32.7}$$

where $\dim_{k(s)} H^0(C_i, \mathscr{O}_{C_i}) = r$ for every i or

$$B_1 \overset{2r}{\relbar\joinrel\relbar} C_1 \overset{2r}{\relbar\joinrel\relbar} \cdots \overset{2r}{\relbar\joinrel\relbar} C_{n-1} \overset{2r}{\relbar\joinrel\relbar} C_n \tag{2.32.8}$$

where $\dim_{k(s)} H^0(C_i, \mathscr{O}_{C_i}) = 2r$ for $1 \leq i < n$ and $\dim_{k(s)} H^0(C_n, \mathscr{O}_{C_n}) = r$. Working through the exact sequences

$$0 \to \mathscr{O}_{C_i}(-C_i \cap C_{i+1}) \to \mathscr{O}_{C_i + \cdots + C_n} \to \mathscr{O}_{C_{i+1} + \cdots + C_n} \to 0$$

we conclude that $H^0(\mathscr{O}_{C_1 + \cdots + C_n}) = H^0(\mathscr{O}_{C_n})$. From (2.32.4) and (10.4) we infer that $R^1 f_* \mathscr{O}_T(-\Delta_T) = 0$, hence $f^*: \mathscr{O}_S \to H^0(\mathscr{O}_{C_1 + \cdots + C_n})$ is surjective. This map factors through $k(s)$, thus $k(s) \simeq H^0(\mathscr{O}_{C_1 + \cdots + C_n}) \simeq H^0(\mathscr{O}_{C_n})$. Therefore $r = 1$ and, in both cases, $(B_T \cdot \Delta_T) = 2$. $\qquad\square$

As a corollary, using the description of the deformations of nodes given in (1.41.2) we obtain the following, which gives a complete description of the codimension 1 behavior of fibers in locally stable families.

Corollary 2.33 *Notation and assumptions as in* (2.31). *Assume in addition that B is a Cartier divisor. Then either B and S are both regular at s, or B has a node at s and S is either regular or has a double point of embedding dimension 3 at s.*

Furthermore, if S is defined over an algebraically closed field k of characteristic 0, then the completion of $\mathcal{O}_{s,S}$ is isomorphic to $k[[x, y, z]]/(xy + z^n)$ and $\hat{B} = (z = 0)$.   □

The different

Definition 2.34 (Different I)   Let S be a regular surface and $B \subset S$ a regular curve. The classical adjunction formula says that $K_B \sim (K_S + B)|_B$. If S is only normal, then each singularity of (S, B) leads to a correction term, called the *different*. For later purposes, we also study the different when (S, B) is not log canonical and we take into account the influence of another $\mathbb{Q}$-divisor B'. Since the singularities of B can be now quite complicated, we compute everything on the normalization $\bar{B} \to B$.

Let S be a normal surface and $B \subset S$ a reduced curve with normalization $\bar{B} \to B$. Let B' be a $\mathbb{Q}$-divisor that has no components in common with B.

Let $f: T \to S$ be a log resolution of (S, B) with exceptional curves C_i and B_T, $B'_T \subset T$ the birational transforms of B, B'. Note that $B_T \simeq \bar{B}$. As in (2.26.2) there is a unique $\Delta(S, T, B + B') := \sum d_j C_j$ such that

$$(\Delta(S, T, B + B') \cdot C_i) = -((K_T + B_T + B'_T) \cdot C_i) \quad \forall i. \quad (2.34.1)$$

Define the different as

$$\mathrm{Diff}_{\bar{B}}(B') := (B'_T + \Delta(S, T, B + B'))|_{B_T}. \quad (2.34.2)$$

We see in (2.35.1) that this is independent of the choice of f. The different is a $\mathbb{Q}$-divisor on $\bar{B}$ and we can write it as

$$\mathrm{Diff}_{\bar{B}}(B') = \sum \delta(\bar{p}, B')[\bar{p}] \quad (2.34.3)$$

where p runs through the points of B that are singular on B or on S or are contained in B' and $\bar{p} \in \bar{B}$ denotes any preimage of p. Our main interest is in computing the coefficients $\delta(\bar{p}, B')$ and relating them to properties of $(S, B + B')$.

Higher dimensional generalizations will be given in Section 4.1.

The fundamental properties of the different are summarized next.

Proposition 2.35 *The different has the following properties.*

(1) $\mathrm{Diff}_{\bar{B}}(B')$ *is independent of the choice of* f.
(2) $\mathrm{Diff}_{\bar{B}}(B') = \mathrm{Diff}_{\bar{B}}(0) + (B' \cdot \bar{B})$.
(3) $\mathrm{Diff}_{\bar{B}}(0)$ *is effective and it is zero if and only if B is regular and S is regular along B*.
(4) *If* $K_S + B + B'$ *is Cartier then* $\mathrm{Diff}_{\bar{B}}(B')$ *is a* $\mathbb{Z}$-*divisor.*
(5) *If* $(S, B + B')$ *is numerically lc at* $p \in S$, *then* $\delta(\bar{p}, B')$ *is the negative of the smallest discrepancy of* $(S, B + B')$ *above p*.
(6) *If* $(S, B + B')$ *is not numerically lc at p then* $\delta(\bar{p}, B') > 1$.

Proof Let $q \in T$ be a point. If $q \neq \bar{p}$ then $\delta(\bar{p}, B')$ is clearly unchanged if we blow-up q. If $q = \bar{p}$ then this is an easy computation, proving (1).

Define $\Theta(B') = \sum c_i C_i$ by the formulas $(\Theta(B') \cdot C_i) = -(B'_T \cdot C_i) \forall i$. Then clearly $\Delta(S, T, B + B') = \Delta(S, T, B) + \Theta(B')$, thus

$$\delta(\bar{p}, B') = \delta(\bar{p}, 0) + (\bar{B} \cdot (B' + \Theta(B')))_{\bar{p}}.$$

If B' is $\mathbb{Q}$-Cartier then the last term is the coefficient of $\bar{p}$ in the pull-back of B' to $\bar{B}$. If B' is not $\mathbb{Q}$-Cartier, this is the usual definition of this intersection number of Mumford (1961), proving (2).

Let $f: T \to S$ be the minimal resolution of (S, B) as in (2.25). By (2.26.4) then $\Delta(0)$ is effective, hence $\mathrm{Diff}_{\bar{B}}(0)$ is effective and it is zero if and only if (S, B) is canonical. Thus (2.29) implies (3).

If $K_S + B + B'$ is Cartier then $K_T + B_T + B'_T + \Delta(B') \sim f^*(K_S + B + B)$ is a $\mathbb{Z}$-divisor and so is $\mathrm{Diff}_{\bar{B}}(B')$, proving (4).

Write $\Delta = \sum d_i C_i$, where $d_i \geq 0$ by (2.26.4). Assume that $(S, B + B')$ is not log terminal (resp. not log canonical) at $p \in S$, that is, $d_r \geq 1$ (resp. $d_r > 1$) for some $C_r \subset f^{-1}(p)$. Let C_s be the exceptional curve passing through $\bar{p}$. We prove in (2.36) that $d_s \geq 1$ (resp. $d_s > 1$) also holds. Since

$$(B_T \cdot (B'_T + \Delta))_{\bar{p}} \geq (B_T \cdot d_s C_s)_{\bar{p}} \geq d_s \cdot \deg \bar{p},$$

this proves (6) and also (5) in case $(S, B + B')$ is not log terminal.

Thus assume that $(S, B + B')$ is log terminal and consider the extended dual graph of the minimal resolution of $(S, 0)$. Write $\Delta + B + B' = \sum(1 - \varepsilon(D_i))D_i$. Note that $\varepsilon(D_i) \geq 0$ for every i since $(S, B + B')$ is log terminal and $\varepsilon(B_T) = 0$. We prove in (2.37) that $\varepsilon(*)$ is a convex function on the dual graph, hence its minimum on all exceptional curves is achieved at the curve intersecting B_T. This proves (5). $\qquad\square$

Higher dimensional versions of the following result play an important role in the general theory, see Kollár and Mori (1998, section 5.4) or Section 4.4.

Proposition 2.36 (Connectedness theorem for surfaces)   *Let (Γ, B) be an extended dual graph and write $\Delta + B = \sum d_i D_i$. Then the set of vertices $\{D_i : d_i \geq 1\}$ form a connected subgraph.*

Proof Let $D_0, \ldots, D_m$ be a connected chain of vertices such that $d_0, d_m \geq 1$. We prove that $d_i \geq 1$ for all $0 < i < m$. Let $\sum^*$ denote summation for $i = 1, \ldots, m - 1$. Write $K + B + \Delta = K + \sum^* d_i D_i + \Delta'$ and compute the following intersection numbers

$$
\begin{aligned}
((\sum^*(d_i - 1)D_i) \cdot D_j) &= -((K + \Delta') \cdot D_j) - ((\sum^* D_i) \cdot D_j) \\
&= -((K + D_j) \cdot D_j) - ((\Delta' + \sum^*_{i \neq j} D_i) \cdot D_j) \qquad (2.36.1) \\
&= -\deg \omega_{D_j} - ((\Delta' + \sum^*_{i \neq j} D_i) \cdot D_j).
\end{aligned}
$$

If $2 \leq j \leq m - 2$, then D_{j-1}, D_{j+1} intersect D_j, each contributing at least r_j (as in (2.26)) hence the last term is $\geq 2r_j$. For $j = 1$ the curve D_2 contributes $\geq r_1$ and the curve $d_0 C_0$ in Δ' contributes $\geq d_0 r_1 \geq r_1$. Similarly for $j = m$. Thus

$$
((\sum_{i=1}^{m-1}(d_i - 1)D_i) \cdot D_j) \leq 0 \quad \text{for } j = 1, \ldots, m - 1,
$$

hence, by (3.29.1), $\sum_{i=1}^{m-1}(d_i - 1)D_i \geq 0$ and $\gg 0$ if $d_0 > 1$ or $d_m > 1$.   □

Proposition 2.37 (Convexity of discrepancies)   *Let (Γ, B) be an extended, log canonical dual graph without (-1)-curves (2.26) and write $\Delta + B = \sum(1 - \varepsilon_i)D_i$. Let D_0, D_1, D_2 be any 3 curves such that $(D_0 \cdot D_1) \neq 0$ and $(D_2 \cdot D_1) \neq 0$. Then $\varepsilon_1 \leq \frac{1}{2}(\varepsilon_0 + \varepsilon_2)$.*

Proof Write $(D_0 \cdot D_1) = a_0 r_1$ and $(D_2 \cdot D_1) = a_2 r_1$. Then (2.36.1) gives that

$$
\begin{aligned}
\varepsilon_0 a_0 r_1 - \varepsilon_1 a_1 r_1 + \varepsilon_2 a_2 r_1 &= ((\varepsilon_0 D_0 + \varepsilon_1 D_1 + \varepsilon_2 D_2) \cdot D_1) \\
&\geq -2r_1 + a_0 r_1 + a_2 r_1.
\end{aligned}
$$

This rearranges to

$$
\varepsilon_1 \leq \frac{1}{c_1}(\varepsilon_0 + \varepsilon_2) - \frac{1}{c_1}((1 - \varepsilon_0)(a_0 - 1) + (1 - \varepsilon_2)(a_2 - 1)).
$$

The last term is positive since $0 \leq \varepsilon_i \leq 1$ and $a_i \geq 1$. If D_1 is not a (-1)-curve then $c_1 \geq 2$ and we are done.   □

Corollary 2.38   *Let $(S, b_1 B_1 + b_2 B_2)$ be a lc surface pair and $s \in B_1 \cap B_2$ a point. Then*

(1) $\mathrm{mld}(s, S, b_1 B_1 + b_2 B_2) \leq 1 - b_1 + 1 - b_2$ *and*
(2) $\mathrm{mld}(s, S, b_1 B_1 + b_2 B_2) \leq \frac{1}{2}(1 - b_1 + 1 - b_2)$ *if $s \in S$ is singular.*

Proof If s is a regular point, let E be the exceptional curve of the blow-up $B_s S$. Then

$$1 + a(E, S, b_1 B_1 + b_2 B_2)$$
$$= 2 - b_1 \operatorname{mult}_s B_1 - b_2 \operatorname{mult}_s B_2 \leq 1 - b_1 + 1 - b_2.$$

If s is a singular point then $\operatorname{mld}(s, S, b_1 B_1 + b_2 B_2)$ is the minimum of the ε_i defined in (2.37) on exceptional curves over s. Thus (2) follows from the convexity property of the ε_i. □

2.3 Ramified covers

In this section we study finite ramified morphisms between normal schemes and investigate how the singularities of the source compare to the singularities of the target.

For the purposes of Chapter 5 we also need to understand ramified morphisms between demi-normal schemes; a concept to be defined in (5.1). For simplicity, we consider only morphisms that are étale over codimension 1 generic points of the singular locus. This restriction is satisfied in our applications, but the general case is needed in some other contexts.

Anyone wishing to study the normal case first, should just ignore every occurrence of "demi" in this section.

Definition 2.39 (Ramified covers)   A finite morphism of (demi-)normal schemes $\pi \colon \tilde{X} \to X$ is called a *ramified cover* of degree m if there is a dense open subset $U \subset X$ that contains every codimension 1 point of $\operatorname{Sing} X$ such that the restriction $\pi_U \colon \tilde{U} \to U$ is étale and has degree m. In this case there is an open subscheme $j \colon X^0 \hookrightarrow X$ whose complement has codimension ≥ 2 such that π is finite and flat of degree m over X^0. Indeed, we can take $X^0 = U \cup X^{\mathrm{reg}}$ where $X^{\mathrm{reg}} \subset X$ is the open subset of regular points. Set $\tilde{X}^0 = \pi^{-1}(X^0)$ and $\pi^0 \colon \tilde{X}^0 \to X^0$ the induced map. Since $\tilde{X}$ is S_2,

$$j_*(\pi^0_* \mathcal{O}_{\tilde{X}^0}) = \pi_* \mathcal{O}_{\tilde{X}}.$$

In particular, $\pi \colon \tilde{X} \to X$ is uniquely determined by the finite, flat morphism $\pi^0 \colon \tilde{X}^0 \to X^0$.

If $\pi_U \colon \tilde{U} \to U$ is Galois with Galois group G then the G action on $\tilde{U}$ extends to a G-action on $\tilde{X}$. The action is free on $\tilde{U}$.

Conversely, let $\tilde{X}$ be a (demi-)normal scheme with the action of a finite group G. The geometric quotient $X := \tilde{X}/G$ exists by (9.29). Let $\tilde{U} \subset \tilde{X}$ be the largest open set on which the G-action is free. Then $\tilde{U} \to \tilde{U}/G$ is étale. (We could take this as the definition of a free action.) Thus if $\tilde{U}$ contains every codimension 1 point of $\operatorname{Sing} X$ then $\tilde{X} \to X = \tilde{X}/G$ is a ramified cover of X.

Let $g\colon Y \to X$ be a morphism with Y (demi-)normal. The pull back $\tilde{X} \times_X Y \to Y$ defines a finite, flat, ramified cover over $g^{-1}(X^0)$. Thus, if $Y \setminus g^{-1}(X^0)$ has codimension ≥ 2, then there is a unique ramified cover $\tilde{Y} \to Y$ that agrees with $\tilde{X} \times_X Y \to Y$ over $g^{-1}(X^0)$. There is always a morphism $\tilde{Y} \to \tilde{X} \times_X Y$ which is an isomorphism if and only if $\tilde{X} \times_X Y$ is S_2.

2.40 (Pull-back and norm of divisors) The pull-back of a Weil divisor B on X by π can be defined as follows.

Take any Weil divisor B on X, restrict it to X^0 as in (2.39), pull it back and then extend uniquely to a Weil divisor $\tilde{B} =: \pi^*B$ on $\tilde{X}$. If B is Cartier, then π^*B is Cartier on $\tilde{X}$ and agrees with the usual pull-back.

To go the other way, for a Weil divisor $\tilde{B}$ on $\tilde{X}$, its *push-forward* $\pi_*\tilde{B}$ or *norm* $\operatorname{norm}_\pi \tilde{B}$ is a Weil divisor on X. To define the norm, assume first that $\tilde{B}$ is a Cartier divisor. Choose an open cover $X = \cup_i X_i$ such that $\tilde{B} \cap \pi^{-1}(X_i)$ is a principal divisor defined by the local equation $(g_i = 0)$. Then $\operatorname{norm}_\pi \tilde{B}$ is defined by the local equations $(\operatorname{norm}_\pi g_i = 0)$. In general, there is a closed subset $Z \subset X$ of codimension ≥ 2 such that $\tilde{B}$ is a Cartier divisor on $\tilde{X} \setminus \pi^{-1}(Z)$. Thus $\operatorname{norm}_\pi \tilde{B}$ is defined as a Cartier divisor on $X \setminus Z$, which then uniquely extends to a Weil divisor on X.

Note that $\operatorname{norm}_\pi \pi^*B = (\deg \tilde{X}/X) \cdot B$. We have thus proved the following.

Claim 2.40.1 The pull-back and the norm take $\mathbb{Q}$-Cartier $\mathbb{Z}$-divisors (resp. $\mathbb{Q}$-divisors) to $\mathbb{Q}$-Cartier $\mathbb{Z}$-divisors (resp. $\mathbb{Q}$-divisors). $\qquad\square$

2.41 (Hurwitz formula) (cf. Hartshorne, 1977, section IV.2) Let $g\colon X' \to X$ be a ramified cover of a (demi-)normal scheme.

The *ramification divisor* of g is defined as

$$R(g) = \sum_{F \subset X'} R(F)[F] := \sum_{F \subset X'} \operatorname{length}_{k(F)}(\Omega_{X'/X})_F[F], \qquad (2.41.1)$$

where the summation is over all prime divisors of X' and $\Omega_{X'/X}$ denotes the sheaf of relative differentials. If $r(F)$ denotes the ramification index of g along F then $R(F) \geq r(F) - 1$. The ramification is called *tame* along F if $R(F) = r(F) - 1$. The latter holds if and only if $\operatorname{char} k(F)$ does not divide $r(F)$.

The support of $g(R(g))$ is called the *branch divisor* of g. The Hurwitz formula compares the canonical class of X with the canonical class of X'. If ω_X is locally free, the formula says that $\omega_{X'}$ is canonically isomorphic to $g^*\omega_X(R)$. Note that ω_X is locally free outside a codimension 2 subset $Z \subset X$, so the locally free case gives a general isomorphism $\omega_{X'} \simeq (g^*\omega_X)^{**}(R)$ which is frequenty written additively as $K_{X'} \sim g^*K_X + R$. The latter form, however, hides the canonical isomorphism.

More generally, if Δ is any $\mathbb{Q}$-divisor on X, then the $\mathbb{Q}$-divisor $g^*\Delta - R$ makes sense and then we have a canonical $\mathbb{Q}$-linear equivalence

$$K_{X'} + g^*\Delta - R \sim_\mathbb{Q} g^*(K_X + \Delta). \tag{2.41.2}$$

Thus $K_X + \Delta$ is $\mathbb{Q}$-Cartier if and only if $K_{X'} + g^*\Delta - R$ is.

In general $g^*\Delta - R$ need not be effective, but there are three important special cases when it is effective and the pull-back formula is very simple.

The first is when $R = 0$, that is, when g is unramified in codimension 1. Then

$$K_{X'} + g^*\Delta \sim_\mathbb{Q} g^*(K_X + \Delta). \tag{2.41.3}$$

The second is when g is tamely ramified and $\Delta = B + \Delta_1$ where B is a reduced divisor whose support contains the branch divisor of g. Then $g^*(B)$ contains the support of R and $g^*B = R + \operatorname{red} g^*B$. Thus we obtain the pull-back formula

$$K_{X'} + \operatorname{red} g^*B + g^*\Delta_1 \sim_\mathbb{Q} g^*(K_X + B + \Delta_1). \tag{2.41.4}$$

More generally, assume that for every $D_i \subset X$, the coefficient of D_i in Δ is at least $1 - \frac{1}{r_i}$ for some $r_i \geq \sup_j\{e_{ij}\}$ where D'_{ij} are the irreducible components of g^*D_i and e_{ij} denotes the ramification index along D'_{ij}. We can then write $\Delta = \sum_i (1 - \frac{1}{r_i})D_i + \Delta'$ where $\Delta' \geq 0$. This gives the formula

$$K_{X'} + \sum_{ij}\left(1 - \tfrac{e_{ij}}{r_i}\right)D'_{ij} + g^*\Delta' \sim_\mathbb{Q} g^*\left(K_X + \sum_i\left(1 - \tfrac{1}{r_i}\right)D_i + \Delta'\right). \tag{2.41.5}$$

In particular, if $\Delta' = 0$ and $e_{ij}|r_i$ for every i, j then every coefficient of the pull-back also has the form $1 - \frac{1}{r_{ij}}$ for some integer r_{ij}.

This is even nicer if $Y \to X$ is Galois with group $G \subset \operatorname{Aut}(Y, \Delta_Y)$ where $\Delta_Y = \sum_i(1 - \frac{1}{m_i})D_i^Y$. Then

$$K_Y + \sum_i\left(1 - \tfrac{1}{m_i}\right)D_i^Y \sim_\mathbb{Q} q^*\left(K_X + \sum_j\left(1 - \tfrac{1}{e_j m_j}\right)D_j^X\right) \tag{2.41.6}$$

where $1 - \frac{1}{m_j}$ is the coefficient of any preimage D_i^Y of D_j^X and e_j is the ramification index of q along D_i^Y.

2.42 (Discrepancies and finite morphisms) (Reid, 1980)   Let $g\colon X' \to X$ be a ramified cover of degree m of (demi-)normal schemes. Let Δ be a $\mathbb{Q}$-divisor on X and Δ' a $\mathbb{Q}$-divisor on X' such that there is a canonical $\mathbb{Q}$-linear equaivalence $K_{X'} + \Delta' \sim_\mathbb{Q} g^*(K_X + \Delta)$ as in (2.41).

Let $f\colon Y \to X$ be a proper birational morphism and $E \subset Y$ an exceptional divisor such that Y is normal at the generic point of Y. Let $Y' \to Y$ be the normalization of Y in the function field of X' and $F \subset Y'$ a divisor dominating

E. We have a diagram

$$
\begin{array}{ccc}
F \subset Y' & \xrightarrow{\ f'\ } & X' \\
\downarrow \quad \downarrow h & & \downarrow g \\
E \subset Y & \xrightarrow{\ f\ } & X.
\end{array}
\tag{2.42.1}
$$

Near the generic point of F we compute that

$$
\begin{aligned}
K_{Y'} &\sim_{\mathbb{Q}} f'^{*}(K_{X'} + \Delta') + a(F, X', \Delta')F \\
&\sim_{\mathbb{Q}} f'^{*}g^{*}(K_X + \Delta) + a(F, X', \Delta')F \\
&\sim_{\mathbb{Q}} h^{*}f^{*}(K_X + \Delta) + a(F, X', \Delta')F, \quad \text{and} \\
K_{Y'} &\sim_{\mathbb{Q}} h^{*}K_Y + R(F)F \\
&\sim_{\mathbb{Q}} h^{*}f^{*}(K_X + \Delta) + a(E, X, \Delta)h^{*}E + R(F)F \\
&\sim_{\mathbb{Q}} h^{*}f^{*}(K_X + \Delta) + (r(F)a(E, X, \Delta) + R(F))F.
\end{aligned}
$$

All these are canonical $\mathbb{Q}$-linear equaivalences, which shows that

$$
a(F, X', \Delta') + 1 = r(F)(a(E, X, \Delta) + 1) + (R(F) + 1 - r(F)). \tag{2.42.2}
$$

Since $R(F) + 1 \geq r(F)$ this implies that

$$
a(F, X', \Delta') + 1 \geq r(F)(a(E, X, \Delta) + 1). \tag{2.42.3}
$$

Conversely, if $R(F) + 1 = r(F)$ then

$$
a(E, X, \Delta) + 1 = \frac{1}{r(F)}(a(F, X', \Delta') + 1) \geq \frac{1}{\deg g}(a(F, X', \Delta') + 1). \tag{2.42.4}
$$

Finally note that these considerations apply to all possible divisors E and F. Given any divisor E over X, we get a divisor F over X' from the diagram (2.42.1). The converse follows from (2.22).

We can summarize these estimates as follows.

Corollary 2.43 *Notation and assumptions as above. Then*

(1) $\operatorname{discrep}(X', \Delta') \geq \operatorname{discrep}(X, \Delta)$. *In particular, if (X, Δ) is klt (resp. lc) then so is (X', Δ').*

(2) *Assume in addition that* char $k = 0$ *(or* $\deg g <$ char k, *or* X'/X *is Galois and* char $k \nmid \deg g$*). Then*

$$
\operatorname{discrep}(X', \Delta') + 1 \geq \operatorname{discrep}(X, \Delta) + 1 \geq \frac{1}{\deg g}(\operatorname{discrep}(X', \Delta') + 1).
$$

In particular, (X, Δ) is klt (resp. plt, lc) if and only if (X', Δ') is klt (resp. plt, lc). □

Warning Neither implication holds for dlt pairs.

Next we consider ramified covers with cyclic Galois group. These are easy to construct and especially useful in the study of slc pairs.

2.44 (μ_m-covers) Let $\pi\colon \tilde{X} \to X$ be a ramified cover with Galois group μ_m (9.49) where m is invertible on X.

Since μ_m is reductive, its action decomposes $\pi_* \mathcal{O}_{\tilde{X}}$ into a sum of eigensheaves L_i, each of rank 1. Multiplication gives maps $L_1^{\otimes i} \to L_i$, hence there are divisors D_i such that

$$\pi_* \mathcal{O}_{\tilde{X}} = \sum_{i=0}^{m-1} L_1^{[i]}(D_i) \tag{2.44.1}$$

where $L_1^{[i]}$ denotes the reflexive ith power (1.2). The $i \equiv 0 \mod m$ eigensubsheaf is isomorphic to $\mathcal{O}_X$, hence we get an isomorphism $\gamma\colon L_1^{[m]}(D_m) \simeq \mathcal{O}_X$.

Since L_1 tends to be negative, we usually choose $L := L_1^{[-1]}$ as our basic sheaf and $D := D_m$ as the key divisor. Then γ corresponds to a section s of $L^{[m]}$ whose zero divisor is D.

Conversely, let X be a (demi-)normal scheme, L a divisorial sheaf on X and s a section of $L^{[m]}$ for some $m > 0$ that does not vanish along any codimension 1 point of $\operatorname{Sing} X$. Set $D := (s)$ and $\Delta := \frac{1}{m} D$. The section s can be identified with an isomorphism

$$\gamma_s\colon L^{[-m]}(m\Delta) = L^{[-m]}(D) \simeq \mathcal{O}_X. \tag{2.44.2}$$

This in turn defines an algebra structure on

$$\mathcal{O}_X + L^{[-1]} + \cdots + L^{[-(m-1)]},$$

where, for $i + j < m$ the multiplication $L^{[-i]} \times L^{[-j]} \to L^{[-(i+j)]}$ is the tensor product and for $i + j \geq m$ we compose the tensor product with the isomorphisms

$$L^{[-(i+j)]} = L^{[-(i+j-m)]} \otimes L^{[-m]} \xrightarrow{1 \otimes \gamma_s} L^{[-(i+j-m)]} \otimes \mathcal{O}_X = L^{[-(i+j-m)]}.$$

The spectrum of this algebra gives $\tilde{X}$ over $X \setminus D$, but it is usually quite singular over D since we have not yet found the correct divisors D_i. By the universal property of the normalization, D_i is the largest divisor such that $(L^{[-i]}(D_i))^{[m]} \subset \mathcal{O}_X$. That is, $D_i = \lfloor \frac{i}{m} D \rfloor = \lfloor i \Delta \rfloor$ and so

$$\tilde{X} = \operatorname{Spec}_X(\mathcal{O}_X + L^{[-1]}(\lfloor \Delta \rfloor) + \cdots + L^{[-(m-1)]}(\lfloor (m-1)\Delta \rfloor)). \tag{2.44.3}$$

We frequently write $\tilde{X} =: X[L, \sqrt[m]{s}]$ to emphasize its dependence on L and s. Alternatively, let I_s be the ideal sheaf in $\sum_{i=0}^{\infty} L^{[-i]}(\lfloor i\Delta \rfloor)$ generated by $\phi - \gamma_s(\phi)$ where ϕ is any local section of $L^{[-m]}(\lfloor m\Delta \rfloor)$. Then

$$X[L, \sqrt[m]{s}] = \operatorname{Spec}_X(\sum_{i=0}^{\infty} L^{[-i]}(\lfloor i\Delta \rfloor))/I_s. \tag{2.44.4}$$

Duality for finite morphisms (cf. Kollár and Mori (1998, 5.68)) now gives that

$$\pi_* \omega_{\tilde{X}} = \omega_X + \omega_X \hat{\otimes} L(-\lfloor \Delta \rfloor) + \cdots + \omega_X \hat{\otimes} L^{[m-1]}(-\lfloor (m-1)\Delta \rfloor). \quad (2.44.5)$$

Note also that the double dual of $\pi_* \pi^*(L(-\lfloor \Delta \rfloor))$ equals

$$L(-\lfloor \Delta \rfloor) + \mathscr{O}_X + L^{[-1]}(\lfloor \Delta \rfloor) + \cdots + L^{[-(m-2)]}(\lfloor (m-2)\Delta \rfloor),$$

which shows that

$$(\pi^*(L(-\lfloor \Delta \rfloor)))^{**} \simeq \mathscr{O}_{\tilde{X}}. \quad (2.44.6)$$

2.45 (Normal forms of μ_m-covers)   There are several ways to change L and s without changing the corresponding μ_m-cover.

First of all, if $(i, m) = 1$ then the same cover is constructed if we think of the ith summand as the basic divisorial sheaf. That is using $L^i(\lfloor -i\Delta \rfloor)$ and the isomorphism

$$(L^{[i]}(\lfloor -i\Delta \rfloor))^{[m]} = L^{[mi]}(m\lfloor -i\Delta \rfloor) \simeq \mathscr{O}_X(mi\Delta + m\lfloor -i\Delta \rfloor). \quad (2.45.1)$$

Second, if D_0 is any divisor then replacing L by $L(-D_0)$ and Δ by $\Delta - D_0$ gives the same μ_m-cover. Thus we can always assume that $\lfloor \Delta \rfloor = 0$.

Finally, there is the choice of the isomorphism $s \colon \mathscr{O}_X \simeq L^{[m]}(-m\Delta)$. Given two such isomorphisms s_i, their quotient $u := s_1/s_2$ is a unit in $\mathscr{O}_X$. If $u = v^m$ is an mth power, then acting by v on L shows that the two μ_m-covers are isomorphic. Thus we should think of s as an element

$$\bar{s} \in H^0(X, \mathscr{O}_X)^{\otimes m} \setminus \mathrm{Isom}_X(\mathscr{O}_X(m\Delta), L^{[m]}). \quad (2.45.2)$$

Different choices of $\bar{s}$ can result in different covers. For instance, if $C = \mathbb{A}^1 \setminus \{0\}$ with coordinate x then $C[\mathscr{O}_C, \sqrt[m]{1}]$ is the reducible plane curve $y^m = 1$ while $C[\mathscr{O}_C, \sqrt[m]{x}]$ is the irreducible plane curve $y^m = x$.

If the residue characteristics do not divide m, then $X[\mathscr{O}_X, \sqrt[m]{s_1/s_2}] \to X$ is étale and the two μ_m-covers $X[L, \sqrt[m]{s_1}]$ and $X[L, \sqrt[m]{s_2}]$ become isomorphic after pulling back to $X[\mathscr{O}_X, \sqrt[m]{u}]$. In particular, they have isomorphic étale covers.

However, they can be quite different if the characteristic divides m. For $m = p$ the above example gives $C[\mathscr{O}_C, \sqrt[p]{1}]$ which is the nonreduced plane curve $y^p = 1$ while $C[\mathscr{O}_C, \sqrt[p]{x}]$ is the smooth plane curve $y^p = x$.

We can summarize these discussions as follows.

Corollary 2.46   *Let X be a (demi-)normal scheme and $U \subset X$ an open subset which contains every codimension 1 point of* Sing X. *Assume that m is invertible on X. Then there is a natural one-to-one correspondence between the following 3 sets.*

(1) *Étale Galois covers $\tilde{U} \to U$ plus an isomorphism* Gal$(\tilde{U}/U) \simeq \mu_m$.
(2) *Ramified Galois covers $\tilde{X} \to X$ whose branch divisor is in $X \setminus U$ plus an isomorphism* Gal$(\tilde{X}/X) \simeq \mu_m$.

(3) *Triples* $(L, \Delta, \bar{s})$ *where*

    (a) *L is a divisorial sheaf on X,*

    (b) *Δ is a $\mathbb{Q}$-divisor whose support is in $X \setminus U$ such that $\lfloor \Delta \rfloor = 0$ and $m\Delta$ is a $\mathbb{Z}$-divisor and*

    (c) *$\bar{s} \in H^0(X, \mathcal{O}_X)^{\otimes m} \setminus \mathrm{Isom}_X(\mathcal{O}_X(m\Delta), L^{[m]})$.*   □

Example 2.47 The description of cyclic covers can be made even more explicit if $L \simeq \mathcal{O}_X$. We can write the isomorphism (2.44.2) as $L^{-m} \simeq \mathcal{O}_X(-m\Delta)$. Let w be a generating section of $L^{-1} \simeq \mathcal{O}_X$ and $f \in \mathcal{O}_X$ a defining equation for $m\Delta$. Then we get the equation $(w^m = f)$. It defines a μ_m-cover of X and we are interested in its normalization. It is, however, not always easy to compute the normalization of a hypersurface. Here are some examples with $X := \mathbb{A}^2_{uv}$, for simplicity in characteristic 0.

(1) Take $\Delta = \frac{a}{m}(u = 0) + \frac{b}{m}(v = 0)$ where $(a, b, m) = 1$. The map $(x, y) \mapsto (x^m, y^m, x^a y^b)$ shows that the cyclic quotient singularity $\mathbb{A}^2_{xy}/\frac{1}{m}(b, -a)$ (3.17) is the normalization of $(w^m = u^a v^b)$.

(2) Take $\Delta_1 = \frac{1}{2}(u = 0) + \frac{2}{3}(v = 0) + \frac{4}{5}(u + v = 0)$. We need to compute the normalization of $(w^{30} = u^{15} v^{10}(u + v)^6)$. We claim that it is $(x^2 + y^3 = z^5) \subset \mathbb{A}^3$ with μ_{30} action $\frac{1}{30}(15, 10, 6)$. This is seen by setting $u = x^2$, $v = y^3$, $w = xyz$. (This looks like a degree 6 cover, but it is birational.) Here (X, Δ_1) is klt and the cover is the Du Val singularity of type E_8 (3.26), hence canonical.

(3) Take $\Delta_2 = \frac{1}{2}(u = 0) + \frac{2}{3}(v = 0) + \frac{5}{6}(u + v = 0)$. We need to compute the normalization of $(w^6 = u^3 v^2(u + v))$. We claim that it is the quotient of $(x^2 + y^3 = z^6) \subset \mathbb{A}^3$ by the μ_6 action $\frac{1}{6}(3, 2, 1)$. This is seen by setting $u = x^2$, $v = y^3$, $w = xyz$.

    Note that $\mathbb{A}^3/\frac{1}{6}(3, 2, 1)$ can be thought of as the cone over $\mathbb{P}^2(3, 2, 1) \subset \mathbb{P}^6$ embedded by $|\mathcal{O}(6)|$. Thus our normalization is a cone over a degree 6 elliptic curve in $\mathbb{P}^5$. Here (X, Δ_2) is lc but not klt and the cover is a simple elliptic singularity (3.27) hence log canonical but not klt.

(4) Let $\pi \colon X \to \mathbb{A}^n$ be a μ_m-cover ramified only along the coordinate hyperplanes $D_i := (x_i = 0)$ with ramification index r_i. Note that $r_i | m$ and X is irreducible if the lcm of the r_i is m. Then $K_X = \pi^*(K_{\mathbb{A}^n} + \sum(1 - \frac{1}{r_i})D_i)$, hence $(X, 0)$ is klt by (2.43). It is also clear that the mth power map $\tau_m \colon \mathbb{A}^n \to \mathbb{A}^n$ given by $(x_1, \dots, x_n) \mapsto (x_1^m, \dots, x_n^m)$ factors through π. Since τ_m is Galois with group μ_m^n, we see that X is also a quotient of $\mathbb{A}^n$ by an action of μ_m^n/μ_m (which is non-canonically isomorphic to μ_m^{n-1}).

2.48 (Local properties of μ_m-covers) Given X and $(L, \Delta, \bar{s})$ as in (2.46.3), let $\pi \colon \tilde{X} \to X$ be the corresponding μ_m-cover. Write $\Delta = \sum(a_i/m_i)D_i$ where $(a_i, m_i) = 1$, $m_i | m$ and assume that m is invertible on X.

By (9.52), for $x \in X$ the evaluation of the product

$$L^{[-i]}(\lfloor i\Delta \rfloor) \times L^{[i-m]}(\lfloor (m-i)\Delta \rfloor) \to L^{[-m]}(\lfloor m\Delta \rfloor) \otimes k(x) \simeq k(x)$$

is zero, unless $i\Delta$ is a $\mathbb{Z}$-divisor near x and $L^{[-i]}(\lfloor i\Delta \rfloor)$ is locally free.

This implies the following:

(1) The number of preimages of x equals the number of indices $0 \le j < m$ such that $j\Delta$ is a $\mathbb{Z}$-divisor near x and $L^{[-j]}(j\Delta)$ is locally free at x.
(2) π is étale at $x \in X$ if and only if L is locally free at x and $x \notin \mathrm{Supp}\,\Delta$.
(3) The ramification index of π over D_i is m_i.

Definition 2.49 (Index 1 covers)   Let (X, Δ) be a (demi-)normal pair.

The *index* of (X, Δ) at a point $x \in X$, denoted by $\mathrm{index}_x(X, \Delta)$ is the smallest positive integer m such that $m\Delta$ is a $\mathbb{Z}$-divisor and $\omega_X^{[m]}(m\Delta)$ is locally free at x. (If there is no such m, set $\mathrm{index}_x(X, \Delta) = \infty$.) If $\Delta = 0$ then $\mathrm{index}_x(X, 0)$ is also called the index of K_X at x.

For a subset $Z \subset X$, let $\mathrm{index}_Z(X, \Delta)$ be the least common multiple of $\mathrm{index}_x(X, \Delta)$ for all $x \in Z$. We write $\mathrm{index}(X, \Delta) := \mathrm{index}_X(X, \Delta)$

Thus $K_X + \Delta$ is $\mathbb{Q}$-Cartier if and only if $\mathrm{index}(X, \Delta) < \infty$.

Pick a point $x \in X$ and set $m = \mathrm{index}_x(X, \Delta)$. After replacing X with an open neighborhood of x, we may assume that there is an isomorphism $s \colon \mathscr{O}_X \simeq \omega_X^{[m]}(m\Delta)$. Write $\Delta = B + \Delta'$ where B is a $\mathbb{Z}$-divisor and $\lfloor \Delta' \rfloor = 0$. Then $L := \omega_X^{[-1]}(-B)$, Δ' and s determine a μ_m-cover $\pi \colon \tilde{X} \to X$. Since m is the smallest, (2.48.1) implies that $\pi^{-1}(x)$ consists of a single point $\tilde{x}$, hence we get a pointed scheme $(\tilde{x} \in \tilde{X})$. Note that $(\tilde{x} \in \tilde{X})$ depends on s if we work Zariski locally, but it does not depend on s if we work étale locally and $\mathrm{char}\,k(x) \nmid m$. Thus, in the latter case, we can talk about *the* index 1 cover $(\tilde{x} \in \tilde{X}, \tilde{\Delta})$ of $(x \in X, \Delta)$. Here $\tilde{\Delta}$ is a $\mathbb{Z}$-divisor, but it need not be effective unless one of the conditions of (2.41.3–5) holds. From (2.44) we get that

$$\pi_* \mathscr{O}_{\tilde{X}} = \sum_{i=0}^{m-1} \omega_X^{[i]}(iB + \lfloor i\Delta' \rfloor) \quad \text{and} \quad \pi_* \omega_{\tilde{X}}(\tilde{B})$$

$$= \sum_{i=0}^{m-1} \omega_X^{[1-i]}((1-i)B - \lfloor i\Delta' \rfloor).$$

As in (2.44.6), the $i = 1$ summand shows that $\omega_{\tilde{X}}(\tilde{B}) \simeq \mathscr{O}_{\tilde{X}}$. Furthermore, the μ_m-action on $\omega_{\tilde{X}}(\tilde{B}) \otimes k(\tilde{x})$ is the standard representation, hence it is faithful.

While the index 1 cover of a (demi-)normal pair (X, Δ) exists near any point $x \in X$, it seems to be useful only if the resulting divisor $\tilde{\Delta}$ is effective. The main cases are enumerated below.

Proposition 2.50 *In each of the following four cases, taking the index 1 cover gives a natural one-to-one correspondence between the sets described in* (a) *and* (b). *Local is understood in the étale topology and m is always assumed invertible.*

(1) (a) *Local (demi-)normal schemes* $(x \in X)$ *such that* $\mathrm{index}_x X = m$.

   (b) *Local (demi-)normal schemes* $(\tilde{x} \in \tilde{X})$ *such that* $\omega_{\tilde{X}}$ *is locally free with a* μ_m-*action that is free outside a codimension* ≥ 2 *subset and the induced action on* $\omega_{\tilde{X}} \otimes k(\tilde{x})$ *is faithful.*

(2) (a) *Local (demi-)normal pairs* $(x \in X, B)$ *such that* $\mathrm{index}_x(X, B) = m$.

   (b) *Local (demi-)normal pairs* $(\tilde{x} \in \tilde{X}, \tilde{B})$ *such that* $\omega_{\tilde{X}}(\tilde{B})$ *is locally free with a* μ_m-*action that is free outside a codimension* ≥ 2 *subset and the induced action on* $\omega_{\tilde{X}}(\tilde{B}) \otimes k(\tilde{x})$ *is faithful.*

(3) (a) *Local (demi-)normal pairs* $(x \in X, \Delta)$ *where* $\Delta = \sum_i (1 - \frac{1}{r_i})D_i$ *with* $r_i \in \mathbb{N}$ *such that* $\mathrm{index}_x(X, \Delta) = m$.

   (b) *Local (demi-)normal schemes* $(\tilde{x} \in \tilde{X})$ *such that* $\omega_{\tilde{X}}$ *is locally free with a* μ_m-*action that is free on a dense open subset that contains every codimension 1 point of* $\mathrm{Sing}\, \tilde{X}$ *and the induced action on* $\omega_{\tilde{X}} \otimes k(\tilde{x})$ *is faithful.*

(4) (a) *Local (demi-)normal pairs* $(x \in X, B + \Delta)$ *where* $\Delta = \sum_i (1 - \frac{1}{r_i})D_i$ *with* $r_i \in \mathbb{N}$ *such that* $\mathrm{index}_x(X, B + \Delta) = m$.

   (b) *Local (demi-)normal pairs* $(\tilde{x} \in \tilde{X}, \tilde{B})$ *such that* $\omega_{\tilde{X}}(\tilde{B})$ *is locally free with a* μ_m-*action that is free on a dense open subset that contains all generic points of* $\tilde{B}$ *(or of* $\tilde{B} + D_{\tilde{X}}$ *in the demi-normal case where* $D_{\tilde{X}}$ *is the conductor* (5.2.2)) *and the induced action on* $\omega_{\tilde{X}}(\tilde{B}) \otimes k(\tilde{x})$ *is faithful.*

Moreover, in all cases the pair $(X, 0)$ *(resp.* (X, B), (X, Δ), $(X, B + \Delta)$) *is klt (or plt or lc or slc* (5.10)) *if and only if the index 1 cover* $(\tilde{X}, 0)$ *(resp.* $(\tilde{X}, \tilde{B})$, $(\tilde{X}, 0)$, $(\tilde{X}, \tilde{B})$) *is klt (or plt or lc or slc).*

Proof Starting with $(X, B + \Delta)$, the construction of $(\tilde{X}, \tilde{B})$ was done in (2.49) and we also saw that $\omega_{\tilde{X}}(\tilde{B})$ is locally free and induced μ_m-action on $\omega_{\tilde{X}}(\tilde{B}) \otimes k(\tilde{x})$ is faithful.

The pull-back of the canonical class is computed in (2.41) and (2.42) shows the last claim about the properties klt, plt, lc or slc. □

The following two special cases are especially important.

Corollary 2.51 *Let X be a normal variety over a field of characteristic 0.*

(1) *The singularity* $(x \in X)$ *is log terminal if and only if it is a quotient of an index 1 canonical singularity* $(\tilde{x} \in \tilde{X})$ *by a* μ_m-*action that is free outside a codimension* ≥ 2 *subset.*

(2) *The singularity $(x \in X, \Delta)$ where $\Delta = \sum_i (1 - \frac{1}{r_i})D_i$ with $r_i \in \mathbb{N}$ is klt if and only if it is a quotient of an index 1 canonical singularity $(\tilde{x} \in \tilde{X})$ by a μ_m-action.* $\square$

Two special aspects of the above covering constructions are worth mentioning.

Remark 2.52

(1) Let $(x \in X, \sum_i a_i D_i)$ be a klt pair such that $a_i \geq \frac{1}{2}$ for every i. Then we can write $\sum_i a_i D_i = \Delta' + \frac{1}{2} \sum_i D_i$ where Δ' is effective. If $K_X + \frac{1}{2} \sum_i D_i$ is $\mathbb{Q}$-Cartier, then one can take the index 1 cover of $(x \in X, \frac{1}{2} \sum_i D_i)$. We get $\pi \colon \tilde{X} \to X$ such that

$$K_{\tilde{X}} + \pi^* \Delta' = \pi^*(K_X + \Delta' + \tfrac{1}{2}\textstyle\sum_i D_i).$$

Moreover, $\tilde{X}$ is canonical and has index 1.

   We stress, again, that this construction works only if $K_X + \frac{1}{2} \sum_i D_i$ is $\mathbb{Q}$-Cartier. We do not know what could replace (2.51.2) in general when the coefficients in Δ are not of the form $1 - \frac{1}{r_i}$.

(2) Another interesting aspect of the covering construction is the extra freedom we get by replacing a projective pair with a suitable cone. For instance, consider a pair $(\mathbb{P}^n, \sum(1 - \frac{1}{r_i})D_i)$. Usually there is no global covering $g \colon Y \to \mathbb{P}^n$ such that $K_Y = g^*(K_{\mathbb{P}^n} + \sum(1 - \frac{1}{r_i})D_i)$.

   By contrast, take the cone $(\mathbb{A}^{n+1}, \sum(1 - \frac{1}{r_i})B_i)$ where B_i is the cone over D_i. Then there is a covering $h \colon X \to \mathbb{A}^{n+1}$ such that $K_X = h^*(K_{\mathbb{A}^{n+1}} + \sum(1 - \frac{1}{r_i})B_i)$. One sees that $X \setminus \{\tilde{0}\}$ is a Seifert bundle (9.50) over $\mathbb{P}^n$.

In some cases this gives us enough freedom to derive interesting global results by going through a local construction; see Kollár (1992, 18.24).

2.4 Log terminal 3-fold singularities

Here we study how index 1 covers can be used to study log canonical singularities of surfaces and log terminal singularities of 3-folds. The traditional classification of log canonical surface singularities over $\mathbb{C}$ proceeds as follows.

2.53 (Log canonical surface singularities II)   Let $(0 \in S)$ be a lc surface singularity over $\mathbb{C}$. If K_S is a Cartier divisor, then every curve appears in the discrepancy divisor Δ (2.26) of its minimal resolution with integral coefficient. Thus, by (2.26.4), either $\Delta = 0$ (and hence Γ is Du Val (3.26)), or $\Delta = \sum C_i$, the sum of all curves in Γ. Thus the system of equations (2.26.3) becomes

$$\#(\text{neighbors of } C_i) = 2 - 2p_a(C_i) \quad \forall i.$$

We have two possibilities. Either Γ consists of a single elliptic or nodal/cuspidal rational curve (as in (3.27)) or Γ consists of smooth rational curves, each with two neighbors (as in (3.27.3)).

By (2.50.1) and (2.43) every lc surface singularity is a cyclic group quotient of a lc surface singularity whose canonical class is Cartier. Thus one can classify all lc surface singularities by describing all cyclic group actions on the singularities listed in (3.26) and (3.27). There are two cases over $\mathbb{C}$:

(1) Log terminal. Since every Du Val singularity is the quotient of $\mathbb{C}^2$ by a finite subgroup of SL(2, $\mathbb{C}$), we see that every log terminal surface singularity is the quotient of $\mathbb{C}^2$ by a finite subgroup of GL(2, $\mathbb{C}$). The classification was carried out in Brieskorn (1968). These singularities are traditionally named after the corresponding subgroup of PGL(2, $\mathbb{C}$). Thus we have the following cases.
   (a) Cyclic, with dual graph as in (3.31.1).
   (b) Dihedral, with dual graph as in (3.40.2).
   (c) Tetrahedral, octahedral or icosahedral with dual graph as in (3.40.3).
(2) Log canonical (but not log terminal). These were worked out in Kawamata (1980). We have two cases.
   (a) Cusp quotients. The group is $\mathbb{Z}/2$ with dual graph as in (3.28).
   (b) Simple elliptic quotients. The group is $\mathbb{Z}/2$, $\mathbb{Z}/3$, $\mathbb{Z}/4$ or $\mathbb{Z}/6$ with dual graph as in (3.28) for $\mathbb{Z}/2$ or (3.39.4) for the other groups.

In positive characteristic this approach has problems when the characteristic divides the index of the canonical class K_S since then the cyclic covering method gives an inseparable cover $\tilde{S} \to S$. For partial results, see Kawamata (1994) and Kawamata (1999b). Note that quotients of $\mathbb{A}^2$ by $\mathbb{Z}/p$ can be very pathological in characteristic p, see (3.25) and Artin (1975, 1977) for examples.

The main steps of this approach work in any dimension and the construction of the index 1 cover $(0 \in \tilde{X}) \to (0 \in X)$ as in (2.49) reduces the problem to two distinct steps.

- Classification of those log terminal singularities $(0 \in X)$ for which K_X is Cartier. Note that these singularities are in fact canonical (2.8.2).
- Classification of cyclic group actions on the above singularities.

Using this, Reid (1980) developed a fairly complete description of terminal, canonical and log terminal singularities in dimension 3. We discuss this next.

Canonical singularities with K_X Cartier

Toward the first step of the classification, we have the following.

Lemma 2.54 (Reid, 1980)   *Let $(0 \in X)$ be a canonical singularity such that K_X is Cartier. Let $0 \in H \subset X$ be a general hypersurface section and $g\colon H' \to H$ any resolution. Then*

$$m_{0,H} \cdot \omega_H \subset g_*\omega_{H'} \subset \omega_H.$$

Proof   If $\dim X = 2$ then H is an isolated curve singularity and H' is its normalization.

If $\dim X \geq 3$ then we use that X is CM (2.88), so H is also CM. Since H is a general hypersurface section, H is smooth in codimension 1, hence normal. There is always an injection $g_*\omega_{H'} \hookrightarrow \omega_H$ whose image is independent of the choice of H' (2.3). Thus we are allowed to make a convenient choice of H'.

Let $B_0 X \to X$ be the blow-up of the origin and $f\colon Y \to X$ a resolution dominating $B_0 X$. Then $f^*m_{0,X} \subset \mathscr{O}_Y$ is an ideal sheaf which defines a Cartier divisor $E \subset Y$. Let $H' := f_*^{-1}H \subset Y$ be the birational transform of H by f. Then $H' + E = f^*H$ and H' is smooth since the linear system $|H'|$ is free. Since $(0 \in X)$ is canonical, $\omega_Y = f^*\omega_X(F)$, where F is effective. Thus, by adjunction,

$$\omega_{H'} = \omega_Y(H')|_{H'} = f^*(\omega_X(H))(F - E)|_{H'} \supset f^*\omega_H \otimes \mathscr{O}_Y(-E).$$

Therefore $f_*\omega_{H'} \supset \omega_H \cdot f_*\mathscr{O}_Y(-E) = m_{0,H} \cdot \omega_H$. $\square$

Note that (2.54) encompasses only two possibilities:

Case 2.54.1 $g_*\omega_{H'} = \omega_H$. That is, H itself is canonical by (2.3.2).

If $\dim H = 1$ then H is smooth and so X is smooth. If $\dim H \geq 2$ then we can apply (2.54) to H and continue until we end up with the next case.

If $\dim H = 2$, we can also use the classification of Du Val singularities (3.26). We obtain that H has multiplicity 2 and embedding dimension 3, hence X has multiplicity 2 and embedding dimension 4 (or X is smooth). Such a singularity $(0 \in X)$ is called a *compound Du Val* or *cDV* singularity. It follows from (2.56) that the pair (X, H) is canonical.

If $(0 \in X)$ is an isolated singularity, then (2.5) implies that $(0 \in X)$ is in fact terminal. If $(0 \in X)$ is not an isolated singularity, the situation is more complicated. It was analyzed in Reid (1983) using a detailed knowledge of deformations of Du Val singularities. Higher dimensional generalizations of this method seem unlikely.

Case 2.54.2 $g_*\omega_{H'} = m_{0,H} \cdot \omega_H$.

Let us start with the case $\dim H = 1$. From the exact sequence

$$0 \to \mathscr{O}_H \to \mathscr{O}_{H'} \to \mathscr{O}_{H'}/\mathscr{O}_H \to 0$$

we get

$$0 \to \omega_{H'} \to \omega_H \to \mathscr{E}xt^1(\mathscr{O}_{H'}/\mathscr{O}_H, \omega_H) \to 0.$$

By local duality on the Gorenstein curve H we see that

$$1 = \dim_k \omega_H / \omega_{H'} = \dim_k \mathcal{E}xt^1(\mathcal{O}_{H'}/\mathcal{O}_H, \omega_H) = \dim_k \mathcal{O}_{H'}/\mathcal{O}_H.$$

Thus H has either a node or a cusp at the origin. If H has a node then X has a type A Du Val singularity. If H has cusp then H can be a type D or E Du Val singularity, but it could be worse. For instance $(x^2 + y^3 + z^r = 0)$ is not even lc if $r \geq 7$. Thus the method comes close to giving the correct answer but needs extra work at the end.

The situation is similar if $\dim H = 2$. Write $K_{H'} = f^* K_H + E_H$. Then $f_* \mathcal{O}_{H'}(E_H) = m_{0,H}$. If H is log canonical, that is, every exceptional divisor appears in E_H with coefficient ≥ -1, then clearly $f_* \mathcal{O}_{H'}(E_H) \supset m_{0,H}$. Our problem is that the converse fails: there are many singularities for which $f_* \mathcal{O}_{H'}(E_H) = m_{0,H}$ yet E contains an exceptional divisor with large negative coefficient.

Nonetheless, for surfaces, Laufer (1977) and Reid (1976) developed a rather complete classification of these so-called elliptic singularities.

Using this, Reid (1980) shows how to construct a minimal model of a 3-fold X from the canonical model of X. Later development turned this picture around: Mori's program constructs the minimal model first; we get the canonical model from the minimal model. However, the method of Siu (2008) and Cascini and Lazić (2010) yields the canonical model first.

Group actions on canonical singularities

Let $(0 \in X)$ be a canonical singularity with ω_X locally free and G a finite group acting on X. We are interested in reversing the index 1 cover construction, hence we assume that the action is fixed point free outside a codimension ≥ 2 subset. From (2.42) we see that the quotient X/G is log terminal.

It seems harder to tell when a quotient is canonical or terminal. To analyze this question, observe that the G-action on X gives a G-action on ω_X. (Note that if one thinks of $\mathcal{O}_X(K_X)$ as a sheaf associated to a linear equivalence class then one naturally gets a G-action only up to multiplication by constants. It is crucial here that ω_X is more functorial than $\mathcal{O}_X(K_X)$.) This then induces an action of G on $\omega_X \otimes k(0)$, hence a character $\chi: G \to \mathbb{G}_m$.

If χ is trivial, then any nonzero vector v gives a G-invariant element in $\omega_X \otimes k(0)$. Pick any section ϕ of ω_X such that $\phi(0) = v$. Then

$$\Phi := \tfrac{1}{|G|} \sum_{g \in G} g^* \phi$$

is a G-invariant section of ω_X such that $\Phi(0) = v$. Thus Φ is a local generator of ω_X and it descends to a local generator of $\omega_{X/G}$. Thus X/G is canonical by (2.3.2).

If χ is nontrivial, let $H = \ker \chi$. By the above argument X/H is canonical, its canonical class is Cartier and $X/G = (X/H)/(G/H)$.

We are thus reduced to the most interesting case when $\chi \colon G \to \mathbb{G}_m$ is injective. Thus G is cyclic; the very case we started with.

There does not seem to be an easy way to decide which quotients by cyclic groups are canonical or terminal. The case when X is smooth is treated in Section 3.2.

Example 2.55 Let $f(z, t)$ be a power series without multiple factors and not divisible by z. Fix natural numbers n, a with $(a, n) = 1$. We claim that, using the notation of (3.19.2),

$$X := (xy + f(z^n, t) = 0)/\tfrac{1}{n}(1, -1, a, 0)$$

is a terminal 3-fold singularity such that K_X has index n (2.49).

It turns out that the above examples give the sole infinite series of terminal 3-fold singularities. There are five other cases, each having index ≤ 4. For their construction and proofs that there are no other cases see Mori (1985), Reid (1987) and Kollár and Mori (1998). More generally, 3-dimensional canonical quotients of hypersurfaces are classified in Hayakawa and Takeuchi (1987).

By assumption $f(z^n, t) = 0$ defines an isolated curve singularity, hence $(xy + f(z^n, t) = 0)$ is an isolated 3-fold singularity. Thus if E is an exceptional divisor over X and center$_X$ E is not the origin, then $a(E, X) > 0$.

In order to understand what happens over the origin, set $H := (xy + f(0, t) = 0)/\tfrac{1}{n}(1, -1, 0)$. By assumption $(xy + f(0, t) = 0)$ is a Du Val singularity and a local generator of its dualizing sheaf is $y^{-1} dy \wedge dt$. This is invariant under the μ_n-action, hence H is canonical.

A local generator of the dualizing sheaf of $(xy + f(z^n, t) = 0)$ is $y^{-1} dy \wedge dz \wedge dt$, hence $z^{-1} y^{-1} dy \wedge dz \wedge dt$ is μ_n-invariant. Thus it descends to a local generator of $\omega_X(H)$. By (2.56) we conclude that (X, H) is canonical.

By (2.5) this implies that $a(E, X) > 0$ if center$_X$ $E \subset H$ and we are done.

The following result of Stevens (1998) was the first general theorem about inversion of adjunction. Now this is a special case of the theory in Section 4.1, but its simple proof is worth seeing.

Proposition 2.56 *Let $(0 \in X)$ be a normal singularity. Let $0 \in H \subset X$ be a divisor such that $H \sim -K_X$ and H is canonical. Then (X, H) is canonical.*

Proof By adjunction (cf. Hartshorne (1977, II.8.20)), we have an exact sequence

$$0 \to \omega_X \to \omega_X(H) \overset{r}{\to} \omega_H.$$

(We know that r is surjective over the open set where H is Cartier; we will prove that it is everywhere surjective.)

Let $f: Y \to X$ be a log resolution of (X, H) and set $H' = f_*^{-1} H$. On Y we have the exact sequence

$$0 \to \omega_Y \to \omega_Y(H') \to \omega_{H'} \to 0.$$

Push this forward by f_*. Since $R^1 f_* \omega_Y = 0$ (cf. (10.32)), we obtain the following commutative diagram with exact rows:

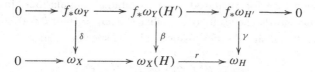

Since H is canonical, γ is an isomorphism. Thus r is surjective and by Nakayama's lemma we conclude that $f_* \omega_Y(H') = \omega_X(H)$. Arguing as in (2.3.2) we conclude that (X, H) is canonical. □

2.5 Rational pairs

One of the early encouraging results about canonical singularities was the quite subtle proof by Elkik (1981) that they are rational.

The aim of this section is two-fold. First, we give rather general results that connect CM sheaves and rationality on X with various vanishing results on a resolution of X (2.72) and (2.74). Then we start to develop a notion of rational pairs (2.80). This concept seems to capture the rationality properties of dlt pairs (2.88).

The proofs rely on general Grothendieck duality theory. Readers who are not familiar with it and are willing to believe (2.88) can skip the rest of this section.

Note that everything before (2.86) works in arbitrary characteristic.

2.57 (Rational and Cohen–Macaulay singularities)   Being *Cohen–Macaulay*, or *CM*, is a somewhat technical but very useful condition for schemes and coherent sheaves; see Matsumura (1986, section 17) and Hartshorne (1977, pp. 184–186). Roughly speaking the basic cohomology theory of CM coherent sheaves works very much like the cohomology theory of locally free sheaves on a smooth variety. Indeed, let X be a projective scheme of pure dimension n over a field k and $\mathscr{F}$ a coherent sheaf on X. By the Noether normalization theorem, there is a finite morphism $\pi: X \to \mathbb{P}^n$. Then

- $H^i(X, \mathscr{F}) = H^i(\mathbb{P}^n, \pi_* F)$ and
- $\mathscr{F}$ is CM if and only if $\pi_* \mathscr{F}$ is locally free.

One can use these to develop the duality theory for projective schemes, see Kollár and Mori (1998, section 5.5).

Let X be a normal variety over a field of characteristic 0. We say that X has *rational* singularities if and only if $R^i f_* \mathcal{O}_Y = 0$ for $i > 0$ for every resolution of singularities and $f : Y \to X$. It is actually enough to check this for one resolution, or even for one proper birational $f : Y \to X$ such that Y has rational singularities. Rational implies CM by (2.82). See Kollár and Mori (1998, section 5.1) for an introduction to rational singularities. In positive characteristic one also needs to assume that $R^i f_* \omega_Y = 0$ for $i > 0$, see (2.76).

If X has rational singularities, using the Leray spectral sequence

$$H^i(X, R^j f_* \mathcal{O}_Y) \Rightarrow H^{i+j}(Y, \mathcal{O}_Y)$$

we obtain that $H^i(X, \mathcal{O}_X) = H^i(Y, \mathcal{O}_Y)$. More generally, for any vector bundle E on X

$$H^i(X, \mathcal{E}) = H^i(Y, f^* \mathcal{E}). \tag{2.57.1}$$

Thus cohomology computations on a variety with rational singularities can be reduced to computations on a smooth variety.

In particular, if X is a projective variety over a field of characteristic 0 with rational singularities and $\mathcal{L}$ is an ample line bundle on X then a variant of Kodaira's vanishing theorem (Kollár and Mori, 1998, 2.64) implies that

$$H^i(X, \mathcal{L}^{-1}) = 0 \quad \text{for } i < \dim X. \tag{2.57.2}$$

CM sheaves

Definition 2.58 Let X be a noetherian scheme, $x \in X$ a (not necessarily closed) point, and $\mathcal{F}$ a coherent sheaf on X. The *dimension* and *codimension* of the closed subscheme $\overline{\{x\}} \subseteq X$ will be denoted by $\dim x$ and $\operatorname{codim}_X x$ respectively. The *dimension* of $\mathcal{F}$ is the dimension of its support: $\dim \mathcal{F} := \dim \operatorname{Supp} \mathcal{F}$. The local dimension, denoted by $\dim_x$ is understood on the local scheme (X, x) and it is equal to $\dim \mathcal{F}_x$, the dimension of $\mathcal{F}_x$ as an $\mathcal{O}_{X,x}$-module. The *depth of $\mathcal{F}$ at x*, denoted by $\operatorname{depth}_x \mathcal{F}$, is defined as the depth of $\mathcal{F}_x$ as an $\mathcal{O}_{X,x}$-module. Recall that the depth of a finitely generated module M over a noetherian local ring $(A, \mathfrak{m})$ is the maximal length of a regular sequence, that is, a sequence $x_1, \ldots, x_r \in \mathfrak{m}$ such that x_1 is not a zero divisor on M and x_i is not a zero divisor on $M/(x_1, \ldots, x_{i-1})M$ for all $i = 2, \ldots, r$.

A non-zero coherent sheaf $\mathcal{F}$ is said to satisfy *Serre's condition S_m* if

$$\operatorname{depth}_x \mathcal{F} \geq \min(m, \dim_x \mathcal{F}) \quad \text{for all } x \in X.$$

Notice that if X is contained in another noetherian scheme Y, then $\mathscr{F}$ satisfies Serre's condition S_m regarded as a sheaf on X if and only if it satisfies S_m regarded as a sheaf on Y.

Let $x \in X$ be a closed point. A coherent sheaf $\mathscr{F}$ is called *Cohen–Macaulay* (*CM* for short) at x if $\mathscr{F}_x$ is a Cohen–Macaulay module over $\mathscr{O}_{X,x}$ (Matsumura, 1986, section 17), that is, the depth of $\mathscr{F}_x$ as an $\mathscr{O}_{X,x}$-module is equal to its dimension. Similarly, $\mathscr{F}$ is *Cohen–Macaulay* (*CM* for short) if it is CM at x for all closed points $x \in \operatorname{Supp} \mathscr{F}$. In particular, X is CM if so is $\mathscr{O}_X$. Also, as with the S_m property above, if X is a closed subscheme of Y then $\mathscr{F}$ is CM as an $\mathscr{O}_X$-sheaf if and only if it is CM as an $\mathscr{O}_Y$-sheaf. Finally, X is called *Gorenstein* if $\mathscr{O}_X$ is CM and ω_X is an invertible sheaf.

Recall that according to Grothendieck's vanishing theorem (see Grothendieck (1967, section 3) or Bruns and Herzog (1993, 3.5.7))

$$H_x^i(X, \mathscr{F}) = 0 \quad \text{for all } i > \dim_x \mathscr{F} \text{ or } i < \operatorname{depth}_x \mathscr{F}.$$

In particular, $\mathscr{F}$ is CM at x if and only if

$$H_x^i(X, \mathscr{F}) = 0 \quad \text{for all } i \neq \dim_x \mathscr{F}.$$

Lemma 2.59 *If $\mathscr{F}$ is CM at x, then $\operatorname{Supp} \mathscr{F}$ is equidimensional in a neighborhood of x. In particular, if $\mathscr{F}$ is CM and $\operatorname{Supp} \mathscr{F}$ is connected, then $\operatorname{Supp} \mathscr{F}$ is equidimensional.*

Proof Let $d_{\min}$ and $d_{\max}$ denote the smallest and largest dimension of the components of $\operatorname{Supp} \mathscr{F}$ containing x. Then by Bruns and Herzog (1993, 1.2.13),

$$\operatorname{depth} \mathscr{F}_x \leq d_{\min} \leq d_{\max} = \dim_x \operatorname{Supp} \mathscr{F} = \dim \mathscr{F}_x.$$

If $\mathscr{F}_x$ is CM, then the two sides of this inequality are equal, forcing that $d_{\min} = d_{\max}$. □

Lemma 2.60 *Let $x \in X$ be a point and*

$$0 \to \mathscr{F}' \to \mathscr{F} \to \mathscr{F}'' \to 0$$

a sequence of coherent sheaves on X which is exact at x. Let $r \in \mathbb{N}$,

(1) *If* $\operatorname{depth} \mathscr{F}_x' \geq r$ *and* $\operatorname{depth} \mathscr{F}_x'' \geq r$ *at x, then* $\operatorname{depth} \mathscr{F}_x \geq r$ *at x.*
(2) *If* $\operatorname{depth} \mathscr{F}_x \geq r$ *and* $\operatorname{depth} \mathscr{F}_x'' \geq r - 1$ *at x, then* $\operatorname{depth} \mathscr{F}_x' \geq r$ *at x.*
(3) *If* $\operatorname{depth} \mathscr{F}_x \geq r$ *and* $\operatorname{depth} \mathscr{F}_x' \geq r + 1$ *at x, then* $\operatorname{depth} \mathscr{F}_x'' \geq r$ *at x.*

Proof Follows from the long exact sequence of local cohomology at x. □

Corollary 2.61 *Let $D \subset X$ be a subscheme of pure codimension 1. If X is S_m, then $\mathscr{O}_D$ is S_{m-1} if and only if $\mathscr{O}_X(-D)$ is S_m.* □

Corollary 2.62   *Let $x \in X$ be a closed point and*

$$0 \to \mathscr{F}' \to \mathscr{F} \to \mathscr{F}'' \to 0$$

a sequence of coherent sheaves on X which is exact at x. Assume that $\operatorname{Supp}\mathscr{F}'$ *and* $\operatorname{Supp}\mathscr{F}$ *are equidimensional of dimension d in a neighborhood of x.*

(1) *If $\mathscr{F}'$ and $\mathscr{F}''$ are CM at x, and $\dim_x \mathscr{F}'' = d$, then $\mathscr{F}$ is also CM at x.*
(2) *If $\mathscr{F}$ and $\mathscr{F}''$ are CM at x, and $d - 1 \le \dim_x \mathscr{F}'' \le d$, then $\mathscr{F}'$ is also CM at x.*
(3) *If $\mathscr{F}'$ and $\mathscr{F}$ are CM and $\dim_x \mathscr{F}'' < d$, then $\mathscr{F}''$ is also CM at x. It follows that in this case $\dim_x \mathscr{F}'' = d - 1$.*

Corollary 2.63   *Assume that X is CM and let $D \subset X$ be a subscheme of pure codimension 1. Then $\mathscr{O}_D$ is CM if and only if $\mathscr{O}_X(-D)$ is CM.* $\square$

CM criteria and the ω-dual

Assumption 2.64   In this section we need to use the *dualizing complex* $\omega_X^{\bullet}$ of a scheme X and Grothendieck duality theory. See Hartshorne (1966), Altman and Kleiman (1970) and Conrad (2000) for introductions to duality theory in derived categories.

For the rest of this section every scheme we work with will be assumed to admit a dualizing complex. This applies, for instance, if X is of finite type over a field. More generally, by Kawasaki (2002, 1.4), this holds if and only if X can be locally embedded as a closed subscheme into a Gorenstein scheme.

A scheme X is CM if and only if $\omega_X^{\bullet} \simeq_{qis} \omega_X[n]$ where $\simeq_{qis}$ denotes quasi-isomorphism, see Conrad (2000, 3.5.1). Thus for CM schemes we can essentially identify the dualizing complex with the dualizing sheaf. Thus the dualizing complex is mostly useful for schemes that are not (or are not known to be) CM.

It is possible to prove the main theorem (2.88) without using Grothendieck duality (cf. Kollár (1997, section 11)), but the proof below is shorter, more general, and we feel that it shows the essential features more clearly.

The (total) right derived functor of a left exact functor $F \colon \mathsf{A} \to \mathsf{B}$ between abelian categories, if it exists, is denoted by $\mathscr{R}F \colon D(\mathsf{A}) \to D(\mathsf{B})$ and $R^i F$ is short for $h^i \circ \mathscr{R}F$ compare (6.1).

Grothendieck's duality theory is very powerful and concise, but this makes it sometimes difficult to see how special cases follow from the general theory. We need the following three results that are discussed in Kovács (2011b, section 3).

Lemma 2.65 *Let X be as in (2.64) and let $X_x \simeq \operatorname{Spec} \mathscr{O}_{X,x}$ denote the local scheme of X at x. Then X_x admits a dualizing complex and*

$$\omega_{X_x}^{\bullet} \simeq \omega_X^{\bullet} \otimes \mathscr{O}_{X,x}[-\dim x].$$

Proof As the statement is local we may assume that X is embedded into a Gorenstein scheme as a closed subscheme by Kawasaki (2002, 1.4). Let $j: X \hookrightarrow Y$ be such an embedding, $N = \dim Y$ and $m = \operatorname{codim}_Y x = \dim \mathscr{O}_{Y,x}$. Then by Grothendieck duality and because Y is CM and j is a closed embedding,

$$\omega_X^{\bullet} \simeq_{qis} \mathcal{RHom}_X(\mathscr{O}_X, \omega_X^{\bullet}) \simeq_{qis} \mathcal{R}j_* \mathcal{RHom}_X(\mathscr{O}_X, \omega_X^{\bullet})$$
$$\simeq_{qis} \underbrace{\simeq_{qis}}_{\text{Grothendieck duality}} \mathcal{RHom}_Y(\mathcal{R}j_*\mathscr{O}_X, \underbrace{\omega_Y^{\bullet}}_{\text{dualizing complex}}) \simeq_{qis} \mathcal{RHom}_Y(\mathscr{O}_X, \underbrace{\omega_Y}_{\text{dualizing sheaf}})[N].$$

Then taking cohomology and localizing at x gives that

$$\omega_X^{\bullet} \otimes \mathscr{O}_{X,x}[-\dim x] \simeq_{qis} (\omega_X^{\bullet})_x[-\dim x]$$
$$\simeq_{qis} (\mathcal{RHom}_Y(\mathscr{O}_X, \omega_Y)[N - \dim x])_x \simeq_{qis} \mathcal{RHom}_{Y_x}(\mathscr{O}_{X,x}, \omega_{Y,x}[m])$$
$$\simeq_{qis} \mathcal{RHom}_{Y_x}(\mathcal{R}(j_x)_*\mathscr{O}_{X,x}, \omega_{Y,x}^{\bullet}) \simeq_{qis} \mathcal{R}(j_x)_* \mathcal{RHom}_{X_x}(\mathscr{O}_{X,x}, \omega_{X,x}^{\bullet}) \simeq_{qis} \omega_{X,x}^{\bullet}$$

$\square$

The following is essentially Grothedieck's vanishing theorem in a somewhat different presentation than usual.

Proposition 2.66 *Let $x \in X$ be a point and $\mathscr{F}$ a coherent sheaf on X. Let $d = \dim_x \mathscr{F} + \dim x$ and $t = \operatorname{depth}_x \mathscr{F} + \dim x$. Then*

$$\left(\mathcal{E}xt_X^{-i}(\mathscr{F}, \omega_X^{\bullet}) \right)_x = 0 \qquad \text{for } i > d \text{ and } i < t.$$

Furthermore, if $\mathscr{F}_x \neq 0$, then

$$\left(\mathcal{E}xt_X^{-d}(\mathscr{F}, \omega_X^{\bullet}) \right)_x \neq 0, \quad \text{and} \quad \left(\mathcal{E}xt_X^{-t}(\mathscr{F}, \omega_X^{\bullet}) \right)_x \neq 0,$$

Proof We may obviously assume that $\mathscr{F} \neq 0$. Localization is exact and commutes with the $\mathcal{H}om$ functor, so $(\mathcal{E}xt_X^{-i}(\mathscr{F}, \omega_X^{\bullet}))_x \simeq \mathcal{E}xt_{X_x}^{-i}(\mathscr{F}_x, (\omega_X^{\bullet})_x)$ and then the latter group is isomorphic to $\mathcal{E}xt_{X_x}^{\dim x - i}(\mathscr{F}_x, \omega_{X_x}^{\bullet})$ by (2.65). This is the Matlis dual of $H_x^{i - \dim x}(Y, \mathscr{F})$ by Hartshorne (1966, V.6.2). Therefore we obtain that

$$\left(\mathcal{E}xt_X^{-i}(\mathscr{F}, \omega_X^{\bullet}) \right)_x = 0 \quad \Leftrightarrow \quad H_x^{i - \dim x}(Y, \mathscr{F}) = 0$$

and since both $\operatorname{depth}_x \mathscr{F}$ and $\dim_x \mathscr{F}$ remain the same over Y, the statement follows from Grothendieck's theorem (Bruns and Herzog, 1993, 3.5.7). $\square$

Corollary 2.67   *Under the same conditions and using the same notation as in (2.66) one has that $\mathscr{F}$ is CM at $x \in X$ if and only if*

$$\left(\mathcal{E}xt_X^{-i}(\mathscr{F}, \omega_X^{\bullet})\right)_x = 0 \qquad \text{for } i \neq d = \dim_x \mathscr{F} + \dim x.$$

Next we will apply the above general statements in somewhat more special situations.

Definition 2.68   Let $\mathscr{F}$ be a coherent sheaf on X with equidimensional support of dimension $d = \dim \mathscr{F}$. Note that in this case $\dim_x \mathscr{F} + \dim x = d$ for any $x \in X$. We define the *ω-dual* of $\mathscr{F}$ to be the coherent sheaf

$$\mathscr{d}(\mathscr{F}) = \mathscr{d}_X(\mathscr{F}) := \mathcal{E}xt_X^{-d}(\mathscr{F}, \omega_X^{\bullet}) = h^{-d}(R\mathcal{H}om_X(\mathscr{F}, \omega_X^{\bullet})).$$

Notice that if X is CM and $\mathscr{F}$ is locally free, then this agrees with the usual dual of $\mathscr{F}$ twisted by the dualizing sheaf: $\mathscr{d}(\mathscr{F}) \simeq \mathcal{H}om(\mathscr{F}, \omega_X) \simeq \mathscr{F}^* \otimes \omega_X$. In fact, something similar holds in general as shown in the next lemma.

Lemma 2.69   *Let $\mathscr{F}$ be a coherent sheaf on X with equidimensional support of dimension $d = \dim \mathscr{F}$. Further let $Z \subseteq X$ be a subscheme that contains the support of $\mathscr{F}$. Then $\mathscr{d}(\mathscr{F}) = \mathscr{d}_X(\mathscr{F}) \simeq \mathscr{d}_Z(\mathscr{F}) = \mathcal{E}xt_Z^{-d}(\mathscr{F}, \omega_Z^{\bullet})$. In particular, $\mathscr{d}(\mathscr{F})$ is coherent and if $\dim Z = d$, then $\mathscr{d}(\mathscr{F}) \simeq \mathcal{H}om_Z(\mathscr{F}, \omega_Z)$. (This is ω_Z and not $\omega_Z^{\bullet}$!)*

Proof   By Grothendieck duality and the definition of ω_Z,

$$\mathscr{d}(\mathscr{F}) \simeq \mathcal{E}xt_X^{-d}(\mathscr{F}, \omega_X^{\bullet}) \simeq h^{-d}(R\mathcal{H}om_X(\mathscr{F}, \omega_X^{\bullet}))$$
$$\simeq h^{-d}(R\mathcal{H}om_Z(\mathscr{F}, \omega_Z^{\bullet})) \simeq \mathcal{E}xt_Z^{-d}(\mathscr{F}, \omega_Z^{\bullet}).$$

If $\dim Z = d$, then $\mathcal{E}xt_Z^{-d}(\mathscr{F}, \omega_Z^{\bullet}) \simeq \mathcal{H}om_Z(\mathscr{F}, \omega_Z)$. Since $\mathscr{F}$ is coherent, this also implies that so is $\mathscr{d}(\mathscr{F})$. $\square$

The following is a direct consequence of (2.67).

Corollary 2.70   *Let $\mathscr{F}$ be a sheaf as in (2.68). Then $\mathscr{F}$ is CM at $x \in X$ if and only if*

$$(R\mathcal{H}om_X(\mathscr{F}, \omega_X^{\bullet}))_x \simeq_{qis} \mathscr{d}(\mathscr{F})_x[d].$$

Corollary 2.71   *Let $\mathscr{F}$ be a CM sheaf. Then $\mathscr{d}(\mathscr{F})$ is also CM, and*

$$R\mathcal{H}om_X(\mathscr{F}, \omega_X^{\bullet}) \simeq_{qis} \mathscr{d}(\mathscr{F})[d].$$

Furthermore, $\operatorname{Supp} \mathscr{d}(\mathscr{F}) = \operatorname{Supp} \mathscr{F}$ *and* $\mathscr{d}(\mathscr{d}(\mathscr{F})) \simeq \mathscr{F}$.

A special case of this was proved in Kollár and Mori (1998, 5.70). The fact that it is indeed a special case follows from (2.69).

Proof It follows directly from (2.70) that $\mathcal{R}\mathcal{H}om_X(\mathcal{F}, \omega_X^{\bullet}) \simeq_{qis} d(\mathcal{F})[d]$ and then since $\omega_X^{\bullet}$ is dualizing, it also follows that

$$\mathcal{R}\mathcal{H}om(d(\mathcal{F}), \omega_X^{\bullet}) \simeq_{qis} \mathcal{R}\mathcal{H}om(\mathcal{R}\mathcal{H}om(\mathcal{F}, \omega_X^{\bullet})[-d], \omega_X^{\bullet}) \simeq_{qis} \mathcal{F}[d].$$

This in turn implies that $\mathrm{Supp}\, d(\mathcal{F}) = \mathrm{Supp}\, \mathcal{F}$ and $d(d(\mathcal{F})) \simeq \mathcal{F}$, and, by (2.70) again, that $d(\mathcal{F})$ is CM. □

Theorem 2.72 *Let* $f\colon \widetilde{X} \to X$ *be a proper morphism and* $\mathcal{G}$ *a CM sheaf with equidimensional support of dimension* $d = \dim \mathcal{G}$ *on* $\widetilde{X}$. *Assume that there exist two integers* $a, b \in \mathbb{Z}$ *such that* $R^i f_* \mathcal{G} = 0$ *for* $i \neq a$ *and* $R^j f_*(d(\mathcal{G})) = 0$ *for* $j \neq b$. *Then* $\mathcal{F} := R^a f_* \mathcal{G}$ *is a CM sheaf on* X *of dimension* $d + a - b$ *with* $d(\mathcal{F}) \simeq R^b f_*(d(\mathcal{G}))$.

Proof By assumption,

$$\mathcal{R}\mathcal{H}om_X(\mathcal{F}, \omega_X^{\bullet}) \simeq_{qis} \mathcal{R}\mathcal{H}om_X(Rf_*\mathcal{G}[-a], \omega_X^{\bullet})$$
$$\simeq Rf_*\mathcal{R}\mathcal{H}om_{\widetilde{X}}(\mathcal{G}, \omega_{\widetilde{X}}^{\bullet})[a] \simeq Rf_*(d(\mathcal{G}))[d + a] \simeq R^b f_*(d(\mathcal{G}))[d + a - b].$$

Then the statement follows from Theorem 2.70. □

Corollary 2.73 *Let* $f\colon \widetilde{X} \to X$ *be a proper morphism and* $\mathcal{G}$ *a CM sheaf with equidimensional support of dimension* $d = \dim \mathcal{G}$ *on* $\widetilde{X}$ *such that* $R^i f_* \mathcal{G} = 0$ *and* $R^i f_*(d(\mathcal{G})) = 0$ *for* $i > 0$. *Then* $\mathcal{F} := f_* \mathcal{G}$ *is a CM sheaf on* X *of dimension* d *with* $d(\mathcal{F}) \simeq f_*(d(\mathcal{G}))$. □

Vanishing theorems

The following is a generalization of Kovács (2000a, theorem 1).

Theorem 2.74 *Let* $f\colon Y \to X$ *be a proper birational morphism of pure dimensional schemes and* $\mathcal{G}$ *a CM sheaf on* Y. *Assume that*

(1) $\mathrm{Supp}\, \mathcal{G} = Y$,
(2) $R^i f_* d(\mathcal{G}) = 0$ *for* $i > 0$ *and*
(3) *the natural map* $\varrho\colon f_*\mathcal{G} \to Rf_*\mathcal{G}$ *admits a left-inverse* $\varrho'\colon Rf_*\mathcal{G} \to f_*\mathcal{G}$, *that is,* $\varrho' \circ \varrho$ *is the identity of* $f_*\mathcal{G}$.

Then $f_* \mathcal{G}$ *is a CM sheaf on* X *and* $R^i f_* \mathcal{G} = 0$ *for* $i > 0$.

(Strictly speaking the automorphism $\varrho' \circ \varrho$ lives in the derived category of coherent $\mathcal{O}_X$-modules, and so it is an auto-quasi-isomorphism, but since $f_*\mathcal{G}$ is a sheaf, h^0 of the derived category auto-quasi-isomorphism induces an honest sheaf automorphism of $f_*\mathcal{G}$.)

Proof   Consider the morphisms ϱ and ϱ'

$$f_*\mathcal{G} \underset{\simeq}{\overset{\varrho}{\Longrightarrow}} Rf_*\mathcal{G} \overset{\varrho'}{\longrightarrow} f_*\mathcal{G},$$

and apply the functor $R\mathcal{H}om(\_\_, \omega_X^\bullet)$;

$$R\mathcal{H}om(f_*\mathcal{G}, \omega_X^\bullet) \underset{\simeq}{\longrightarrow} R\mathcal{H}om(Rf_*\mathcal{G}, \omega_X^\bullet) \longrightarrow R\mathcal{H}om(f_*\mathcal{G}, \omega_X^\bullet).$$

$$(2.74.4)$$

Applying Grothendieck duality, (2.71) and (2) to the middle term yields that

$$R\mathcal{H}om_X(Rf_*\mathcal{G}, \omega_X^\bullet) \simeq_{qis} Rf_* R\mathcal{H}om_Y(\mathcal{G}, \omega_Y^\bullet)$$
$$\simeq_{qis} Rf_* d(\mathcal{G})[n] \simeq_{qis} f_* d(\mathcal{G})[n].$$

This implies that the automorphism of $h^i(R\mathcal{H}om(f_*\mathcal{G}, \omega_X^\bullet))$ induced by (4) factors through 0 for $i \neq d$. Therefore

$$R\mathcal{H}om_X(f_*\mathcal{G}, \omega_X^\bullet) \simeq_{qis} \mathcal{H}[n]$$

for some sheaf $\mathcal{H}$ and we obtain that the induced automorphism of $\mathcal{H}$ factors through $f_*(d(\mathcal{G}))$:

$$\mathcal{H} \underset{\simeq}{\overset{\alpha}{\Longrightarrow}} f_*(d(\mathcal{G})) \overset{\beta}{\longrightarrow} \mathcal{H}.$$

Since $f_*(d(\mathcal{G}))$ is torsion-free and both α and β are generically isomorphisms, it follows that they are isomorphisms everywhere. In other words, we conclude that

$$R\mathcal{H}om_X(f_*\mathcal{G}, \omega_X^\bullet) \simeq_{qis} \mathcal{H}[n] \simeq f_* d(\mathcal{G})[n] \simeq_{qis} Rf_* d(\mathcal{G})[n].$$

Finally consider the following sequence of isomorphisms:

$$f_*\mathcal{G} \simeq_{qis} R\mathcal{H}om_X(R\mathcal{H}om_X(f_*\mathcal{G}, \omega_X^\bullet), \omega_X^\bullet)$$
$$\simeq_{qis} R\mathcal{H}om_X(Rf_* d(\mathcal{G})[n], \omega_X^\bullet) \simeq_{qis} Rf_* R\mathcal{H}om_Y(d(\mathcal{G})[n], \omega_Y^\bullet)$$
$$\simeq_{qis} Rf_* R\mathcal{H}om_Y(R\mathcal{H}om_Y(\mathcal{G}, \omega_Y^\bullet), \omega_Y^\bullet) \simeq_{qis} Rf_*\mathcal{G}.$$

It follows that $R^i f_*\mathcal{G} = 0$ for $i > 0$. Finally, the statement that $f_*\mathcal{G}$ is CM follows from (2.73). $\qquad\square$

We will only use the following special case of (2.74) in the sequel.

Corollary 2.75   *Let* $f: Y \to X$ *be a proper birational morphism and* $\mathcal{L}$ *an invertible sheaf on* Y*. Assume that*

(1) Y *is Gorenstein of dimension n,*
(2) $R^i f_*(\omega_Y \otimes \mathcal{L}^{-1}) = 0$ *for* $i > 0$ *and*

(3) *the natural morphism* $\varrho\colon f_*\mathscr{L} \to \mathcal{R}f_*\mathscr{L}$ *admits a left-inverse* ϱ': $\mathcal{R}f_*\mathscr{L} \to f_*\mathscr{L}$, *that is,* $\varrho' \circ \varrho$ *is the identity of* $f_*\mathscr{L}$.

Then $f_*\mathscr{L}$ *is a CM sheaf and* $R^i f_*\mathscr{L} = 0$ *for* $i > 0$. $\qquad\qquad\square$

It is through the condition (2) that the characteristic 0 assumption enters into many of the applications. See Section 10.3 for many known cases.

Rational pairs

First we recall the definition of rational singularities and then define an analogous notion for pairs.

Definition 2.76 Let X be a variety. A resolution of singularities $\phi\colon Y \to X$ is called a *rational resolution* if

(1) $\mathscr{O}_X \simeq \phi_*\mathscr{O}_Y$, that is, X is normal,
(2) $R^i\phi_*\mathscr{O}_Y = 0$ for $i > 0$ and
(3) $R^i\phi_*\omega_Y = 0$ for $i > 0$.

X is said to have *rational singularities* if it has a rational resolution.

It turns out that if X has a rational resolution then every resolution is rational. In characteristic 0 this is relatively easy (cf. Kollár and Mori (1998, section 5.1)) but quite hard in positive characteristic Chatzistamatiou and Rülling (2009).

In characteristic 0 (3) follows from the Grauert-Riemenschneider vanishing theorem; a special case of (10.32) with f birational and $L = \Delta = 0$. Thus it is enough (and customary) to define rational singularities by assuming only (1) and (2).

The following alternative description is due to Kempf *et al.* (1973), see also Kollár and Mori (1998, 5.10).

Proposition 2.77 *Let X be a variety and $\phi\colon Y \to X$ a resolution of singularities. Then ϕ is a rational resolution if and only if the following conditions are satisfied:*

(1) *X is CM,*
(2) *$\phi_*\omega_Y \simeq \omega_X$ and*
(3) *$R^i\phi_*\omega_Y = 0$ for $i > 0$.*

As in (2.76), in characteristic zero (3) always holds, so it is enough to assume (1) and (2). In particular, using (2.3), in characteristic zero CM + canonical implies rational. One of our main aims is to show that in fact canonical implies CM, hence canonical implies rational (2.88).

Proof of (2.77)   Let $n = \dim X$. The condition that X is CM is equivalent to $\omega_X^\bullet \simeq_{qis} \omega_X[n]$, while (2) and (3) together are equivalent to $\omega_X \simeq_{qis} R\phi_*\omega_Y$. As Y is also CM, we obtain that the conditions (1–3) are equivalent to (3) and

$$\omega_X^\bullet \simeq_{qis} R\phi_*\omega_Y^\bullet.$$

Applying $R\mathcal{H}om_X(\underline{\hspace{1em}}, \omega_X^\bullet)$ and Grothendieck duality shows that this is equivalent to (3) and

$$\mathcal{O}_X \simeq_{qis} R\phi_*\mathcal{O}_Y.$$

These in turn are the conditions (2.76.1–3).   □

Next we generalize the notion of rational singularities to pairs.

Definition 2.78   Let X be a normal variety and $D \subset X$ a reduced divisor. Let $\phi\colon Y \to X$ be a resolution such that $(Y, D_Y := \phi_*^{-1}D)$ is snc. Then $\phi\colon (Y, D_Y) \to (X, D)$ is called a *rational resolution* if

(1) $\mathcal{O}_X(-D) \simeq \phi_*\mathcal{O}_Y(-D_Y)$,
(2) $R^i\phi_*\mathcal{O}_Y(-D_Y) = 0$ for $i > 0$
(3) $R^i\phi_*\omega_Y(D_Y) = 0$ for $i > 0$.

Formally this is very much like (2.76) but there are significant differences.

Consider an snc pair (X^0, D^0) and observe that it has many non-rational resolutions. To find these, let $Z^0 \subset X^0$ be any stratum (1.7). By explicit computation we see that $B_{Z^0}X^0 \to X^0$ is not a rational resolution. In fact, as a very special case of (2.87), we see that a resolution $\phi^0\colon (Y^0, D_Y^0) \to (X^0, D^0)$ is rational if and only if ϕ^0 is an isomorphism over the generic point of every stratum of (X^0, D^0). To get the best notion, we need to add one more property.

Definition 2.79   Let X be a normal variety and $D \subset X$ a reduced divisor. A resolution $f\colon Y \to X$ is called *thrifty* if $(Y, D_Y := f_*^{-1}D)$ is snc and the following equivalent conditions hold.

(1) f is an isomorphism over the generic point of every stratum of $\mathrm{snc}(X, D)$ and f is an isomorphism at the generic point of every stratum of (Y, D_Y).
(2) $\mathrm{Ex}(f)$ does not contain any stratum of (Y, D_Y) and $f(\mathrm{Ex}(f))$ does not contain any stratum of $\mathrm{snc}(X, D)$.

In characteristic 0, thrifty resolutions exist and any two of them can be dominated by a third thrifty resolution (10.45.2). If (X, D) is dlt then the first requirement in (1) implies the second. The concept of thrifty resolutions may seem artificial, but it is dictated by (2.87).

Definition 2.80   Let X be a normal variety and $D \subset X$ a reduced divisor. The pair (X, D) is called *rational* if it has a thrifty rational resolution.

In characteristic 0 this implies that every other thrifty resolution is rational (2.86). In positive characteristic this is not known.

Remark 2.81

(1) Our definition is different from the one in Schwede and Takagi (2008) in several ways. A key point is that we want snc pairs to be rational.
(2) It can happen that (X, D) is rational but X is not. This may appear strange at first but it is frequently an advantage. To get such examples, let Z be a smooth projective variety and L a sufficiently ample line bundle on $Z \times \mathbb{P}^1$. Let X be the affine cone $C_a(Z \times \mathbb{P}^1, L)$ and $D \subset X$ the cone over $Z \times$ (point). Then (X, D) is rational but in general X is not even CM. We see, however, that $\mathscr{O}_X(-D)$ is CM (2.82).
(3) By our definition, the pair $(\mathbb{A}^3, (x^2 = y^2 z))$ is not rational but Kollár and Shepherd-Barron (1988) argues that it should be considered "rational." It would be very useful to extend our definition to cover such cases. Also, a really good notion of rationality should be preserved by quotients by finite groups. Finding the correct definition is very much connected with the resolution problem (10.51).

Our definition of rational pairs includes all dlt pairs, as we will see in (2.87). (Note that $(\mathbb{A}^3, (x^2 = y^2 z))$ is not dlt and being dlt is not preserved by quotients by finite groups.)

Next we consider various versions of (2.77) for pairs. The situation is more complicated in general but it becomes very similar if D is a Cartier divisor.

Proposition 2.82 (Kovács, 2011b) *Let $\phi: (Y, D_Y) \to (X, D)$ be a rational resolution. Then $\mathscr{O}_X(-D)$ is CM.*

Proof Apply (2.73) to $\mathscr{G} = \mathscr{O}_Y(-D_Y)$. Then $R^i \phi_* \mathscr{G} = 0$ by (2.78.2) and $R^i f_*(\omega_Y(D_Y)) = 0$ for $i > 0$ by (2.78.3) since $\omega_Y(D_Y) = d(\mathscr{G})$. Therefore $\mathscr{O}_X(-D) = \phi_* \mathscr{G}$ is a CM sheaf. $\qquad\square$

Proposition 2.83 *Let (X, D) be a reduced pair and $\phi: Y \to X$ a resolution such that $(Y, D_Y := \phi_*^{-1} D)$ is snc. Then ϕ is a rational resolution if and only if*

(1) $\mathscr{R}\mathcal{H}om(\mathscr{O}_X(-D), \omega_X^{\bullet}) \simeq_{qis} \phi_* \omega_Y(D_Y)[n]$ *and*
(2) $R^i \phi_* \omega_Y(D_Y) = 0$ *for $i > 0$.*

Proof As (2) is the same as (2.76.3), we may assume that it holds. Let $n = \dim X$. Assuming (2) ϕ is a rational resolution if and only if

$$\mathscr{R}\mathcal{H}om_X(\mathscr{O}_X(-D), \omega_X^{\bullet}) \simeq_{qis} \mathscr{R}\mathcal{H}om_X(\mathscr{R}\phi_* \mathscr{O}_Y(-D_Y), \omega_X^{\bullet}),$$

and (1) is equivalent to

$$\mathcal{R}\mathcal{H}om(\mathcal{O}_X(-D), \omega_X^{\bullet}) \simeq_{qis} \mathcal{R}\phi_*\omega_Y(D_Y)[n].$$

On the other hand, Grothendieck duality (2.80) implies that

$$\mathcal{R}\mathcal{H}om_X(\mathcal{R}\phi_*\mathcal{O}_Y(-D_Y), \omega_X^{\bullet})$$
$$\simeq_{qis} \mathcal{R}\phi_*\mathcal{R}\mathcal{H}om_Y(\mathcal{O}_Y(-D_Y), \omega_Y^{\bullet}) \simeq_{qis} \mathcal{R}\phi_*\omega_Y(D_Y)[n], \quad (2.83.3)$$

so the desired equivalence follows.   □

Proposition 2.84   *Let* (X, D) *be a reduced pair and* $\phi: Y \to X$ *a resolution such that* $(Y, D_Y := f_*^{-1}D)$ *is snc. Assume that D is a Cartier divisor. Then* ϕ *is a rational resolution if and only if the following conditions are satisfied.*

(1) *X is CM,*
(2) $\omega_X(D) \simeq \phi_*\omega_Y(D_Y)$ *and*
(3) $R^i\phi_*\omega_Y(D_Y) = 0$ *for* $i > 0$.

Proof   Let $n = \dim X$. By (2.83),

$$\omega_X^{\bullet} \simeq_{qis} \mathcal{R}\mathcal{H}om(\mathcal{O}_X(-D), \omega_X^{\bullet}) \otimes \mathcal{O}_X(-D) \simeq_{qis} (\phi_*\omega_Y(D_Y) \otimes \mathcal{O}_X(-D))[n],$$

so $h^i(\omega_X^{\bullet}) = 0$ for $i \neq -n$ and hence X is CM. Then (2.83.1) implies (2) and therefore it follows that if ϕ is a rational resolution, then the conditions (1–3) hold.

Conversely, if D is Cartier and X is CM, then (1–3) trivially implies (2.83.1), so the statement follows.   □

Remark 2.85   Let $\phi: (Y, D_Y) \to (X, D)$ be a thrifty rational resolution. By looking at the higher direct images of the sequences

$$0 \longrightarrow \mathcal{O}_Y(-D_Y) \longrightarrow \mathcal{O}_Y \longrightarrow \mathcal{O}_{D_Y} \longrightarrow 0 \qquad \text{and}$$
$$0 \longrightarrow \omega_Y \longrightarrow \omega_Y(D_Y) \longrightarrow \omega_{D_Y} \longrightarrow 0$$

we see that if two of the following three statements hold, then so does the third:

(1) (X, D) is a rational pair.
(2) X has rational singularities.
(3) $\mathcal{R}\phi_*\mathcal{O}_{D_Y} = \mathcal{O}_D$ and $R^i\phi_*\omega_{D_Y} = 0$ for $i > 0$.

Note that $\phi|_{D_Y}: D_Y \to D$ is what one should call a thrifty semi-resolution of D (10.59) and then (3) should be the definition of *semi-rational* singularities.

One can then also define semi-rational pairs. As we noted before, it would be more useful to develop this theory by working with the nc locus, not just the snc locus.

Rational pairs in characteristic zero

For the rest of the section everything is defined in characteristic 0.

Corollary 2.86 *If (X, D) has a thrifty rational resolution then every thrifty resolution is rational.*

Proof Assume first that (X, D) is snc. Then the identity is a thrifty rational resolution. Let $\phi: (Y, D_Y) \to (X, D)$ be a thrifty resolution. Then $\phi^*(K_X + D) \sim K_Y + D_Y - B$ where B is effective and exceptional, hence (2.84.2) is clear. Since ϕ is birational on every stratum of (Y, D_Y), (2.84.3) is implied by (10.38.1). Thus ϕ is rational by (2.84).

Now to the general case. Let $\phi_i: (Y_i, D_i) \to (X, D)$ be two thrifty resolutions and assume that ϕ_1 is rational. By (10.45.2) there is a third thrifty resolution $\phi_3: (Y_3, D_3) \to (X, D)$ that dominates both, that is, we have morphisms $g_i: Y_3 \to Y_i$. Since (Y_i, D_i) are snc, we know that the g_i are rational. The Leray spectral sequence

$$\mathcal{R}(\phi_i \circ g_i)_* = \mathcal{R}(\phi_i)_* \circ \mathcal{R}(g_i)_*$$

shows that ϕ_i is rational if and only if $\phi_i \circ g_i$ is. Applying this in one direction we see that $\phi_1 \circ g_1$ is rational. Since $\phi_2 \circ g_2 = \phi_1 \circ g_1$, using the same in the reverse direction shows that ϕ_2 is also rational. $\square$

The following theorem ties together the three concepts of "especially nice" resolutions of dlt pairs.

Theorem 2.87 *Let (X, Δ) be dlt and $f: Y \to X$ a resolution such that $(Y, f_*^{-1}\lfloor \Delta \rfloor)$ is snc. The following are equivalent:*

(1) *Every exceptional divisor of f has discrepancy > -1.*
(2) *f is a thrifty resolution of $(X, \lfloor \Delta \rfloor)$.*
(3) *f is a rational resolution of $(X, \lfloor \Delta \rfloor)$.*

Proof Assume (2) and let E be an exceptional divisor. If $a(E, X, \Delta) = -1$ then center$_X$ E is a stratum of $(X, \lfloor \Delta \rfloor)$ by definition of dlt and (2.10). There is no such E if f is thrifty, proving (1).

Still assuming (2) we will show that it implies (3). By (2.86) it is enough to show that there is a thrifty and rational resolution. Thus let $f: (Y, \Delta_Y) \to (X, \Delta)$ be a thrifty log resolution. Since (2) $\Rightarrow$ (1), every exceptional divisor appears in Δ_Y with coefficient < 1, so we can write $\Delta_Y = D_Y + A - B$ where A is exceptional with $\lfloor A \rfloor = 0$ and B is exceptional and integral. Since

$$B - D_Y \sim_{\mathbb{Q}} K_Y - f^*(K_X + \Delta) + A,$$

we see that $R^i f_* \mathcal{O}_Y(B - D_Y) = 0$ for $i > 0$ by (10.32).

B is an effective exceptional divisor, so that the composition,

$$\mathscr{O}_X(-D) \xrightarrow{\;\;\;\;} \underbrace{Rf_*\mathscr{O}_Y(-D_Y) \xrightarrow{\;\;\;\;} Rf_*\mathscr{O}_Y(B-D_Y)}_{\simeq_{qis}} \xrightarrow{\;\simeq_{qis}\;} f_*\mathscr{O}_Y(B-D_Y)$$

is a quasi-isomorphism. This shows that (2.75.3) holds.

Since f is thrifty, it is birational on all strata of (Y, D_Y), hence $R^i f_* \omega_Y(D_Y) = 0$ for $i > 0$ by (10.38.1), proving (2.75.2). Since Y is smooth, (2.75.1) holds. Thus (2.75) implies that $R^i f_* \mathscr{O}_Y(-D_Y) = 0$ for $i > 0$, proving (3).

Next we show that (3) implies (1). Choose a rational resolution f and assume to the contrary that $E \subset Y$, the union of all exceptional divisors with discrepancy -1, is not empty. By localizing at a generic point of $f(E)$ we may assume that $f(E)$ is a closed point. Since (X, Δ) is dlt, $f(E)$ is a smooth point and we may assume that X is smooth, $D := \Delta$ is a reduced snc divisor and $K_X + D \sim 0$. Thus we can write

$$K_Y + D_Y + E - B \sim f^*(K_X + D) \sim 0$$

where B is an effective $\mathbb{Z}$-divisor. Thus $\mathscr{O}_Y(-D_Y) \simeq \omega_Y(E - B)$ and for $n = \dim X$ we get surjections

$$R^{n-1} f_* \mathscr{O}_Y(-D_Y) = R^{n-1} f_* \omega_Y(E - B) \twoheadrightarrow H^{n-1}(E, \omega_E(-B|_E))$$
$$\twoheadrightarrow H^{n-1}(E, \omega_E).$$

The last term is nonzero which contradicts the assumption that f is a rational resolution. Thus (3) implies (1).

Finally we show that (1) implies (2). Since $(X\Delta)$ is dlt, being thrifty is a property of the snc locus $\mathrm{snc}(X, \lfloor \Delta \rfloor)$, thus we may assume that $\Delta = \sum_{i=1}^{n} D_i$ is a $\mathbb{Z}$-divisor and $(X, \sum D_i)$ is snc. We use induction on n, the number of irreducible components of $\lfloor \Delta \rfloor$. By induction and localization we may assume that $\dim X = n$, $x := \cap_i D_i$ is a closed point and f is thrifty over $X \setminus \{x\}$. We need to prove that f is an isomorphism near $x \in X$.

If $n = \dim X = 2$ then every resolution is either an isomorphism over $x = D_1 \cap D_2$ (hence thrifty) or it factors through the blow-up of x (giving an exceptional divisor with discrepancy -1).

Assume next that $n = \dim X \geq 3$. Set $D_i' := f_*^{-1} D_i$ and $f_i := f|_{D_i'}$. Then $f_i \colon D_i' \to D_i$ is a resolution and $(D_i', (f_i)_*^{-1} \sum_{j \neq i} D_i \cap D_j)$ is snc.

Claim 2.87.1 f_i has no exceptional divisors of discrepancy -1.

Proof   By the adjunction formula, the fact that f has no exceptional divisors of discrepancy -1, and (2.10.1), such divisors would arise as irreducible components of $D_i' \cap D_j'$ for some $j \neq i$, hence we need to show that $(f_i)_*^{-1}(D_i \cap D_j) = D_i' \cap D_j'$. Since $D_i' \cap D_j'$ is smooth, the latter holds if and only if $D'_i \cap D'_j \to D_i \cap D_j$ has connected fibers. Equivalently, if and only if $\mathscr{O}_{D_i \cap D_j} = f_* \mathscr{O}_{D'_i \cap D'_j}$. To see the last equality, consider the exact Koszul sequence

$$0 \to \mathscr{O}_Y(-D'_i - D'_j) \to \mathscr{O}_Y(-D'_i) \oplus \mathscr{O}_Y(-D'_j) \to \mathscr{O}_Y \to \mathscr{O}_{D'_i \cap D'_j} \to 0.$$

By induction on the number of irreducible components and since we are in the case $n \geq 3$, we see that f is a thrifty resolution of $(X, D_i + D_j)$ and (X, D_i) for all i, j. Since we have already proved that (2) implies (3) it follows that

$$R^r f_* \mathscr{O}_Y(-D'_i - D'_j) = R^r f_* \mathscr{O}_Y(-D'_i) = 0 \quad \text{for } r > 0.$$

Thus the push-forward is an exact sequence

$$0 \to \mathscr{O}_X(-D_i - D_j) \to \mathscr{O}_X(-D_i) \oplus \mathscr{O}_X(-D_j) \to \mathscr{O}_X \xrightarrow{\partial} f_* \mathscr{O}_{D'_i \cap D'_j} \to 0.$$

Note that ∂ factors as $\partial \colon \mathscr{O}_X \twoheadrightarrow \mathscr{O}_{D_i \cap D_j} \hookrightarrow f_* \mathscr{O}_{D'_i \cap D'_j}$, thus $\mathscr{O}_{D_i \cap D_j} = f_* \mathscr{O}_{D'_i \cap D'_j}$ as needed. (Alternatively, we could have applied the Connectedness theorem (Kollár and Mori, 1998, 5.48) to $D'_i \to (D_i, D_i \cap D_j)$.) □

Thus, by induction, $f_i \colon D'_i \to D_i$ is an isomorphism over a neighborhood of $x \in D_i$ and hence $\sum_i D'_i \to \sum_i D_i$ is an isomorphism over the same neighborhood.

Set $y := \cap_i D'_i$. The tangent space $T_{y,Y}$ is isomorphic to the tangent space of $\sum_i D'_i$. Thus $f_* \colon T_{y,Y} \to T_{x,X}$ is an isomorphism and hence so is f. This shows that f is an isomorphism near $x \in X$ and therefore (2) follows. □

The main application of the results of this section is the following; see Kollár and Mori (1998, 5.25) and Fujino (2009b, 4.14)).

Corollary 2.88 *Let X be a scheme of finite type over a field of characteristic 0, D a reduced $\mathbb{Z}$-divisor and L a $\mathbb{Q}$-Cartier $\mathbb{Z}$-divisor. Assume that (X, Δ) is dlt for some effective $\mathbb{Q}$-divisor Δ and $D \leq \lfloor \Delta \rfloor$. Then*

(1) *$\mathscr{O}_X$ is CM,*
(2) *$\mathscr{O}_X(-D - L)$ is CM,*
(3) *$\omega_X(D + L)$ is CM,*
(4) *if $D + L$ is effective then $\mathscr{O}_{D+L}$ is CM and*
(5) *(X, D) is rational.*

Proof Let us start with the cases when $L = 0$. Then (2) is a restatement of (2.82) and (1) is the $D = 0$ special case. These imply (3) by duality (2.71). The exact sequence

$$0 \to \mathscr{O}_X(-D) \to \mathscr{O}_X \to \mathscr{O}_D \to 0,$$

and (2.62) imply (4).

By (3.71) there exists an effective $\mathbb{Q}$-divisor Δ' such that $\lfloor \Delta' \rfloor = 0$ and $(X, D + \Delta')$ is dlt. Replacing Δ with $D + \Delta'$ we may assume that $D = \lfloor \Delta \rfloor$. Thus (5) follows from (2.87).

Finally we consider the case when $L \neq 0$. First assume that $mL \sim 0$ for some $m > 0$. Let $\pi \colon X' \to X$ be the corresponding cyclic cover (2.44). Then $(X', \pi^*(D))$ is dlt (cf. Kollár and Mori (1998, 5.20)) and $\mathscr{O}_X(-D - L)$ is a direct summand of $\pi_* \mathscr{O}_{X'}(-\pi^* D)$. Thus $\mathscr{O}_X(-D - L)$ is CM if $\mathscr{O}_{X'}(-\pi^* D)$ is (by Kollár and Mori, 1998, 5.4) (cf. (2.73)). Now observe that the assumption $mL \sim 0$ always holds locally on X and being CM is a local property. Thus everything reduces to the $L = 0$ case.　　　　　　　　　　　　□

3

Examples

We studied log canonical surface singularities in Section 2.2 and gave examples of typical log terminal 3-fold singularities in Section 2.4. In Section 2.5 we proved that dlt singularities are rational in any dimension.

The aim of this chapter is to show, by many examples, that the above results are by and large optimal: log canonical singularities get much more complicated in dimension 3 and even terminal singularities are likely non-classifiable in dimension 4.

A first set of examples of the various higher dimensional singularities occurring in the minimal model program are given in Section 3.1. These are rather elementary, mostly cones, but they already illustrate how subtle log canonical pairs can be.

We consider in greater detail quotient singularities in Section 3.2. This is a quite classical topic but with many subtle aspects. These are some of the simplest log terminal singularities in any dimension but they are the most likely to come up in applications.

Section 3.3 gives a rather detailed classification of log canonical surface singularities. Strictly speaking, not all of Section 3.3 is needed for the general theory, but it is useful and instructive to have a thorough understanding of a class of concrete examples. Our treatment may be longer than usual, but it applies in positive and mixed characteristics as well.

More complicated examples of terminal and log canonical singularities are constructed in Section 3.4. The treatment closely follows Kollár (2011b).

In Section 3.5 we study perturbations of the boundary Δ of an lc pair (X, Δ). We show that in some cases the singularity cannot be improved by changing Δ. We also give examples of flat deformations $\{X_t : t \in \mathbb{C}\}$ where (X_0, Δ_0) is klt for some Δ_0 but (X_t, Δ_t) is not klt for $t \neq 0$, no matter how one chooses Δ_t.

Assumptions In Section 3.3. we work with arbitrary excellent surfaces, in the other sections with varieties over perfect fields. The examples of Section 3.4 are worked out only over $\mathbb{C}$.

3.1 First examples: cones

The simplest examples of terminal, canonical, etc. singularities are given by cones. (See (3.8) for our conventions on cones.) Cones are rather special, but illustrate many of the difficulties that appear when dealing with these singularities.

In most of the subsequent examples we use the next lemma when X is smooth. Then the assumptions that X be terminal (resp. canonical . . .) are all satisfied. The proof is also much easier in this case.

Lemma 3.1 *Let X be normal, projective, L an ample Cartier divisor on X and $C_a(X, L)$ the corresponding affine cone (3.8). Let Δ be an effective $\mathbb{Q}$-divisor on X and $\Delta_{C_a(X,L)}$ the corresponding $\mathbb{Q}$-divisor on $C_a(X, L)$. Assume that $K_X + \Delta \sim_{\mathbb{Q}} r \cdot L$ for some $r \in \mathbb{Q}$. Then $(C_a(X, L), \Delta_{C_a(X,L)})$ is*

(1) *terminal if and only if $r < -1$ and (X, Δ) is terminal,*
(2) *canonical if and only if $r \le -1$ and (X, Δ) is canonical,*
(3) *klt if and only if $r < 0$ (that is, $-(K_X + \Delta)$ is ample) and (X, Δ) is klt,*
(4) *dlt if and only if either $r < 0$ and (X, Δ) is dlt or $(X, \Delta) \simeq (\mathbb{P}^n, (\prod x_i = 0))$ and the cone is $(\mathbb{A}^{n+1}, (\prod x_i = 0))$,*
(5) *lc if and only if $r \le 0$ (that is, $-(K_X + \Delta)$ is nef) and (X, Δ) is lc.*

If we allow X to be non-normal, then $(C_a(X, L), \Delta_{C_a(X,L)})$ is

(6) *semi-dlt (5.19) if and only if either $r < 0$ and X is semi-dlt or, for some $m \le n$, the cone is $((x_0 \cdots x_m = 0) \subset \mathbb{A}^{n+1}, (x_{m+1} \cdots x_n = 0))$.*
(7) *semi-log-canonical (5.10) if and only if $r \le 0$ and X is semi-log-canonical.*

Proof We see in (3.14.4) that $K_{C_a(X,L)} + \Delta_{C_a(X,L)}$ is $\mathbb{Q}$-Cartier if and only if $K_X + \Delta \sim_{\mathbb{Q}} r \cdot L$ for some $r \in \mathbb{Q}$.

As in (3.8), let $p: BC_a(X, L) \to C_a(X, L)$ be the blow-up of the vertex of the cone with exceptional divisor $E \simeq X$. Then, by (3.14.4),

$$K_{BC_a(X,L)} + \Delta_{BC_a(X,L)} + (1+r)E \sim_{\mathbb{Q}} 0 \sim_{\mathbb{Q}} p^*(K_{C_a(X,L)} + \Delta_{C_a(X,L)}).$$

Thus, by (2.23),

$$\operatorname{discrep}(C_a(X, L), \Delta_{C_a(X,L)})$$
$$= \min\{1 + r, \operatorname{discrep}(BC_a(X, L), \Delta_{C_a(X,L)+(1+r)E})\}.$$

The various conditions on r are now clear. Next note that the projection

$$\pi : (BC_a(X, L), \Delta_{BC_a(X,L)}) \to (X, \Delta)$$

is smooth and E is a section. Thus, by (2.14),

$$\operatorname{discrep}(BC_a(X, L), \Delta_{C_a(X,L)+E}) \ge \operatorname{discrep}(X, \Delta).$$

The rest of the assertions (1–5) are clear.

The non-normal cases (6–7) are equivalent with the corresponding assertions about their normalizations. ☐

Both of the next two results are proved by an easy case analysis.

Corollary 3.2 *Let B be a smooth curve and L an ample line bundle on B. Then $C_a(B, L)$ is*

(1) *terminal if and only if $g(B) = 0$ and $\deg L = 1$ ($C_a(B, L)$ is an affine plane),*
(2) *canonical if and only if $g(B) = 0$ and $\deg L \leq 2$ ($C_a(B, L)$ is an affine plane or a quadric cone),*
(3) *log terminal if and oly if $g(B) = 0$ ($C_a(B, L)$ is a cone over a rational normal curve) and*
(4) *log canonical if and only if $g(B) = 1$ or (3) holds.* ☐

The easiest higher dimensional case is the following:

Corollary 3.3 *Let $X \subset \mathbb{P}^n$ be a smooth complete intersection of hypersurfaces of degrees $(d_1, \ldots, d_m)$. Then $C_a(X) = C_a(X, \mathcal{O}_X(1))$ and it is*

(1) *terminal if and only if $\sum d_i < n$,*
(2) *canonical if and only if $\sum d_i \leq n$,*
(3) *log canonical if and only if $\sum d_i \leq n + 1$.* ☐

In many applications, one needs to pay special attention to two properties. It is easy to decide when a cone considered in (3.1) is CM:

Corollary 3.4 *Let X be a smooth, projective variety over a field of characteristic 0 and L an ample line bundle on X.*

(1) *If $-K_X$ is ample (that is, X is Fano) then $C_a(X, L)$ is CM and has rational singularities.*
(2) *If $-K_X$ is nef (for instance, X is Calabi–Yau), then*
    (a) *$C_a(X, L)$ is CM if and only if $H^i(X, \mathcal{O}_X) = 0$ for $0 < i < \dim X$, and*
    (b) *$C_a(X, L)$ has rational singularities iff $H^i(X, \mathcal{O}_X) = 0$ for $0 < i \leq \dim X$.*

Proof By assumption $\omega_X \otimes L^{-m}$ is anti-ample for $m \geq 1$, and even for $m = 0$ if $-K_X$ is ample. Hence, by Kodaira vanishing and by Serre duality, $H^i(X, L^m) = 0$ for $i > 0$ and $m \geq 0$ in case (1) and $m > 0$ in case (2). Thus in (3.11) we only need the vanishing of $H^i(X, L^0) = H^i(X, \mathcal{O}_X)$ for $0 < i < \dim X$. This proves (1) and (2a) while (2b) follows from (3.13). ☐

Remark 3.5 Kodaira's vanishing theorem does not hold in positive characteristic, but it is possible that it holds for Fano varieties. That is, we do not know any example of a smooth projective variety X and an ample line bundle L on X such that $-K_X$ is ample and $H^i(X, L^{-1}) \neq 0$ for some $i < \dim X$.

Partial results for $i = 1$ are in Kollár (1996, section II.6) and Biswas and dos Santos (2012).

The following examples are proved by a straightforward application of (3.4).

Example 3.6 (Log canonical but not CM)   The following examples show that log canonical singularities need not be rational, not even CM. By (2.88), none of these are log terminal. The claims below follow directly from (3.4) and hold for every ample line bundle L.

(1) A cone over an Abelian variety A is CM if and only if dim $A = 1$.
(2) A cone over a K3 surface is CM but not rational.
(3) A cone over an Enriques surface is CM and rational.
(4) A cone over a smooth Calabi–Yau complete intersection is CM but not rational.

While not related to any of our questions, the following example shows that one of the standard constructions, the taking of the index 1 cover, can also lead to more complicated singularities.

Example 3.7 (Index 1 covers)   Let $x \in X$ be a normal singularity over $\mathbb{C}$ and assume that K_X is $\mathbb{Q}$-Cartier. Then there is a smallest $m > 0$ such that mK_X is Cartier near x. By replacing X with a smaller neighborhood of x, we may assume that $mK_X \sim 0$. As in Kollár and Mori (1998, 2.49) or (2.49), there is a corresponding (ramified) finite, cyclic cover $\pi\colon \tilde{X} \to X$, called the *canonical cover* or *index 1 cover* of X such that $K_{\tilde{X}} = \pi^* K_X$ is Cartier.

For the singularities of the MMP, $\tilde{X}$ is usually simpler than the original singularity X. Nonetheless, here are four examples to show that $\tilde{X}$ can be cohomologically more complicated than X; see also Singh (2003).

1. Let S be a K3 surface with a fixed point free involution τ. Thus $T := S/\tau$ is an Enriques surface. A cone $C_a(T)$ over the Enriques surface T is CM and rational. By (3.14) the canonical class of $C_a(T)$ is not Cartier but $2K_{C_a(T)}$ is Cartier. Its canonical cover is a cone over the K3 surface S which is CM but not rational.

2. Let S be a K3 surface with a fixed point free involution τ as above. Then τ acts as multiplication by -1 on $H^2(S, \mathcal{O}_S)$. By the Künneth formula, (τ, τ) acts as multiplication by -1 on $H^2(S \times S, \mathcal{O})$ and as multiplication by $1 = (-1) \cdot (-1)$ on $H^4(S \times S, \mathcal{O})$. Set $X := (S \times S)/(\tau, \tau)$. Then $h^i(X, \mathcal{O}_X) = 0$ for $0 < i < 4$, hence a cone over X is CM. Its canonical cover is a cone over $S \times S$ which is not CM.

3. Let A be the Jacobian of the hyperelliptic curve $y^2 = x^5 - 1$ and τ the automorphism induced by $(x, y) \mapsto (\varepsilon x, y)$ where ε is a 5th root of unity. The space of holomorphic 1-forms is spanned by $dx/y, xdx/y$. Thus τ acts on $H^1(A, \mathcal{O}_A)$ with eigenvalues $\varepsilon, \varepsilon^2$ and on $H^2(A, \mathcal{O}_A) = \wedge^2 H^1(A, \mathcal{O}_A)$ with

eigenvalue ε^3. Thus τ has only isolated fixed points. Set $S := A/\langle\tau\rangle$. Then $H^i(S, \mathscr{O}_S) = 0$ for $i = 1, 2$, hence a cone over S is a rational singularity. Its canonical cover is a cone over A, which is not CM.

4. Let S_m be a K3 surface and σ_m an automorphism of order m whose fixed point set is finite and such that σ_m acts as multiplication by ε_m on $H^2(S_m, \mathscr{O}_{S_m})$ where ε_m is a primitive mth root of unity. (Such S_m and σ_m exist for $m \in \{5, 7, 11, 17, 19\}$ by Kondō (1992).) By the Künneth formula, (σ_m, σ_m) acts as multiplication by ε_m on $H^2(S_m \times S_m, \mathscr{O})$ and as multiplication by σ_m^2 on $H^4(S_m \times S_m, \mathscr{O})$. Set $X_m := (S_m \times S_m)/(\sigma_m, \sigma_m)$. Then $h^i(X_m, \mathscr{O}_{X_m}) = 0$ for $0 < i \leq 4$, hence a cone over X_m is rational.

By (3.14) the canonical class of $C_a(X_m)$ is not Cartier but $m K_{C_a(X_m)}$ is Cartier. Its canonical cover is a cone over $S_m \times S_m$ which is not even CM.

Auxiliary results on cones

3.8 (Cones) Let $X \subset \mathbb{P}^n$ be a projective scheme over a field k and $f_1, \ldots, f_s \in k[x_0, \cdots, x_n]$ generators of its homogeneous ideal.

The *classical affine cone* over X is the variety $C_a(X) \subset \mathbb{A}^{n+1}$ defined by the same equations ($f_1 = \cdots = f_s = 0$). The closure of the classical affine cone is the *classical projective cone* over X, denoted by $C_p(X) \subset \mathbb{P}^{n+1}(x_0:\cdots:x_n:x_{n+1})$. It is defined by the same equations ($f_1 = \cdots = f_s = 0$) as X. (Thus x_{n+1} does not appear in any of the equations.)

Assume that X is normal (resp. S_2). Then $C_a(X)$ is normal (resp. S_2) if and only if $H^0(\mathbb{P}^n, \mathscr{O}_{\mathbb{P}^n}(d)) \to H^0(X, \mathscr{O}_X(d))$ is onto for every $d \geq 0$ (cf. Hartshorne (1977, exercise II.5.14)). Therefore, it is better to define cones for any ample line bundle as follows.

Let X be a projective scheme with an ample line bundle L. The *affine cone* over X with conormal bundle L is

$$C_a(X, L) := \operatorname{Spec}_k \sum_{m \geq 0} H^0(X, L^m),$$

and the *projective cone* over X with conormal bundle L is

$$C_p(X, L) := \operatorname{Proj}_k \sum_{m \geq 0} (\sum_{r=0}^{m} H^0(X, L^r) \cdot x_{n+1}^{m-r}).$$

Note that if $X \subset \mathbb{P}^N$ and $L = \mathscr{O}_X(1)$ then there is a natural finite morphism $C_p(X, \mathscr{O}_X(1)) \to C_p(X)$ which is an isomorphism away from the vertex. If X is normal then $C_p(X, \mathscr{O}_X(1))$ is the normalization of $C_p(X)$ but $C_p(X, \mathscr{O}_X(1)) \to C_p(X)$ is an isomorphism only if

$$H^0(\mathbb{P}^N, \mathscr{O}_{\mathbb{P}^N}(m)) \twoheadrightarrow H^0(X, \mathscr{O}_X(m)) \quad \text{is onto } \forall\, m.$$

In particular, the $m = 0$ case shows that X has to be connected.

An advantage of the general notion is that if X is normal (resp. S_2) then $C_a(X, L)$ and $C_p(X, L)$ are also normal (resp. S_2).

In the affine case a natural partial resolution (a weighted blow-up of the vertex) is given by

$$p: BC_a(X, L) := \operatorname{Spec}_X \sum_{m \geq 0} L^m \to C_a(X, L).$$

It is an isomorphism over the *punctured affine cone* $C_a^*(X, L) := C_a(X, L) \setminus \{\text{vertex}\}$. The exceptional divisor of p is $E \simeq X$ and $C_a^*(X, L) \simeq BC_a(X, L) \setminus E$. In the projective case we have

$$p: BC_p(X, L) := \operatorname{Proj}_X \sum_{m \geq 0} (\sum_{r=0}^{m} L^r \otimes \mathcal{O}_X^{m-r})$$

$$= \operatorname{Proj}_X S(L + \mathcal{O}_X) \to C_p(X, L).$$

Aside   Assume that we have a flat family of varieties $X_t \subset \mathbb{P}^n$. Flatness is equivalent to requiring that $h^0(X_t, \mathcal{O}_{X_t}(d))$ be independent of t for $d \gg 1$; see Hartshorne (1977, III.9.9). On the other hand, the projective cones $C_p(X_t)$ form a flat family if and only if $\sum_{i=0}^{d} h^0(X_t, \mathcal{O}_{X_t}(i))$ is independent of t for $d \gg 1$. Since the $h^0(X_t, \mathcal{O}_{X_t}(i))$ are upper semi continuous functions of t, this implies that $h^0(X_t, \mathcal{O}_{X_t}(i))$ is independent of t for every $i \geq 0$. This is a source of many interesting examples.

3.9 (Deformation to the cone over a hyperplane section)   Let X be projective, L ample with a nonzero section $s \in H^0(X, L)$ and zero set $H := (s = 0)$. Consider the deformation $\pi: Y \to \mathbb{A}_t^1$ where

$$Y := (s - t x_{n+1} = 0) \subset C_p(X, L) \times \mathbb{A}_t^1.$$

If $t \neq 0$ then we can use $x_{n+1} = t^{-1} s$ to eliminate x_{n+1} and obtain that $Y_t \simeq X$. If $t = 0$ then the extra equation becomes $(s = 0)$, thus we get the fiber $Y_0 = C_p(X, L) \cap (s = 0)$. How does this compare with the cone $C_p(H, L|_H)$?

The answer is given by the $m = 1$ case of the next result.

Proposition 3.10   *Let X be a projective scheme over a field k with ample line bundle L and $H \subset X$ a Cartier divisor such that $\mathcal{O}_X(H) \simeq L^m$ for some $m > 0$. The following are equivalent*

(1) $C_p(H, L|_H)$ *is a subscheme of* $C_p(X, L)$.
(2) $H^1(X, L^d) = 0$ *for every $d \geq -m$.*

Proof   Let $s \in H^0(X, L^m)$ be a section defining H and consider the beginning of the exact sequence

$$0 \to H^0(X, L^{d-m}) \xrightarrow{s} H^0(X, L^d) \xrightarrow{r_d} H^0(H, L^d|_H)$$
$$\to H^1(X, L^{d-m}) \to H^1(X, L^d). \tag{3.10.3}$$

The restriction maps r_d: $H^0(X, L^d) \to H^0(H, L^d|_H)$ give a natural morphism $C_p(H, L_H) \to C_p(X, L)$ which is an embedding if and only if the r_d are all surjective.

If $H^1(X, L^{d-m}) = 0$ for every d then the r_d are all surjective.

Conversely, assume that $H^1(X, L^{d-m}) \neq 0$ for some $d \geq 0$. By Serre vanishing, the set of such values d is bounded from above. Thus we can find a $d \geq 0$ such that $H^1(X, L^{d-m}) \neq 0$ but $H^1(X, L^d) = 0$. Then (3.10.3) shows that r_d is not surjective hence $C_p(H, L|_H)$ is not a subscheme of $C_p(X, L)$. $\square$

By iterating this argument, we obtain the following.

Corollary 3.11 *Let X be a projective, CM variety and L an ample line bundle on X. Then $C_a(X, L)$ is CM if and only if $H^i(X, L^m) = 0$ $\forall m$, $\forall 0 < i < \dim X$.* $\square$

Remark 3.12 A sheaf version in terms of local cohomologies is the following. As in (3.15) let F be a coherent sheaf on X without 0-dimensional embedded points and $C_a(F, L)$ the corresponding sheaf on $C_a(X, L)$. Let $v \in C_a(X, L)$ be the vertex. Then, for $i \geq 2$,

$$H^i_v(C_a(X, L), C_a(F, L)) \simeq \sum_{m \in \mathbb{Z}} H^{i-1}(X, F \otimes L^m).$$

Thus $C_a(F, L)$ satisfies Serre's condition S_r if and only if $H^i(X, F \otimes L^m) = 0$ for all $m \in \mathbb{Z}$ and $0 < i \leq r - 2$. For $F = \mathscr{O}_X$ we recover (3.11).

Proposition 3.13 *Let X be projective variety with rational singularities and L an ample line bundle on X. Then the cone $C_a(X, L)$ has rational singularities if and only if $H^i(X, L^m) = 0$ $\forall m$, $\forall 0 < i \leq \dim X$.*

Proof Let p: $BC_a(X, L) \to C_a(X, L)$ be the blow-up of the vertex with exceptional divisor $E \simeq X$. Since $BC_a(X, L)$ is a $\mathbb{P}^1$-bundle over X, it has rational singularities. As noted in (2.57), $C_a(X, L)$ has rational singularities if and only if $R^i p_* \mathscr{O}_{BC_a(X,L)} = 0$ and $R^i p_* \omega_{BC_a(X,L)} = 0$ for $i > 0$.

Let $I \subset \mathscr{O}_{BC_a(X,L)}$ be the ideal sheaf of E. By construction,

$$\mathscr{O}_{BC_a(X,L)}/I^m \simeq \mathscr{O}_X + L + \cdots + L^{m-1},$$

hence, by the Theorem on Formal Functions,

$$R^i p_* \mathscr{O}_{BC_a(X,L)} = \sum_{m \geq 0} H^i(X, L^m) \quad \text{for } i > 0. \tag{3.13.1}$$

By (3.14.4) $K_{BC_a(X,L)} \sim \pi^* K_X - E$, hence a similar argument and Serre duality show that, for $i > 0$

$$R^i p_* \omega_{BC_a(X,L)} = \sum_{m > 0} H^i(X, \omega_X \otimes L^m) = \sum_{m < 0} H^{\dim X - i}(X, L^m) \tag{3.13.2}$$

Now (3.13.1) and (3.13.2) together complete the proof. $\square$

Proposition 3.14   *Assume that X is normal, projective and L is an ample line bundle on X. Then*

(1) $\mathrm{Pic}(C_a(X, L)) = 0$ *and*
(2) $\mathrm{Cl}(C_a(X, L)) = \mathrm{Cl}(X)/\langle L \rangle$.

Let Δ_X be a $\mathbb{Q}$-divisor on X. By pull-back, we get $\mathbb{Q}$-divisors $\Delta_{C_a^*(X,L)}$ on the punctured affine cone $C_a^*(X, L)$, $\Delta_{C_a(X,L)}$ on $C_a(X, L)$ and $\Delta_{BC_a(X,L)}$ on $BC_a(X, L)$. Assume that $K_X + \Delta_X$ is $\mathbb{Q}$-Cartier. Then

(3) $K_{BC_a(X,L)} + \Delta_{BC_a(X,L)} \sim \pi^*(K_X + \Delta_X) - E$ *where $\pi\colon BC_a(X, L) \to X$ is projection from the vertex and $E \simeq X$ is the exceptional divisor of $p\colon BC_a(X, L) \to C_a(X, L)$.*
(4) $m(K_{C_a(X,L)} + \Delta_{C_a(X,L)})$ *is Cartier iff $\mathscr{O}_X(mK_X + m\Delta_X) \simeq L^{rm}$ for some $r \in \mathbb{Q}$. If this holds then*

$$K_{BC_a(X,L)} + \Delta_{BC_a(X,L)} + (1 + r)E \sim_{\mathbb{Q}} 0.$$

Proof   By construction, $\pi\colon BC_a(X, L) \to X$ is an $\mathbb{A}^1$-bundle over $E \simeq X$, hence $\mathrm{Cl}(BC_a(X, L)) = \mathrm{Cl}(X)$ and $\mathrm{Pic}(BC_a(X, L)) = \mathrm{Pic}(X)$. If M is any line bundle on $C_a(X, L)$ then $p^*M|_E$ is trivial, hence p^*M is trivial and so is M, proving (1).

The class group of $C_a(X, L)$ is the same as the class group of the punctured affine cone $C_a^*(X, L) \simeq BC_a(X, L) \setminus E$. The latter is computed by the exact sequence

$$\mathbb{Z}[E] \to \mathrm{Cl}(BC_a(X, L)) \to \mathrm{Cl}(C_a^*(X, L)) \to 0.$$

Since $\mathscr{O}_{BC_a(X,L)}(E)|_E \simeq L^{-1}$, we obtain (2).

The projection $\pi\colon C_a^*(X, L) \to X$ is a $\mathbb{G}_m$-bundle. If t is a coordinate on $\mathbb{G}_m = \mathrm{Spec}\, k[t, t^{-1}]$ then the 1-form dt/t is independent of the choice of t since $d(ct)/ct = dt/t$. Thus there is a natural linear equivalence

$$K_{BC_a(X,L)}(E) \sim \pi^* K_X.$$

Therefore $K_{BC_a(X,L)} + \Delta_{BC_a(X,L)} + E \sim \pi^*(K_X + \Delta_X)$.

Combining with the earlier results, we see that $m K_{C_a(X,L)} + m\Delta_{C_a(X,L)}$ is Cartier if and only if $m K_{C_a^*(X,L)} + m\Delta_{C_a^*(X,L)}$ is trivial if and only if $\mathscr{O}_X(mK_X + m\Delta_X) \simeq L^{rm}$ for some $r \in \mathbb{Q}$. Then $\pi^*(K_X + \Delta_X) \sim_{\mathbb{Q}} -rE$, proving (4). $\square$

3.15 (Cones of sheaves)   Let F be a coherent sheaf on X without 0-dimensional embedded points. Then

$$\sum_{m \in \mathbb{Z}} H^0(X, F \otimes L^m)$$

is a coherent module over $\sum_{m \geq 0} H^0(X, L^m)$. Thus it corresponds to a coherent sheaf $C_a(F, L)$ on the affine cone $C_a(X, L)$.

As before, it is easy to see that $C_a(F, L)$ is CM if and only if

$$H^i(X, F \otimes L^m) = 0 \quad \forall m, \ \forall 0 < i < \dim X. \tag{3.15.1}$$

As an example, consider the case when $S := (xy - zt = 0) \subset \mathbb{P}^3$. The affine cone is $X := C_a(S, \mathscr{O}_S(1)) = (xy - zt = 0) \subset \mathbb{A}^4$, Let $A := (x = z = 0)$ and $B := (x = t = 0)$ be two planes on X.

Claim 3.15.2 $\mathscr{O}_X(nA + mB)$ is CM if and only if either $n = m$ (and then $\mathscr{O}_X(nA + mB)$ is locally free) or $|n - m| = 1$ (in which case $\mathscr{O}_X(nA + mB)$ is not locally free).

Proof   Let $L_A := (x = z = 0)$ and $L_B := (x = t = 0)$ the corresponding lines. Then $\mathscr{O}_X(nA + mB)$ is the cone over $\mathscr{O}_S(nL_A + mL_B)$ and $\mathscr{O}_S(1) \sim L_A + L_B$. By (3.15.1) we need to check when

$$H^1(S, \mathscr{O}_S((n - r)L_A + (m - r)L_B)) = 0 \quad \forall r.$$

By the Künneth formula,

$$
\begin{aligned}
&H^1(S, \mathscr{O}_S((n - r)L_A + (m - r)L_B)) \\
&= H^1(\mathbb{P}^1, \mathscr{O}(n - r)) \otimes H^0(\mathbb{P}^1, \mathscr{O}(m - r)) \\
&\quad \oplus H^0(\mathbb{P}^1, \mathscr{O}(n - r)) \otimes H^1(\mathbb{P}^1, \mathscr{O}(m - r)).
\end{aligned}
$$

If $|n - m| \geq 2$ then we get a nonzero term for $r = \max\{n, m\}$. Otherwise we always get zero.   □

3.16 (Singularities of Cox rings)   An interesting generalization of cones is given by the *Cox rings* of varieties. Let X be a projective variety and assume for simplicity that $\mathrm{Pic}(X)$ is torsion free. Then the Cox ring of X is defined as

$$\mathrm{Cox}(X) := \sum_{L \in \mathrm{Pic}(X)} H^0(X, L),$$

where multiplication is given by the natural multiplication maps $H^0(X, L_1) \times H^0(X, L_2) \to H^0(X, L_1 \otimes L_2)$. If $\mathrm{Pic}(X) = \mathbb{Z}[L]$ then the spectrum of the Cox ring is exactly the cone $C_a(X, L)$.

In general, $\mathrm{Cox}(X)$ is not even finitely generated. However, if (X, Δ) is $\mathbb{Q}$-Fano and klt then $\mathrm{Cox}(X)$ is finitely generated and its spectrum has log terminal singularities Brown (2011). For the converse and for generalizations see Fujino and Takagi (2011), Gongyo *et al.* (2012) and Kawamata and Okawa (2012).

3.2 Quotient singularities

The singularities $(0 \in \mathbb{C}^n / G)$ for some finite subgroup $G \subset \mathrm{GL}(n, \mathbb{C})$ form an interesting subclass of log terminal singularities. In dimension 2 every log terminal singularity is a quotient and in dimension 3 one can still view quotient singularities as "typical" among terminal singularities, (though not any more among canonical or log terminal singularities). For instance, every 3-dimensional terminal singularity can be deformed to 3-dimensional terminal quotient singularities (Reid, 1987, 9.4) while terminal quotient singularities are rigid in all dimensions Schlessinger (1971). Another good illustration of this is Reid's plurigenus formula (Reid, 1987, section 10).

In higher dimensions quotient singularities are rather special, but they occur frequently in applications.

3.17 (Quotient singularities)   Let G be a finite group acting on a scheme X and X/G the corresponding quotient (9.29). Let $\bar{x} \in X/G$ be a point, $x_1, \ldots, x_m$ its preimages in X and $G_i \subset G$ the stabilizer of x_i. The subgroups G_i are conjugate to each other and the natural map $X/G_i \to X/G$ is étale at the image of x_i. Thus, as far as the singularities of X/G are concerned, we need to understand only neighborhoods of fixed points.

Assume now that we are over a field k and $x \in X(k)$ is a G-fixed point. $\hat{\mathscr{O}}_{x,X}$ is a complete local k-algebra with maximal ideal m and with a G-action. If $\operatorname{char} k \nmid |G|$ then G-representations are completely reducible, thus $m \to m/m^2$ has a G-equivariant splitting. This gives a G-equivariant map from the polynomial ring $k[m/m^2]$ to R. (The simplest example of a nonlinearizable action is $t \mapsto \frac{t}{1+t}$ which has order p in characteristic p. See (3.25) and Artin (1975, 1977) for more examples when $\operatorname{char} k$ divides $|G|$.)

We claim that if $x \in X(k)$ is a nonsingular point then we get a G-equivariant isomorphism $k[[m/m^2]] \simeq \hat{\mathscr{O}}_{x,X}$. Indeed, the homomorphism $k[m/m^2] \to R$ gives surjections $S^i(m/m^2) \twoheadrightarrow m^i/m^{i+1}$. If R is regular, then the two sides have the same dimension, so the maps are isomorphisms.

Furthermore, a power series is G-invariant if and only if each of its homogeneous parts are G-invariant. Thus we get that taking invariants commutes with completion. This implies the following.

Claim 3.17.1   Over a field k of characteristic zero every quotient singularity $(x \in X)$ where $x \in X(k)$ is formally isomorphic to $(0 \in \mathbb{A}_k^n/G)$ for some $G \subset \operatorname{GL}(n, k)$.

It is not hard to see that $(x \in X)$ and $(0 \in \mathbb{A}_k^n/G)$ have isomorphic Henselizations (4.38). A special case is treated in (3.32).

Remark 3.17.2   One should note that a subgroup $G \subset \operatorname{GL}(n, k)$ gives two natural actions on the affine space $\mathbb{A}_k^n$ and on the polynomial ring $k[x_1, \ldots, x_n]$. If $g = (g_{ij}) \in G$, then one action is matrix multiplication

$$(x_1, \ldots, x_n)^t \mapsto (g_{ij}) \cdot (x_1, \ldots, x_n)^t.$$

On the other hand, if we think of $k[x_1, \ldots, x_n]$ as functions on $\mathbb{A}_k^n$, then the natural action is pulling back by the G-action. This corresponds to

$$(x_1, \ldots, x_n)^t \mapsto (g_{ij})^{-1} \cdot (x_1, \ldots, x_n)^t,$$

which is not an action of G but of the opposite group.

3.18 (Classification of quotient singularities)   For general introductions, see Benson (1993) or Smith (1995).

Let $G \subset \mathrm{GL}(n, k)$ be a finite group, again assuming char $k \nmid |G|$. An element $1 \neq g \in \mathrm{GL}(n, k)$ is called a *pseudo-reflection* if all but one of its eigenvalues are 1.

By a theorem proved originally via a classification by Shephard and Todd (1954) and later more conceptually by Chevalley (1955), if $G \subset \mathrm{GL}(n, k)$ is generated by pseudo-reflections then $\mathbb{A}^n/G \simeq \mathbb{A}^n$.

For any $G \subset \mathrm{GL}(n, k)$, the subgroup $G_{pr} \subset G$ generated by pseudo-reflections is normal, and so

$$\mathbb{A}^n/G = (\mathbb{A}^n/G_{pr})/(G/G_{pr}) \simeq \mathbb{A}^n/(G/G_{pr}).$$

Thus it is enough to classify quotients by subgroups containing no pseudo-reflections. In this case G acts freely outside a codimension 2 subset $F \subset \mathbb{A}^n$ (and F is a union of linear spaces). Thus $\pi \colon \mathbb{A}^n \to \mathbb{A}^n/G$ is unramified in codimension 1 hence

$$K_{\mathbb{A}^n} = \pi^* K_{\mathbb{A}^n/G}. \tag{3.18.1}$$

By (2.43) this implies that quotient singularities are log terminal. So, by (2.88), quotient singularities are rational in characteristic 0. (For a simple direct proof of the latter see Brieskorn (1968).) The positive characteristic cases are treated in Chatzistamatiou and Rülling (2009, theorem 2).

Furthermore, $\mathbb{A}^n \setminus F$ is simply connected by (3.58), hence $\mathbb{A}^n \setminus F$ is the universal cover of $\mathbb{A}^n/G \setminus \pi(F)$. (If char $k = p$ then we restrict ourselves to prime-to-p covers and if k is not algebraically closed, we restrict to those étale covers whose normal extension over $\mathbb{A}^n$ has no residue field extension over $0 \in \mathbb{A}^n$.) Thus we obtain the following.

Theorem 3.18.1   Let k be a field and $G_1, G_2 \subset \mathrm{GL}(n, k)$ two subgroups containing no pseudo-reflections. If char $k \nmid |G_i|$, then the following are equivalent:

(a) $\mathbb{A}^n/G_1 \simeq \mathbb{A}^n/G_2$,
(b) $\hat{\mathbb{A}}^n/G_1 \simeq \hat{\mathbb{A}}^n/G_2$,
(c) G_1 and G_2 are conjugate in $\mathrm{GL}(n, k)$. $\qquad\qquad\square$

3.19 (Cyclic quotients)   If $G \simeq \mu_m$ is cyclic of order m, k is algebraically closed and char $k \nmid m$ then the G-action can be diagonalized. Fixing a generator

ε, a primitive mth root of unity , we can write it as

$$(x_1, \ldots, x_n) \mapsto (\varepsilon^{a_1} x_1, \ldots, \varepsilon^{a_n} x_n) \tag{3.19.1}$$

where ε is a primitive mth root of unity. The a_i are determined modulo m and there are no pseudo-reflections if and only if for every prime $\ell | m$, at most $n - 2$ of the a_i are divisible by ℓ. We usually abbreviate the quotient by this action as

$$\mathbb{A}^n / \tfrac{1}{m}(a_1, \ldots, a_n). \tag{3.19.2}$$

If $(b, m) = 1$ then this singularity is isomorphic to $\mathbb{A}^n / \tfrac{1}{m}(ba_1, \ldots, ba_n)$. Thus it is very easy to decide when two cyclic quotient singularities are isomorphic.

The description can be simplified if $n = 2$. In this case m is relatively prime to both a_1, a_2. We can thus choose $ba_1 \equiv 1 \mod m$ to get the normal form

$$\mathbb{A}^2 / \tfrac{1}{m}(1, q) \quad \text{where} \quad (q, m) = 1. \tag{3.19.3}$$

Two such singularities $\mathbb{A}^2 / \tfrac{1}{m}(1, q)$ and $\mathbb{A}^2 / \tfrac{1}{m}(1, q')$ are isomorphic if and only if $q \equiv q' \mod m$ or $qq' \equiv 1 \mod m$.

Algebraically, a μ_m-action is a $\mathbb{Z}/m\mathbb{Z}$-valued grading of $k[x_1, \ldots, x_n]$ given by

$$w(x_1^{i_1} \cdots x_n^{i_n}) = a_1 i_1 + \cdots + a_n i_n \in \mathbb{Z}/m\mathbb{Z}. \tag{3.19.4}$$

Then the μ_m-invariants are spanned by the monomials M with $w(M) = 0$. Thus

$$\mathbb{A}^n / \tfrac{1}{m}(a_1, \ldots, a_n) = \operatorname{Spec}_k k[x_1, \ldots, x_n]_{(w=0)}. \tag{3.19.5}$$

Note that the right hand side of (3.19.5) makes sense even if char k divides m; we adopt it as the definition of the left hand side in this case. It is consistent with the usual definition of μ_m as the subgroup scheme of mth roots of unity.

In any given case it is easy to work out the minimal generators of the algebra of invariants, but it seems impossible to do it in a systematic way. If $n = 2$ and we use the normal form (3.19.3) then three of the invariants are $u := x_1^m$, $v := x_2^m$ and $w = x_1^{m-q} x_2$. These show that

$$\mathbb{A}^2 / \tfrac{1}{m}(1, q) \quad \text{is the normalization of} \quad (w^m = u^{m-q} v) \subset \mathbb{A}^3. \tag{3.19.6}$$

Note that $(w^m = u^{m-q} v)$ is normal if and only if $m - q = 1$, that is, when we have a Du Val singularity of type A_{m-1} (3.26).

Definition 3.20 For $g \in \mathrm{U}(n) \subset \mathrm{GL}(n, \mathbb{C})$, its eigenvalues (with multiplicity) can be written as $e^{2\pi i r_1}, \ldots, e^{2\pi i r_n}$ where $0 \leq r_i < 1$. Following Ito and Reid (1996) and Reid (2002), we define the *age* of g as $\operatorname{age}(g) := r_1 + \cdots + r_n$.

If k is a field of characteristic 0 and $g \in \mathrm{GL}(n, k)$ has finite order, we can choose an embedding $k \hookrightarrow \mathbb{C}$ and conjugate g into $\mathrm{U}(n)$ to obtain its age. If g has order m, the value we get actually depends on which mth root of unity we identify with $e^{2\pi i/m}$. However, changing to another root of unity has the same effect as changing to a power of g. Thus if ε is a primitive mth root of unity and the eigenvalues of g are $\varepsilon^{c_1}, \ldots, \varepsilon^{c_n}$ for $0 \le c_i < m$ then $\sum c_i = m \operatorname{age}(g^r)$ for some $(r, m) = 1$.

Thus the set $\{\operatorname{age}(g) : g \in G\}$ is well defined, even in positive characteristic.

Note that $t \mapsto \operatorname{diag}(e^{2\pi i t r_1}, \ldots, e^{2\pi i t r_n})$ is a geodesic connecting the identity to g in the natural invariant metric of $\mathrm{U}(n)$. Thus the age is essentially the distance of g from the identity. (More precisely, the usual distance is $2\pi \cdot \min\{\operatorname{age}(g), \operatorname{age}(g^{-1})\}$ corresponding to the shorter of two ways going around the circle to get to g and it is more natural to take the shortest circle $t \mapsto \operatorname{diag}(e^{2\pi i t}, 1, \ldots, 1)$ to have length 2π.)

Thus the element of smallest age in a group $G \subset \mathrm{U}(n)$ is the (non-identity) element in G that is closest to the identity. If $G \subset \mathrm{SU}(n)$, it is also the length of the shortest closed geodesic in $\mathrm{SU}(n)/G$.

Theorem 3.21 (Reid–Tai criterion) (Reid, 1980) *Let $G \subset \mathrm{GL}(n, k)$ be a finite group that contains no pseudo-reflections such that char $k \nmid |G|$. The following are equivalent.*

(1) $\mathbb{A}^n/G$ *is canonical (resp. terminal).*
(2) $\mathbb{A}^n/C$ *is canonical (resp. terminal) for every cyclic subgroup $C \subset G$.*
(3) $\operatorname{age}(g) \ge 1$ *(resp. $\operatorname{age}(g) > 1$) for every $1 \ne g \in G$.*

Proof Let E be an exceptional divisor over $\mathbb{A}^n_\mathbf{x}$ that is point-wise fixed by a (necessarily) cyclic subgroup $\operatorname{Fix}(E)$. Using (2.22), there is a unique exceptional divisor F over $\mathbb{A}^n/G$ dominated by E and every exceptional divisor over $\mathbb{A}^n/G$ arises this way. We compute in (2.42.4) that

$$a(F, \mathbb{A}^n/G) + 1 = \frac{a(E, \mathbb{A}^n) + 1}{|\operatorname{Fix}(E)|}.$$

Thus $\mathbb{A}^n/G$ is canonical (resp. terminal) if and only if $a(E, \mathbb{A}^n) \ge |\operatorname{Fix}(E)| - 1$ (resp. $\ge |\operatorname{Fix}(E)|$) for every E. Since this involves only cyclic subgroups of G, we see that (1) and (2) are equivalent.

To see that (3) $\Rightarrow$ (1) let E be an exceptional divisor over $\mathbb{A}^n_\mathbf{x}$ that is point-wise fixed by a cyclic subgroup $\langle g \rangle$ of order m. At a general point of E choose local coordinates $y_1, \ldots, y_n$ such that $E = (y_1 = 0)$, $g^* y_1 = \varepsilon y_1$ is a g-eigenfunction and the y_i are g-invariant. We can thus write $g^* x_i = y_1^{b_i} u_i$ where the u_i are units. Thus $g^*(dx_1 \wedge \cdots \wedge dx_n)$ vanishes to order $\ge -1 + \sum b_i$ along E, hence $a(E, \mathbb{A}^n_\mathbf{x}) \ge -1 + \sum b_i$. If $g^* x_i = \varepsilon^{c_i} x_i$ and $0 \le c_i < m$ then $b_i \equiv c_i \mod m$. Thus $\sum b_i \ge \sum c_i = m \operatorname{age}(g)$.

The above proof also gives the converse once we show that for every g there is an exceptional divisor F_g that is fixed by g such that $a(F_g, \mathbb{A}^n_\mathbf{x}) \le -1 + m \operatorname{age}(g)$.

To see this, consider the cyclic subgroup $\frac{1}{m}(a_1, \ldots, a_n)$. Assume that $a_1, \ldots, a_n$ have a common prime divisor ℓ. If $\ell \mid m$ then the action is not faithful and if $(\ell, m) = 1$ then $\frac{1}{m}(a_1/\ell, \ldots, a_n/\ell)$ has smaller age. Thus we may assume that $a_1, \ldots, a_n$ are relatively prime.

Consider the weighted blow-up $\mathbb{A}_{\mathbf{x}}^n$ with weights $(a_1, \ldots, a_n)$. (See Kollár *et al.* (2004, 6.38) for a quick introduction to weighted blow-ups.) A local chart is

$$g: \mathbb{A}_{\mathbf{y}}^n / \frac{1}{a_1}(1, -a_2, \ldots, -a_n) \to \mathbb{A}_{\mathbf{x}}^n \quad \text{with} \quad g^* x_1 = y_1^{a_1}, \ g^* x_i = y_1^{a_i} y_i \ (i > 1).$$

Here $\mathbb{A}_{\mathbf{y}}^n$ is a degree a_1 ramified cover of the chart, but, if the $a_1, \ldots, a_n$ are relatively prime, the covering is generically unramified along the exceptional divisor ($y_1 = 0$). Hence we can compute the discrepancy of $F := (y_1 = 0)$ on $\mathbb{A}_{\mathbf{y}}^n$ instead of on the quotient. Since

$$g^*(dx_1 \wedge \cdots \wedge dx_n) = a_1 y_1^{(-1 + \sum a_i)} \cdot dy_1 \wedge \cdots \wedge dy_n,$$

we conclude that $a(F, \mathbb{A}_{\mathbf{x}}^n) = -1 + \sum a_i$. $\quad\square$

For any given representation it is relatively easy to decide if the Reid–Tai condition is satisfied or not. It is, however, difficult to get a good understanding of all representations that satisfy it. For instance, it is quite tricky to determine all 3-dimensional representations of cyclic groups that satisfy the Reid–Tai condition (Morrison and Stevens (1984); Morrison (1985); Reid (1987)) and the analogous question in dimension 4 is still open (Mori *et al.*, 1988).

By contrast, it seems that a "typical" representation of a "typical" nonabelian group satisfies the Reid–Tai condition, and hence defines a terminal qutient singularity. In order to make this precise, let $G \subset \mathrm{GL}(V)$ be a finite subgroup and $g \in G$ an element such that $0 < \mathrm{age}(g) < 1$. If the conjugacy class of g does not generate the full group G, it must generate a normal subgroup H of G. After classifying the cases for which the conjugacy class of g generates H, all finite subgroups intermediate between H and its normalizer give further examples.

If V is reducible, then for every irreducible factor V_i of V on which g acts nontrivially, $0 < \mathrm{age}(g|_{V_i}) < 1$. Motivated by these reduction steps, $G \subset \mathrm{GL}(V)$ is called a *basic non-RT* subgroup if the conjugacy class of any $g \in G$ with $0 < \mathrm{age}(g) < 1$ generates G. (And there is at least one element with $0 < \mathrm{age}(g) < 1$.)

The following result was proved in Kollár and Larsen (2009) when the dimension is large enough. The optimal form was proved in Guralnick and Tiep (2010, theorem 1.7). For examples and related questions see Kollár and Larsen (2009, section 5).

Theorem 3.22 *Let $G \subset \mathrm{GL}(V)$ be a basic non-RT subgroup. If $\dim V \geq 5$ then G is projectively equivalent to a reflection group.*

That is, there is a subgroup $G' \subset \mathrm{GL}(V')$ generated by pseudo-reflections and an isomorphism $\mathrm{PGL}(V) \to \mathrm{PGL}(V')$ mapping the image of G in $\mathrm{PGL}(V)$ isomorphically to the image of G' in $\mathrm{PGL}(V')$. $\square$

The reflection groups are classified in Shephard and Todd (1954), but it is not always easy to decide if a group is projectively equivalent to a reflection group or not. The following gives a more precise answer.

Theorem 3.23 (Guralnick and Tiep, 2010, corollary 1.6) *Let $G \subset \mathrm{GL}(V)$ be a finite irreducible, primitive, tensor indecomposable subgroup containing no pseudo-reflections. Assume that $n = \dim V \geq 11$ and V/G is not terminal. Then one of the following holds.*

(1) *G contains $Z(G) \times A_{n+1}$ as a subgroup of index ≤ 2, with A_{n+1} acting on $\mathbb{C}_{\mathbf{x}}^{n+1}$ by permutations and $V = (\sum x_i = 0) \subset \mathbb{C}_{\mathbf{x}}^{n+1}$.*
(2) *All elements of age < 1 are central.* $\square$

Note that for any $G \subset \mathrm{GL}(V)$ and mth root of unity ε, the subgroup $\langle G, \varepsilon\mathbf{1}\rangle \subset \mathrm{GL}(V)$ is finite and the quotient is not terminal if $m \geq \dim V$. These explain the second possibility in (3.23).

3.24 (Isolated quotient singularities)   Let $G \subset \mathrm{GL}(n, \mathbb{C})$ be a finite subgroup without pseudo-reflections. Then $\mathbb{C}^n/G$ is an isolated singularity if and only if 1 is not an eigenvalue of g for every $1 \neq g \in G$.

This turns out to be a surprisingly restrictive condition that was much studied in connection with the *Clifford–Klein space form problem.*

Every finite subgroup of $\mathrm{GL}(n, \mathbb{C})$ is conjugate to a subgroup of $\mathrm{U}(n)$, hence we may assume that $G \subset \mathrm{U}(n)$. Thus G acts on the unit sphere $\mathbb{S}^{2n-1} \subset \mathbb{C}^n$ by isometries and $\mathbb{C}^n/G$ is an isolated singularity if and only if the G-action has no fixed points.

Conversely, by a theorem of Frobenius–Schur (Wolf, 1967, 4.7.3), any finite, fixed point free subgroup of isometries of an odd-dimensional sphere arises this way.

The full classification, due mostly to Zassenhaus and Vincent, is lengthy (Wolf, 1967, chapter 6) and the resulting groups turn out to be quite special. Most of them are solvable; the only such simple group is the icosahedral one.

Quotient singularities are much more complicated when the characteristic divides the order of the group. The only case that is fully understood is $\mathbb{Z}/2\mathbb{Z}$-quotients of surfaces in characteristic 2, but this is already quite subtle.

3.25 ($\mathbb{Z}/2\mathbb{Z}$-quotients in characteristic 2) (Artin, 1975)   Let k be a field of characteristic 2 and consider the ring

$$k[[x, y, z]]/(z^2 + abz + a^2y + b^2x)$$

where $a, b \in (x, y)k[[x, y]]$ are relatively prime. It is obtained as the ring of invariants of the regular local ring

$$k[[x, y, u, v]]/(u^2 + au + x, v^2 + bv + y)$$

by the $\mathbb{Z}/2\mathbb{Z}$-action $\sigma(x, y, u, v) = (x, y, u + a, v + b)$ by setting $z = u(v + b) + v(u + a)$.

Even when a, b are powers of the variables x, y, these singularities are quite complicated. For instance $(z^2 + xyz + x^2y + y^2x = 0)$ is of type D_4 and $(z^2 + yx^2z + y^3 + x^5 = 0)$ is of type E_8. $(z^2 + x^2y^2z + x^4y + y^4x = 0)$ is a cusp singularity and $(z^2 + x^iy^jz + x^{2i}y + y^{2j}x = 0)$ is not even log canonical for $i, j \geq 3$.

The main result of Artin (1975) says that every $\mathbb{Z}/2\mathbb{Z}$-action with an isolated fixed point arises this way.

3.3 Classification of log canonical surface singularities

Instead of giving a direct classification of germs of log canonical pairs ($y \in Y, B$), we aim to develop a description of numerically log canonical extended dual graphs.

This approach is independent of the characteristic. A disadvantage is that it is not always easy to go from the dual graph to the actual surface germ ($y \in Y$). Fortunately, it turns out that for our applications the extended dual graph contains all the necessary information.

First we classify dual graphs of canonical singuarities and list key examples of lc singularities. Then we study the dual graphs combinatorially, essentially by trying to solve the system of equations (2.26.3). We are able to get a good understanding of how the solutions change if we change the dual graph. Ultimately, we end up with a short list.

Example 3.26 (Du Val singularities)   Consider case (2.29.2) when the residue field $k(s)$ is algebraically closed. If the exceptional curves are E_i, then their self-intersections are computed from $(E_i \cdot E_i) = (K_T + E_i) \cdot E_i = 2p_a(E_i) - 2$. By (10.1), $(E_i \cdot E_i) < 0$, hence $p_a(E_i) = 0$, $E_i \simeq \mathbb{P}^1$ and $(E_i \cdot E_i) = -2$. Since the intersection form is negative definite, this implies that $(E_i \cdot E_j) \in \{0, 1\}$ for $i \neq j$.

The classification of such dual graphs is easy to do by hand and it has been done many times from different points of view. The first time probably as part of the classification of root systems of simple Lie algebras; giving rise to their names.

The corresponding singularities are called *Du Val* singularities or *rational double points* or *simple* surface singularities. See Kollár and Mori (1998, section 4.2) or Durfee (1979) for more information. (The dual graphs are correct in every characteristic. The equations below are correct in characteristic zero; see Artin (1977) about positive characteristic.)

A_n: $x^2 + y^2 + z^{n+1} = 0$, with $n \geq 1$ curves in the dual graph:

$$2 \text{---} 2 \text{---} \cdots \text{---} 2 \text{---} 2$$

D_n: $x^2 + y^2 z + z^{n-1} = 0$, with $n \geq 4$ curves in the dual graph:

$$
\begin{array}{c}
2 \\
| \\
2 \text{---} 2 \text{---} \cdots \text{---} 2 \text{---} 2
\end{array}
$$

E_6: $x^2 + y^3 + z^4 = 0$, with 6 curves in the dual graph:

$$
\begin{array}{c}
2 \\
| \\
2 \text{---} 2 \text{---} 2 \text{---} 2 \text{---} 2
\end{array}
$$

E_7: $x^2 + y^3 + yz^3 = 0$, with 7 curves in the dual graph:

$$
\begin{array}{c}
2 \\
| \\
2 \text{---} 2 \text{---} 2 \text{---} 2 \text{---} 2 \text{---} 2
\end{array}
$$

E_8: $x^2 + y^3 + z^5 = 0$, with 8 curves in the dual graph:

$$
\begin{array}{c}
2 \\
| \\
2 \text{---} 2 \text{---} 2 \text{---} 2 \text{---} 2 \text{---} 2 \text{---} 2
\end{array}
$$

If $k(s)$ is not algebraically closed, we have a few more twisted forms. These can be classified by the same method, as in Lipman (1969, section 24), or one can use the general results of (3.31).

B_n or twisted A_n: $2 \Longrightarrow (2 \cdot 2) \Longrightarrow \cdots \Longrightarrow (2 \cdot 2)$

C_n or twisted D_n: $2 \text{---} \cdots \text{---} 2 \Longrightarrow (2 \cdot 2)$

G_2 or twisted D_4: $2 \Longrightarrow (3 \cdot 2)$

F_4 or twisted E_6: $2 \text{---} 2 \Longrightarrow (2 \cdot 2) \Longrightarrow (2 \cdot 2)$

First examples of log canonical singularities

Next we discuss the main examples of surface singularities that are lc but not lt.

Example 3.27 (Numerically elliptic/cusp singularities)   By definition, these are the singularities $(s \in S, B)$ where $B = 0$ and every exceptional curve appears in $\Delta := \Delta(s \in S, B)$ with coefficient 1.

These are the only lc surface singularities that are not rational (2.28).

One should treat elliptic/cusp singularities as one class, but traditionally they have been viewed as two distinct types. For these singularities one can write the equations (2.26.1) as

$$\deg \omega_{C_i} = -\textstyle\sum_{i \neq j}(C_i \cdot C_j).$$

The usual approach distinguishes two solutions:

(1) Γ consists of a single curve C with $\deg \omega_C = 0$. These are called numerically (simple) *elliptic*. If we are over an algebraically closed field, then there are three possibilities for C:
   (a) a smooth elliptic curve (called a *simple elliptic singularity*),
   (b) a nodal rational curve, or
   (c) a cuspidal rational curve (three more blow-ups show that this is not lc).
(2) Γ consists of at least two curves C_i with $\deg \omega_{C_i} < 0$ for every i. If C has only nodes then these, together with the case (1b), are called *cusp*[1] singularities. There are two degenerate cases: two curves can be tangent at a point or three curves can meet in one point. These are numerically log canonical but not log canonical.

The distinction between the two cases is made mostly for historic reasons, and over a nonclosed field k the two concepts are mixed together. Indeed, we can have a cusp over $\bar{k}$ where the Galois group of $\bar{k}/k$ acts transitively on the exceptional curves. In that case, we have a numerically elliptic singularity over k but a cusp over $\bar{k}$.

For a cusp the dual graph is a circle of conics

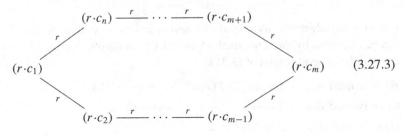

$$(3.27.3)$$

---

[1] Cusps of curves and cusps of surfaces are quite unrelated. Usually there is no danger of confusion.

where $n \geq 3$. The $n = 1, 2$ cases are somewhat special. For $n = 2$ we have two curves which intersect at two distinct points. For $n = 1$ we get a nodal rational curve; this was already considered above.

There is also a further degeneration when the circle folds on itself, giving the dual graph

$$(r \cdot c_1) \xrightarrow{2r} (2r \cdot c_1) \xrightarrow{2r} \cdots \xrightarrow{2r} (2r \cdot c_{n-1}) \xrightarrow{2r} (r \cdot c_n) \quad (3.27.4)$$

In both cases, the intersection form is negative definite if at least one of the $c_i \geq 3$, as shown by the vector $\sum C_i$ (10.3.4.2).

If the residue field $k(s)$ is algebraically closed then $r = 1$. If $k(s)$ is perfect, then passing to the algebraic closure we still have a cusp. Its dual graph is now a circle of smooth rational curves of length rn.

Example 3.28 Here we list the dual graphs of those lc pairs $(s \in S, B)$ that are not plt, B is reduced and S has a rational singularity at s. The main series is constructed as follows.

We start with a chain $c_1 \textrm{------} \cdots \textrm{------} c_n$ and at both end we attach one of the following

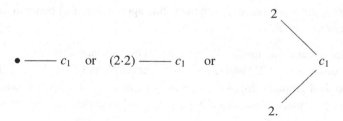

Here $c_1 = 1$ is allowed. We get six different types all together. For any $c_1, \ldots, c_n \geq 2$, with at least one $c_i \geq 3$, the resulting graphs are negative definite.

The main examples are

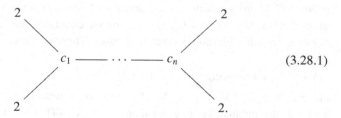

$$(3.28.1)$$

(When the characteristic is different from 2, the corresponding singularity is a $\mathbb{Z}/2$-quotient of a cusp (for $n \geq 2$) or of a simple elliptic singularity (for $n = 1$).)

Over nonclosed fields, there is one more series obtained from the above by a left-right symmetry of order 2. We start with a chain

$$(2 \cdot c_1) =\!=\!= \cdots =\!=\!= (2 \cdot c_{n-1}) =\!=\!= (c_n) =$$

and at the left end we attach one of the following

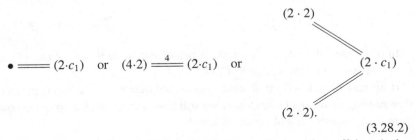

$$(3.28.2)$$

In all of these cases, the original curves C_i appear in Δ with coeffcient 1, the newly added curves with coeffcient $\frac{1}{2}$.

Plan of the classification

In what follows, we classify lc surface germs $(s \in S, B)$. We focus on the dual graph, following the method of Alexeev (Kollár, 1992, section 3). Besides working for arbitrary excellent schemes, this approach is also better at dealing with the boundary B.

3.29 (Classification method)   Let $(s \in S, B)$ be a surface singularity, $f: T \to S$ a resolution as in (2.25) with exceptional curves C_i and $\Gamma = \Gamma(s \in S, B)$ the extended dual graph (2.26) of the curves $\sum C_i$ and $f_*^{-1} B$. More generally, we consider all connected, numerically log canonical extended dual graphs (Γ, B).

In what follows, we consider the equations (2.26.3)

$$(\Delta \cdot C_i) = (C_i \cdot C_i) - \deg \omega_{C_i} - (f_*^{-1} B \cdot C_i) \qquad \forall i.$$

where $\Delta = \sum_j d_j C_j$ and try to find conditions on the right hand side which force $d_j > 1$ for some j. This would then show that (Γ, B) is not lc.

We will identify various small subsets of the equations which lead to a contradiction, no matter what the remaining equations are. After four such steps we are left with only a handful of cases. These will then be studied more carefully.

We use the following form of (10.3.4):

Claim 3.29.1   If $(\Delta' \cdot C_i) \leq (\Delta'' \cdot C_i)$ for every i then $\Delta' \geq \Delta''$. Furthermore, if one of the inequalities is strict then $\Delta' \gg \Delta''$. (That is, $\Delta' - \Delta''$ contains every curve with positive coefficient.) $\square$

We use three applications of this claim.

3.29.2 (Estimating Δ)   If we have $\Delta' = \sum d'_j C_j$ such that

$$(\Delta' \cdot C_i) \geq (C_i \cdot C_i) - \deg \omega_{C_i} - (f_*^{-1} B \cdot C_i) \qquad (3.29.2i)$$

for every i then $\Delta(\Gamma, B) \geq \Delta'$ and $\Delta(\Gamma, B) \gg \Delta'$ if one of the inequalities (3.29.2i) is strict.   □

Next we see how $\Delta(\Gamma, B)$ changes if we change B. Note that changing B changes the notion of minimality, but this does not matter for the inequalities.

3.29.3 (Changing B)   If $B' \lneq B$ then $\Delta(\Gamma, B') \ll \Delta(\Gamma, B)$.   □

The most important comparison is the following.

3.29.4 (Passing to a subgraph)   Let (Θ, B) be a graph obtained from (Γ, B) by deleting some vertices and edges. Set $\Delta(\Gamma, B) = \sum_{j \in \Gamma} d_j C_j$ and let $(C_i \cdot C_j)_\Theta$ denote the intersection number in Θ. Then

$$\left(\sum_{j \in \Theta} d_j C_j \cdot C_i\right)_\Theta \leq (C_i \cdot C_i) - \deg \omega_{C_i} - (B \cdot C_i) - \left(\sum_{j \notin \Theta} d_j C_j \cdot C_i\right).$$

Thus if (Γ, B) is log minimal then $\Delta(\Theta, B) \leq \sum_{j \in \Theta} d_j C_j$ and in fact

$$\Delta(\Theta, B) \ll \sum_{j \in \Theta} d_j C_j,$$

unless $d_j = 0$ for every $j \notin \Theta$. By (2.26.4) this happens iff (Γ, B) is canonical, in which case (Θ, B) is also canonical.

In particular, if (Θ, B) is not log terminal then (Γ, B) is not log canonical.   □

3.30 (First steps of the classification)   Let (Γ, B) be a connected, negative definite, log minimal, log canonical extended dual graph. We develop a rough classification in a series of steps.

Claim 3.30.1   Assume that (Γ, B) contains a curve C_i such that $\deg \omega_{C_i} \geq 0$. Then $\deg \omega_{C_i} = 0$ and $(\Gamma, B) = (\{C_i\}, 0)$. These are the numerically simple elliptic singularities (3.27).

Proof   The coefficient of C_i in $\Delta(\{C_i\}, 0)$ equals $1 - \frac{\deg \omega_{C_i}}{(C_i \cdot C_i)}$. If $\deg \omega_{C_i} > 0$ this is bigger than 1. If $\deg \omega_{C_i} = 0$ then $(\{C_i\}, 0)$ is lc but not lt. Thus, by (3.29.3–4), adding new vertices or increasing B yields a non-lc graph.   □

Reduction 3.30.2   From now on we assume that $\deg \omega_{C_i} = -2r_i$ for every i where $r_i := h^0(C_i, \mathcal{O}_{C_i})$. That is, each C_i is a conic (10.6).

Claim 3.30.3   Assume that (Γ, B) contains two curves C_i, C_j such that $(C_i \cdot C_j) > \max\{r_i, r_j\}$. Then $(C_i \cdot C_j) = 2r_i = 2r_j$ and $(\Gamma, B) = (\{C_i, C_j\}, 0)$. These are cusp singularities with two curves (3.27).

Proof   Note that $(C_i \cdot C_j) \geq 2 \cdot \max\{r_i, r_j\}$ since r_i, r_j divide $(C_i \cdot C_j)$. Thus

$$((C_i + C_j) \cdot C_i) = (C_i \cdot C_i) + (C_j \cdot C_i) \geq (C_i \cdot C_i) + 2r_i$$
$$= (C_i \cdot C_i) - \deg \omega_{C_i}$$

and similarly for $((C_i + C_j) \cdot C_j)$. Thus, by (3.29.2), $(\{C_i, C_j\}, 0)$ is not lt and lc only if $(C_i \cdot C_j) = 2r_i = 2r_j$. By (3.29.3–4), adding new vertices or increasing B yields a non-lc graph.   □

Reduction 3.30.4   From now on we assume that, for every $i \neq j$, the intersection number $(C_i \cdot C_j)$ is either 0 or $\max\{r_i, r_j\}$.

Claim 3.30.5   Assume that (Γ, B) contains a subgraph $(\Theta, 0)$ consisting of vertices $C_1, \ldots, C_n$ such that all the intersection numbers $(C_i \cdot C_{i+1})$ (subscript modulo n) are nonzero. Then $(\Theta, 0)$ is not log terminal hence $(\Gamma, B) = (\Theta, 0)$. Furthermore, $(\Theta, 0)$ is lc if and only if all the other intersection numbers are 0 and $r_1 = \cdots = r_n$. These are the cusp singularities (3.27.3).

Proof   Choose $\Delta' = \sum C_j$. Then, since $(C_i \cdot C_{i+1}) = \max\{r_i, r_{i+1}\}$,

$$(\Delta \cdot C_i) \geq (C_i \cdot C_i) + (C_i \cdot C_{i-1}) + (C_i \cdot C_{i+1}) \geq (C_i \cdot C_i) + 2r_i.$$

Thus (3.29.2) shows that $(\Gamma, 0)$ is not log terminal and it is lc if and only if equality holds in both places.   □

Reduction 3.30.6   From now on we assume that the dual graphs are trees.

Claim 3.30.7   Assume that (Γ, B) contains a subgraph (Θ, B') built up as follows. We start with a chain

$$(r \cdot c_1) \mathrel{\rule[0.5ex]{2em}{0.4pt}} \cdots \mathrel{\rule[0.5ex]{2em}{0.4pt}} (r \cdot c_n)$$

of length $n \geq 1$ and at both ends we attach one of the following

$$\bullet \mathrel{\rule[0.5ex]{2em}{0.4pt}} (r \cdot c_1) \quad \text{or} \quad (s \cdot c_0) \mathrel{\rule[0.5ex]{2em}{0.4pt}} (r \cdot c_1) \quad \text{or} \quad \begin{array}{c} (r \cdot c_0') \\ \diagdown \\ (r \cdot c_1) \\ \diagup \\ (r \cdot c_0'') \end{array}$$

$$(3.30.8)$$

where $s \neq r$ and $c_0, c_0', c_0'' \geq 2$. Then the resulting graph is not log terminal hence $(\Gamma, B) = (\Theta, B')$. Furthermore, (Θ, B') is lc if and only if both ends are among the following: (i) the first type, (ii) the second type with $s = 2r$ and $c_0 = 2$ or with $s = \frac{1}{2}r$, (iii) the third type where $c_0' = c_0'' = 2$.

These are the dual graphs enumerated in (3.28) and in (3.27.4).

Proof We use a Δ' whose coefficients are 1 for the curves $C_1, \ldots, C_n$. For the new curves we choose the coefficient 1 if $s < r$ and $\frac{1}{2}$ if $s > r$ or we are in the last case. $\qquad\square$

Conclusion 3.30.9 The remaining dual graphs contain at most one copy of the three configurations listed in (3.30.8). By (10.10) there is at least one curve for which $r_i = 1$.

We can summarize our results as follows:

Corollary 3.31 *Let S be a normal, excellent surface, $s \in S$ a closed point and B a reduced boundary. Assume that $(s \in S, B)$ is numerically log canonical and let (Γ, B) be its extended dual graph. The either (Γ, B) is listed in (3.27) and (3.28) or one of the following holds.*

(1) *(Cyclic quotient) $\lfloor B \rfloor$ is regular and $(\Gamma, \lfloor B \rfloor)$ is*

$$\bullet \!\!-\!\!-\!\! \circ \!\!-\!\!-\!\! \cdots \!\!-\!\!-\!\! \circ \quad or \quad \circ \!\!-\!\!-\!\! \circ \!\!-\!\!-\!\! \cdots \!\!-\!\!-\!\! \circ.$$

Furthermore, $C_i \simeq \mathbb{P}^1_{k(s)}$ for every i, save when $B = 0$ and there is only one exceptional curve which can be a $k(s)$-irreducible conic.

(2) *(Other quotient) $\lfloor B \rfloor = 0$, $C_i \simeq \mathbb{P}^1_{k(s)}$ for every i and Γ is a tree with three branches:*

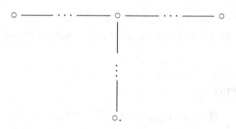

(3) *(Twisted quotient) $\lfloor B \rfloor = 0$ and, for some $1 \leq i < n$ and $r \geq 2$, Γ is*

$$c_1 \!-\!-\! \cdots \!-\!-\! c_i \overset{r}{-\!-\!-} (r \cdot c_{i+1}) \overset{r}{-\!-\!-} \cdots \overset{r}{-\!-\!-} (r \cdot c_n).$$

For the higher dimensional theory the cyclic quotients are the most important, especially when $\lfloor B \rfloor \neq 0$. We describe them in detail next. A full description of the other two classes is given afterwards.

Cyclic quotients

The following theorem explains the name of cyclic quotient singularities.

Theorem 3.32 *Let Y be a normal surface over a field k, B a reduced curve and $y \in Y$ a k-point such that the dual graph of $(y \in Y, B)$ is*

$$\circledast \!-\!-\! c_1 \!-\!-\! c_2 \!-\!-\! \cdots \!-\!-\! c_{n-1} \!-\!-\! c_n \!-\!-\! \circledast$$

where ⊛ *denotes either the empty set or a local branch of B. Then the completion of* $(y \in Y)$ *is isomorphic to the cyclic quotient singularity* (3.19)

$$\widehat{\mathbb{A}}_k^2 / \tfrac{1}{d_n}(1, d_{n-1})$$

where the d_r *are computed as in* (3.33). *B is contained in the image of the coordinate axes.*

We stress that we allow char k to divide d_n. The result also holds for the Henselization (4.38).

Definition 3.33 Given a sequence $c_1, c_2, \ldots$, let $M_r = M_r(c_1, \ldots, c_r)$ denote the intersection form of the dual graph

$$c_1 \text{———} c_2 \text{———} \cdots \text{———} c_{r-1} \text{———} c_r.$$

For instance, for $r = 5$ we get

$$M_5(c_1, \ldots, c_5) = \begin{pmatrix} c_1 & -1 & 0 & 0 & 0 \\ -1 & c_2 & -1 & 0 & 0 \\ 0 & -1 & c_3 & -1 & 0 \\ 0 & 0 & -1 & c_4 & -1 \\ 0 & 0 & 0 & -1 & c_5 \end{pmatrix}$$

Set $d_0 := 1$ and $d_r(c_1, \ldots, c_r) := \det M_r(c_1, \ldots, c_r)$ for $r \geq 1$. Expanding by the last column we see that the d_r satisfy the recurrence relation

$$d_r = c_r d_{r-1} - d_{r-2}. \tag{3.33.1}$$

These in turn imply that

$$M_r \cdot (d_0, d_1, \ldots, d_{r-1})^t = (0, \ldots, 0, d_r)^t. \tag{3.33.2}$$

By induction one also sees that $\gcd(d_r, d_{r-1}) = 1$ for every r and that the d_r are also computed by the following continued fraction

$$\frac{d_r}{d_{r-1}} = c_r - \cfrac{1}{c_{r-1} - \cfrac{1}{c_{r-2} - \cfrac{1}{\ddots \cfrac{1}{c_3 - \cfrac{1}{c_2 - \cfrac{1}{c_1}}}}}} \tag{3.33.3}$$

This makes it possible to enumerate all chains with a given determinant d. For any $0 < e < d$ with $(d, e) = 1$, compute the above continued fraction expansion of d/e. These give all such chains.

For instance, the cases $\det(\Gamma) \in \{2, 3, 4, 5, 6\}$ give the possibilities

$$\begin{aligned}
\det(\Gamma) = 2 &\Leftrightarrow \quad \Gamma \text{ is } 2 \\
\det(\Gamma) = 3 &\Leftrightarrow \quad \Gamma \text{ is } 3 \text{ or } 2 - 2, \\
\det(\Gamma) = 4 &\Leftrightarrow \quad \Gamma \text{ is } 4 \text{ or } 2 - 2 - 2, \\
\det(\Gamma) = 5 &\Leftrightarrow \quad \Gamma \text{ is } 5 \text{ or } 2 - 2 - 2 - 2 \text{ or } 2 - 3 \text{ or } 3 - 2, \\
\det(\Gamma) = 6 &\Leftrightarrow \quad \Gamma \text{ is } 6 \text{ or } 2 - 2 - 2 - 2 - 2.
\end{aligned}$$

3.34 *Proof of (3.32)* Let $f: X \to Y$ be the minimal resolution with exceptional curves $C_1, \ldots, C_n \subset X$. As in (3.33), set

$$d_r := \det M_r(c_1, \ldots, c_r) \quad \text{and} \quad d_r^* := \det M_r(c_n, \ldots, c_{n+1-r}).$$

Note that the C_i are geometrically integral rational curves over k that have a k-point (the intersection point with the adjacent curve). Thus in fact $C_i \simeq \mathbb{P}_k^1$. By working in a suitable étale neighborhood of $(y \in Y)$ with the same residue field (4.38.1), we can pick (nonproper) curves C_0, C_{n+1} such that C_0 intersects only C_1 and C_{n+1} intersects only C_n, both with multiplicity 1. If there are such local branches of B, we choose them to be the C_0, C_{n+1}. Set $d_{-1} = d_{-1}^* = 0$. By (3.33.2), both of the line bundles

$$\mathscr{O}_X(-\textstyle\sum_{i=0}^{n+1} d_{i-1} C_i) \quad \text{and} \quad \mathscr{O}_X(-\textstyle\sum_{i=0}^{n+1} d_{n-i}^* C_i) \tag{3.34.1}$$

have degree zero on the exceptional curves $C_1, \ldots, C_n$. Hence, by (10.9), in a neighborhood of $\mathrm{Ex}(f)$ they are both trivial. Thus, as subsheaves on $\mathscr{O}_X$, they are generated by functions v (resp. u) such that

$$(v) = \textstyle\sum_{i=0}^{n+1} d_{i-1} C_i \quad \text{and} \quad (u) = \textstyle\sum_{i=0}^{n+1} d_{n-i}^* C_i.$$

Set $b := d_n - d_{n-1}$ and consider

$$(u^b v) = \textstyle\sum_{i=0}^{n+1} a_{i-1} C_i \quad \text{where} \quad a_{i-1} = (d_{i-1} + b d_{n-i}^*).$$

Note that the two highest coefficients are $a_n = d_n + b d_{-1} = d_n$ and $a_{n-1} = d_{n-1} + b d_0 = d_n$. Using (3.33.1) for both sequences, we get a recursion

$$a_{r-2} = c_r a_{r-1} - a_r$$

which shows that all the coefficients in $(u^b v)$ are divisible by d_n. Thus, as before, there is a function w such that

$$(w) = \sum_{i=0}^{n+1} \frac{a_{i-1}}{d_n} C_i.$$

Since w^{d_n} and $u^b v$ have the same divisors, the functions u, v, w are related by an equation $w^{d_n} = (\text{invertible}) u^b v$. We can set $v' := (\text{invertible}) v$ to obtain the simpler equation $w^{d_n} = u^b v'$. Thus we obtain a morphism

$$(u, v', w): Y \to \operatorname{Spec} k[u, v', w]/(w^{d_n} = u^b v')$$

which factors through the normalization (3.19.6)

$$\pi\colon Y \to \mathbb{A}^2/\tfrac{1}{d_n}(1, d_{n-1}).$$

We are left to prove that π is étale at y. Since $k[u, v', w]/(w^{d_n} = u^b v')$ has degree d_n over $k[u, v']$, it is enough to prove that $(u, v')\colon Y \to \operatorname{Spec} k[u, v']$ also has degree d_n at y. Equivalently, that the intersection number of their divisors is

$$(u) \cdot (v) = (\textstyle\sum_{i=0}^{n+1} d_{i-1} C_i) \cdot (\textstyle\sum_{i=0}^{n+1} d_{n-i}^* C_i) = d_n.$$

As we noted, C_{n+1} intersects only C_n and (v) has 0 intersection with the curves $C_1, \ldots, C_n$. Thus

$$\begin{aligned}
(\textstyle\sum_{i=0}^{n+1} d_{i-1} C_i) \cdot (\textstyle\sum_{i=0}^{n+1} d_{n-i}^* C_i) &= d_n(C_{n+1} \cdot \textstyle\sum_{i=0}^{n+1} d_{n-i}^* C_i) \\
&= d_n(C_{n+1} \cdot C_n) = d_n.
\end{aligned}$$

This completes the proof of (3.32). $\square$

Cyclic quotients play a key role in the adjunction formulas in Chapter 4. For now, let us examine in detail what we have for surfaces.

3.35 (Adjunction)   Let S be a normal, excellent surface and $B \subset S$ a curve such that (S, B) is lc. By (2.31), B is a nodal curve and by (2.31) and the classification (3.30) and (3.31), there are three types of singularities of the pair (S, B):

(1) (Cyclic, plt) (S, B) is plt at p, B is regular at p and the dual graph of the minimal resolution is

$$\bullet \;\rule[0.5ex]{1.2em}{0.4pt}\; c_1 \;\rule[0.5ex]{1.2em}{0.4pt}\; \cdots \;\rule[0.5ex]{1.2em}{0.4pt}\; c_n.$$

(2) (Cyclic, lc) (S, B) is lc but not plt at p, B has a node at p and the dual graph of the minimal resolution is

$$\bullet \;\rule[0.5ex]{1.2em}{0.4pt}\; c_1 \;\rule[0.5ex]{1.2em}{0.4pt}\; \cdots \;\rule[0.5ex]{1.2em}{0.4pt}\; c_n \;\rule[0.5ex]{1.2em}{0.4pt}\; \bullet$$

or

$$\bullet = (2 \cdot c_1) = \cdots = (2 \cdot c_{n-1}) = (c_n).$$

(3) (Dihedral) (S, B) is lc but not plt at p, B is regular at p and the dual graph of the minimal resolution is

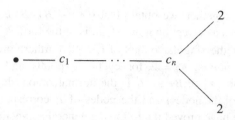

or

$$\bullet \text{\rule{1cm}{0.4pt}} c_1 \text{\rule{1cm}{0.4pt}} \cdots \text{\rule{1cm}{0.4pt}} c_{n-1} \text{\rule{1cm}{0.4pt}} (2 \cdot 2).$$

Assume next for simplicity that B is proper. (See (2.34) for the general case.) Let $f \colon T \to S$ be a minimal log resolution of (S, B) as in (2.25) and $B_T \subset T$ the birational transform of B. Write

$$f^*(K_S + B) \sim_{\mathbb{Q}} K_T + B_T + \sum_p \Delta_p,$$

where Δ_p is supported on $f^{-1}(p)$. By the projection formula,

$$\begin{aligned} ((K_S + B) \cdot B) &= (f^*(K_S + B) \cdot B_T) \\ &= ((K_T + B_T + \sum_p \Delta_p) \cdot B_T) \\ &= \deg \omega_{B_T} + \sum_p (\Delta_p \cdot B_T). \end{aligned}$$

For the cases (3.35.2–3), the divisor Δ_p was computed in (3.30). For both of these, the curves intersecting B_T have coefficient 1 in Δ_p. Next we compute the discrepancy divisor for (3.35.1).

Claim 3.35.1 Consider a pair $(S, B + b'B')$ whose dual graph is

$$\bullet \text{\rule{1cm}{0.4pt}} c_1 \text{\rule{1cm}{0.4pt}} \cdots \text{\rule{1cm}{0.4pt}} c_n \overset{b'}{\text{\rule{1cm}{0.4pt}}} \circledast,$$

where the birational transfrom of B' is denoted by $\circledast$. Then the discrepancy divisor is

$$\Delta = \sum_{i=1}^{n} \left(1 - (1 - b') \frac{d_{i-1}}{\det(\Gamma)} \right) C_i \quad \text{and} \quad B_T \cdot \Delta = 1 - \frac{1 - b'}{\det(\Gamma)}.$$

Proof By the adjunction formula,

$$((K_T + B_T + C_1 + \cdots + C_n + b'B_T') \cdot C_j) = \begin{cases} 0 & \text{if } j < n, \text{ and} \\ -(1 - b') & \text{if } j = n. \end{cases}$$

Thus, using (3.33.2), we conclude that

$$\left(\left(K_T + B_T + C_1 + \cdots + C_n - \frac{1 - b'}{\det(\Gamma)} \sum_{i=1}^{n} d_{i-i} C_i \right) \cdot C_j \right) = 0 \quad \forall i.$$

This gives the formula for Δ. The only curve in Δ that intersects B_T is C_1 and its coefficient is $1 - (1 - b')\frac{d_0}{\det(\Gamma)} = 1 - \frac{1-b'}{\det(\Gamma)}$. □

Putting these together, we obtain that $((K_S + B) \cdot B)$ equals the expected term $\deg \omega_{B_T}$ plus a correction term which is of the form $\sum_p \delta_p \deg[p]$ where the sum is over the singular points of (S, B). Furthermore, $\delta_p = 1 - \frac{1}{\det(\Gamma_p)}$ for cyclic, plt points, $\delta_p = 2$ for cyclic lc points and $\delta_p = 1$ for dihedral points. Since $f|_{B_T}: B_T \to B$ is the normalization, $\deg \omega_{B_T} = \deg \omega_B -$ deg(preimages of the nodes) and the nodes of B correspond to the cyclic lc points. Thus we have proved the following general adjunction formula:

Theorem 3.36 (Adjunction formula I) *Let S be a normal, excellent surface and $B \subset S$ a reduced, proper curve such that (S, B) is lc. Then*

$$((K_S + B) \cdot B) = \deg \omega_B + \sum_{p \text{ cyclic, plt}} \left(1 - \tfrac{1}{\det(\Gamma_p)}\right) \deg[p]$$

$$+ \sum_{p \text{ dihedral}} \deg[p]. \quad \square$$

As stated, (3.36) is only an equality of two numbers. However, by looking at its proof, we see that we have computed the different

$$\mathrm{Diff}_B(0) = \sum_{p \text{ cyclic, plt}} \left(1 - \tfrac{1}{\det(\Gamma(p))}\right)[p] + \sum_{p \text{ dihedral}} [p]. \tag{3.36.1}$$

One fork case

We continue our investigations by working out the remaining two cases of (3.31). We start with (3.31.2), when we have a a tree with one fork.

3.37 Let (Γ, B) be a tree with one fork and all the $r_i = 1$ (hence we drop it from the notation). We further assume that the components of B intersect only the leaves. That is, the dual graph is

$$B_1 \text{——} \Gamma_1 \text{——} c_0 \text{——} \Gamma_2 \text{——} B_2$$

$$|$$

$$\Gamma_3 \tag{3.37.1}$$

$$|$$

$$B_3.$$

The fork is denoted by C_0 and for $i = 1, 2, 3$, the curves on the branches are indexed as

$$(C_0 \text{——} \Gamma_i \text{——} B_i) = (C_0 \text{——} C_1^i \text{——} \cdots \text{——} C_{n(i)}^i \text{——} B_i).$$

The curves $C_{n(i)}^i$ are called *leaves* and the curves C_j^i for $0 < j < n(i)$ are called *intermediate*. Set $c_j^i = -(C_j^i \cdot C_j^i)$ and $\beta_i = (C_{n(i)}^i \cdot B_i)$. We allow two degenerate cases:

(2) some of the Γ_i can be empty and
(3) the self-intersection of the fork C_0 can be -1.

Theorem 3.38 *Notation as above. Then* (3.37.1) *is numerically log canonical (resp. numerically log terminal) if and only if*

$$\frac{1 - \beta_1}{\det(\Gamma_1)} + \frac{1 - \beta_2}{\det(\Gamma_2)} + \frac{1 - \beta_3}{\det(\Gamma_3)} \geq 1 \quad (\text{resp.} > 1). \tag{3.38.1}$$

Proof Let $\Sigma^i := \sum_j C_j^i$ denote the sum of all curves in Γ_i with coefficient 1 and $\Sigma_\Gamma := C_0 + \sum_i \Sigma^i$ the sum of all curves in Γ.

Write the discrepancy divisor as $\Delta = a_0 C_0 + \sum_i \Delta_i$ where Δ_i involves only the curves in Γ_i. Thus $\Sigma_\Gamma - \Delta \sim_{\mathbb{Q}} K_\Gamma + \Sigma_\Gamma + B$ and for every exceptional curve C we have by (2.26.3)

$$((\Sigma_\Gamma - \Delta) \cdot C) = -2 + \#\{\text{neighbors of } C\} + (B \cdot C).$$

In particular,

$$((\Sigma_\Gamma - \Delta) \cdot C) = \begin{cases} 1 & \text{if } C = C_0 \text{ is the fork} \\ 0 & \text{if } C = C_j^i \text{ is intermediate and} \\ \beta_i - 1 & \text{if } C = C_{n(i)}^i \text{ is a leaf.} \end{cases}$$

Using $d_j^i = \det M(c_1^i, \dots, c_j^i)$, consider the $\mathbb{Q}$-divisor

$$D^i := \frac{1}{\det(\Gamma_i)} \left(d_0^i C_1^i + \cdots + d_{n(i)-1}^i C_{n(i)}^i \right),$$

where $\det(\Gamma_i) = d_{n(i)}^i$ is the determinant of Γ_i. Therefore, by (3.33.2),

$$\left(\left(\Sigma_\Gamma - \Delta - \sum_{i=1}^3 (1 - \beta_i) D^i \right) \cdot C \right) = \begin{cases} 1 - \sum_{i=1}^3 \frac{1 - \beta_i}{\det(\Gamma_i)} & \text{if } C \text{ is the fork} \\ 0 & \text{if } C \text{ is any other curve.} \end{cases}$$

$$\tag{3.38.2}$$

Thus, if (3.38.1) holds then $\Sigma_\Gamma - \Delta - \sum_{i=1}^3 (1 - \beta_i) D^i$ has ≤ 0 intersection with every curve. Hence, by (3.29.1),

$$\Delta \leq \Sigma_\Gamma - \sum_{i=1}^3 (1 - \beta_i) D^i \leq \Sigma_\Gamma,$$

showing that (Γ, B) is numerically lc.

If (3.38.1) fails then $\Sigma_\Gamma - \Delta - \sum_{i=1}^3 (1 - \beta_i) D^i$ has ≥ 0 intersection with every curve, hence, by (3.29.1), it is a negative vector. Thus C_0 appears in Δ with coefficient > 1, hence (Γ, B) is not numerically lc. $\square$

List of log canonical surface singularities

Here is a summary of the results proved so far where $(s \in S, B)$ is an lc germ and $k(s)$ is algebraically closed.

3.39 (($s \in S, 0$) not log terminal) Thus $B = 0$.

3.39.1 (Simple elliptic) $\Gamma = \{E\}$ has a single vertex which is a smooth elliptic curve with self-intersection ≤ -1.

3.39.2 (Cusp) Γ is a circle of smooth rational curves, $c_i \geq 2$ and at least one of them with with $c_i \geq 3$. The cases $n = 1, 2$ are somewhat special.

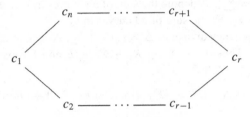

3.39.3 ($\mathbb{Z}/2$-quotient of a cusp or simple elliptic) Γ has two forks and each branch is a single (-2)-curve. We have $c_i \geq 2$.

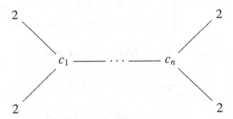

3.39.4 (Other quotients of a simple elliptic) These are coming from (3.38). Since $\beta_i = 0$, the inequalities (3.38.1) give

$$\frac{1}{\det(\Gamma_1)} + \frac{1}{\det(\Gamma_2)} + \frac{1}{\det(\Gamma_3)} = 1.$$

If $\det(\Gamma_i) \geq 2$ for every i, there are only three possibilities for $(\det(\Gamma_1), \det(\Gamma_2), \det(\Gamma_3))$:

($\mathbb{Z}/3$-quotient) (3,3,3)
($\mathbb{Z}/4$-quotient) (2,4,4)
($\mathbb{Z}/6$-quotient) (2,3,6).

The names reflect that, at least in characteristic 0, these are obtained as the quotient of a simple elliptic singularity by the indicated group (2.53).

3.40 (($s \in S, 0$) log terminal)   Here B can be nonzero.

3.40.1 (Cyclic quotient)

$$c_1 \text{———} \cdots \text{———} c_n.$$

3.40.2 (Dihedral quotient)   Here $n \geq 2$ with dual graph

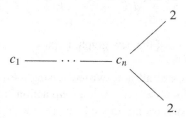

3.40.3 (Other quotients)   The dual graph is as in (3.37.1) with three possibilities for $(\det(\Gamma_1), \det(\Gamma_2), \det(\Gamma_3))$:

(Tetrahedral) $(2,3,3)$
(Octahedral) $(2,3,4)$
(Icosahedral) $(2,3,5)$.

The names reflect that, at least in characteristic 0, these are obtained as the quotient of $\mathbb{C}^2$ by the indicated type of group (2.53). See Brieskorn (1968) for details.

Twisted quotients

Let ($s \in S$) be an lc surface singularity over a perfect field k and $f : Y \to S$ a minimal resolution with dual graph Γ. By passing to the algebraic closure, we get $f_{\bar{k}} : Y_{\bar{k}} \to S_{\bar{k}}$; let $\Gamma_{\bar{k}}$ be the dual graph of the exceptional curves. Then $\Gamma_{\bar{k}}$ is a (possibly disconnected) log canonical dual graph. The Galois group $\mathrm{Gal}(\bar{k}/k)$ acts on the vertices of $\Gamma_{\bar{k}}$ and its orbits are exactly the vertices of Γ. If $C \subset Y$ is an exceptional curve then $\dim_k H^0(C, \mathscr{O}_C)$ equals the number of connected components of $C_{\bar{k}}$. Thus, over a perfect field k, we can obtain all twisted dual graphs by first describing all dual graphs over $\bar{k}$ and then classifying all possible group actions on them. (It is not obvious that every abstract possibility is actually realized, but in our case this turns out to be easy.)

These geometric arguments do not apply for non-perfect residue fields. One can, however, follow this method with abstract dual graphs.

3.41 (Classification of twisted quotients)   Start with a twisted quotient

$$c_1 \xrightarrow{r} \cdots \xrightarrow{r} c_i \xrightarrow{r} (r \cdot c_{i+1}) \xrightarrow{r} \cdots \xrightarrow{r} (r \cdot c_n).$$

Create a new graph by replacing each of the vertices $C_{i+1}, \ldots, C_n$ with r new vertices $C^a_{i+1}, \ldots, C^a_n$ for $a = 1, \ldots, r$. The new intersection numbers are all zero except $(C_i \cdot C_{i+1}^a) = 1$ and $(C_j^a \cdot C_{j+1}^a) = 1$ for $j > i$. The symmetric group on r letters acts on the new dual graph.

It is easy to see that if $\Delta = \sum d_j C_j$ is the discrepancy divisor of the old graph then

$$\Delta' := \sum_{j=1}^{i} d_j C_j + \sum_{j=i+1}^{n} d_j (\textstyle\sum_{a=1}^{r} C_j^a)$$

is the discrepancy divisor of the new graph. In particular, the new graph is lc if and only if the old one is.

Thus we need to look at all non-twisted lc singularities that we have just classified and list all possible symmetric group actions on them. This is quite easy to do, hopefully we have not missed any. Besides the cases already listed in (3.27.4) and (3.28.2), we get the following.

3.41.1 (Twisted $\mathbb{Z}/2$-quotients of simple elliptic)

$$(c_0) \xrightarrow{\ 4\ } (4 \cdot 2) \quad \text{or} \quad 2 \xrightarrow{\hspace{1cm}} (c_0) \xrightarrow{\ 3\ } (3 \cdot 2).$$

3.41.2 (Twisted $\mathbb{Z}/3$-quotients of simple elliptic)   With a triple twist we get

$$(c_0) \xrightarrow{\ 3\ } (3 \cdot 3) \quad \text{or} \quad (c_0) \xrightarrow{\ 3\ } (3 \cdot 2) \xrightarrow{\ 3\ } (3 \cdot 2).$$

With a double twist we get

$$3 \xrightarrow{\hspace{0.6cm}} (c_0) \xrightarrow{\ 2\ } (2 \cdot 3) \quad \text{or} \quad 3 \xrightarrow{\hspace{0.6cm}} (c_0) \xrightarrow{\ 2\ } (2 \cdot 2) \xrightarrow{\ 2\ } (2 \cdot 2) \quad \text{or}$$

$$2 \xrightarrow{\hspace{0.6cm}} 2 \xrightarrow{\hspace{0.6cm}} (c_0) \xrightarrow{\ 2\ } (2 \cdot 3) \quad \text{or} \quad 2 \xrightarrow{\hspace{0.6cm}} 2 \xrightarrow{\hspace{0.6cm}} (c_0) \xrightarrow{\ 2\ } (2 \cdot 2) \xrightarrow{\ 2\ } (2 \cdot 2).$$

3.41.3 (Twisted $\mathbb{Z}/4$-quotients of simple elliptic)

$$2 \xrightarrow{\hspace{0.6cm}} (c_0) \xrightarrow{\ 2\ } (2 \cdot 2) \xrightarrow{\ 2\ } (2 \cdot 2) \xrightarrow{\ 2\ } (2 \cdot 2) \quad \text{or} \quad 2 \xrightarrow{\hspace{0.6cm}} (c_0) \xrightarrow{\ 2\ } (2 \cdot 4).$$

3.41.4 (Twisted tetrahedral)

$$2 \xrightarrow{\hspace{0.6cm}} (c_0) \xrightarrow{\ 2\ } (2 \cdot 3) \quad \text{or} \quad 2 \xrightarrow{\hspace{0.6cm}} (c_0) \xrightarrow{\ 2\ } (2 \cdot 2) \xrightarrow{\ 2\ } (2 \cdot 2).$$

3.41.5 (Twisted D_4)

$$(c_0) \xrightarrow{\ 3\ } (3 \cdot 2).$$

3.41.6 (Twisted dihedral)

$$(2 \cdot 2) \xrightarrow{\ 2\ } c_1 \xrightarrow{\hspace{0.6cm}} \cdots \xrightarrow{\hspace{0.6cm}} c_n.$$

3.41.7 (Twisted cyclic)

$$(c_0) \xrightarrow{\ 2\ } (2 \cdot c_1) \xrightarrow{\ 2\ } \cdots \xrightarrow{\ 2\ } (2 \cdot c_n).$$

Boundary with coefficients $\geq \frac{1}{2}$

Here we classify all log canonical surface pairs (S, B) where $B = \sum \beta_i B_i$ and

- either $\beta_i \in \{1, \frac{1}{2}\}$ for every i,
- or $\beta_i \geq \frac{2}{3}$ for every i.

This method essentially also classifies all pairs (S, B) such that $\beta_i \geq \frac{1}{2}$ for every value of i. Indeed, in this case we can replace $(S, \sum \beta_i B_i)$ by $(S, \sum \beta_i^* B_i)$ where $\beta_i^* = \frac{1}{2} \lfloor 2\beta_i \rfloor$ is the "half integral round down."

The main remaining issue is the following. Given a lc pair $(S, \sum \beta_i^* B_i)$, for which values of $\beta_i \geq \beta_i^*$ is $(S, \sum \beta_i B_i)$ also log canonical? In almost all instances our formulas give the answer, but some case analysis is left undone.

As before, we consider the dual graph of the log minimal resolution, and we show that the B_i can be replaced by (-2)-curves such that the resulting new dual graph is still lc. We then read off the classification of the pairs (S, B) from the classification of all lc dual graphs.

We consider the case when all the exceptional curves are geometrically integral. The general case can be obtained from this by listing all symmetric groups actions on such dual graphs as in (3.41).

The problems with small coefficients already appear in dimension 2. Let S be a smooth or log terminal surface and C an arbitrary effective curve on S. It is easy to see from the definition that $(S, \varepsilon C)$ is klt for all $0 \leq \varepsilon \ll 1$. Thus a classification of all possible lc pairs (S, B) would include a classification of all curve singularities on smooth or log terminal surfaces. This is an interesting topic in its own right but not our main concern here.

For a pair (S, C), the optimal bound on ε depends on S and C in a quite subtle way. See Kollár *et al.* (2004, section 6.5) for the case when S is smooth. The general case has not yet been worked out.

For several reasons, the most important fractional coefficients are 1 and $1 - \frac{1}{n}$ (4.4). These are all $\geq \frac{1}{2}$, exactly the value where our classification works.

3.42 (Replacing B by (-2)-curves) Let $(\Gamma, B = \frac{1}{2} \sum_j B_j)$ be an extended lc dual graph where we do not assume that the B_j are distinct.

Usually B_j intersects exactly one curve $C_{a(j)}$ and the intersection is transverse. However, B_j may intersect $\sum C_i$ at a singular point or it may be tangent to some C_i. Keeping these in mind, we construct a new dual graph (Γ^*, B^*) as follows.

For each i, write $2(C_i \cdot B) = 2m(i) + b(i)$ where $b(i) \in \{0, 1\}$. We introduce $m(i)$ new vertices $B_{i1}, \ldots, B_{i,m(i)}$ and, if $b(i) = 1$, another new vertex C_{i1}. The intersection numbers are unchanged for the old vertices, $(C_i \cdot B_{ij}) = 1$, $c_{i1} = -(C_{i1}^2) = 2$ and $(C_i \cdot C_{i1}) = 1$. No other new intersections.

Claim 3.42.1   If (Γ, B) is lc and minimal then (Γ^*, B^*) is also lc and minimal.

Proof Let $\Delta = \sum_{i=1}^{n} d_i C_i$ be the discrepancy divisor and set $\Delta^* := \Delta + \frac{1}{2} \sum C_{i1}$. Then

$$(\Delta^* \cdot C) = \begin{cases} (\Delta \cdot C_i) + \frac{1}{2} b(i) & \text{if } C = C_i, \text{ and} \\ d_i + \frac{1}{2}(C_{i1}^2) \le 1 + \frac{1}{2}(-2) = 0 & \text{if } C = C_{i1}. \end{cases}$$

Note that $(\Delta \cdot C_i) = 2 - c_i - (C_i \cdot B)$ is a half integer. If it is negative, then $(\Delta \cdot C_i) + \frac{1}{2} \le 0$ and if $(\Delta \cdot C_i) = 0$ then $b(i) = 0$ hence $(\Delta \cdot C_i) + \frac{1}{2} b(i) = 0$. Thus Δ^* has negative intersection with every curve so Γ^* is negative definite by (10.3.4). If Γ contains a (-1)-curve C_j then $(C_j \cdot B) \ge 1$. Thus $(C_j \cdot B^*) \ge 1$ and so (Γ^*, B^*) is minimal. Furthermore, by (3.29.1), $\Delta(\Gamma^*, B^*) \le \Delta^*$ and so (Γ^*, B^*) is also lc. □

3.43 (Log canonical, B contains a half integral divisor)   We distinguish two main cases.

3.43.1 (Two fork or bullet case)   The extended dual graph is a chain where at both ends we attach one of the following

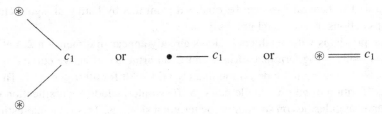

where $\circledast$ can be a (-2)-curve, a component of B with coefficient $\ge \frac{1}{2}$ or it can be missing. Here $c_1 = 1$ is allowed if all the $\circledast$ are components of B.

3.43.2 (One fork case)   In all other cases without triple intersections, the dual graph is

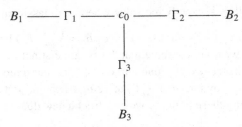

where B_i appears in B with coefficient β_i. We allow the degenerate possibility that some Γ_i is empty. In the triple intersection point case, we blow-up the triple intersection point to get the above graph with $c_0 = 1$.

Next we enumerate these possibilities by the number of components of B. We use the inequality (3.38.1) where, for every i, either $\beta_i \ge \frac{1}{2}$ or $\det(\Gamma_i) \ge 2$.

3.43.2.1 (One component)   Here $\beta_3 \geq \frac{1}{2}$ and $\det(\Gamma_1)$, $\det(\Gamma_2) \geq 2$.

(a) If $\det(\Gamma_3) = 1$ then (3.38.1) gives $\beta_3 \leq \frac{1}{\det(\Gamma_1)} + \frac{1}{\det(\Gamma_2)}$. Depending on what $(\det(\Gamma_1), \det(\Gamma_2))$ is, we get the following cases.
   (i) (2,2): We have a dihedral point as in (3.35); these are log canonical for any $\beta_3 \leq 1$.
   (ii) (2,3): Then $\beta_3 \leq \frac{5}{6}$. There are two dual graphs:

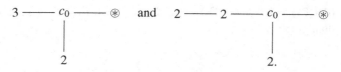

The one on the left with $c_0 = 1$ gives the embedded resolution of the planar cusp. Note that these two cases have no symmetries, hence we do not get any new such cases when the residue field is not algebraically closed.
   (iii) (2,4): Then $\beta_3 \leq \frac{3}{4}$.
   (iv) (2,5): Then $\beta_3 \leq \frac{7}{10}$.
   (v) (2,6) or (3,3): Then $\beta_3 \leq \frac{2}{3}$.
   (vi) In all other cases $\beta_3 \leq \frac{9}{14} < \frac{2}{3}$.
(b) $((\det(\Gamma_1), \det(\Gamma_2)) = (2, 2)$. For any $\det(\Gamma_3)$, we have a dihedral point as in (3.35); these are log canonical for any $\beta_3 \leq 1$.
(c) $\det(\Gamma_1) \geq 3$ and $\det(\Gamma_2)$, $\det(\Gamma_3) \geq 2$. This implies that $\beta_3 \leq \frac{2}{3}$.

3.43.2.2 (Two components)   We have $\beta_1, \beta_3 \geq \frac{1}{2}$ and $\det(\Gamma_2) \geq 2$. Again there are three cases.

(a) $(\det(\Gamma_1), \det(\Gamma_2), \det(\Gamma_3)) = (1, d_2, 1)$ with $d_2 \geq 2$. This implies that $\beta_1 + \beta_2 \leq \frac{3}{2}$ and the dual graph is

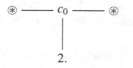

(b) $(\det(\Gamma_1), \det(\Gamma_2), \det(\Gamma_3)) = (1, d_2, d_3)$ with $d_2, d_3 \geq 2$. If $\beta_i \geq \frac{2}{3}$ then $d_2 = d_3 = 2$ and the dual graph is

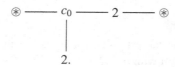

(c) If $\det(\Gamma_i) \geq 2$ for every i then $\beta_1 + \beta_3 \leq 1$ and the dual graph is

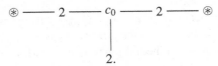

3.43.2.3 (Three components)   All $\beta_i \geq \frac{1}{2}$. There are three possibilities.

(a) $(\det(\Gamma_1), \det(\Gamma_2), \det(\Gamma_3)) = (1, 1, 1)$. Then $\beta_1 + \beta_2 + \beta_3 \leq 2$ and the dual graph is

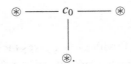

The case when $c_0 = 1$ corresponds to 3 smooth curves meeting at a smooth point of the surface.

(b) $(\det(\Gamma_1), \det(\Gamma_2), \det(\Gamma_3)) = (1, 1, d_3)$ with $d_3 \geq 2$. This replicates part of (3.43.1).

(c) $(\det(\Gamma_1), \det(\Gamma_2), \det(\Gamma_3)) = (2, 1, 2)$. Then $\beta_1 = \beta_2 = \beta_3 = \frac{1}{2}$. This gives the dual graph

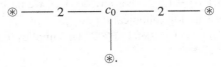

3.43.2.4 (Triple intersection)   These are the cases when B passes through the intersection point of two curves C_j. The procedure of (3.42) tells us to attach (-2)-curves to both of these C_j. This is a strong restriction, and all such cases have been enumerated above. We just need to look at the cases when the components of B are attached to a pair of intersecting curves. This can happen in only a few cases. A separate treatment of these possibilities is needed, however, since not all these cases are lc. We have encountered a similar problem in (3.27).

It is better to blow-up the intersection point and view this as a one fork case with $c_0 = 1$. We get only the following possibilities

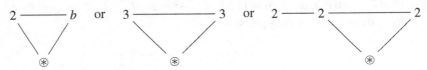

where $2 \leq b \leq 5$. The largest value for β_2 is $\frac{2}{3}$ which appears only for the configuration on the left with $b = 2$.

The following consequences of this classification will be very useful.

Corollary 3.44 *Let $(s \in S)$ be a normal, excellent surface singularity and C, C_i curve germs passing through s.*

(1) *If (S, cC) is lc for some $c > \frac{5}{6}$ then (S, C) is also lc. In particular, C is either regular or has a node at s. For $c = \frac{5}{6}$ the only exceptions are listed in (3.43.2.1aii).*
(2) *If $(S, c_1C_1 + c_2C_2)$ is lc for some $c_1 + c_2 > \frac{3}{2}$ then $(S, C_1 + C_2)$ is also lc. In particular, $C_1 + C_2$ has a node at s. For $c_1 + c_2 = \frac{3}{2}$ the only exceptions are listed in (3.43.2.2a), (3.43.1) and (3.43.2.2b).* $\square$

Corollary 3.45 *Let $(s \in S)$ be a normal, excellent surface singularity and C, C_i curve germs passing through s. Assume that $(S, C + \sum_{i=1}^{m} c_iC_i)$ is lc and $c_i \geq \frac{1}{2}$ for every i. Then we have one of the following cases.*

(1) *(S, C) not cyclic or not plt (3.35.1). Then $m = 0$ and $\mathrm{Diff}_C = [s]$.*
(2) *$m = 2$. Then $c_1 = c_2 = \frac{1}{2}$ and $\mathrm{Diff}_C(\frac{1}{2}C_1 + \frac{1}{2}C_2) = [s]$.*
(3) *(S, C) cyclic, plt and $m = 1$. Then*

$$\mathrm{Diff}_C(c_1C_1) = \left(1 - \frac{1 - c_1}{\det \Gamma}\right)[s].$$
$\square$

Note that, at least in characteristic 0, these can be proved quicker by using the index 1 cover as in (2.53) and then a simple analysis of curve singularities on a smooth surface. (See, for instance, Kollár *et al.* (2004, section 6.5).)

3.4 More examples

In Section 3.1 we used a very simple recipe to construct log canonical singularities: take a cone over a variety with lc singularities whose canonical class is either negative or trivial.

Here we introduce a slight twist as in Kollár (2011b). We start with a reducible variety Z with snc singularities and construct a smooth (or mildly singular) variety Y that contains Z as a Cartier divisor with negative normal bundle L^{-1}. A theorem of Artin (1970) says that Z can be contracted (analytically or as an algebraic space) and we get a singular point $(0 \in X)$. One can also think of $(0 \in X)$ as a deformation of the affine cone $C_a(X, L)$. If the canonical class of Z is sufficiently negative (resp. trivial), we get new examples of terminal (resp. log canonical) singularities. It turns out that reducible varieties with snc singularities can be much more involved than smooth varieties, thus we get substantially more complicated examples of terminal and log canonical singularities.

In principle this method is not surprising since every singularity $(0 \in X)$ can be thought of as a deformation of its tangent cone $(0 \in TC_0 X)$ (maybe with some embedded points), but frequently the tangent cone $TC_0 X$ has much worse singularities than X. (For example, for canonical surface singularities (3.26), the tangent cone is a smooth conic for A_1, a pair of lines for $A_n : n \geq 2$ and a double line for the D_n, E_n cases.)

First we discuss the general method and then work out in detail the following two series of examples whose proofs are given in (3.67) and (3.64).

Theorem 3.46 *For every $r \geq 4$ there exist germs of isolated terminal 4-fold singularities $(0 \in X_r)$ over $\mathbb{C}$ such that K_{X_r} is Cartier, $X_r \setminus \{0\}$ is simply connected, the class group $\mathrm{Cl}(X_r)$ is trivial and the embedding dimension of X_r is r.*

Theorem 3.47 *Let F be a connected 2-manifold without boundary. Then there are germs of isolated log canonical 3-fold singularities $(0 \in X = X(F))$ over $\mathbb{C}$ such that for any resolution $p: Y \to X$ we have $R^1 p_* \mathcal{O}_Y \simeq H^1(F, \mathbb{C})$, $\pi_1(Y) \simeq \pi_1(F)$ and $\pi_1(X \setminus \{0\})$ is an extension of $\pi_1(F)$ by a cyclic group.*

3.48 (Previous examples) By Reid (1980), every 3-fold terminal singularity is a quotient of a hypersurface double point by a cyclic group action. This description can be developed into a complete list; see Reid (1987) or Section 2.4.

For some time it has been an open question whether, in some sense, terminal singularities in higher dimensions also form an essentially bounded family. The first relevant counter examples are in an unpublished note of Brown and Reid (2010).

In the log canonical examples of Section 3.1 we had dim $R^1 p_* \mathcal{O}_Y \leq$ dim $X -$ 1 and $\pi_1(Y)$ contained a finite index Abelian subgroup.

It is also worthwhile emphasizing the difference between the log canonical and the dlt cases. Let $(0 \in X, \Delta)$ be a germ of a dlt pair and $p: Y \to X$ any resolution of singularities. Then $R^i p_* \mathcal{O}_Y = 0$ for $i > 0$ by (2.88) and $\pi_1(Y) = 1$ by Kollár (1993b) and Takayama (2003).

Singularities with prescribed exceptional divisors

Every smooth variety Z can be realized as the exceptional set of a resolution of an isolated singular point $(0 \in X)$; one can simply take X to be a cone over Z. More generally, given any scheme Z of dimension n, one can ask if there is a normal, isolated singularity $(0 \in X)$ of dimension $n + 1$ with a resolution

$$
\begin{array}{ccc}
Z \subset Y & & \text{such that } Y \setminus Z \simeq X \setminus \{0\}. \\
\downarrow \quad \downarrow & & \\
0 \in X & &
\end{array}
$$

An obvious restriction is that Z should have only hypersurface singularities, but this is not sufficient. For instance, consider $Z = (xy = 0) \subset \mathbb{P}^3$ and let Y be any smooth 3-fold containing Z. Then the normal bundle of the line $L := (x = y = 0) \subset Z$ in Y is $\mathscr{O}_{\mathbb{P}^1}(1) + \mathscr{O}_{\mathbb{P}^1}(1)$, thus $L \subset Y$ deforms in every direction. Hence Z is not contractible; it cannot even be a subscheme of the exceptional set of a resolution of an isolated 3-fold singularity. Other obstructions are discussed in (1.9).

It turns out, however, that if we allow Y to be (very mildly) singular, then one can obtain such a $(0 \in X)$. The final construction is in (3.53), but the key ingredient is the following.

Proposition 3.49 *Let P be a smooth variety over a field of characteristic 0 and $Z \subset P$ a subscheme of codimension n that is a local complete intersection. Let L be a line bundle on P such that $L(-Z)$ is generated by global sections. Let $Z \subset Y \subset P$ be the complete intersection of $(n-1)$ general sections of $L(-Z)$. Set*

$$\pi \colon B_{(-Z)}Y := \mathrm{Proj}_Y \sum_{m=0}^{\infty} \mathscr{O}_Y(mZ) \to Y.$$

(The non-standard notation indicates that we blow up not the ideal sheaf of Z but its inverse in the class group.) Then:

(1) *$\pi_*^{-1}Z$ is isomorphic to Z and it is a Cartier divisor in $B_{(-Z)}Y$.*
(2) *The exceptional set of π has codimension 2 in $B_{(-Z)}Y$.*
(3) *$\omega_{B_{(-Z)}Y} \simeq \pi^*\omega_Y \simeq \pi^*(\omega_P \otimes L^{n-1}|_Y)$ and the normal bundle of $\pi_*^{-1}Z$ in $B_{(-Z)}Y$ is $\omega_Z \otimes \omega_P^{-1} \otimes L^{1-n}$.*

Proof The claims (1–2) are étale local and, once they are established, (3) follows from the adjunction formula. Thus we may assume that $P =: A$ is affine and $Z \subset A$ is a complete intersection of codimension n defined by $f_1 = \cdots = f_n = 0$. Y is defined by a system of equations

$$\begin{array}{ccc}
h_{1,1}f_1 & + \cdots + & h_{1,n}f_n & = 0 \\
\vdots & & \vdots & \\
h_{n-1,1}f_1 & + \cdots + & h_{n-1,n}f_n & = 0.
\end{array} \qquad (3.49.4)$$

Let $H = (h_{ij})$ be the matrix of the system and H_i the submatrix obtained by removing the ith column. Let $\tilde{Y} \subset A \times \mathbb{P}^1_{st}$ be the subscheme defined by the equations

$$s \cdot f_i = (-1)^i \cdot t \cdot \det H_i \quad \text{for } i = 1, \dots, n. \qquad (3.49.5)$$

Set $Y_0 := \tilde{Y} \cap (s \neq 0)$ and $Y' \subset A \times \mathbb{P}^1_{st}$ the closure of Y_0.

We claim that Y_0 is a complete intersection, there is a natural map $\sigma \colon B_{(-Z)}Y \to Y'$ and σ is an isomorphism over Y_0. (In fact, Y' is CM by Kollár (2011b), hence σ is an isomorphism, but we do not need this.)

Let $\tilde{p}\colon \tilde{Y} \to Y$ and $p\colon Y' \hookrightarrow Y$ be the projections. Let $(\mathbf{x}, s, t) \in \tilde{Y}$ be a point such that rank $H(\mathbf{x}) = n - 1$ and $s \neq 0$. Then the $f_i(\mathbf{x})$ are the (unique up to scalar) solutions of the equations (3.49.4). Thus $\tilde{p}\colon Y_0 \to Y$ is an isomorphism outside the set $Z \cap (\text{rank } H(\mathbf{x}) < n - 1)$.

Thus the exceptional set $\mathrm{Ex}(p)$ is a $\mathbb{P}^1$-bundle over $Z \cap (\text{rank } H(\mathbf{x}) < n - 1)$. By (3.52) we conclude that $\dim \mathrm{Ex}(p) = \dim Z - 1 = \dim Y - 2$. Thus Y' has codimension n in $A \times \mathbb{P}^1_{st}$. Therefore $Y_0 \subset Y'$ is a complete intersection, hence CM.

From the equations (3.49.5) we see that $Z' := Y' \cap (t = 0) \simeq Z$ and Z' is a p-ample Cartier divisor. This gives injections $p_* \mathscr{O}_{Y'}(mZ') \hookrightarrow \mathscr{O}_Y(mZ)$ hence a morphism $\sigma\colon B_{(-Z)}Y \to Y'$ which is an isomorphism over the points where Y' is S_2.   □

If Z has hypersurface singularities, then, possibly after shrinking A, we may assume that $df_2, \ldots, df_n$ are independent at every point of Z. Thus the equations (3.49.5) can be used to eliminate $(n - 1)$ variables and we end up with one local equation for $B_{(-Z)}Y$:

$$f_1 = -t \cdot \det H_1 \tag{3.49.6}$$

where H_1 is a general $(n - 1) \times (n - 1)$ matrix in the $\mathbf{x}$-variables.

Assume next that Z has simple normal crossing singularities and let $W \subset Z$ be any stratum. Then $\det H_1$ is nonzero at a general point of W and vanishes with multiplicity 1 outside a codimension 4 subset of W by (3.52). We have thus proved the following.

Proposition 3.50 *Notation and assumption as in* (3.49). *Assume in addition that Z has snc singularities and L is sufficiently ample. Then*

(1) *At a general point of every stratum of Z, the local equation of $\pi_*^{-1}Z \subset B_{(-Z)}Y$ is $(x_1 \cdots x_m = t = 0) \subset (x_1 \cdots x_m = t)$.*
(2) *Every positive dimensional stratum of Z contains a point where the local equation of $\pi_*^{-1}Z \subset B_{(-Z)}Y$ is $(x_1 \cdots x_m = t = 0) \subset (x_1 \cdots x_m = tx_{m+1})$.*
(3) *If $\dim Z \leq 3$ then these are the only singular points along $\pi_*^{-1}Z$. In particular, $B_{(-Z)}Y$ has only terminal singularities along $\pi_*^{-1}Z$ and $B_{(-Z)}Y \setminus \pi_*^{-1}Z$ is smooth.*   □

Remark 3.51 For $\dim Z \geq 4$ a complete description of the singularities of $B_{(-Z)}Y$ is given in Kollár (2012a). One can choose local coordinates, denoted by (x_i, y_{ij}, t), such that the local equation of $B_{(-Z)}Y$ is

$$x_1 \cdots x_m = t \cdot \det \begin{pmatrix} y_{11} & \cdots & y_{1n} \\ \vdots & & \vdots \\ y_{n1} & \cdots & y_{nn} \end{pmatrix}. \tag{3.51.1}$$

From this we conclude the folowing.

(2) $B_{(-Z)}Y$ has terminal singularities along $\pi_*^{-1}Z$.
(3) If dim $Z \geq 4$ then Sing $B_{(-Z)}Y$ has codimension 5 in $B_{(-Z)}Y \setminus \pi_*^{-1}Z$ and
(4) If dim $Z \leq 10$ then $B_{(-Z)}Y \setminus \pi_*^{-1}Z$ has only cA-type singularities.

3.52 (Determinantal varieties) We have used the following properties of determinantal varieties. These are quite easy to prove directly; see Harris (1995, 12.2 and 14.16) for a more general case.

Let V be a smooth, affine variety over a field of characteristic 0, $V_i \subset V$ a finite set of smooth, affine subvarieties and $\mathcal{L} \subset \mathcal{O}_V$ a finite dimensional base point free linear system. Let $H_{n,m} = (h_{ij})$ be an $n \times m$ matrix whose entries are general elements in $\mathcal{L}$. Then for every i

(1) the singular set of $V_i \cap (\det H_{n,n} = 0)$ has codimension 4 in V_i and
(2) the set $V_i \cap (\text{rank } H_{n,m} \leq r)$ has codimension $(n-r)(m-r)$ in V_i.

Let now Z be a projective, connected, local complete intersection scheme of pure dimension n and L an ample line bundle on Z. A large multiple of L embeds Z into $P := \mathbb{P}^N$ for some N. Applying (3.49) we get $Z \simeq \pi_*^{-1}Z \subset B_{(-Z)}Y$ where the normal bundle of $\pi_*^{-1}Z$ is very negative. Thus $\pi_*^{-1}Z$ can be contracted (analytically or as an algebraic space) and we get a singular point $(0 \in X)$ Artin (1970). Its properties are summarized next.

Theorem 3.53 *Let Z be a projective, connected, local complete intersection scheme of pure dimension n and L an ample line bundle on Z. Then for $m \gg 1$ there are germs of normal singularities $(0 \in X = X(Z, L, m))$ with a partial resolution*

$$
\begin{array}{ccc}
Z \subset Y & & \text{where } Y \setminus Z \simeq X \setminus \{0\} \\
\downarrow \quad \downarrow \pi & & \\
0 \in X & &
\end{array}
$$

such that

(1) *Z is a Cartier divisor in Y,*
(2) *the normal bundle of Z in Y is $\omega_Z \otimes L^{-m}$,*
(3) *if Z is snc then*
    (a) *Y has only terminal singularities and (3.50.1–2) hold,*
    (b) *if dim $Z \leq 4$ then $(0 \in X)$ is an isolated singular point.* $\square$

Next we consider various properties of these $(0 \in X)$.

Proposition 3.54 *Let* $(0 \in X)$ *be a normal singularity with a partial resolution*

$$Z \subset Y \qquad \text{where } Y \setminus Z \simeq X \setminus \{0\}.$$
$$\downarrow \quad \downarrow \pi$$
$$0 \in X$$

Let $\hat{Y}$ denote the completion of Y along Z and $\hat{X}$ the completion of X at 0.

Assume that Z is a reduced Cartier divisor with normal bundle L^{-1} where L is ample and that either $-K_Z$ is nef or L is sufficiently ample. Then:

(1) $R^i \pi_* \mathscr{O}_Y \simeq H^i(Z, \mathscr{O}_Z)$.
(2) *The restriction* $\operatorname{Pic}(\hat{Y}) \to \operatorname{Pic}(Z)$ *is an isomorphism.*
(3) *If Y is smooth then* $\operatorname{Cl}(\hat{X}) = \operatorname{Pic}(\hat{Y})/\langle[Z_i] : i \in I\rangle$ *where the* $\{Z_i : i \in I\}$ *are the irreducible components of Z.*
(4) *Assume in addition that Z is snc, Y is smooth at the generic point of every stratum of Z and $K_Z \sim_{\mathbb{Q}} -rL$ for some $r > 0$. Then*
   (a) *K_X is $\mathbb{Q}$-Cartier and $(0 \in X)$ is lc.*
   (b) *If $r \geq 1$ then $(0 \in X)$ is canonical.*
   (c) *If $r > 1$ and $X \setminus \{0\}$ is terminal then $(0 \in X)$ is terminal.*

Proof Let $I_Z \subset \mathscr{O}_Y$ be the ideal sheaf of $Z \subset Y$. Then the completion of $R^i \pi_* \mathscr{O}_Y$ equals the inverse limit of the groups $H^i(Y, \mathscr{O}_Y/I_Z^m)$, hence it is enough to prove that the maps

$$H^i\left(Y, \mathscr{O}_Y/I_Z^{m+1}\right) \to H^i\left(Y, \mathscr{O}_Y/I_Z^m\right) \to \cdots \to H^i(Z, \mathscr{O}_Z)$$

are all isomorphisms. For each of these we have an exact sequence

$$H^i(Z, L^m) \to H^i\left(Y, \mathscr{O}_Y/I_Z^{m+1}\right) \to H^i\left(Y, \mathscr{O}_Y/I_Z^m\right) \to H^{i+1}(Z, L^m)$$

and $H^i(Z, L^m) = 0$ by Kodaira vanishing (10.42) or by Serre vanishing.

A similar argument proves (2) using the exact sequence

$$0 \to L^m \to \left(\mathscr{O}_Y/I_Z^{m+1}\right)^* \to \left(\mathscr{O}_Y/I_Z^m\right)^* \to 1.$$

which gives

$$H^1(Z, L^m) \to \operatorname{Pic}\left(\operatorname{Spec}_Y \mathscr{O}_Y/I_Z^{m+1}\right) \to \operatorname{Pic}\left(\operatorname{Spec}_Y \mathscr{O}_Y/I_Z^m\right).$$

If Y is smooth then $\operatorname{Cl}(\hat{Y}) = \operatorname{Pic}(\hat{Y})$ which implies (3).

Finally, by adjunction, $(K_Y - (r-1)Z)|_Z \sim_{\mathbb{Q}} 0$, hence $K_Y - (r-1)Z \sim_{\mathbb{Q}} 0$ by (2). Thus $K_X \sim \pi_*(K_Y - (r-1)Z)$ is $\mathbb{Q}$-Cartier, $K_Y \sim_{\mathbb{Q}} \pi^* K_X + (r-1)Z$ and the rest of (4) follows from (1.42). $\qquad\square$

Proposition 3.55 *Let Z be a connected, reduced, proper, local complete intersection variety that is a Cartier divisor in a normal analytic space Y such that Z is a deformation retract of Y. For each irreducible component $Z_i \subset Z$ let $\gamma_i \subset Y \setminus Z$ be a small loop around Z_i.*

(1) *Assume that* $\text{codim}_Z(Z \cap \text{Sing } Y) \geq 2$. *Then the* γ_i *(and their conjugates) generate* $\ker[\pi_1(Y \setminus Z) \to \pi_1(Y)]$.

(2) *Assume in addition that* Z *is a normal crossing variety and every codimension 1 stratum of* Z *contains a node of* Y *as in* (3.50.2). *Then* $\ker[\pi_1(Y \setminus Z) \to \pi_1(Y)]$ *is cyclic and is generated by any of the* γ_i.

Proof Let $\varrho^0 : (Y \setminus Z)^\sim \to Y \setminus Z$ be an étale cover that is trivial on all the γ_i. This means that ϱ^0 extends to an étale cover of $Y \setminus (Z \cap \text{Sing } Y)$. It then extends to an étale cover of Y by (3.58). This proves (1).

Next let $\varrho^0 : (Y \setminus Z)^\sim \to Y \setminus Z$ be an étale cover and $\varrho : Y^\sim \to Y$ its largest étale extension.

Assume that there is at least one point $z \in Z$ such that ϱ has a section over a neighborhood of z. We claim that then ϱ has a global section, hence it is trivial.

First we use the existence of nodes. In local coordinates we have

$$(Z \subset Y) \simeq (t = 0) \subset (x_1 x_2 = t x_3) \subset \mathbb{C}^{n+1}_{(x_1,\ldots,x_n,t)}.$$

Note that $(x_1 x_2 = t x_3) \setminus (t = 0) \sim \mathbb{C}^* \times \mathbb{C}^{n-1}$, thus, in this neighborhood, the two loops γ_1 around $(t = x_1 = 0)$ and γ_2 around $(t = x_2 = 0)$ are homotopic. Thus any étale cover of $(x_1 x_2 = t x_3) \setminus (t = 0)$ that is locally trivial along $(t = x_1 = 0)$ is trivial. Hence ϱ is an étale cover outside the codimension ≥ 2 strata of Z.

Finally, since Y has hypersurface singularities, if $W \subset Y$ has codimension ≥ 3 then any étale cover of $Y \setminus W$ extends to an étale cover of Y by (3.58). $\square$

Remark 3.56 Note that the seemingly artificial condition (3.55.2) is necessary and (3.55) may fail if Y is smooth.

As an example, let S be a resolution of a rational surface singularity ($0 \in T$) with exceptional curve $E \subset S$. Take $Z := E \times \mathbb{P}^1 \hookrightarrow S \times \mathbb{P}^1 =: Y$.

Then Y is simply connected but $\pi_1(Y \setminus Z) \simeq \pi_1(S \setminus E)$ is infinite, non-abelian as soon as T is not a quotient singularity (Brieskorn, 1968).

The presence of nodes along the double locus of Z (3.49.10), which at first seemed to be a blemish of the construction, thus turned out to be of great advantage to us.

3.57 The kernel of $[\pi_1(Y \setminus Z) \to \pi_1(Y)]$ in (3.55) can be infinite cyclic; for instance this happens if Z is an abelian variety.

However, if $\pi_1(Z) = 1$ and there is an (analytic or algebraic) morphism that contracts Z to a point then $\pi_1(Y \setminus Z)$ is finite. To see this note that since $\pi_1(Y \setminus Z)$ is abelian, it is enough to show that γ_i has finite order in $H_1(Y \setminus Z, \mathbb{Z})$. We can now repeatedly cut Y by hyperplanes until it becomes a smooth surface, hence a resolution of a normal surface singularity. The latter case is computed in Mumford (1961).

An especially simple situation is when $\omega_Z \simeq L^r$ for some r. We can choose L to be non-divisible in $\mathrm{Pic}(Z)$. Then the normal bundle of Z in Y is L^{r-m}, thus $\mathscr{O}_Y(-Z)$ is divisible by $(m - r)$ in $\mathrm{Pic}(Y)/\langle [Z_i] : i \in I \rangle$. As in Section 2.3 we can thus take a degree $(m - r)$ cyclic cover of $X \setminus \{0\}$ and replace $Z \subset Y$ with another diagram

$$
\begin{array}{ccc}
Z & \subset & \tilde{Y} \\
\downarrow & & \downarrow{\scriptstyle\pi} \\
 & 0 \in & \tilde{X}
\end{array}
$$

where $\tilde{Y} \setminus Z \simeq \tilde{X} \setminus \{0\}$, the normal bundle of Z is L^{-1} and $\tilde{X} \setminus \{0\}$ is simply connected.

The singularities of $\tilde{Y}$ are, however, a little worse than before. At double points of Z, the original local equations $(x_1 x_2 = t)$ or $(x_1 x_2 = t x_3)$ become

$$(x_1 x_2 = s^{m-r}) \quad \text{or} \quad (x_1 x_2 = s^{m-r} x_r).$$

We have used the following result due to Grothendieck (1968, X.3.4) in the algebraic case and Hamm (1971) in the topological case; see also Goresky and MacPherson (1988, section II.1.3).

Theorem 3.58 *Let Y be a scheme and $W \subset Y$ a closed subscheme. Then the natural map $\pi_1(Y \setminus W) \to \pi_1(Y)$ is an isomorphism in both of the following cases.*

(1) *(Zariski–Nagata purity theorem) Y is regular and $\mathrm{codim}_Y W \geq 2$.*
(2) *Y is a local complete intersection and $\mathrm{codim}_Y W \geq 3$.*   □

Construction of log canonical 3-fold singularities

We start by constructing certain reducible snc surfaces. Clearly the construction and its properties were known to the authors of Friedman and Morrison (1983), though only some of it is there explicitly.

Proposition 3.59 *For every connected 2-manifold without boundary F there are (many) connected algebraic surfaces $Z = Z(F)$ with snc singularities such that*

(1) *$K_Z \sim 0$ if F is orientable and $2K_Z \sim 0$ if F is not orientable,*
(2) *$h^i(Z, \mathscr{O}_Z) = h^i(F, \mathbb{C})$ and*
(3) *$\pi_1(Z) \simeq \pi_1(F)$.*

We first describe the irreducible components of these surfaces and then explain how to glue them together.

3.60 (Rational surfaces with an anticanonical cycle) Fix $m \geq 1$ and let Z be a rational surface such that $-K_Z$ is linearly equivalent to a length m cycle of rational curves $C_1, \ldots, C_m$. One can get such surfaces by starting with three lines in $\mathbb{P}^2$, then repeatedly blowing up an intersection point and adding the exceptional curve to the collection of curves $\{C_j\}$.

Let L be an ample line bundle on Z and consider the sequence

$$0 \to L(-\textstyle\sum C_j) \to L \to L|_{\sum C_j} \to 0.$$

Since $-\sum C_j \sim K_Z$, we see that $H^1(Z, L(-\sum C_j)) = 0$. Thus we have a surjection

$$H^0(Z, L) \twoheadrightarrow H^0(\textstyle\sum C_j, L|_{\sum C_j}).$$

Since $\mathrm{Pic}^0(\sum C_j) \simeq \mathbb{G}_m$, $L|_{\sum C_j}$ has a section which has exactly one zero on each C_j (not counting multiplicities). Thus there is a divisor $A_L \in |L|$ such that $A_L \cap C_j$ is a single point for every j.

We would like to choose L such that $\deg(L|_{C_j})$ is independent of j. This is not always possible, but it can be arranged as follows.

Assume that the self intersections $(C_j \cdot C_j)$ are ≤ -2 for every j. (This can be achieved by blowing up points on the C_j if necessary.) Then, for any j, the intersection form of the other curves is negative definite. Thus, as in (10.3), if H_j is any ample divisor on Z then there is an effective linear combination

$$H'_j := H_j + \textstyle\sum_{i \neq j} a_i C_i$$

such that $(H'_j \cdot C_i) = 0$ for $i \neq j$ and $(H'_j \cdot C_j) > 0$. Set

$$H := \textstyle\sum_j \frac{1}{(H'_j \cdot C_j)} \cdot H'_j.$$

H is an ample $\mathbb{Q}$-divisor that has degree 1 on every C_j. A suitable multiple gives the required line bundle L.

Let next $P \subset \mathbb{R}^2$ be a convex polygon with vertices $v_1, \ldots, v_m$ and sides $S_i = [v_i, v_{i+1}]$. We map P into the algebraic surface Z as follows.

We map the vertex v_i to the point $C_i \cap C_{i-1}$. We can think of $C_i \simeq \mathbb{CP}^1$ as a sphere with $C_i \cap C_{i+1}$ as north pole, $C_i \cap C_{i-1}$ as south pole and the unique point $A_L \cap C_i$ as a point on the equator. We map the side S_i to a semicircle in $C_i \simeq \mathbb{CP}^1 \sim S^2$ whose midpoint is $A_L \cap C_i$. Since $\pi_1(Z) = 1$, this mapping of the boundary of P extends to $\tau \colon P \to Z(\mathbb{C})$ (whose image could have self-intersections).

3.61 (Gluing rational surfaces with anticanonical cycles) Let F be a (connected) topological surface and T a triangulation of F.

Dual to T is a subdivision of F into polygons P_i such that at most three polygons meet at any point.

For each polygon P_i with sides $S_1^i, \ldots, S_{m_i}^i$ (in this cyclic order) choose a rational surface Z_i with an anticanonical cycle of rational curves $C_1^i, \ldots, C_{m_i}^i$ (in this cyclic order) and a map $P_i \hookrightarrow Z_i(\mathbb{C})$ as in (3.60).

If the sides S_j^i and S_k^l are identified on F by an isometry $\phi_{jk}^{il} : S_j^i \to S_k^l$, then there is a unique isomorphism $\Phi_{jk}^{il} : C_j^i \to C_k^l$ extending ϕ_{jk}^{il}.

These gluing data define a surface $Z = \cup_i Z_i$ (this is a very simple special case of (9.21)) and the maps τ_i glue to $\tau \colon F \to Z(\mathbb{C})$. Since only three polygons meet at any vertex, we get an snc surface. The curves H_i glue to an ample Cartier divisor H on Z.

We claim that τ induces an isomorphism $\pi_1(F) \to \pi_1(Z(\mathbb{C}))$. This is clear for the 1-skeleton where each 1-cell in F is replaced by a $\mathbb{CP}^1 \sim S^2$.

As we attach each polygon P_i to the 1-skeleton, we kill an element of the fundamental group corresponding to its boundary. On each rational surface Z_i with anticanonical cycle $\sum_j C_j^i$ we have a surjection

$$\pi_2(Z_i, \textstyle\sum_j C_j^i) \twoheadrightarrow \pi_1(\textstyle\sum_j C_j^i),$$

thus as we attach Z_i we kill the same element of the fundamental group. Thus $\pi_1(F) \simeq \pi_1(Z(\mathbb{C}))$. The statement about the homology groups is proved in (3.63).

Definition 3.62 Let X be an snc veriety over $\mathbb{C}$ with irreducible components $\{X_i : i \in I\}$. The combinatorics of X is encoded in its *dual complex* (or *dual graph*). It is a CW-complex whose vertices are labeled by the irreducible components of X and for every stratum $Z \subset \cap_{i \in J} X_i$ we attach a $(|J| - 1)$-dimensional cell. Note that for any $j \in J$ there is a unique irreducible component of $\cap_{i \in J \setminus \{j\}} X_i$ that contains Z; this specifies the attaching map. (The dual complex is usually not a simplicial complex since the intersection of two simplices could be a union of several faces but it is an unordered Δ-complex in the terminology of Hatcher (2002, p. 534).)

Lemma 3.63 (Friedman and Morrison, 1983, pp. 26–27) *Let $X = \cup_{i \in I} X_i$ be an snc scheme over $\mathbb{C}$ with dual complex T. For $J \subset I$ set $X_J := \cap_{i \in J} X_i$. Assume that $H^r(X_J, \mathcal{O}_{X_J}) = 0$ for every $r > 0$ and for every $J \subset I$. Then $H^r(X, \mathcal{O}_X) = H^r(T, \mathbb{C})$ for every r.*

Proof Fix an ordering of I. It is not hard to check that there is an exact complex

$$0 \to \mathcal{O}_X \to \textstyle\sum_i \mathcal{O}_{X_i} \to \textstyle\sum_{i<j} \mathcal{O}_{X_{ij}} \to \cdots$$

where the kth term is $\sum_{|J|=k} \mathcal{O}_{X_J}$. If $i \in J$ then the map $\mathcal{O}_{X_{J \setminus i}} \to \mathcal{O}_{X_J}$ is the natural restriction with a plus (resp. minus) sign if i is in odd (resp. even) position in the ordering of J.

Thus the cohomology of $\mathcal{O}_X$ is also the hypercohomology of the rest of the complex $\sum_i \mathcal{O}_{X_i} \to \sum_{i<j} \mathcal{O}_{X_{ij}} \to \cdots$. This is computed by a spectral sequence whose E_1 term is

$$\sum_{|J|=q} H^p(X_J, \mathcal{O}_{X_J}) \Rightarrow H^{p+q}(X, \mathcal{O}_X).$$

By assumption the only nonzero terms have $p = 0$, hence $H^q(X, \mathcal{O}_X)$ is identified with the qth homology group of

$$\cdots \to \sum_{|J|=q} H^0(X_J, \mathcal{O}_{X_J}) \to \sum_{|J|=q+1} H^0(X_J, \mathcal{O}_{X_J}) \to \cdots. \quad (3.63.1)$$

Dual to the triangulation of T there is an open cover $\{T_J : J \subset I\}$ where T_J is the union of the open stars of the simplices corresponding to X_J. With the previous sign conventions, the complex

$$0 \to \mathbb{C}_T \to \sum_i \mathbb{C}_{T_i} \to \sum_{i<j} \mathbb{C}_{T_{ij}} \to \cdots$$

is exact, hence $H^q(T, \mathbb{C}_T)$ is identified with the qth homology group of

$$\cdots \to \sum_{|J|=q} H^0(T_J, \mathbb{C}) \to \sum_{|J|=q+1} H^0(T_J, \mathbb{C}) \to \cdots. \quad (3.63.2)$$

The complexes (3.63.1) and (3.63.2) agree, proving the claim. $\square$

3.64 *Proof of (3.47)* Start with a connected 2-manifold F and construct $Z = Z(F)$ as in (3.59). Then use (3.53) to obtain $Z \subset Y$ and $(0 \in X)$. The required properties of $(0 \in X)$ are verified in (3.54–3.55), the only slight problem is that Y is not smooth. Rather, by (3.50), at finitely many points the local equation of $Z \subset Y$ is

$$(x_1 x_2 = t = 0) \subset (x_1 x_2 = t x_3) \subset \mathbb{A}^4.$$

This can be resolved without changing either $R^1 p_* \mathcal{O}$ or the fundamental group by blowing up either of the local irreducible components of Z. Globally such a resolution can be obtained by ordering the irreducible components of Z and then blowing them up in that order. $\square$

Construction of terminal 4-fold singularities

Let W be a smooth Fano 3-fold of index 2. That is, W is smooth and there is an ample line bundle L such that $-K_W \sim L^2$. Then the cone $C_a(W, L)$ is a terminal singularity by (3.1).

Such smooth Fano 3-folds have been classified; they give examples only up to embedding dimension 10.

As a generalization, one can try to look for Fano 3-folds of index 2 with normal crossing singularities. Thus the irreducible components of its normalization are normal crossing pairs (W_i, S_i) with an ample divisor H_i such that $-(K_{W_i} + S_i) \sim 2H_i$.

The first examples of such pairs that come to mind are $(\mathbb{P}^3, \mathbb{P}^2 + \mathbb{P}^2)$, $(\mathbb{P}^3, \mathbb{P}^1 \times \mathbb{P}^1)$ and $(\mathbb{Q}^3, \mathbb{P}^1 \times \mathbb{P}^1)$. These are all the examples where the underlying variety W is also Fano with Picard number 1. A general classification was started in Maeda (1986) and extended to higher dimensions in Fujita (2011). We are mainly interested in an infinite sequence of index 2 log Fano pairs $(P_r, \mathbb{P}^1 \times \mathbb{P}^1)$ where P_r is a $\mathbb{P}^2$-bundle over $\mathbb{P}^1$.

Any 2 of these examples can be glued together to get infinitely many families of Fano 3-folds of index 2 with normal crossing singularities. Then we apply (3.49) and (3.53–3.55) to conclude the proof of (3.46).

Example 3.65 For $r \geq 0$ set

$$\pi \colon P_r := \mathbb{P}_{\mathbb{P}^1}(E_r) \to \mathbb{P}^1 \quad \text{where} \quad E_r := \mathscr{O}_{\mathbb{P}^1} \oplus \mathscr{O}_{\mathbb{P}^1} \oplus \mathscr{O}_{\mathbb{P}^1}(r).$$

Write $\mathscr{O}_{P_r}(a, b) := \mathscr{O}_{P_r}(a) \otimes \pi^* \mathscr{O}_{\mathbb{P}^1}(b)$. Since $\pi_* \mathscr{O}_{P_r}(a, b) = S^a E_r \otimes \mathscr{O}_{\mathbb{P}^1}(b)$, we see that $\mathscr{O}_{P_r}(a, b)$ is ample if and only if $a > 0$ and $b > 0$. Let $S_r \subset P_r$ be the surface corresponding to the unique section of $\mathscr{O}_{P_r}(1, -r)$. Note that $K_{P_r} \sim \mathscr{O}_{P_r}(-3, r - 2)$, $S_r \simeq \mathbb{P}_{\mathbb{P}^1}(\mathscr{O}_{\mathbb{P}^1} + \mathscr{O}_{\mathbb{P}^1}) \simeq \mathbb{P}^1 \times \mathbb{P}^1$ and

$$-(K_{P_r} + S_r) \sim \mathscr{O}_{P_r}(2, 2) \quad \text{is ample on } P_r.$$

For any $a, b \geq 0$ there are natural isomorphisms

$$H^0(P_r, \mathscr{O}_{P_r}(a, b)) = H^0(\mathbb{P}^1, S^a E_r \otimes \mathscr{O}_{\mathbb{P}^1}(b))$$

and $S^a E_r \otimes \mathscr{O}_{\mathbb{P}^1}(b)$ naturally decomposes as the sum of line bundles of the form $\mathscr{O}_{\mathbb{P}^1}(cr + b)$ where $0 \leq c \leq a$. This makes it easy to compute the spaces $H^0(P_r, \mathscr{O}_{P_r}(a, b))$ and to show the following.

(1) For $a, b \geq 0$ the restriction maps

$$H^0(P_r, \mathscr{O}_{P_r}(a, b)) \to H^0(S_r, \mathscr{O}_S(a, b)) \quad \text{are surjective.}$$

(2) For $a_i, b_i \geq 0$ the multiplication maps

$$H^0(P_r, \mathscr{O}_{P_r}(a_1, b_1)) \otimes H^0(P_r, \mathscr{O}_{P_r}(a_2, b_2))$$
$$\to H^0(P_r, \mathscr{O}_{P_r}(a_1 + a_2, b_1 + b_2))$$

are surjective.

3.66 (Fano 3-folds of index 2) We consider in detail two series of examples.

1. Fix $r \geq 0$ and let Z_r be obtained from (P_r, S_r) and $(\mathbb{P}^3, S_0)$ by an isomorphism $S_r \simeq \mathbb{P}^1 \times \mathbb{P}^1 \simeq S_0$. Then ω_{Z_r} is ample and isomorphic to L_r^{-2} where L_r is ample.

One can check that $h^0(Z_r, L_r) = r + 6$ and the algebra $\sum_{m \geq 0} H^0(Z_r, L_r^m)$ is generated by $H^0(Z_r, L_r)$.

2. Fix $r, s \geq 0$ and let Z_{rs} be obtained from (P_r, S_r) and (P_s, S_s) by an isomorphism $S_r \simeq \mathbb{P}^1 \times \mathbb{P}^1 \simeq S_s$. Then $\omega_{Z_{rs}}$ is ample and isomorphic to L_{rs}^{-2}

where L_{rs} is ample. Using (3.65.1–2) we compute that $h^0(Z_{rs}, L_{rs}) = r + s + 8$ and the algebra $\sum_{m \geq 0} H^0(Z_{rs}, L_{rs}^m)$ is generated by $H^0(Z_{rs}, L_{rs})$.

For $r \in \{0, 1\}$, the (Z_r, L_r) series should give degenerations of smooth Fano 3-folds. The simplest is $r = 0$. Take $\mathbb{P}^1 \times \mathbb{P}^2$ and embed it into $\mathbb{P}^5$ by $\mathcal{O}(1, 1)$. Under this embedding, $\mathbb{P}^1 \times \mathbb{P}^1 \subset \mathbb{P}^1 \times \mathbb{P}^2$ becomes a quadric; it is thus contained in a 3-plane H^3. The union of $\mathbb{P}^1 \times \mathbb{P}^2$ and of the 3-plane gives Z_0. It is a $(2, 2)$ complete intersection.

The construction of X_r (as in (3.53)) can also be realized by taking the cone over Z_r and then deforming the (reducible) cone by taking high-order perturbations of the two quadratic defining equations.

Putting these together with (3.53) and (3.57) we get the following more precise version of (3.46).

Proposition 3.67 *Let (Z, L) be any of the pairs from (3.66). Then there are 4-dimensional, normal, isolated singularities $(0 \in X)$ with a partial resolution $\pi: (Z \subset Y) \to (0 \in X)$ such that*

(1) *Y has only canonical singularities of type cA.*
(2) *$Z \subset Y$ is a Cartier divisor and its normal bundle is L^{-1}.*
(3) *$(0 \in X)$ is terminal, K_X is Cartier, $X \setminus \{0\}$ is simply connected.*
(4) *The embedding dimension of $(0 \in X)$ is $h^0(Z, L)$.*
(5) *$\mathrm{Cl}(X_r) = 0$ for the series X_r and $\mathrm{Cl}(X_{rs}) \simeq \mathbb{Z}$ for the series X_{rs}.*

Proof Everything was already done save the computation of the class group of X. We apply (3.54.3), which needs a resolution of singularities of Y. Our set-up is simple enough that this can be done explicitly. □

3.68 (Explicit resolution) Let Y be a 4-fold and $Z \subset Y$ a Cartier divisor. Assume that Z is the union of two smooth components $Z = Z_1 \cup Z_2$ and along the intersection $S := Z_1 \cap Z_2$ in suitable local analytic coordinates $[Z_1 \cup Z_2 \subset Y]$ is isomorphic to

$$[(x_1 = s = 0) \cup (x_2 = s = 0) \subset (x_1 x_2 = s^m)] \quad \subset \mathbb{A}^5 \quad \text{or}$$
$$[(x_1 = s = 0) \cup (x_2 = s = 0) \subset (x_1 x_2 = s^m x_3)] \quad \subset \mathbb{A}^5.$$

Let $C \subset S$ be the curve defined locally by $(x_1 = x_2 = x_3 = s = 0)$.

We can resolve the singularities by iterating the following steps.

(1) If $m \geq 3$, we blow-up S. We get two exceptional divisors, both are $\mathbb{P}^1$-bundles over S with two disjoint sections. In the above local coordinates the equation changes to $(x_1 x_2 = s^{m-2})$ or $(x_1 x_2 = s^{m-2} x_3)$.
(2) If $m = 2$, we blow up S; the resulting 4-fold is smooth. We get one exceptional divisor, which is a conic bundle over S with two disjoint sections. It is isomorphic to a $\mathbb{P}^1$-bundle over S blown up along C (contained in one of the sections).

(3) If $m = 1$, we blow-up the component, call it N, that intersects Z_1. (This component is Z_2 if and only if there are no previous blow-ups.) The birational transform of Z_1 is isomorphic to Z_1 and the birational transform of N is isomorphic to N blown up along C.

Thus at the end we have a chain of smooth 3-folds

$$E_0 := Z_1, E_1, \ldots, E_{m-1}, E_m := Z_2$$

such that

(4) the only intersections are $S_i := E_i \cap E_{i+1} \simeq S$ for $0 \le i < m$.
(5) $E_2, \ldots, E_{m-1}$ are $\mathbb{P}^1$-bundles over S with two disjoint sections.
(6) E_1 is a $\mathbb{P}^1$-bundle over S blown up along C if $m \ge 2$ and Z_2 blown up along C if $m = 1$.

By taking the cohomology of the exact sequence

$$0 \to \mathbb{Z}_{\cup_i E_i} \to \sum_{i=0}^{m} \mathbb{Z}_{E_i} \to \sum_{i=0}^{m-1} \mathbb{Z}_{S_i} \to 0$$

we get an exact sequence

$$\sum_{i=0}^{m-1} H^1(S_i, \mathbb{Z}) \to H^2(\cup_i E_i, \mathbb{Z}) \to \sum_{i=0}^{m} H^2(E_i, \mathbb{Z}) \to \sum_{i=0}^{m-1} H^2(S_i, \mathbb{Z}).$$

Assume now that

$$H^1(S, \mathbb{Z}) = 0 \quad \text{and} \quad H^2(Z_2, \mathbb{Z}) \to H^2(S, \mathbb{Z}) \quad \text{is surjective.} \qquad (3.68.7)$$

Setting $h^i(\ ) := \operatorname{rank} H^i(\ , \mathbb{Z})$ we get that

$$h^2(\cup_i E_i) = \sum_{i=0}^{m} h^2(E_i) - \sum_{i=0}^{m-1} h^2(S_i)$$
$$= h^2(Z_1) + h^2(Z_2) - h^2(S) + h^2(C) + (m - 1).$$

In (3.66) $S \simeq \mathbb{P}^1 \times \mathbb{P}^1$ and C is connected, thus the formula becomes

$$h^2(\cup_i E_i) = h^2(Z_1) + h^2(Z_2) + (m - 2).$$

Therefore, by (3.54.3), the class group of X, obtained by contracting $Z \subset Y$ to a point, satisfies

$$\operatorname{rank} \operatorname{Cl}(X) \le h^2(Z_1) + h^2(Z_2) + (m - 2) - (m + 1) = h^2(Z_1) + h^2(Z_2) - 3.$$
$$(3.68.8)$$

Thus we see that for the series $(0 \in X_r)$ we get $\operatorname{Cl}(X_r) = 0$ but for the series $(0 \in X_{rs})$ we get $\operatorname{Cl}(X_{rs}) \simeq \mathbb{Z}$. $\square$

Remark 3.69 The series X_{rs} can also be constructed as follows. Set

$$Y_{rs} := \mathbb{P}_{\mathbb{P}^1}(\mathcal{O}_{\mathbb{P}^1} \oplus \mathcal{O}_{\mathbb{P}^1} \oplus \mathcal{O}_{\mathbb{P}^1}(r) \oplus \mathcal{O}_{\mathbb{P}^1}(s)).$$

This visibly contains both P_r, P_s as divisors and $K_{Y_{rs}} + P_r + P_s \sim \mathcal{O}_{Y_{rs}}(-2, -2)$. We can now take the affine cone $C_a(Y_{rs}, \mathcal{O}_{Y_{rs}}(1, 1))$ and inside it the cone $C_a(P_r + P_s)$. Perturbing the equation of the cone $C_a(P_r + P_s)$ as in Kollár and Mori (1998, 2.43) we get our varieties X_{rs}.

Since $\mathrm{Cl}(X_{rs}) = \mathbb{Z}$, the singularities X_{rs} have a small modification; now we can see this explicitly. The cone $C_a(Y_{rs})$ has a small resolution determined by the pencil of the 4-planes that are the cones over the fibers of $Y_{rs} \to \mathbb{P}^1$.

By contrast, we believe that one cannot realize the series Z_r as hypersurfaces in a cone over a smooth variety.

Remark 3.70 Although these examples show that terminal 4-fold singularities do not form an "essentially bounded family" in the most naive sense, it should not be considered a final answer.

There are several ways to "simplify" a given terminal singularity $(0 \in X)$.

First, if $X \setminus \{0\}$ has finite fundamental group, one can pass to the universal cover. Since finite group actions on a given singularity are frequently not too hard to understand, it makes sense to concentrate on those singularities $(0 \in X)$ for which $X \setminus \{0\}$ is simply connected.

Another simplification is given by the $\mathbb{Q}$-factorial model $X^{\mathrm{qf}} \to X$ (1.37). Note that $X^{\mathrm{qf}} = X$ if and only if $\mathrm{Cl}(X) = 0$.

This suggests that the most basic terminal singularities $(0 \in X)$ of dimension n are those that satisfy both $\pi_1(X \setminus \{0\}) = 1$ and $\mathrm{Cl}(X) = 0$.

Finally, one might argue that any collection of examples constructed in a simple uniform way forms an "essentially bounded family." The complete list of snc Fano 3-folds of index 2 (and also many higher dimensional cases) is in Fujita (2011). Thus these examples do form an "essentially bounded family."

3.5 Perturbations and deformations

Starting with an lc pair (X, Δ), we would like to understand ways of changing X or Δ, and hopefully improving the singularities of the pair in the process.

First we look at perturbations of Δ. By (2.5), we need to decrease Δ if we want to increase the discrepancy. This is not possible if $\Delta = 0$ to start with, but for most lc pairs (X, Δ) such that $(X, 0)$ is klt outside $\mathrm{Supp}\,\Delta$, there is a divisor Δ' such that (X, Δ') is klt. (If Δ is $\mathbb{Q}$-Cartier then $\Delta' := (1 - \varepsilon)\Delta$ works.) However, the examples (3.72) show that this is not always the case.

Then we ask if every deformation $\{X_t : t \in \mathbb{C}\}$ of $X_0 = X$ is again lc. If the boundary Δ moves in a "sufficiently nice" family $\{\Delta_t : t \in \mathbb{C}\}$, then this is indeed the case by (4.10). Here we are interested in cases where Δ does not move with X_t. The examples (3.75) show that usually one cannot choose any Δ_t that makes (X_t, Δ_t) lc, not even if we do not assume any relationship between Δ and the Δ_t.

In fact, we get flat families of 3-fold singularities $\{Y_t : t \in \mathbb{C}\}$ such that Y_0 is log terminal but, for $t \neq 0$, there is no effective $\mathbb{Q}$-divisor Δ_t on Y_t such that (Y_t, Δ_t) is lc.

Perturbing the boundary

We start with a positive result.

3.71 (Perturbation method (cf. Kollár and Mori (1998, 2.43)))   Let X be an affine variety over a field of characteristic 0 and D an effective Weil divisor. Then the linear system $|D|$ is infinite dimensional and its base locus is exactly the set of points where D is not Cartier.

Choose a finite dimensional subsystem $D \in |L| \subset |D|$. Let $f: X' \to X$ be a log resolution such that $f^*|L| = B + |L'|$ where $|L'|$ is base point free and the base locus B is snc.

Pick any $m > 0$ and general members $D'_1, \ldots, D'_m \in |L'|$. Then $mD \sim f(D'_1) + \cdots + f(D'_m)$.

Assume that we have a pair (X, Δ) and write $\Delta = \Delta' + \Delta''$ as the sum of two effective $\mathbb{Q}$-divisors. Choose r such that $r\Delta''$ is a $\mathbb{Z}$-divisor. Applying the above method to $D := r\Delta''$, we get a divisor

$$\Delta''_{rm} := \tfrac{1}{rm} f(D'_1) + \cdots + \tfrac{1}{rm} f(D'_m) \sim_{\mathbb{Q}} \Delta''.$$

As we go from (X, Δ) to $(X, \Delta' + \Delta''_{rm})$, the discrepancies of the f-exceptional divisors can only increase since $rm\Delta''$ is a special member of the linear system $|L'|$, thus its support may contain f-exceptional divisors.

Hence, if (X, Δ) is lc or klt then so is $(X, \Delta' + \Delta''_{rm})$. If (X, Δ) is dlt and $\lfloor \Delta' \rfloor = 0$ then choose $|L|$ such that its base locus does not contain any stratum of $\mathrm{snc}(X, \Delta)$. Then $(X, \Delta' + \Delta''_{rm})$ is klt.

If (X, Δ) is quasi-projective, we can apply the same method to $(X, (1 - \varepsilon)\Delta + \varepsilon\Delta + H)$ where H is a general very ample divisor and $0 < \varepsilon \ll 1$.

This way we can move the support of the boundary away from any smooth point and make the coefficients of the boundary as small as we wish. However, as the next examples show, it is not always possible to eliminate lc centers by perturbations.

Example 3.72 (No klt perturbations)   Let (X, Δ_X) be an affine cone over any of the following surfaces (S, Δ_S). Then (X, Δ_X) is lc but there is no effective $\mathbb{Q}$-divisor Δ' such that (X, Δ') is klt.

(1) Let $E_3 \subset \mathbb{P}^2$ be a smooth cubic and $Q \subset E_3$ a set of at least 10 points. Set $S := B_Q \mathbb{P}^2$ and let $\Delta_S = E$ be the birational transform of E_3. Then $K_S + \Delta_S \sim 0$ and $E^2 < 0$.

(2) Let $C_6 \subset \mathbb{P}^2$ be a rational sextic with 10 nodes and $Q \subset C_6$ the set of nodes. Set $S := B_Q \mathbb{P}^2$ and let $\Delta_S := \frac{1}{2}E$ where E is the birational transform of C_6. Then $K_S + \frac{1}{2}\Delta_S \sim 0$ and $E^2 = -4$.

We prove these claims in two steps. First we check a special property of S (3.73) and then we see in (3.74) that the latter implies that there are no klt perturbations.

Proposition 3.73 *Let Y be a smooth, proper variety and $E \subset Y$ an irreducible $\mathbb{Q}$-Cartier divisor such that $K_Y + cE \sim_\mathbb{Q} 0$ for some $0 < c \leq 1$ and $E|_E$ is not $\mathbb{Q}$-linearly equivalent to an effective divisor.*

Let Δ be an effective $\mathbb{Q}$-divisor on Y such that $K_Y + \Delta \sim_\mathbb{Q} 0$. Then $\Delta = cE$. In particular, $-K_Y$ is not $\mathbb{Q}$-linearly equivalent to an effective divisor plus an ample divisor.

Proof Write $\Delta = (c - \gamma)E + \Delta'$ where Δ' is effective, its support does not contain E and $\gamma \in \mathbb{Q}$. Then $\Delta' \sim_\mathbb{Q} \gamma E$ which shows that $\gamma \geq 0$. Restricting to E we get $\gamma(E|_E) \sim_\mathbb{Q} \Delta'|_E$. This is a contradiction, unless $\gamma = 0$ and $\Delta = cE$. $\qquad\square$

Lemma 3.74 *Let L be an ample line bundle on a normal variety Y and $C_a(Y, L)$ the affine cone over Y. Let Δ be an effective $\mathbb{Q}$-divisor on Y.*

(1) *If $(C_a(Y, L), \Delta)$ is lc then $-K_Y$ is $\mathbb{Q}$-linearly equivalent to an effective divisor.*

(2) *If $(C_a(Y, L), \Delta)$ is klt then $-K_Y$ is $\mathbb{Q}$-linearly equivalent to an effective divisor plus an ample divisor.*

Proof Blowing up the vertex gives a partial resolution $\pi \colon BC_a(Y, L) \to C_a(Y, L)$ (3.8) whose exceptional divisor is $E \simeq Y$ with normal bundle L^{-1}. We can write

$$K_{BC_a(Y,L)} + (1 - c)E + \Delta' \sim_\mathbb{Q} \pi^*(K_{C_a(Y,L)} + \Delta) \sim_\mathbb{Q} 0$$

where $c = 1 + a(E, C_a(Y, L), \Delta) \geq 0$ if $(C_a(Y, L), \Delta)$ is lc and $c > 0$ if $(C_a(Y, L), \Delta)$ is klt. Thus $K_{BC_a(Y,L)} + E \sim_\mathbb{Q} cE - \Delta'$ and, restricting to E we obtain that

$$K_Y = K_E \sim_\mathbb{Q} -cL - (\Delta'|_E). \qquad\square$$

Non-lc deformations

Example 3.75 1. Take cones over the two families $\{X_{5,t}\}$ and $\{X_{6,t}\}$ from (3.76.1–2).

We get flat families of 3-fold singularities $\{C_a(X_{n,t}): t \in \mathbb{C}\}$ such that $C_a(X_{n,0})$ is log terminal for $n = 5$ and log canonical for $n = 6$ but, for $t \neq 0$, there is no effective $\mathbb{Q}$-divisor Δ_t on $C_a(X_{n,t})$ such that $(C_a(X_{n,t}), \Delta_t)$ is lc. This follows from (3.74.1) and (3.76.1–2).

2. Let S be a smooth, projective surface and $E \sim -K_S$ a smooth, irreducible curve such that $(E^2) < 0$. (For instance, (S, E) can be as in (3.72.1).) Let $p \in S$ be a point, $S_p := B_p S$ the blow-up with projection $\pi_p: S_p \to S$ and $E_p \subset S_p$ the birational transform of E. If $p \in E$ then (S_p, E_p) is still lc and $K_{S_p} + E_p \sim 0$.

On the other hand, if p is not on E, we claim that there is no effective $\mathbb{Q}$-divisor Δ_p on S_p such that $K_{S_p} + \Delta_p \sim_{\mathbb{Q}} 0$. Indeed, this would imply that $K_S + (\pi_p)_* \Delta_p \sim_{\mathbb{Q}} 0$ hence $(\pi_p)_* \Delta_p = E$ by (3.73). But $p \notin E$, a contradiction.

By taking cones over these surfaces, we get a flat family of 3-fold singularities $C_a(S_p, H_p)$ such that

$$(C_a(S_p, H_p), C_a(E_p, H_p|_{E_p})) \quad \text{is lc if } p \in E,$$

but if $p \notin E$, then $(C_a(S_p, H_p), \Delta_p)$ is not lc, no matter what Δ_p is.

Example 3.76 For $n \geq 4$ we construct projective, rational surfaces with quotient singularities $X_n = X_{n,0}$ whose minimal resolution is the blow-up of $\mathbb{P}^2$ in $\binom{n}{2}$ points in special position. Then we describe flat deformations $\{X_{n,t} : t \in \mathbb{C}\}$ where, for $t \neq 0$, $X_{n,t}$ is the blow-up of $\mathbb{P}^2$ in $\binom{n}{2}$ points in general position.

These give several interesting examples.

(1) ($n = 5$ case): $X_5 = X_{5,0}$ is a del Pezzo surface with quotient singularities, but $X_{5,t}$ is not del Pezzo for $t \neq 0$. Even more, for $t \neq 0$ there is no effective $\mathbb{Q}$-divisor Δ_t on $X_{5,t}$ such that $-(K_{X_{5,t}} + \Delta_t)$ is nef.

(2) ($n = 6$ case): Here $2K_{X_{6,0}} \sim 0$ but $K_{X_{6,t}} \not\sim 0$ for $t \neq 0$. Even more, there is no effective $\mathbb{Q}$-divisor Δ_t on $X_{6,t}$ such that $-(K_{X_{6,t}} + \Delta_t)$ is nef.

(3) ($n \geq 7$ case): Here $K_{X_{n,0}}$ is ample but $X_{n,t}$ is a smooth rational surface for $t \neq 0$. Thus $K_{X_{n,t}}$ is not even pseudo-effective.

(4) ($n = 4$ case): The deformation is not very interesting. $X_{4,0}$ is a cubic surface with four singular points and $X_{4,t}$ is a smooth cubic surface.

We start with the surfaces $X_{n,0}$.

Construction 3.76.1 For $n \geq 3$ let $L_n \subset \mathbb{P}^2$ be the union of n general lines. Let $P \subset \mathbb{P}^2$ be the $\binom{n}{2}$ intersection points and $p: B_P \mathbb{P}^2 \to \mathbb{P}^2$ the blow-up. Let E_n denote the sum of all exceptional curves and $L'_n \subset B_P \mathbb{P}^2$ the birational transform of L_n. We check that $p^* \mathcal{O}_{\mathbb{P}^2}(n-1)(-E_n)$ is generated by global sections and

$$H^i(B_P\mathbb{P}^2, (p^* \mathcal{O}_{\mathbb{P}^2}(n-1)(-E_n))^{\otimes m}) = 0 \quad \text{for } i, m > 0.$$

The Stein factorization of the map given by the global sections of $p^* \mathcal{O}_{\mathbb{P}^2}(n-1)(-E_n)$ gives a morphism $q: B_P\mathbb{P}^2 \to X_n$. Furthermore,

a) X_n has n cyclic quotient singularities of the form $\mathbb{A}^2/\frac{1}{n-2}(1,1)$ (3.19.2),
b) $X_3 \simeq \mathbb{P}^2$ and X_4 is a cubic surface in $\mathbb{P}^3$,
c) X_5 is a log Del Pezzo surface,
d) $2K_{X_6} \sim 0$ and
e) X_n has ample canonical class for $n \geq 7$.

Proof Note that $L'_n \subset B_P\mathbb{P}^2$ is a union of n disjoint smooth rational curves, each with self-intersection $2 - n$ and $p^* \mathcal{O}_{\mathbb{P}^2}(n-1)(-E_n)$ has zero intersection number with L'_n.

The sum of any $(n-1)$ lines in L_n pulls back to a divisor in $|p^* \mathcal{O}_{\mathbb{P}^2}(n-1)|$ which contains E_n, giving a divisor in $|p^* \mathcal{O}_{\mathbb{P}^2}(n-1)(-E_n)|$. The intersection of all these divisors is empty, thus the linear system $|p^* \mathcal{O}_{\mathbb{P}^2}(n-1)(-E_n)|$ is base point free. Thus (the Stein factorization of) the map given by $|p^* \mathcal{O}_{\mathbb{P}^2}(n-1)(-E_n)|$ gives a morphism $q: B_P\mathbb{P}^2 \to X_n$ such that $p^* \mathcal{O}_{\mathbb{P}^2}(n-1)(-E_n) \simeq q^* M_n$ for some ample line bundle M_n on X_n.

The sum of all these sections of $p^* \mathcal{O}_{\mathbb{P}^2}(n-1)(-E_n)$ is supported on $E_n + L'_n = p^{-1}(L_n)$. Thus if $C \subset B_P\mathbb{P}^2$ is any irreducible curve different from the lines in L'_n then $\deg_C p^* \mathcal{O}_{\mathbb{P}^2}(n-1)(-E_n) > 0$. Thus q is birational, and the lines L'_n are the only q-exceptional curves. For $n \geq 4$ we get n singular points which are cyclic quotient singularities of the form $\mathbb{A}^2/\frac{1}{n-2}(1,1)$ (3.32). Using the theorem of formal functions, we see that $R^i q_* \mathcal{O}_{B_P\mathbb{P}^2} = 0$ for $i > 0$.

Since $K_{\mathbb{P}^2} \sim_{\mathbb{Q}} -\frac{3}{n} L_n$, we can write

$$K_{B_P\mathbb{P}^2} \sim_{\mathbb{Q}} -\frac{3}{n} L'_n + \left(1 - \frac{6}{n}\right) E_n.$$

Thus

$$K_{X_n} = q_* K_{B_P\mathbb{P}^2} \sim_{\mathbb{Q}} \left(1 - \frac{6}{n}\right) q_* E_n.$$

We can further write

$$q^* K_{X_n} \sim_{\mathbb{Q}} \left(1 - \frac{6}{n}\right) \left[E_n + \frac{n-1}{n-2} L'_n\right]$$
$$\sim_{\mathbb{Q}} \frac{n-6}{n-2} \cdot [p^* \mathcal{O}_{\mathbb{P}^2}(n-1)(-E_n)] \sim_{\mathbb{Q}} \frac{n-6}{n-2} \cdot q^* M_n.$$

This shows that K_{X_n} is ample for $n \geq 7$, numerically trivial for $n = 6$ and anti-ample for $n < 6$.

As we saw, $R^i q_* \mathcal{O}_{B_P\mathbb{P}^2} = 0$ for $i > 0$, thus

$$H^i(B_P\mathbb{P}^2, (p_0^* \mathcal{O}_{\mathbb{P}^2}(n-1)(-E_{Q_0}))^{\otimes m}) = H^i(X_n, M_n^{\otimes m}).$$

Since $M_n \sim_{\mathbb{Q}} K_{X_n} + \frac{4}{n-2} M_n$ the Kodaira vanishing theorem (10.32) implies that $H^i(X_n, M_n^{\otimes m}) = 0$ for $i, m > 0$. □

Next we study the blow-up of $\mathbb{P}^2$ at points in general position $r: B_R\mathbb{P}^2 \to \mathbb{P}^2$. We need to understand ample line bundles on these surfaces. The conjecture of Nagata (1965) says that, for $|R| \geq 9$, the line bundle $r^*\mathcal{O}_{\mathbb{P}^2}(m_0)(-m_1 E_R)$ is ample if and only if its self-intersection is positive; that is, when $m_0^2 - m_1^2|R| > 0$. This is still open but we need only some special cases.

Claim 3.76.2   Let $R \subset \mathbb{P}^2$ be points in general position and $r: B_R\mathbb{P}^2 \to \mathbb{P}^2$ the blow-up with exceptional curve E_R. Then

a) $r^*\mathcal{O}_{\mathbb{P}^2}(m)(-E_R)$ is ample if $|R| < m^2$,
b) $r^*\mathcal{O}_{\mathbb{P}^2}(30)(-9E_R)$ is ample if $|R| \leq 10$.

Proof   In order to prove (a) it is enough to find one point set $R \subset \mathbb{P}^2$ such that $r^*\mathcal{O}_{\mathbb{P}^2}(m)(-E_R)$ is ample. Pick general degree m curves C_1, C_2 such that every member of the pencil $|C_1, C_2|$ is irreducible and choose $R \subset C_1 \cap C_2$.
We can then write

$$r^*\mathcal{O}_{\mathbb{P}^2}(m)(-E_R) \sim_{\mathbb{Q}} \tfrac{1}{2}(C_1' + C_2'),$$

where C_i' denotes the birational transform of C_i. Note that the C_i' lie in a pencil with $m^2 - |R| > 0$ base points all of whose members are irreducible. Using the Nakai–Moishezon criterion of ampleness we conclude that $r^*\mathcal{O}_{\mathbb{P}^2}(m)(-E_R)$ is ample.

To see (b) note that for any point $p_i \in R$ there is a cubic C_i passing through the other nine points. The sum of their birational transforms gives a divisor in $|r^*\mathcal{O}_{\mathbb{P}^2}(30)(-9E_R)|$. The self-intersection is 10. If the points are very general, then there is no effective curve D in $\mathbb{P}^2$ such that $\mathrm{Supp}(C \cap D) \subset R$. Using the Nakai–Moishezon criterion of ampleness we conclude that $r^*\mathcal{O}_{\mathbb{P}^2}(30)(-9E_R)$ is ample. □

Corollary 3.76.3   Let $R \subset \mathbb{P}^2$ be ≥ 10 points in general position and $r: B_R\mathbb{P}^2 \to \mathbb{P}^2$ the blow-up. Then minus the canonical class of $B_R\mathbb{P}^2$ is not pseudo-effective.

Proof   It is enough to prove this for the blow-up of 10 points. We compute that $K_{B_R\mathbb{P}^2} \cdot (r^*\mathcal{O}_{\mathbb{P}^2}(30)(-9E_R)) = 0$. By Kleiman's ampleness criterion this implies that $K_{B_R\mathbb{P}^2}$ is not pseudo-effective. □

Next we construct a deformation of $X_{n,0}$ by moving the set of intersection points $P \subset \mathbb{P}^2$ into general position.
Fix $n \geq 4$ and let $Q_t \subset \mathbb{P}^2$ be a flat family of $\binom{n}{2}$ points such that $Q_0 = P$ as in (3.76.5) and Q_t is general for $t \neq 0$. Note that $\binom{n}{2} < (n-1)^2$ holds thus

$p_t^* \mathscr{O}_{\mathbb{P}^2}(n-1)(-E_{Q_t})$ is ample for $t \neq 0$ by (3.76.6a) where $p_t \colon B_{Q_t}\mathbb{P}^2 \to \mathbb{P}^2$ denotes the blow-up.

We saw that $(p_0^* \mathscr{O}_{\mathbb{P}^2}(n-1)(-E_{Q_0}))^{\otimes m}$ is generated by global sections and has no higher cohomologies. Thus the same holds for t in an open neighborhood of $0 \in \mathbb{C}$. Therefore

$$h^0(B_{Q_t}\mathbb{P}^2, (p_t^* \mathscr{O}_{\mathbb{P}^2}(n-1)(-E_{Q_t}))^{\otimes m})$$

is independent of t for $m \gg 1$. Using Hartshorne (1977, III.9.9) we obtain:

Claim 3.76.4   For $n \geq 4$ there is a flat family of surfaces $\{X_{n,t} : t \in \mathbb{C}\}$ such that $X_{n,0} \simeq X_n$ and $X_{n,t} \simeq B_{Q_t}\mathbb{P}^2$. □

4

Adjunction and residues

This chapter starts the main part of the book where we aim to understand the difference between klt pairs and general lc pairs.

Once the machinery of Mori's program was developed, it became clear that while klt pairs behave nearly as well as smooth varieties, many of the theorems and methods fail for lc pairs. Thus a separate, detailed study of lc pairs became necessary.

Furthermore, lc pairs turn out to be very useful in applications. Time and again, the solution to a question involving a smooth variety X relied on introducing an auxiliary $\mathbb{Q}$-divisor Δ such that (X, Δ) is lc but not klt and then using the resulting exceptional divisor with discrepancy -1 to solve the original problem; see (8.29–8.34) for such examples.

The key advantage of lc pairs is that divisors with discrepancy -1 allow induction on the dimension. This relies on what was classically called *adjunction* or the *Poincaré residue map*. The study of these forms the technical core of the book and a good understanding of the origins and aims of these methods is very helpful.

Let S be a smooth hypersurface in a smooth variety X. The classical adjunction theorem computes the canonical class of S by the formula $K_S \sim (K_X + S)|_S$. It is even better to view this linear equivalence as a computation of residues, which gives a *canonical* isomorphism

$$\mathcal{R} : \omega_X(S)|_S \simeq \omega_S.$$

We call it the *Poincaré residue isomorphism*. It has been long understood that, if X, S are singular, one should add a correction term, called the *different*, to these formulas. Then the classical adjunction, extended to pairs $(X, S + \Delta)$, becomes

$$K_S + \mathrm{Diff}_S \Delta \sim_{\mathbb{Q}} (K_X + S + \Delta)|_S.$$

For a pair (X, Δ_X) a divisor with discrepancy -1 may appear only on a suitable modification $g: (Y, \Delta_Y) \to (X, \Delta_X)$. Thus adjunction naturally takes place for a pair $S \subset Y$ where $a(S, X, \Delta) = -1$, but we would need to understand it as a relationship for the pair $Z := \operatorname{center}_X S \subset X$. These are called the *log canonical centers* of (X, Δ_X). There are only finitely many log canonical centers on any given lc pair and the study of these lc centers holds the key to understanding lc pairs.

There are three intertwined topics running through the theory.

- Understanding adjunction for divisors.
- Studying the geometry of log canonical centers.
- Relating adjunction on exceptional divisors to properties of the log canonical centers.

The first of these topics is treated in Section 4.1. It was realized by Shokurov (1992) that there is a very close relationship between the singularities of the pairs $(X, S + \Delta)$ and $(S, \operatorname{Diff}_S \Delta)$. For example, Kawakita (2007) proved that $(X, S + \Delta)$ is lc in a neighborhood of S if and only if $(S, \operatorname{Diff}_S \Delta)$ is lc. The optimal version (4.9) is an important tool that allows induction on the dimension in many cases.

We show in Section 4.2 that, with minor modifications, the divisorial case is sufficient to deal with all log canonical centers in the dlt case.

The case when (X, Δ) is not dlt is much harder; it occupies the rest of the chapter. After an overview in Section 4.3, we discuss in detail the two new concepts that help us understand adjunction and log canonical centers. These are *crepant log structures* and *sources* of log canonical centers.

Recall, (1.36), that for any lc pair (X, Δ) there is a proper birational morphism $g: (X^{\mathrm{dlt}}, \Delta^{\mathrm{dlt}}) \to (X, \Delta)$ such that $(X^{\mathrm{dlt}}, \Delta^{\mathrm{dlt}})$ is dlt and

$$K_{X^{\mathrm{dlt}}} + \Delta^{\mathrm{dlt}} \sim_{\mathbb{Q}} g^*(K_X + \Delta).$$

Then we start with a log canonical center $Z \subset X$, find a divisor $S \subset X^{\mathrm{dlt}}$ that dominates Z, use adjunction for $S \subset X^{\mathrm{dlt}}$ and finally employ the resulting map $g|_S: S \to Z$ to relate adjunction on $S \subset X^{\mathrm{dlt}}$ to properties of $Z \subset X$.

Axiomatizing the properties of (the Stein factorization of) the maps $g|_S: S \to Z$ leads to the concept of crepant log structures. These were introduced in Kollár and Kovács (2010), building on earlier work of Ambro (2003). They are studied in Section 4.4.

The approach above can be considerably improved by replacing S with a minimal lc center of $(X^{\mathrm{dlt}}, \Delta^{\mathrm{dlt}})$ that dominates Z. Usually there are many different choices, but they turn out to be birational to each other. This leads to the concepts of *sources* and *springs* of log canonical centers, introduced in Kollár (2011d). These are discussed in Section 4.5.

Assumptions   In Section 4.1 we work with arbitrary schemes. Starting with Section 4.2 we work with schemes (or algebraic spaces) that are of finite type over a base scheme S that is essentially of finite type over a field of characteristic 0. One would like to allow more general base schemes, for instance spectra of complete local rings, but there are some technical difficulties, primarily with the vanishing theorems of Section 10.3.

4.1 Adjunction for divisors

Adjunction is a classical method that relates the canonical class of a variety and the canonical class of a divisor. It is a very useful tool that allows lifting information from divisors to the ambient variety and facilitates induction on the dimension.

Definition 4.1 (Poincaré residue map I)   Let X be a smooth variety (or a regular scheme as in (1.5)) and $S \subset X$ a smooth (or regular) divisor. Let

$$\mathcal{R}_S : \omega_X(S) \to \omega_S$$

(or $\mathcal{R}_{X \to S}$ if the choice of X is not clear) denote the *Poincaré residue map*. It can be defined in two equivalent ways.

(Local definition.) If X is a smooth variety then at a point $s \in S \subset X$ choose local coordinates $x_1, \ldots, x_n$ such that $S = (x_1 = 0)$. Then

$$\frac{dx_1}{x_1} \wedge dx_2 \wedge \cdots \wedge dx_n$$

is a local generator of $\omega_X(S) = \Omega_X^n(\log S)$. Set

$$\mathcal{R}_S \left(f \cdot \frac{dx_1}{x_1} \wedge dx_2 \wedge \cdots \wedge dx_n \right) := f|_S \cdot dx_2 \wedge \cdots \wedge dx_n. \quad (4.1.1)$$

It is easy to check that $\mathcal{R}_S$ is independent of the local coordinates.

(Global definition.) If X is a regular scheme, we view ω_X and ω_S as dualizing sheaves as in (1.6). Then $\omega_S = \mathcal{E}xt_X^1(\mathcal{O}_S, \omega_X)$. By applying $\mathcal{H}om(\ , \omega_X)$ to the exact sequence

$$0 \to \mathcal{O}_X(-S) \to \mathcal{O}_X \to \mathcal{O}_S \to 0$$

we get a long exact sequence

$$\mathcal{H}om_X(\mathcal{O}_X, \omega_X) \to \mathcal{H}om_X(\mathcal{O}_X(-S), \omega_X) \to \mathcal{E}xt_X^1(\mathcal{O}_S, \omega_X) \to \mathcal{E}xt_X^1(\mathcal{O}_X, \omega_X)$$

and the last term is zero since $\mathcal{O}_X$ is locally free. Thus we get the usual short exact sequence

$$0 \to \omega_X \to \omega_X(S) \xrightarrow{\mathcal{R}_S} \omega_S \to 0. \quad (4.1.2)$$

Note that this sequence is exact whenever X is CM and $S \subset X$ has pure codimension 1. If, in addition, $\omega_X(S)$ is locally free near S then

$$\omega_X(S)|_S = \omega_S. \qquad (4.1.3)$$

(Note that ω_S can be locally free even if $\omega_X(S)$ is not locally free. As an example, take $X = (xy - zt = 0) \subset \mathbb{A}^4$ and $S = (x = z = 0)$.) In many of our applications $K_X + S$ and K_S are not Cartier but $m(K_X + S)$ and mK_S are Cartier for some $m > 0$. By taking tensor powers, we get maps

$$\mathcal{R}_S^{\otimes m}: (\omega_X(S))^{\otimes m} \to \omega_S^{\otimes m},$$

but we really would like to get a corresponding map between locally free sheaves

$$\omega_X^{[m]}(mS) := ((\omega_X(S))^{\otimes m})^{**} \stackrel{?}{\dashrightarrow} \left(\omega_S^{\otimes m}\right)^{**} =: \omega_S^{[m]}. \qquad (4.1.4)$$

As we saw in (2.34), no such map exists in general, not even if X is a normal surface. One needs a correction term, called the different.

   Our next aim is to extend the definition of the different (and of the Poincaré residue map) to singular schemes in all dimensions. This does not seem always possible and in (4.2) we tried to list the minimal set of assumptions. They are somewhat numerous, but rather mild and satisfied in two important special cases: if X is normal and $S' \to S$ is the normalization of a divisor and if $(X, S + \Delta)$ is lc and $S' = S$.

Definition 4.2 (Different II)   Consider schemes and divisors as in (1.5) satisfying the following additional conditions.

(1) X is a reduced, pure dimensional scheme.
(2) $S \subset X$ is a reduced subscheme of pure codimension 1 and X is regular at all generic points of S.
(3) Δ is a $\mathbb{Q}$-divisor on X such that X is regular at all generic points of Δ and no irreducible component of S is contained in $\mathrm{Supp}\,\Delta$.
(4) For some $m > 0$, the rank 1 sheaf $\omega_X^{[m]}(mS + m\Delta)$ is locally free at all codimension 1 points of S.
(5) S' is a reduced, pure dimensional scheme such that $\omega_{S'}$ is locally free in codimension 1.
(6) $\pi: S' \to S$ is a finite birational morphism (1.11).

   By assumption, there is a closed subscheme $Z \subset S$ of codimension 1 such that S and X are both regular at every point of $S \setminus Z$, $\pi: (S' \setminus \pi^{-1}Z) \to (S \setminus Z)$ is an isomorphism, $\omega_{S'}$ is locally free on $S' \setminus \pi^{-1}Z$ and $\mathrm{Supp}\,\Delta \cap S \subset Z$.

Thus the Poincaré residue map (4.1) gives an isomorphism

$$\mathcal{R}_S : \omega_X(S)|_{(S \setminus Z)} \simeq \omega_{(S \setminus Z)}.$$

Since $\Delta \cap (S \setminus Z) = \emptyset$, taking mth power gives an isomorphism

$$\mathcal{R}_S^m : \pi^* \omega_X^{[m]}(mS + m\Delta)|_{(S' \setminus \pi^{-1} Z)} \simeq \omega_{S'}^{[m]}|_{(S' \setminus \pi^{-1} Z)}.$$

By (4–5) there is a dense, open subset $S^0 \subset S'$ such that $\operatorname{codim}_{S'}(S' \setminus S^0) \geq 2$ and both $\omega_X^{[m]}(mS + m\Delta)|_{S'}$ and $\omega_{S'}^{[m]}$ are locally free on S^0. Therefore

$$\mathcal{H}om_{S'}\left(\pi^* \omega_X^{[m]}(mS + m\Delta), \omega_{S'}^{[m]}\right)$$

is a rank 1 sheaf, locally free on S^0 and $\mathcal{R}_S^m$ defines a rational section. Hence there is a unique (not necessarily effective) divisor $\Delta_{S'}$ which is Cartier on S^0 such that

$$\mathcal{R}_{S^0}^m : (\pi^* \omega_X^{[m]}(mS + m\Delta))|_{S^0} \simeq \omega_{S^0}^{[m]}(\Delta_{S'}|_{S^0}). \tag{4.2.7}$$

We formally divide by m and define the *different* of Δ on S' as the $\mathbb{Q}$-divisor

$$\operatorname{Diff}_{S'}(\Delta) := \tfrac{1}{m} \Delta_{S'}. \tag{4.2.8}$$

We write the formula (4.2.7) in terms of $\mathbb{Q}$-divisors as

$$(K_X + S + \Delta)|_{S'} \sim_{\mathbb{Q}} K_{S'} + \operatorname{Diff}_{S'}(\Delta). \tag{4.2.9}$$

As in (2.6), the formula (4.2.9) has the disadvantage that it indicates only that the two sides are $\mathbb{Q}$-linearly equivalent, whereas (4.2.7) is a *canonical* isomorphism.

Note that if $K_X + S$ and Δ are both $\mathbb{Q}$-Cartier, then $\operatorname{Diff}_{S'}(0)$ is defined and

$$\operatorname{Diff}_{S'}(\Delta) = \operatorname{Diff}_{S'}(0) + \Delta|_S, \tag{4.2.10}$$

but in general the individual terms on the right do not make sense.

Example 4.3   As an illustration we compute the case $X = (xy - z^n = 0) \subset \mathbb{A}^3$ and $S = (y = z = 0)$. Note that y vanishes along S with multiplicity n while z vanishes along S with multiplicity 1 (but it also vanishes along the $(x = z = 0)$-line).

A local generator of ω_X is given by $x^{-1} dx \wedge dz$. Thus $\omega_X^n(nS)$ is locally free and a local generator is

$$\frac{1}{y} \cdot \frac{(dx \wedge dz)^n}{x^n} = \frac{(dx \wedge dz)^n}{x^{n-1} z^n} = (-1)^n \cdot \frac{(dz)^n}{z^n} \wedge \frac{(dx)^n}{x^{n-1}}.$$

Thus, up to sign, the residue of this form is $x^{-(n-1)}(dx)^n$, hence

$$\operatorname{Diff}_S(0) = \left(1 - \tfrac{1}{n}\right)[0].$$

This is consistent with (3.36).

4.4 (Coefficients $1 - \frac{1}{m}$)   If we start with the classical case, that is with lc pairs $(X, S + D)$ where S, D are reduced $\mathbb{Z}$-divisors, then $\mathrm{Diff}_S D$ contains divisors whose coefficient is either 1 or $1 - \frac{1}{m}$ for some $m \in \mathbb{N}$ by (3.36.1).

Remarkably, if we allow Δ to have coefficients of the form $1 - \frac{1}{m}$ for some $m \in \mathbb{N}$, then $\mathrm{Diff}_S \Delta$ has coefficients $1 - \frac{1}{mn}$ for some $n \in \mathbb{N}$ by (3.45). This makes it very pleasant and important to work with pairs (X, Δ) where the coefficients of Δ are in the set

$$\mathbf{T} := \left\{ 1 - \tfrac{1}{2}, 1 - \tfrac{1}{3}, 1 - \tfrac{1}{4}, \ldots, 1 \right\}.$$

As an added benefit, if (X, Δ) is such and $G \subset \mathrm{Aut}(X, \Delta)$ is a finite subgroup then the boundary divisor of the quotient $(X, \Delta)/G$ also has coefficients in $\mathbf{T}$ by (2.41.6).

Proposition 4.5   *Notation and assumptions as in (4.2). Let $\pi \colon \bar{S} \to S$ be the normalization and $V \subset \bar{S}$ a prime divisor. Then:*

(1) *If Δ is effective then $\mathrm{Diff}_{\bar{S}}(\Delta)$ is effective and $\mathrm{coeff}_V \mathrm{Diff}_{\bar{S}}(\Delta) = 0$ if and only if X, S are both regular at the generic point of V and $V \not\subset \mathrm{Supp}\,\Delta$.*
(2) *If Δ is effective then $\mathrm{coeff}_V \mathrm{Diff}_{\bar{S}}(\Delta) \le 1$ if and only if $(X, S + \Delta)$ is lc at the generic point of $\pi(V)$.*
(3) *If $(X, S + \Delta)$ is lc and Δ is effective then $\mathrm{Diff}_{\bar{S}}(\Delta)$ is a boundary.*
(4) *If S is Cartier outside a codimension 3 subset then $\mathrm{Diff}_S(\Delta) = \Delta|_S$.*
(5) *If $K_X + S$ and D are both Cartier outside a codimension 3 subset then $\mathrm{Diff}_S D$ is a $\mathbb{Z}$-divisor and $(K_X + S + D)|_S \sim K_S + \mathrm{Diff}_S D$.*
(6) *If $(X, S + \Delta)$ is lc and Δ is effective then $\mathrm{Diff}_{\bar{S}}(\Delta) + K_{\bar{S}/S} = \pi^* \mathrm{Diff}_S(\Delta)$.*

Proof   Note that the different involves only divisors on X and on $\bar{S}$, hence its computation involves only points of codimension 2 on X. Therefore we can localize at the generic point of $\pi(V)$ and assume that $\dim X = 2$.

For surfaces the different was computed in (2.35) and (1–5) all hold. If $(X, S + \Delta)$ is an lc surface and Δ is effective then, by (2.31), at any point $s \in S$, the curve S is either regular or it has a node and $s \not\subset \mathrm{Supp}\,\Delta$. In the first case π is an isomorphism near s and (6) is clear. If S has a node then $\mathrm{coeff}_s \mathrm{Diff}_S(\Delta) = 0$ and $\mathrm{Diff}_{\bar{S}}(\Delta)$ is the sum of the two preimages of the node (2.35). Thus $\mathrm{Diff}_{\bar{S}}(\Delta) = -K_{\bar{S}/S}$, proving (6). $\qquad\square$

Proposition 4.6 (Crepant maps and the different)   *Let $\phi \colon (X_1, S_1 + \Delta_1) \dashrightarrow (X_2, S_2 + \Delta_2)$ be a crepant birational map (2.23) such that ϕ induces a birational map $S_1 \dashrightarrow S_2$. Then the induced map*

$$\phi|_{\bar{S}_1} \colon (\bar{S}_1, \mathrm{Diff}_{\bar{S}_1} \Delta_1) \dashrightarrow (\bar{S}_2, \mathrm{Diff}_{\bar{S}_2} \Delta_2) \quad \textit{is also crepant.}$$

Proof By assumption there is a pair $(Y, S_Y + \Delta_Y)$ and a commutative diagram

$$
\begin{array}{ccc}
 & (Y, S_Y + \Delta_Y) & \\
 & h_1 \swarrow \qquad \searrow h_2 & \\
(X_1, S_1 + \Delta_1) & \cdots\cdots\overset{\phi}{\cdots\cdots} \to & (X_2, S_2 + \Delta_2)
\end{array}
$$

where the h_i are crepant. It is thus enough to show that the morphisms

$$
h_i|_{\bar S_Y} : (\bar S_Y, \mathrm{Diff}_{\bar S_Y} \Delta_Y) \to (\bar S_i, \mathrm{Diff}_{\bar S_i} \Delta_i)
$$

are crepant. Since $K_Y + S_Y + \Delta_Y \sim_{\mathbb{Q}} h_i^*(K_{X_i} + S_i + \Delta_i)$, by restriction we obtain that

$$
K_{\bar S_Y} + \mathrm{Diff}_{\bar S_Y}(\Delta_Y) \sim_{\mathbb{Q}} (h_i|_{\bar S_Y})^*(K_{\bar S_i} + \mathrm{Diff}_{\bar S_i}(\Delta_i)).
$$

It remains to show that

$$
(h_i|_{\bar S_Y})_* \mathrm{Diff}_{\bar S_Y}(\Delta_Y) = \mathrm{Diff}_{\bar S_i}(\Delta_i).
$$

This can be checked at codimension 1 points of $\bar S_i$. By localization we can thus assume that $\dim X_i = 2$ and this case was done in (2.35.1). $\qquad\square$

4.7 (Different and discrepancies) Let $(X, S + \Delta)$ be a pair and $f: (Y, S_Y + \Delta_Y) \to (X, S + \Delta)$ a log resolution. Then (4.6) gives that

$$
K_{S_Y} + \Delta_Y|_{S_Y} \sim_{\mathbb{Q}} f_S^*(K_{\bar S} + \mathrm{Diff}_{\bar S}(\Delta)) \quad \text{and} \quad \mathrm{Diff}_{\bar S}(\Delta) = (\bar f_S)_*(\Delta_Y|_{S_Y}).
$$
$$(4.7.1)$$

If the existence of a log resolution is not known, we can still use (4.7.1) if we focus on finitely many divisors at a time. Indeed, let E_S be any divisor over $\bar S$. By (2.22), there is a sequence of blow-ups $f: Y \to X$ such that $\mathrm{center}_{S_Y} E_S$ is a divisor on S_Y and both S_Y and Y are regular at the generic point of $\mathrm{center}_{S_Y} E_S$.

(4.7.2) In order to get additional information, note that by further blowing up we may assume that $f_*^{-1}\Delta$ is disjoint from S_Y and if E_i is an exceptional divisor of f that intersects S_Y then $\mathrm{center}_X E_i \subset S$. (Note, however, that if E_i is f-exceptional, $E_i \cap S_Y$ need not be f_S-exceptional.) Then (2.35) and (4.7.1) give the following.

Claim 4.7.3 Let $f: Y \to X$ be a birational morphism and $S_Y := f_*^{-1}S$ the birational transform of S.

a) Let $E \subset Y$ be a divisor and E_S any irreducible component of $E \cap S_Y$. Then
$a(E_S, \bar S, \mathrm{Diff}_{\bar S}(\Delta_Y)) \leq a(E, X, S + \Delta)$.
b) Let $F \subset S_Y$ be a divisor such that S_Y is regular at the generic point of F. Let F_X be the divisor obtained by blowing up $F \subset Y$. Then
$a(F, \bar S, \mathrm{Diff}_{\bar S}(\Delta_Y)) = a(F_X, X, S + \Delta)$ and $\mathrm{center}_S F = \mathrm{center}_X F_X$.

c) If $f: Y \to (X, S + \Delta)$ is a log resolution then

$$\text{totaldiscrep}(\bar{S}, \text{Diff}_{\bar{S}}(\Delta))$$
$$= \min_i \{a(E_i, X, S + \Delta) : E_i \text{ is } f\text{-exceptional and } E_i \cap S_Y \neq \emptyset\}.$$

$\square$

The condition $E_i \cap S_Y \neq \emptyset$ in (4.7.3c) is rather difficult to control since it is not birationally invariant. In order to get something that is visibly birational in nature, we consider the following variants of the discrepancy (2.9).

Definition 4.7.4   Let (X, Δ) be a pair and $Z \subset X$ a closed subscheme. Define

$$\text{discrep(center} \subset Z, X, \Delta) \quad := \inf_E \{a(E, X, \Delta) : \text{center}_X E \subset Z\}, \quad \text{and}$$
$$\text{discrep(center} \cap Z \neq \emptyset, X, \Delta) := \inf_E \{a(E, X, \Delta) : \text{center}_X E \cap Z \neq \emptyset\},$$

where, in both cases, E runs through the set of all exceptional divisors over X satisfying the indicated restrictions. Both of these have a totaldiscrep version.

Finally, if $\mathcal{E}$ is a set of divisors over X then inserting the conditions $E \notin \mathcal{E}$ means that we take the infimum over all divisors that are not in $\mathcal{E}$. For instance, if $[S]$ denotes the set of divisors corresponding to the irreducible components of some $S \subset X$, then

$$\text{totaldiscrep}(E \notin [S], \text{center} \cap S \neq \emptyset, X, \Delta)$$

denotes the infimum of all $a(E, X, \Delta)$, where E runs through all divisors over X, save the irreducible components of S, whose center has nonempty intersection with S.

Using this notation, the above arguments established the following:

Lemma 4.8 (Easy adjunction)   *Let X be a normal scheme as in (1.5) and S a reduced divisor on X. Let Δ be an effective $\mathbb{Q}$-divisor that has no irreducible components in common with S. Assume that $K_X + S + \Delta$ is $\mathbb{Q}$-Cartier and let $\bar{S} \to S$ denote the normalization.*

Then for every divisor E_S over $\bar{S}$ there is a divisor E_X over X such that

$$a(E_X, X, S + \Delta) = a(E_S, \bar{S}, \text{Diff}_{\bar{S}}(\Delta)) \quad \text{and} \quad \text{center}_X E_X = \text{center}_S E_S.$$

In particular,

$$\text{totaldiscrep}(\bar{S}, \text{Diff}_{\bar{S}}(\Delta)) \geq \text{discrep(center} \subset S, X, S + \Delta)$$
$$\geq \text{totaldiscrep}(E \notin [S], \text{center} \cap S \neq \emptyset, X, S + \Delta).$$

Note that if $(X, S + \Delta)$ is plt then S is normal in codimension 1 by (2.31) hence normal in characteristic 0 by (2.88). Thus, in these cases, we do not worry about the normalization and write $(S, \text{Diff}_S(\Delta))$ instead.

The above lemma (or similar results and conjectures) is frequently referred to as *adjunction* if we assume something about X and obtain conclusions about

S, or *inversion of adjunction* if we assume something about S and obtain conclusions about X.

The main theorem in the subject (4.9) asserts that the inequalities in (4.8) are equalities, at least in characteristic 0. It was conjectured in Shokurov (1992, 3.3) and extended to the current form in Kollár (1992, 17.3). The hard part is to establish the converse: we assume something about the lower dimensional pair $(S, \mathrm{Diff}_S(\Delta))$ and we want to conclude something about the higher dimensional pair $(X, S + \Delta)$.

The surface case was proved in (2.35).

For applications the following two special cases are especially important.

(1) Kollár (1992, 17.4) If totaldiscrep$(S, \mathrm{Diff}_S(\Delta)) > -1$ then
$$\mathrm{totaldiscrep}(E \notin [S], \mathrm{center} \cap S \neq \emptyset, X, S + \Delta) > -1.$$
(2) Kawakita (2007) If totaldiscrep$(\bar{S}, \mathrm{Diff}_{\bar{S}}(\Delta)) \geq -1$ then
$$\mathrm{totaldiscrep}(\mathrm{center} \cap S \neq \emptyset, X, S + \Delta) \geq -1.$$

Both of these have been proved in characteristic 0 without using any form of the MMP and these have been important tools in higher dimensional birational geometry. The simplest proof of (1) is in Kollár *et al.* (2004, section 6.4).

It was also observed in Kollár (1992, 17.9–12) that (4.9) is implied by the full MMP for dlt pairs and this is essentially the proof that we present below.

Theorem 4.9 (Adjunction) *Let X be a normal variety over a field of characteristic 0 and S a reduced divisor on X with normalization $\bar{S} \to S$. Let Δ be an effective $\mathbb{Q}$-divisor that has no irreducible components in common with S. Assume that $K_X + S + \Delta$ is $\mathbb{Q}$-Cartier. Then*

$$\mathrm{totaldiscrep}(\bar{S}, \mathrm{Diff}_{\bar{S}}(\Delta)) = \mathrm{discrep}(\mathrm{center} \subset S, X, S + \Delta)$$
$$= \mathrm{totaldiscrep}(E \notin [S], \mathrm{center} \cap S \neq \emptyset, X, S + \Delta).$$

In particular,

(1) *If $(\bar{S}, \mathrm{Diff}_{\bar{S}}(\Delta))$ is klt then $(X, S + \Delta)$ is plt in a neighborhood of S.*
(2) *If $(\bar{S}, \mathrm{Diff}_{\bar{S}}(\Delta))$ is lc then $(X, S + \Delta)$ is lc in a neighborhood of S.*

Furthermore, the following more precise result holds.

(3) *Assume that $(X, S + \Delta)$ is lc and let E be a divisor over X such that $a(E, X, S + \Delta) < 0$. Let $Z_S \subset \bar{S}$ be an irreducible subvariety that dominates an irreducible component of $S \cap \mathrm{center}_X E$. Then there is a divisor E_S over $\bar{S}$ such that $a(E_S, \bar{S}, \mathrm{Diff}_{\bar{S}}(\Delta)) \leq a(E, X, S + \Delta)$ and $\mathrm{center}_{\bar{S}} E_S = Z_S$.*

Warning

1. If $(\bar{S}, \mathrm{Diff}_{\bar{S}}(\Delta))$ is terminal (canonical, plt or dlt) then $(X, S + \Delta)$ is usually *not* terminal (canonical, plt or dlt) in a neighborhood of S.

2. The assumption that $K_X + S + \Delta$ be $\mathbb{Q}$-Cartier is quite important. Let C be a smooth projective curve. By taking a cone over $C \times \mathbb{P}^n$, we get a normal variety X^{n+2} which contains a smooth divisor $S \simeq \mathbb{A}^{n+1}$. There is only an isolated singular point at the vertex and, for $n \geq 1$, $\mathrm{Diff}_S(0)$ is the zero divisor. In particular, $(S, \mathrm{Diff}_S(0))$ is even terminal. Nonetheless, if $g(C) \geq 2$ then (X, S) is not lc. In fact, (X, Δ) is never lc, no matter how we choose a $\mathbb{Q}$-divisor Δ.

3. In (4.9.3) the assumption $a(E, X, S + \Delta) < 0$ may not be necessary but our method of proof definitely fails in general. This is connected with the conjecture on the lower semi-continuity of minimal log discrepancies (7.13).

As a first application of (4.9) we obtain that discrepancies behave well in flat families. The last claim was originally proved in Kawamata (1999a) and Nakayama (1986) without using MMP.

Corollary 4.10 *Let X be a normal variety over a field of characteristic 0 and Δ an effective $\mathbb{Q}$-divisor on X such that $K_X + \Delta$ is $\mathbb{Q}$-Cartier. Let $f : X \to B$ be a flat, proper morphism to a smooth curve such that every irreducible component of Δ dominates B. Set $X_b := f^{-1}(b)$ with normalization $\bar{X}_b \to X_b$. Then:*

(1) *The function $b \mapsto \mathrm{totaldiscrep}(\bar{X}_b, \mathrm{Diff}_{\bar{X}_b}(\Delta))$ is lower semi-continuous.*
(2) *The set $\{b \in B : (\bar{X}_b, \mathrm{Diff}_{\bar{X}_b}(\Delta))$ is lc$\}$ is open in B.*
(3) *The set $\{b \in B : (X_b, \Delta_b)$ is klt$\}$ is open in B.*
(4) *The set $\{b \in B : (X_b, \Delta_b)$ is canonical (resp. terminal)$\}$ is open in B.*

Proof Fix $0 \in B$ and set $c = \mathrm{totaldiscrep}(X_0, \mathrm{Diff}_{\bar{X}_0}(\Delta))$. By (4.9), there is an open neighborhood $X_0 \subset U \subset X$ such that $\mathrm{totaldiscrep}(E \notin [X_0], U, X_0 + \Delta|_U) \geq c$.

Let $f : Y \to X$ be a log resolution of $(X, X_0 + \Delta)$ and write $K_Y + \Delta_Y \sim_{\mathbb{Q}} f^*(K_X + X_0 + \Delta)$. For general $b \in B$, X_b is contained in U and, by Bertini's theorem, $Y_b := f^{-1}(X_b) \to X_b$ is a log resolution of $(X_b, \Delta|_{X_b})$ and $K_{Y_b} + \Delta_Y|_{Y_b} \sim_{\mathbb{Q}} f^*(K_{X_b} + \Delta|_{X_b})$. Thus $\mathrm{totaldiscrep}(\bar{X}_b, \mathrm{Diff}_{\bar{X}_b}(\Delta)) \geq c$, proving (1) and (2).

In the klt cases X_b is normal and by (4.5), the restriction of Δ is the same as the different. This shows (3).

A more delicate argument is needed for (4). By (3) we may assume that, possibly after shrinking B, $(X, X_b + \Delta)$ is plt for every $b \in B$.

Assume that (X, Δ) is not terminal (resp. canonical). Then there is an exceptional divisor E over X such that $a(E, X, \Delta) \leq 0$ (resp. < 0). If center$_X$ $E \subset X_b$ then $a(E, X, X_b + \Delta) \leq -1$ which is impossible. Thus E dominates B and so $a(E, X, X_b + \Delta) = a(E, X, \Delta)$. By (1.38) there is a $\mathbb{Q}$-factorial dlt model $\pi : X' \to X$ whose sole exceptional divisor is E. Since $\pi^{-1}(X_b)$ is a Cartier divisor, $\pi^{-1}(X_b) \cap E$ is a divisor in $\pi^{-1}(X_b)$; let E_b be any of its irreducible components. By (4.7.3a)

$$a(E_b, X_b, \Delta_b) \leq a(E, X, X_b + \Delta) = a(E, X, \Delta).$$

Since E is exceptional over X and dominates B, $\dim X_b \cap \text{center}_X E \leq \dim X_b - 2$ for every $b \in B$, hence E_b is exceptional over X_b. Thus (X_0, Δ_0) is also not terminal (resp. canonical). □

4.11 Proof of (4.9) All assertions concern a neighborhood of S in X, thus we may assume from now on that every irreducible component of Δ intersects S and we can ignore all exceptional divisors whose center in X is disjoint from S.

Note that $\text{totaldiscrep}(\bar{S}, \text{Diff}_{\bar{S}}(\Delta)) \leq 0$ (since every divisor on $\bar{S}$ has discrepancy ≤ 0) and $\text{totaldiscrep}(E \notin [S], X, S + \Delta) \leq 0$ as shown by blowing up a divisor on S along which S and X are smooth. Thus both sides in (4.9) are in $\{-\infty\} \cup [-1, 0]$.

Assume next that $(X, S + \Delta)$ is plt and let E be a divisor over X such that $-1 < a(E, X, S + \Delta) < 0$. By (1.39) there is a projective, birational morphism $g: X^E \to X$ such that E is an exceptional divisor on X^E and $E = g^{-1}(g(E))$. In particular, every irreducible component of $S \cap g(E)$ is dominated by an irreducible component of $g_*^{-1}S \cap E$. Since X^E is $\mathbb{Q}$-factorial, the latter are divisors over S and by (4.7.3a) their discrepancy is $\leq a(E, X, S + \Delta)$. This proves the main equalities and also (4.9.3) in the plt case.

It remains to show that

(1) if $(X, S + \Delta)$ is not plt then $(\bar{S}, \text{Diff}_{\bar{S}}(\Delta))$ is not klt,
(2) if $(X, S + \Delta)$ is not lc then $(\bar{S}, \text{Diff}_{\bar{S}}(\Delta))$ is also not lc and
(3) if $(X, S + \Delta)$ is lc then (4.9.3) holds.

We could try to argue as above, but this would need (1.39) in the non-dlt case (which is known but we have not proved yet) or in the non-lc case (which does not hold in general). There are two ways to go around this problem. First we see what a small modification of the above method gives. Then we prove an even more general result that includes (4.9.2).

Special case 4.11.4 Assume that S is $\mathbb{Q}$-Cartier. (This is the only case that we used in (4.10) and will use later in this book.)

Apply (1.34) to get $g: X' \to X$ where all exceptional divisors have disrepancy ≤ -1. Since S is $\mathbb{Q}$-Cartier, so is g^*S. If $g^*S \neq g_*^{-1}S$ then there is a g-exceptional divisor $E \subset g^*S$ such that $E \cap g_*^{-1}S \neq \emptyset$. As before we see that $(\bar{S}, \text{Diff}_{\bar{S}}(\Delta))$ is not klt.

If $g^*S = g_*^{-1}S$ then $g_*^{-1}S$ intersects every divisor, and again we conclude that $(\bar{S}, \text{Diff}_{\bar{S}}(\Delta))$ is not klt. This completes (4.11.1).

Next assume that $(X, S + \Delta)$ is not lc. Pick $0 < \eta \ll 1$ and apply (1.36) to $(X, (1 - \eta)S + \Delta)$ with $c = 1$ to get $g: X_\eta \to X$ with g-exceptional divisors

E_i. Note that as we replace S by $(1 - \eta)S$, we increase the discrepancy of every exceptional divisor whose center is contained in S. Thus we can choose η such that no g-exceptional divisor with center in S has discrepancy -1 and some have discrepancy < -1.

Now we argue as before. If $g^*S \neq g_*^{-1}S$ then there is a g-exceptional divisor with center in S that intersects $g_*^{-1}S$ and we are done.

If $g^*S = g_*^{-1}S$ then $g_*^{-1}S$ intersects every divisor. Since $(X, S + \Delta)$ is not lc, there is at least one divisor whose discrepancy is < -1 and we are again done.

General case 4.11.5   If S is not $\mathbb{Q}$-Cartier, the above proofs do not work. However, both (4.11.1–2) were proved without using minimal models. The proof of (4.11.1) given in Kollár and Mori (1998, 5.50) is short and quite self-contained. The proof of (4.11.2) in Kawakita (2007) is an elementary but quite mysterious computation.

Finally, the proof of (4.11.3) will be completed in (7.5).   □

The following result of Hacon (2012) is a generalization of (4.9.2). In order to state it, recall the notation $\Delta_{>1}$, $\Delta_{\leq 1}$ introduced in (1.3).

Theorem 4.12   *Let X be a normal variety over a field of characteristic 0 and Δ_X a boundary on X such that $K_X + \Delta_X$ is $\mathbb{Q}$-Cartier. Let $f: (Y, \Delta_Y) \to (X, \Delta_X)$ be a log resolution where $K_Y + \Delta_Y \sim_{\mathbb{Q}} f^*(K_X + \Delta_X)$.*

Assume that there is an irreducible divisor $E \subset Y$ such that $\operatorname{coeff}_E \Delta_Y = 1$ and E is disjoint from $\operatorname{Supp}((\Delta_Y)_{>1})$. Then

(1)   *(X, Δ_X) is lc in a neighborhood of $f(E)$, equivalently,*
(2)   *$f(E)$ is disjoint from $f(\operatorname{Supp}(\Delta_Y)_{>1})$.*

If Δ_Y is effective, then we can be even more general.

Theorem 4.13   *Let (Y, Δ) be a pair over a field of characteristic 0 such that Δ is effective and $(Y, \Delta_{\leq 1} + \operatorname{red}\Delta_{>1})$ is dlt. Let $f: Y \to X$ be a projective morphism with connected fibers such that $K_Y + \Delta \sim_{f,\mathbb{Q}} 0$. Let $Z \subset Y$ be an lc center (4.15) of $(Y, \Delta_{\leq 1} + \operatorname{red}\Delta_{>1})$ that is disjoint from $\operatorname{Supp}(\Delta_{>1})$. Then*

(1)   *(Y, Δ) is dlt in a neighborhood of $f^{-1}(f(Z))$, equivalently,*
(2)   *$f(Z)$ is disjoint from $f(\operatorname{Supp}(\Delta_{>1}))$.*

Proof   Assume to the contrary that there is a point $x \in f(Z) \cap f(\operatorname{Supp}(\Delta_{>1}))$. The problem is étale local on X, thus we may assume that $k(x)$ is algebraically closed and $D_i \cap f^{-1}(x)$ is connected for every irreducible component D_i of $\Delta_{\geq 1}$ (4.38).

If $f^{-1}(x) \cap \sum D_i$ is disconnected then we apply (4.37) to $(Y, \Delta - \lfloor\Delta\rfloor)$ with $D := \lfloor\Delta\rfloor$. Thus (Y, Δ) is plt in a neighborhood of $f^{-1}(x)$ and we are done.

If $f^{-1}(x) \cap \sum D_i$ is connected then, up to reindexing, let $D_1, \ldots, D_r$ be a shortest chain of divisors such that

(3) each D_i has coefficient 1 in Δ,
(4) $\Delta_{>1} \cap D_1 \cap f^{-1}(x) \neq \emptyset$,
(5) $D_i \cap D_{i+1} \cap f^{-1}(x) \neq \emptyset$ for $i = 1, \ldots, r-1$ and
(6) $D_r \cap Z \cap f^{-1}(x) \neq \emptyset$.

We claim that (the Stein factorization of) $(D_1, \mathrm{Diff}^*_{D_1} \Delta) \to X$ gives a lower dimensional counter example to (4.13). The only question is to find an lc center Z_1 of $(D_1, \mathrm{Diff}^*_{D_1} \Delta)$ that is disjoint from $D_1 \cap \Delta_{>1}$.

Note that if $i \geq 2$ then $\Delta_{>1} \cap D_i \cap f^{-1}(x) = \emptyset$ since otherwise we could omit D_1 from the chain. Thus if $r \geq 2$ then any irreducible component of $D_1 \cap D_2$ gives such a Z_1. If $r = 1$ then any irreducible component of $D_1 \cap Z$ works.

We complete the proof by induction on $\dim Y$. $\square$

4.14 **Proof of (4.13) $\Rightarrow$ (4.12)**   If Δ_Y is effective then $f: (Y, \Delta_Y) \to X$ satisfies the assumptions of (4.13) with $Z := E$. We will run a suitable MMP to get rid of the negative part of Δ_Y while keeping track of E.

We can replace Y by any other log resolution dominating it, thus, by (1.12.1), we may assume that there is an effective f-exceptional divisor F such that $-F$ is f-ample. The conclusions are local on X, thus we may assume that X is quasi-projective. Choose H_X ample on X such that $-F + f^*H_X$ is ample on Y and let H be an effective $\mathbb{Q}$-divisor such that $H \sim_{\mathbb{Q}} -F + f^*H_X$ and $(Y, \Delta_Y + H)$ is also snc. Write $\Delta_Y = \Delta_h + \Delta_v$ where $\Delta_h := f_*^{-1}\Delta_X$ is the horizontal part and Δ_v is f-exceptional. Let $\Theta = \sum_i E_i$ be the sum of all f-exceptional divisors. Note that $\Theta = \Delta_v - A + B$ where A, B are effective, $\mathrm{Supp}\,A = \mathrm{Supp}\,(\Delta_v)_{>1}$ and $\mathrm{Supp}\,B = \mathrm{Supp}\,(\Delta_v)_{<1}$. Choose $0 < \varepsilon$ to be smaller than any of the coefficients appearing in B.

By (1.30.5) we can run an MMP for $(Y, \Delta_h + \Theta + \varepsilon H)$ over X. We get a series of contractions and flips $\phi_i: Y^i \dashrightarrow Y^{i+1}$ starting with $Y^0 = Y$ and ending with a minimal model

$$f^m: \left(Y^m, \Delta_h^m + \Theta^m + \varepsilon H^m\right) \to X.$$

We prove by induction on r that

(1.r) E^r is disjoint from $\Delta_{>1}^r$ and
(2.r) E^r is not contracted by ϕ_r.

We start with (1.0) which is our assumption. Assume that (1.r) holds. Note that

$$K_{Y^r} + \Delta_h^r + \Theta^r + \varepsilon H^r \sim_{f^r, \mathbb{Q}} B^r + \varepsilon H^r - A^r,$$

the divisors B^r, H^r, A^r are effective, none of them contains E^r and Supp $A^r =$ Supp $(\Delta_v^r)_{>1}$ is disjoint from E^r by (1.r). Thus

$$(K_{Y^r} + \Delta_h^r + \Theta^r + \varepsilon H^r)\big|_{E^r} \sim_{f^r,\mathbb{Q}} \text{(effective divisor)},$$

hence E^r is not contracted by ϕ_r, proving (2.r).

Assume that (2.r-1) holds, thus E^r is a divisor on X^r. By our assumption $(E, \text{Diff}_E^* \Delta_Y)$ is lc, hence $(\bar{E}^r, \text{Diff}_{\bar{E}^r}^* \Delta_Y^r)$ is also lc by (4.6) and (2.23.4) where $\pi \colon \bar{E}^r \to E^r$ is the normalization. Since Y^r is $\mathbb{Q}$-factorial, if E^r is not disjoint from $\Delta_{>1}^r$ then there is a divisor $V^r \subset \bar{E}^r$ such that $\pi(V^r) \subset \Delta_{>1}^r$. Thus V^r appears in $\text{Diff}_{\bar{E}^r}^* \Delta_Y^r$ with coefficient > 1, unless $\pi(V^r)$ is also contained in $\Delta_{<0}$ (2.32). This is, however, impossible since the irreducible components of $E^r + \Delta_{>1}^r + \Delta_{<0}^r$ all appear in Θ^r with coefficient $= 1$. The latter is dlt, hence at most two of its irreducible components can contain $\pi(V^r)$ (2.32).

At the end we get that

$$K_{Y^m} + \Delta_h^m + \Theta^m + \varepsilon H^m \sim_{f^m,\mathbb{Q}} B^m + \varepsilon H^m - A^m \sim_{f^m,\mathbb{Q}} B^m - \varepsilon F^m - A^m$$

is f^m-nef and f^m-exceptional. Thus $A^m + \varepsilon F^m - B^m$ is effective by (1.17). Since ε is smaller than any of the coefficients appearing in B, this implies that $B^m = 0$.

Thus $f^m \colon (Y^m, \Delta_Y^m) \to X$ satisfies the assumptions of (4.13) with $Z := E^m$. □

4.2 Log canonical centers on dlt pairs

For an lc pair (X, Δ) it is especially interesting and useful to study exceptional divisors with discrepancy -1. These divisors, and their centers on X, play a crucial role in the inductive treatment of lc pairs.

Definition 4.15 Let (X, Δ) be a pair with Δ not necessarily effective. We say that an irreducible subvariety $Z \subset X$ is a *log canonical center* or *lc center* of (X, Δ) if (X, Δ) is lc at the generic point of Z and there is a divisor E over X such that $a(E, X, \Delta) = -1$ and center$_X E = Z$.

We sometimes call a (not necessarily closed) point an lc center if its closure is an lc center.

Let $f \colon Y \to X$ be a proper birational morphism and write $K_Y + \Delta_Y \sim_{\mathbb{Q}} f^*(K_X + \Delta)$ as in (2.6). If (X, Δ) is lc, then, by (2.23), the lc centers of (X, Δ) are exactly the images of the lc centers of (Y, Δ_Y).

If X is smooth and Δ is snc, then the lc centers of (X, Δ) are exactly the strata of $\Delta_{=1}$. This follows from (2.11), or see (4.16) for a more general case.

Warning Traditionally X was not considered to be a log canonical center of (X, Δ). However, in many contexts it is very convenient to view X

itself as a log canonical center, thereby avoiding trivial exceptions to many theorems.

Log canonical centers in the dlt case

The lc centers of a dlt pair behave very much like lc centers of an snc pair.

Theorem 4.16 (Fujino, 2007, section 3.9) *Let (X, Δ) be a dlt pair and $D_1, \ldots, D_r$ the irreducible divisors that appear in Δ with coefficient 1.*

(1) *The s-codimensional lc centers of (X, Δ) are exactly the irreducible components of the various intersections $D_{i_1} \cap \cdots \cap D_{i_s}$ for some $\{i_1, \ldots, i_s\} \subset \{1, \ldots, r\}$.*

(2) *Every irreducible component of $D_{i_1} \cap \cdots \cap D_{i_s}$ is normal and of pure codimension s.*

(3) *Let $Z \subset X$ be any lc center. Assume that D_i is $\mathbb{Q}$-Cartier for some i and $Z \not\subset D_i$. Then every irreducible component of $D_i|_Z$ is also $\mathbb{Q}$-Cartier.*

Proof Let E be a divisor over X such that $a(E, X, \Delta) = -1$ and $Z = \text{center}_X E$. By localizing at the generic point of Z, we may assume that Z is a closed point of X. By the dlt assumption, X is smooth at Z and Δ is snc. If $\dim_Z X = n$, there are snc divisors $B_1, \ldots, B_n$ through Z such that $\Delta = \sum a_i B_i$ for some $0 \le a_i \le 1$ (where we can ignore the components of Δ that do not pass through Z). Set $\Delta' := \sum B_i$. Then $a(E, X, \Delta') \ge -1$ by (2.7) and $a(E, X, \Delta) > a(E, X, \Delta') \ge -1$ by (2.5) unless $a_i = 1$ for every i. Thus every B_i appears in Δ with coefficient $= 1$ – hence is one of the D_j – and Z is an irreducible component of the intersection of the corresponding D_j. This proves one direction of (1).

By (2.31) each D_i is smooth in codimension 1 and S_2 (even CM) by (2.88). Thus each D_i is normal. We use induction on s to prove the rest, building on the following result.

Claim 4.16.4 Assume that (X, Δ) is dlt and let $D \subset \lfloor \Delta \rfloor$ be an irreducible component. Then $(D, \text{Diff}_D(\Delta - D))$ is dlt. (The converse fails, for instance for $(\mathbb{A}^3, (z = 0) + (xy = z^m))$.)

Proof By (4.8), $(D, \text{Diff}_D(\Delta - D))$ is lc. Let $Z \subset D$ be a log canonical center of $(D, \text{Diff}_D(\Delta - D))$. Then there is a divisor E_D over D whose center is Z and whose discrepancy is -1. Thus, by (4.8) there is a divisor E_X over X whose center is Z and whose discrepancy is -1. Since (X, Δ) is dlt this implies that (X, Δ) is snc at the generic point of Z. Thus $\text{Diff}_D(\Delta - D) = (\Delta - D)|_D$ and $(D, \text{Diff}_D(\Delta - D))$ is snc at the generic point of Z. □

By (2.62), the support of $\mathscr{O}_{D_i} + \mathscr{O}_{D_j}/\mathscr{O}_{D_i+D_j}$ has pure codimension 2 in X, hence $D_i \cap D_j$ has pure codimension 2 in X. By localizing at a generic point of $D_i \cap D_j$, we see from (2.35) that each irreducible component of $D_i \cap D_j$ is an lc center. This shows the $s = 2$ case of (2). For $s > 2$ we use induction and the equality

$$D_{i_1} \cap \cdots \cap D_{i_s} = (D_{i_1}|_{D_{i_s}}) \cap \cdots \cap (D_{i_{s-1}}|_{D_{i_s}}).$$

Finally, assume that D_i is $\mathbb{Q}$-Cartier and $i \neq s$. Then $D_i|_{D_s}$ is also $\mathbb{Q}$-Cartier, and, since $D_i \cap D_s$ is normal, every irreducible component of $D_i|_{D_s}$ is also $\mathbb{Q}$-Cartier; proving (3). □

The following result compares the fields of definitions of various lc centers.

Proposition 4.17 *Let (X, Δ) be a dlt pair over a field k. Let $K \supset k$ be a field extension and $W_1 \subset W_2 \subset X_K$ lc centers of (X_K, Δ_K).*
Then, if W_1 is defined over k, so is W_2.

Proof Let $D_i \subset \lfloor \Delta \rfloor$ be the irreducible components that contain W_1. By (4.16) each D_i is smooth at the generic point of W_1, hence the K-irreducible component $D_i' \subset (D_i)_K$ that contains W_1 is also defined over k. By (4.16), W_2 is a K-irreducible component of the intersection of some of the D_i'. Since W_2 is also smooth at the generic point of W_1, it is also defined over k. □

4.18 (Higher codimension Poincaré residue maps)   First, let X be a smooth variety and $D_1 + \cdots + D_r$ an snc divisor on X. We can iterate the codimension 1 Poincaré residue maps (4.1) to obtain higher codimension Poincaré residue maps

$$\mathcal{R}_{X \to D_1 \cap \cdots \cap D_r} : \omega_X(D_1 + \cdots + D_r)|_{D_1 \cap \cdots \cap D_r} \simeq \omega_{D_1 \cap \cdots \cap D_r} \qquad (4.18.1)$$

defined by

$$\mathcal{R}_{X \to D_1 \cap \cdots \cap D_r} := \mathcal{R}_{D_1 \cap \cdots \cap D_{r-1} \to D_1 \cap \cdots \cap D_r} \circ \cdots \circ \mathcal{R}_{D_1 \to D_1 \cap D_2} \circ \mathcal{R}_{X \to D_1}.$$

These maps are defined up to sign. It is enough to check this for two divisors. As a local model, take $X := \mathbb{A}^n$ and $D_i := (x_i = 0)$. For any $1 \leq r \leq n$, a generator of $\omega_X(D_1 + \cdots + D_r)$ is given by

$$\sigma := \frac{1}{x_1 \cdots x_r} dx_1 \wedge \cdots \wedge dx_n.$$

If we restrict first to D_1 and then to D_2, we get

$$\mathcal{R}_{X \to D_1 \cap D_2}(\sigma) = \frac{1}{x_3 \cdots x_r} dx_3 \wedge \cdots \wedge dx_n.$$

If we restrict first to D_2 and then to D_1, then we need to interchange dx_1 and dx_2 in the computation, hence we get

$$\mathcal{R}_{X \to D_1 \cap D_2}(\sigma) = \frac{-1}{x_3 \cdots x_r} dx_3 \wedge \cdots \wedge dx_n.$$

Note also that, if $W = D_1 \cap \cdots \cap D_r$ and $Z = W \cap D_{r+1} \cap \cdots \cap D_s$ then, by construction,

$$\mathcal{R}_{X \to Z} = \pm \mathcal{R}_{W \to Z} \circ \mathcal{R}_{X \to W} \qquad (4.18.2)$$

Assume next that X is a normal variety and Δ is a $\mathbb{Q}$-divisor which can be written as

$$\Delta = D_1 + \cdots + D_r + \Delta'$$

where $D_i \subset X$ are irreducible divisors not contained in Supp Δ'. Let $Z^0 \subset \operatorname{snc}(X, \sum D_i) \cap D_1 \cap \cdots \cap D_r$ be an open subset and $\pi_Z \colon Z \to X$ the normalization of its closure. We may assume that Z^0 is disjoint from Supp Δ'.

Assume that $m(K_X + \Delta)$ is a Cartier $\mathbb{Z}$-divisor where $m > 0$ is even. We can apply (4.18.1) to the snc locus of $(X, D_1 + \cdots + D_r)$ to obtain an isomorphism

$$\mathcal{R}_{X \to Z^0}^m : \omega_X^{[m]}(m\Delta)|_{Z^0} \xrightarrow{\simeq} \omega_{Z^0}^{[m]}. \qquad (4.18.3)$$

As in (4.2), this gives a rational section of

$$\mathcal{H}om_Z \big(\pi_Z^* \omega_X^{[m]}(m\Delta), \omega_Z^{[m]} \big)$$

There is thus a unique $\mathbb{Q}$-divisor $\operatorname{Diff}_Z^* \Delta$, called the *different* such that (4.18.3) extends to an isomorphism

$$\mathcal{R}_{X \to Z}^m : \omega_X^{[m]}(m\Delta)|_Z \xrightarrow{\simeq} \omega_Z^{[m]}(m \cdot \operatorname{Diff}_Z^* \Delta). \qquad (4.18.4)$$

Note that this formula simultaneously defines $\mathcal{R}_{X \to Z}^m$ and $\operatorname{Diff}_Z^* \Delta$ and that $\mathcal{R}_{X \to Z}^m$ is defined only for those even values of m for which $m(K_X + \Delta)$ is Cartier.

Remark on the notation   In (2.34) the original Diff_D is set up so that if $\Delta = D + D'$ then the restriction of $K_X + \Delta = K_X + D + D'$ to D is $K_D + \operatorname{Diff}_D D' = K_D + \operatorname{Diff}_D(\Delta - D)$. That is, we first remove D from Δ and then take the different. For higher codimension lc centers $Z \subset X$, it does not make sense to "first remove Z from Δ." Thus we need the new notation Diff^*. In the classical case, $\operatorname{Diff}_D^* \Delta = \operatorname{Diff}_D(\Delta - D)$.

Applying this to the dlt case, we obtain the following.

Theorem 4.19 *Let (X, Δ) be dlt, Δ effective and $Z \subset X$ an lc center. Then:*

(1) $\operatorname{Diff}_Z^* \Delta$ *is an effective $\mathbb{Q}$-divisor and $(Z, \operatorname{Diff}_Z^* \Delta)$ is dlt.*
(2) *If $m > 0$ is even and $m(K_X + \Delta)$ is Cartier then the Poincaré residue map gives an isomorphism*

$$\mathcal{R}_{X \to Z}^m \colon \omega_X^{[m]}(m\Delta)|_Z \xrightarrow{\simeq} \omega_Z^{[m]}(m \cdot \operatorname{Diff}_Z^* \Delta).$$

(3) *If $W \supset Z$ is another lc center of (X, Δ) then Z is also an lc center of $(W, \mathrm{Diff}^*_W \Delta)$ and*

$$\mathrm{Diff}^*_Z \Delta = \mathrm{Diff}^*_Z(\mathrm{Diff}^*_W \Delta).$$

Proof   If Z is a divisor, (1) was proved in (4.5).

By the definition, (4.18.4) and (4.18.2), (3) holds if Z is a divisor on X and W a divisor on Z. Thus, by induction, (3) also holds whenever Z and W are irreducible components of complete intersections of divisors in $\lfloor \Delta \rfloor$. By (4.16) this covers all cases. Now (3) and induction gives (1) and (2) was part of the definition of $\mathrm{Diff}^*_Z \Delta$. □

Complement 4.19.1   The above results also hold for any pair (X, Δ), provided it is dlt at the generic point of Z. The different is then defined on the normalization $\bar{Z}$. If Δ is effective then so is $\mathrm{Diff}^*_{\bar{Z}} \Delta$ and if (X, Δ) is lc then so is $(\bar{Z}, \mathrm{Diff}^*_{\bar{Z}} \Delta)$.

There are two ways to obtain these. First, one can go back and see that the only part of dlt we used was that (X, Δ) is snc at the generic point of Z. Second, we can use (1.36) to get a dlt model $g: (X^m, \Delta^m) \to (X, \Delta)$ that is an isomorphism over the generic point of Z. We have already proved (1–3) for (X^m, Δ^m); by push-forward we get (1–3) for (X, Δ). □

4.3 Log canonical centers on lc pairs

Next we aim to generalize (4.19) to lc pairs. This is a quite lengthy project. The key observation is that that one needs to look behind lc centers in order to understand lc pairs. This leads first to the notion of *hereditary* lc centers and then to *sources* and *springs* of lc centers. Here we give an outline of the arguments, gathered into six main topics.

4.20 (Lc centers as subvarieties)   We saw in (4.16) that the lc centers of a dlt pair are normal and by (4.19) and (2.88) they have rational singularities. Neither of these hold for general lc pairs, but they do not fail by much. We will prove four special properties of lc centers of arbitrary lc pairs.

(1) (Normality) Every minimal (with respect to inclusion) lc center is normal (9.26).
(2) (Intersection) Any intersection of lc centers is also a union of lc centers (4.41.2).
(3) (Seminormality) Every union of lc centers is seminormal (9.26, 10.11).
(4) (Du Bois) Every union of lc centers is Du Bois (6.10).

Originally, (4.20.1–3) were proved directly by rather subtle arguments in Ambro (2003, 2011) and Fujino (2009b). We take a more roundabout approach.

First we show much weaker versions of these assertions. For instance, in (4.41) we show that every minimal lc center is unibranch and that the union of *all* lc centers is seminormal. These are much easier to prove than (1) or (3). Then we set up an inductive framework and use these weaker versions to get the general results.

A disadvantage of this method is that one needs to wait for the general theory to get the basic results on lc centers. For example, (4.20.1) and (4.20.3) will be proved as special cases of (9.26) and (4.20.4) is established in (6.32).

On the other hand, we find this approach shorter at the end and it also works well for semi-log canonical pairs and for the proof of the Du Bois property.

4.21 (Hereditary lc centers)   If all lc centers are normal (as in the dlt case), then they show everything. In general, however, one should start with the lc centers, then look at the lc centers on their normalizations and iterate this process if necessary. These are called *hereditary lc centers* (5.30).

This concept is illustrated by the examples

$$(\mathbb{A}^3, D_1 := (x^2 = y^2 z)) \quad \text{and} \quad (\mathbb{A}^3, D_2 := (x^3 + x^2 z = y^2 z)).$$

The 2-dimensional lc centers are D_1 (resp. D_2)). In both cases, the sole 1-dimensional lc center is the line $(x = y = 0)$. The normalization of D_1 is $\pi \colon \mathbb{A}^2_{st} \to D_1$ given by $x = st, y = s, z = t^2$ and the preimage of the line $(x = y = 0)$ is the line $(s = 0)$. It is smooth and there are no more lc centers. The normalization of D_2 is the cone over the twisted cubic $\operatorname{Spec} \mathbb{C}[s^3, s^2 t, st^2, t^3] \subset \mathbb{A}^4$ and the map is given by $x = s(t^2 - s^2), y = t(t^2 - s^2), z = s^2$. The preimage of the line $(x = y = 0)$ is the pair of intersecting lines $(s \pm t = 0)$. This shows that the origin has to be an lc center. Of course we could easily have seen this by blowing up the origin, but now the underlying geometry reveals it.

Hereditary lc centers play a key role in Chapter 9. They seem to be the right concept to use in the study of finite equivalence relations.

A problem with the above definition of hereditary lc centers is that if we start with an lc pair (X, Δ), then the normalization of an lc center cannot be made into an lc pair in a natural way. We come back to this in (4.23).

4.22 (Sources of lc centers)   Let us start with an example. Consider the lc pair $(X, 0)$ where

$$X := (s(x^3 + y^3 + z^3) + t(3xyz) = 0) \subset \mathbb{P}^1_{st} \times \mathbb{A}^3_{xyz}. \qquad (4.22.1)$$

There is only one lc center $Z := \mathbb{P}^1 \times \{(0, 0, 0)\}$. We compute that $\omega_X|_Z \simeq \mathcal{O}_{\mathbb{P}^1}(-1)$. This is larger than $\omega_{\mathbb{P}^1} \simeq \mathcal{O}_{\mathbb{P}^1}(-2)$ and the different should compensate for this.

Since both ω_X and $\omega_{\mathbb{P}^1}$ are locally free, (4.5.5) would suggest that there should be a special point $p \in Z$ such that the residue map is an isomorphism $\omega_X|_Z \simeq \omega_Z(p)$. At first sight it looks like $(0 : 1)$ might be a special point, where the elliptic curve $E(s, t) := (s(x^3 + y^3 + z^3) + t(3xyz) = 0) \subset \mathbb{P}^2$ becomes reducible.

A closer inspection reveals that there are three other points where $E(s, t)$ becomes reducible; this also happens when $t^3 + s^3 = 0$. Moreover, these four points play a completely symmetric role; the automorphism group of X acts like S_4 on these four points and it has no fixed point on Z. For instance, if ε is a primitive 3rd root of unity then

$$(x, y, z, s, t) \mapsto (x + y + z, x + \varepsilon y + \varepsilon^2 z, x + \varepsilon^2 y + \varepsilon z, s + t, 2s - t)$$

is an automorphism of X which interchanges $E(0, 1)$ with $E(1, -1)$. (See Artebani and Dolgachev (2009) for details about this so called *Hesse pencil*.)

To find the correct solution, we blow-up Z to get $p: B_Z X \to X$. Note that $B_Z X$ is a dlt modification of X. Let $E \subset B_Z X$ denote the exceptional divisor. Then

$$E \simeq (s(x^3 + y^3 + z^3) + t(3xyz) = 0) \subset \mathbb{P}^1_{st} \times \mathbb{P}^2_{xyz}$$

and the first projection $p|_E: E \to Z$ is a family of elliptic curves. Then one gets natural maps

$$p^* \omega_X \simeq \omega_{B_Z X}(E) \quad \text{and} \quad \mathcal{R}_{B_Z X \to E} : \omega_{B_Z X}(E) \to \omega_E. \qquad (4.22.2)$$

These in turn give natural isomorphisms

$$\omega_X|_Z \simeq p_* \omega_E \simeq \omega_Z \otimes p_* \omega_{E/Z}. \qquad (4.22.3)$$

There are two ways to think about the formulas (4.22.2–3). For the traditional approach, see (4.26).

Our choice is to think of $p|_E: E \to Z$ as the basic object naturally associated to the lc center Z of the pair (X, Δ); we call it the *source* of Z and denote it by $\text{Src}(Z, X, \Delta)$. Its key properties are established in Section 4.5.

Before we can study these sources, we need to study the abstract properties of fiber spaces similar to $p|_E: E \to Z$.

4.23 (Crepant log structures)   In general, the above constructions lead to a compound object $p: (X, \Delta) \to Z$ where p is a proper morphism with connected fibers, (X, Δ) is lc and $K_X + \Delta \sim_{p, \mathbb{Q}} 0$. These *crepant log structures* are studied in Section 4.4.

Especially when $\Delta = 0$, these fiber spaces $p: X \to Z$ have been intensely studied in connection with Iitaka's classification program, see Ueno (1975) and Mori (1987) for surveys. These works almost exclusively focused on the postivity properties of the sheaves $p_* \omega_{X/Z}^{[m]}$. By contrast, here we are mainly interested in the log canonical centers of (X, Δ).

Crepant log structures also provide the right level of generality to define hereditary lc centers.

So far we have studied individual lc centers. Now it is time to put all the information together.

4.24 (Log canonical stratification)   For an lc pair (X, Δ) let $S_i^*(X, \Delta)$ be the union of all $\leq i$-dimensional lc centers and set

$$S_i(X, \Delta) := S_i^*(X, \Delta) \setminus S_{i-1}^*(X, \Delta).$$

Thus $S_i(X, \Delta)$ is a locally closed subscheme of X of pure dimension i and X is the disjoint union of the $S_i(X, \Delta)$. (Eventually we prove that the $S_i(X, \Delta)$ are normal and the $S_i^*(X, \Delta)$ are seminormal and Du Bois.) The collection of all the $S_i(X)$ is called the *log canonical stratification* of X. The *boundary* of the stratification is the closed subscheme

$$B(X, \Delta) := S_{\dim X - 1}^*(X, \Delta).$$

Our main interest is in understanding the log canonical stratification but for most arguments one needs two additional structures.

4.24.1 (Hereditary structure)   Frequently one needs information not just about the stratification of X, but also about the preimages of the lc stratification on the normalization of the boundary $\bar{B}(X) \to B(X) \hookrightarrow X$, on the normalization of the boundary of the normalization of the boundary

$$\bar{B}(\bar{B}(X)) \to B(\bar{B}(X)) \hookrightarrow \bar{B}(X) \to B(X) \hookrightarrow X$$

and so on. This is the theme of Sections 5.5 and 9.1.

4.24.2 (Source structure)   For some questions, besides the hereditary structure we also need to keep track of the sources of the lc centers. These considerations play a key role in Sections 5.6 and 5.7.

These topics complete the basic structure theory of lc (and slc) pairs. They focused entirely on divisors with discrepancy -1. Next we study centers of divisors with negative discrepancy.

4.25 (Log centers)   Let E be a divisor over X such that $a(E, X, \Delta) < 0$. Then $\text{center}_X E$ is called a *log center* of (X, Δ). Note that every irreducible component of Δ is a log center of (X, Δ), thus the log centers give a much fuller picture of (X, Δ) than the log canonical centers alone.

The main properties of log centers are proved in Sections 7.1 and 9.2. The guiding principle is that if $a(E, X, \Delta)$ is close to -1 then $\text{center}_X E$ behaves very much like a log canonical center but as $a(E, X, \Delta)$ gets close to 0 there is less and less to say.

4.26 (Historical remark I)   The first approach defining the different and the Poincaré residue map in general was developed in Kawamata (1998) and Kollár

(2007c). Looking at (4.22.3) we accept that the different is zero and think of $p_* \omega_{E/Z}$ as a secondary correction term.

Let (X, Δ) be an lc pair and $Z \subset X$ an lc center. The aim is to define the different and a new divisor class J_Z such that we get a (weaker) Poincaré residue isomorphism

$$\mathcal{WR}_{X \to Z}^m \colon \omega_X^{[m]}(m\Delta)|_Z \xrightarrow{\simeq} \omega_Z^{[m]}(mJ_Z + m\operatorname{Diff}_Z^* \Delta). \qquad (4.26.1)$$

A major difficulty is that while $\operatorname{Diff}_Z^* \Delta$ is an actual divisor, J_Z is only a $\mathbb{Q}$-linear equivalence class and there does not seem to be any sensible way to pick a divisor corresponding to J_Z. Correspondingly, $\mathcal{WR}_{X \to Z}$ is not as canonical as one would wish.

There have been several attempts to correct these shortcomings. A natural candidate is to work with the variation of Hodge structures $R^1 p_* \mathbb{C}_E$ since $p_* \omega_{E/Z}$ is the bottom piece of its Hodge filtration. This works quite well for families of elliptic curves. The higher dimensional case is discussed in Hacon and Xu (2011b).

Also, with this formulation, the two sides of (4.26.1) are not symmetric. It should be possible to work out a symmetric variant, but this seems to involve several technical issues about variations of mixed Hodge structures that are not yet settled.

4.27 (Historical remark II) Let (X, Δ) be an lc pair and $g \colon (Y, \Delta_Y) \to (X, \Delta)$ a log resolution. Let $E \subset Y$ be the union of *all* divisors that appear in Δ_Y with coefficient 1. Set $\Delta_E := (\Delta_Y - E)|_E$.

Note that E dominates the boundary $B(X, \Delta)$ and

$$g|_E \colon (E, \Delta_E) \to B(X, \Delta)$$

acts as a non-normal version of a resolution. Its properties were axiomatized by Ambro (2003) to arrive at the notion of a *quasi-log structure*. Working with quasi-log structures requires subtle use of mixed Hodge structures, but they were used to prove several important properties of lc centers (Ambro, 2003, 2011; Fujino, 2009b).

It was observed in Kollár and Kovács (2010) that many of the technical difficulties involving quasi-log structures go away if instead of an arbitrary log resolution we start with a $\mathbb{Q}$-factorial dlt modification $g \colon (X^{\mathrm{dlt}}, \Delta^{\mathrm{dlt}}) \to (X, \Delta)$ as in (1.36).

Our current treatment is a further simplification. Instead of keeping track of the whole reducible variety (E, Δ_E), we first understand the minimal lc centers of $(X^{\mathrm{dlt}}, \Delta^{\mathrm{dlt}})$ and then the hereditary lc centers of (X, Δ).

Of course, this approach relies on (1.36) and thus on the MMP, which was not available for Ambro (2003).

4.4 Crepant log structures

Definition 4.28   Let Z be a normal variety. A *weak crepant log structure* on Z is a normal pair (X, Δ) together with a proper, surjective morphism $f \colon X \to Z$ such that

(1) $f_* \mathcal{O}_X = \mathcal{O}_Z$, hence f has connected fibers, and
(2) $K_X + \Delta \sim_{f,\mathbb{Q}} 0$.

It is called a *crepant log structure* if, in addition,

(3) (X, Δ) is lc and Δ is effective.

Any lc pair (Z, Δ_Z) with Δ_Z effective has a trivial crepant log structure where $(X, \Delta) = (Z, \Delta_Z)$. Conversely, if f is birational then $(Z, \Delta_Z := f_* \Delta)$ is lc.

An irreducible subvariety $W \subset Z$ is a *log canonical center* or *lc center* of a weak crepant log structure $f \colon (X, \Delta) \to Z$ if and only if it is the image of an lc center $W_X \subset X$ of (X, Δ). We sometimes call a point an lc center if its closure is an lc center. By (4.15), a crepant log structure has only finitely many lc centers.

As in (4.15), it is sometimes convenient to view Z itself as an lc center.

For birational maps, the definitions of log canonical centers etc. give the same for the lc pair $(Z, f_* \Delta)$ and for the crepant log structure $f \colon (X, \Delta) \to Z$.

4.29 (Calabi–Yau pairs)   An lc pair (X, Δ) defined over a field k such that Δ is effective and $K_X + \Delta \sim_{\mathbb{Q}} 0$ is called a *Calabi–Yau pair* or a log Calabi–Yau variety. (This is probably the most general definition, many authors use a more restrictive variant.)

Each Calabi–Yau pair can be viewed as a crepant log structure $f \colon (X, \Delta) \to \operatorname{Spec} k$. Conversely, if $f \colon (X, \Delta) \to Z$ is a crepant log structure then there is an open set $Z^0 \subset Z$ such that the fiber (X_z, Δ_z) is a Calabi–Yau pair for all $z \in Z^0$. Thus a crepant log structure can be viewed as an especially nice compactification of a family of Calabi–Yau pairs over Z^0.

Let (X, Δ) be a dlt Calabi–Yau pair over $\mathbb{C}$ and $W_i \subset X$ the minimal (with respect to inclusion) lc centers. It turns out that

(1) every $(W_i, \Delta_i := \operatorname{Diff}^*_{W_i} \Delta)$ is a klt Calabi–Yau pair,
(2) the various (W_i, Δ_i) are birational to each other and
(3) for sufficiently divisible m there are canonical isomorphisms

$$H^0\big(X, \omega_X^{[m]}(m\Delta)\big) \simeq H^0\big(W_i, \omega_{W_i}^{[m]}(m\Delta_i)\big).$$

Here (1) is easy, (2–3) were proved for surfaces in Shokurov (1992) and then extended and used in many papers (Kollár, 1992; Keel *et al.*, 1994; Fujino, 2000; Kollár and Kovács, 2010; Fujino, 2011). Our results generalize these to arbitrary crepant log structures.

4.30 (Key examples of crepant log structures) Let (Z, Δ_Z) be an lc pair. By (1.36), there is a crepant log structure $f: (X, \Delta_X) \to (Z, \Delta_Z)$ such that X is $\mathbb{Q}$-factorial and (X, Δ_X) is dlt.

Let $Y \subset X$ be an lc center. Consider the Stein factorization

$$f|_Y : Y \xrightarrow{f_Y} Z_Y \xrightarrow{\pi} Z$$

and set $\Delta_Y := \mathrm{Diff}_Y^* \Delta_X$. By (4.16.4), (Y, Δ_Y) is dlt and $f_Y: (Y, \Delta_Y) \to Z_Y$ is a crepant log structure. In general, f_Y is not birational.

A key result of this section (4.42) establishes a very close relationship between lc centers of $f_Y: (Y, \Delta_Y) \to Z_Y$ and lc centers of $f: (X, \Delta_X) \to Z$.

4.31 (Birational weak crepant log structures) Let $f: (X, \Delta) \to Z$ be a (weak) crepant log structure. If f factors as $X \xrightarrow{g} X' \xrightarrow{f'} Z$ where g is birational, then $f': (X', \Delta' := g_* \Delta) \to Z$ is another (weak) crepant log structure. We say that $f: (X, \Delta) \to Z$ *dominates* $f': (X', \Delta') \to Z$.

Conversely, let $f': (X', \Delta') \to Z$ be a (weak) crepant log structure and $g: X \to X'$ a proper birational morphism. Write $K_X + \Delta \sim_{\mathbb{Q}} g^*(K_{X'} + \Delta')$. Then $f := f' \circ g: (X, \Delta) \to Z$ is a (weak) crepant log structure.

By (1.36), every crepant log structure $f: (X, \Delta) \to Z$ is dominated by another crepant log structure $f^*: (X^*, \Delta^*) \to Z$ such that (X^*, Δ^*) is dlt and $\mathbb{Q}$-factorial.

Crepant, birational maps between two (weak) crepant log structures are defined in (2.23).

Let $f: (X, \Delta) \to Z$ be a weak crepant log structure and $f': X' \to Z$ a proper morphism. Assume that there is a birational map $\phi: X \dashrightarrow X'$ such that $f' \circ \phi = f$. Then there is a unique $\mathbb{Q}$-divisor Δ' such that $f': (X', \Delta') \to Z$ is a weak crepant log structure that is birational to $f: (X, \Delta) \to Z$. If ϕ^{-1} has no exceptional divisors, then $\Delta' = \phi_* \Delta$ and hence Δ' is effective if Δ is. Thus we have proved the following.

Claim 4.31.1 Let $f: (X, \Delta) \to Z$ be a (weak) crepant log structure. Let $f': X' \to Z$ be a proper, normal Z-scheme and $\phi: X \dashrightarrow X'$ a birational contraction such that $f = f' \circ \phi$. Then $f': (X', \phi_* \Delta) \to Z$ is a (weak) crepant log structure and ϕ is crepant. □

The following is a weak version of the (semi)normality properties of lc centers discovered by Ambro (2003). The general case will be derived from it in Section 5.5. See (4.41) for other partial results.

Proposition 4.32 *Let $f: (X, \Delta) \to Z$ be a crepant log structure. Let $W \subset Z$ be the union of all lc centers (except Z itself) and $B(W) \subset W$ the union of all non-maximal (with respect to inclusion) lc centers. Then W is seminormal (10.11) and $W \setminus B(W)$ is normal.*

(For the future we note that the seminormality of W gives us conditions (SN) and (HSN) in (9.18).)

Proof   We may assume that (X, Δ) is dlt and $\mathbb{Q}$-factorial.

Let us start with the special case when $\lfloor\Delta\rfloor$ does not dominate Z. Then $W = f(\lfloor\Delta\rfloor)$. Write $\Delta = \lfloor\Delta\rfloor + \Delta'$ and consider the exact sequence

$$0 \to \mathscr{O}_X(-\lfloor\Delta\rfloor) \to \mathscr{O}_X \to \mathscr{O}_{\lfloor\Delta\rfloor} \to 0$$

and its push-forward

$$\mathscr{O}_Z = f_*\mathscr{O}_X \to f_*\mathscr{O}_{\lfloor\Delta\rfloor} \xrightarrow{\delta} R^1 f_*\mathscr{O}_X(-\lfloor\Delta\rfloor).$$

Note that $-\lfloor\Delta\rfloor \sim_{f,\mathbb{Q}} K_X + \Delta'$ and (X, Δ') is klt. Hence by (10.40.1), $R^i f_*\mathscr{O}_X(-\lfloor\Delta\rfloor)$ is torsion free for every i. On the other hand, $f_*\mathscr{O}_{\lfloor\Delta\rfloor}$ is supported on W, hence it is a torsion sheaf. Thus the connecting map δ is zero, hence $\mathscr{O}_Z \twoheadrightarrow f_*\mathscr{O}_{\lfloor\Delta\rfloor}$ is surjective. Since this map factors through $\mathscr{O}_W$, we conclude that $\mathscr{O}_W \twoheadrightarrow f_*\mathscr{O}_{\lfloor\Delta\rfloor}$ is also surjective, hence an isomorphism.

Note that $\lfloor\Delta\rfloor$ has only nodes at codimension 1 points and it is S_2 by (2.88). Thus $\lfloor\Delta\rfloor$ is seminormal by (10.14) and (10.15) implies that W is also seminormal.

To see the normality claim, let $V \subset \lfloor\Delta\rfloor$ be an irreducible component of its non-normal locus. By (4.16.2), V is an irreducible component of the intersection of two irreducible components of $\lfloor\Delta\rfloor$, hence an lc center of (X, Δ). Thus $f(V) \subset X$ is an lc center. Hence either $f(V)$ is an irreducible component of W or $f(V) \subset B(W)$. Thus (10.15.1) implies that $W \setminus B(W)$ is normal.

There are two ways to extend this method to the general case.

First variant   We deal with the dominant components of $\lfloor\Delta\rfloor$ directly.

Write $\Delta = \lfloor\Delta\rfloor + \Delta'$ and $\lfloor\Delta\rfloor = D_1 + D_2$ where D_1 is the sum of those components that dominate Z and D_2 is the sum of those components that do not dominate Z. By (1.38), we may assume that every irreducible component of W is dominated by an irreducible component of D_2, that is, $W = f(D_2)$. As before, we use the exact sequence

$$0 \to \mathscr{O}_X(-D_2) \to \mathscr{O}_Y \to \mathscr{O}_{D_2} \to 0$$

and its push-forward

$$\mathscr{O}_Z = f_*\mathscr{O}_X \to f_*\mathscr{O}_{D_2} \xrightarrow{\delta} R^1 f_*\mathscr{O}_X(-D_2).$$

Note that $-D_2 \sim_{f,\mathbb{Q}} K_X + D_1 + \Delta'$ and $(X, D_1 + \Delta')$ is dlt. We can finish the proof as before if $R^1 f_*\mathscr{O}_Y(-D_2)$ is torsion free. This is not true in general, but (10.40.2) applies if every lc center of $(X, D_1 + \Delta')$ dominates Z. Using (10.45.2) this can be achieved if we replace (X, Δ) by a suitable other model.

Second variant Here we restrict to a dominant component of $\lfloor \Delta \rfloor$ and use induction.

Let $E \subset \lfloor \Delta \rfloor$ be an irreducible component that dominates Z. Let $f_E \colon E \to Z_E$ and $\pi \colon Z_E \to Z$ be the Stein factorization of $f|_E$. Set $\Delta_E := \operatorname{Diff}_E^* \Delta$. Then $f_E \colon (E, \Delta_E) \to Z_E$ is a dlt, crepant log structure. Let $W_E \subset Z_E$ be the union of its lc centers and $B(W_E) \subset W_E$ the union of its non-maximal lc centers. By induction on the dimension, W_E is seminormal and $W_E \setminus B(W_E)$ is normal. Thus,

$$\text{if } W_E = \pi^{-1}(W) \quad \text{and} \quad B(W_E) = \pi^{-1}(B(W)) \qquad (4.32.1)$$

both hold, then (10.26) implies that W is seminormal and $W \setminus B(W)$ is normal. We prove (4.32.1) in (4.42).

As we see there, $W_E \subset \pi^{-1}(W)$ is clear, but the converse is not. Every irreducible component $W^i \subset W$ is dominated by an lc center V^i of (X, Δ), but if V^i is disjoint from E then we do not get any obvious lc center on E. The Connectedness theorem (Kollár, 1992, 17.4) is designed to deal with such issues. Here we need a non-birational version of it. Once that is done in (4.42), the second variant of the proof will also be complete. $\square$

Connectedness theorems

Let $f \colon (X, \Delta) \to Z$ be a dlt, crepant log structure. It is quite useful to understand the connected components of $\lfloor \Delta \rfloor$ and, more generally, connected components of the fibers of $f \colon \lfloor \Delta \rfloor \to Z$.

If f is birational, a complete answer is given by the Connectedness theorem proved in Kollár (1992, 17.4). For somewhat simpler proofs see Kollár and Mori (1998, 5.48) or Kollár *et al.* (2004, section 6.7).

Theorem 4.33 (Connectedness theorem) *Let $g \colon Y \to Z$ be a proper, birational morphism, Y smooth and Δ_Y an snc $\mathbb{Q}$-divisor on Y such that $g_* \Delta_Y$ is effective and $-(K_Y + \Delta_Y)$ is g-nef. Write $\Delta_Y = \Delta_Y^+ - \Delta_Y^-$ where Δ_Y^+, Δ_Y^- are effective and without common irreducible components. Then*

$$g|_{\lfloor \Delta_Y^+ \rfloor} \colon \lfloor \Delta_Y^+ \rfloor \to Z \quad \text{has connected fibers.} \qquad \square$$

Two easy consequences for crepant log structures are the following.

Corollary 4.34 *Let (X, Δ_X) be pair with Δ_X effective and $g \colon (Y, \Delta_Y) \to (X, \Delta_X)$ a crepant log resolution (2.23). Then*

$$g|_{\lfloor \Delta_Y^+ \rfloor} \colon \lfloor \Delta_Y^+ \rfloor \to \text{non-klt}(X, \Delta_X)$$

is surjective and has connected fibers. $\square$

Corollary 4.35   *Let* $f_i\colon (X_i, \Delta_i) \to Z$ *be weak crepant log structures with* Δ_i
effective and $\phi\colon (X_1, \Delta_1) \dashrightarrow (X_2, \Delta_2)$ *a crepant, birational map.*

(1) *There is a bijection between the connected components of the* non-klt(X_i, Δ_i).
(2) *If the* (X_i, Δ_i) *are dlt then there is a bijection between the connected components of the* $\lfloor \Delta_i \rfloor$.
(3) *If* (X_1, Δ_1) *is klt (resp. lc) then* (X_2, Δ_2) *is also klt (resp. lc).*
(4) *If* (X_1, Δ_1) *is plt and* ϕ *does not contract any irreducible component of* $\lfloor \Delta_1 \rfloor$ *then* (X_2, Δ_2) *is also plt.*

Proof   We can choose (Y, Δ_Y) in (4.31.1) to be snc. Let Δ_Y^+ be the effective part of Δ_Y. By (4.34) the fibers of $h_i\colon \lfloor \Delta_Y^+ \rfloor \to$ non-klt(X_i, Δ_i) are connected. Thus the preimage of a connected component of non-klt(X_i, Δ_i) is a connected component of $\lfloor \Delta_Y^+ \rfloor$ proving (1). If the (X_i, Δ_i) are dlt then non-klt$(X_i, \Delta_i) = \lfloor \Delta_i \rfloor$, thus (1) implies (2). Finally (3–4) follow from (2.23.4).   □

If a crepant log structure $f\colon (X, \Delta) \to Z$ is not birational, the fibers of $f\colon \lfloor \Delta \rfloor \to Z$ need not be connected. An easy example is given by $f\colon (\mathbb{P}^1, [0] + [\infty]) \to \operatorname{Spec} k$. Surprisingly, this is essentially the only one. First we give the examples and then we show that there are no other cases.

Definition 4.36   A *standard* $\mathbb{P}^1$*-link* is a $\mathbb{Q}$-factorial pair $(X, D_1 + D_2 + \Delta)$ plus a proper morphism $\pi\colon X \to S$ such that

(1) $K_X + D_1 + D_2 + \Delta \sim_{\mathbb{Q},\pi} 0$,
(2) $(X, D_1 + D_2 + \Delta)$ is plt (in particular, D_1 and D_2 are disjoint),
(3) $\pi\colon D_i \to S$ are both isomorphisms and
(4) every reduced fiber red X_s is isomorphic to $\mathbb{P}^1$.

Let F denote a general smooth fiber. Then $((K_X + D_1 + D_2) \cdot F) = 0$, hence $(\Delta \cdot F) = 0$. That is, Δ is a vertical divisor, the projection gives an isomorphism $(D_1, \operatorname{Diff}_{D_1} \Delta) \simeq (D_2, \operatorname{Diff}_{D_2} \Delta)$ and these pairs are klt by (4.9).

The simplest example of a standard $\mathbb{P}^1$-link is a product

$$(S \times \mathbb{P}^1, S \times \{0\} + S \times \{\infty\} + \Delta_S \times \mathbb{P}^1) \to S$$

for some $\mathbb{Q}$-divisor Δ_S. More generally, assume that the cyclic group μ_m acts on (S, Δ_S) such that its fixed point set has codimension ≥ 2. If ε is a primitive mth root of unity then $(x : y) \mapsto (\varepsilon x : y)$ generates a μ_m action on $\mathbb{P}^1$. Then

$$(S \times \mathbb{P}^1, S \times \{0\} + S \times \{\infty\} + \Delta_S \times \mathbb{P}^1)/\mu_m \to S/\mu_m$$

is a standard $\mathbb{P}^1$-link which is not a product. We see at the end of the proof of (4.37) that every standard $\mathbb{P}^1$-link is locally the quotient of a product (Kollár, 2011d).

For later applications, in the following result we need to keep careful track of the residue fields.

Proposition 4.37 *Let (X, Δ) be a klt pair with Δ effective, D an effective $\mathbb{Z}$-divisor on X and $f: X \to Z$ a proper morphism such that $K_X + \Delta + D \sim_{\mathbb{Q}, f} 0$. Fix a point $z \in Z$ and assume that $f^{-1}(z)$ is connected but $f^{-1}(z) \cap D$ is disconnected (as $k(z)$-schemes).*

(1) *There is an étale morphism $(z' \in Z') \to (z \in Z)$ and a proper morphism $S' \to Z'$ such that $k(z) = k(z')$ and $(X, \Delta + D) \times_Z Z'$ is birational to a standard $\mathbb{P}^1$-link over S'.*

(2) *$(X, \Delta + D)$ is plt in a neighborhood of $f^{-1}(z)$.*

(3) *$f^{-1}(z) \cap D$ has 2 connected components and there is a crepant, birational involution of $(D, \mathrm{Diff}_D \Delta)$ that interchanges these two components.*

Proof By (1.37) we may assume that X is $\mathbb{Q}$-factorial. By passing to an étale neighborhood of $(z \in Z)$ with the same residue field $k(z)$ we may assume that different connected components of $f^{-1}(z) \cap D$ are contained in different connected components of D (4.38.1).

First we show that D dominates Z. Indeed, consider the exact sequence

$$0 \to \mathcal{O}_X(-D) \to \mathcal{O}_X \to \mathcal{O}_D \to 0$$

and its push-forward

$$f_* \mathcal{O}_X \to f_* \mathcal{O}_D \xrightarrow{\delta} R^1 f_* \mathcal{O}_X(-D).$$

Since $-D \sim_{\mathbb{Q}, f} K_X + \Delta$, the sheaf $R^1 f_* \mathcal{O}_X(-D)$ is torsion free by (10.40.1). If D does not dominate Z, then $f_* \mathcal{O}_D$ is a torsion sheaf, hence $\delta = 0$. Therefore $f_* \mathcal{O}_X \twoheadrightarrow f_* \mathcal{O}_D$ is surjective. This is, however, impossible since $\mathcal{O}_D$ has a section that is 0 on one connected component of $f^{-1}(z) \cap D$ and 1 on the other, but every section of $\mathcal{O}_X$ that vanishes at a point of red $f^{-1}(z)$ vanishes identically on red $f^{-1}(z)$.

Since D dominates Z, $K_X + \Delta$ is not pseudo-effective on the generic fiber of f and so, by Birkar *et al.* (2010, 1.3.2), the (X, Δ)-MMP over Z terminates with a Fano contraction $p: (X^*, \Delta^*) \to S$. We prove that the latter is a standard $\mathbb{P}^1$-link.

By (4.35) $p^{-1}(z) \cap D^*$ is still disconnected. By assumption there is an irreducible component, say $D_1^* \subset D^*$ that has positive intersection with the contracted ray. Therefore D_1^* is p-ample. Furthermore, there is another irreducible component, say $D_2^* \subset D^*$ that is disjoint from D_1^*. Let $F_s \subset X^*$ be any fiber of p that intersects D_2^*. Since D_2^* is disjoint from D_1^*, we see that D_2^* does not contain F_s. Thus D_2^* also has positive intersection with the contracted ray, hence D_2^* is also p-ample.

Thus D_1^* and D_2^* both intersect every curve contracted by p and they are disjoint. Thus every fiber of p has dimension 1 and a general fiber F_g is a smooth rational curve. Since $(F_g \cdot (\Delta^* + D^*)) = 2$, we see that D_1^* and D_2^* are sections of p, they both appear in D with coefficient 1 and the rest of $\Delta^* + D^*$ consists of vertical divisors. Since $f^{-1}(z) \cap D^*$ is disconnected, the vertical components of D^* are disjoint from $f^{-1}(z)$. Thus, by shrinking Z, we may assume that $D^* = D_1^* + D_2^*$.

Since p is an extremal contraction, $R^1 p_* \mathcal{O}_{X^*} = 0$ hence $H^1(F_s, \mathcal{O}_{F_s}) = 0$ and so red F_s is a tree of smooth rational curves for every $s \in S$. Both of $D_i^* \cap \mathrm{red}\, F_s$ are points that are common to all irreducible components of red F_s. This is only possible if red $F_s \simeq \mathbb{P}^1_{k(s)}$ for every $s \in S$.

Note that $D_1^* \sim_{p,\mathbb{Q}} D_2^*$, hence, working locally on S we may assume that $D_1^* \sim_{\mathbb{Q}} D_2^*$. If $D_1^* \sim D_2^*$ then these two sections of $\mathcal{O}_{X^*}(D_1^*) \simeq \mathcal{O}_{X^*}(D_2^*)$ give an isomorphism $X^* \simeq S \times \mathbb{P}^1$. Thus there is a $\mathbb{Q}$-divisor Δ_S on S such that

$$(X^*, \Delta^* + D^*) \simeq (S \times \mathbb{P}^1, \Delta_S \times \mathbb{P}^1 + S \times \{(0{:}1)\} + S \times \{(1{:}0)\}).$$

Since (X^*, Δ^*) is klt, so is (S, Δ_S) and thus (4.9.1) implies that $(X^*, \Delta^* + D^*)$ is plt. Thus $(X, \Delta + D)$ is also plt by (4.35.4).

Otherwise pick the smallest $m > 0$ such that $m(D_1^* - D_2^*) \sim 0$. This gives a cyclic cover $\pi \colon \tilde{X}^* \to X^*$ that ramifies along the locus where $D_1^* + D_2^*$ is not Cartier. Thus the preimage of a general fiber F_g consists of m disjoint copies of F_g but the preimage of F_z is the unique degree m cyclic cover ramified at $F_z \cap (D_1^* + D_2^*)$. The previous argument shows that $\tilde{X}^* \simeq \tilde{S} \times \mathbb{P}^1$ where $\tilde{S} \to S$ is the Stein factorization of $\tilde{X}^* \to S$. Thus $(X^*, \Delta^* + D^*)$ is a cyclic quotient of a plt pair hence plt (2.43). $\square$

4.38 A useful subclass of morphisms of pointed schemes is given by those étale morphisms $\pi \colon (z' \in Z') \to (z \in Z)$ such that π^* induces an isomorphism $k(z) \simeq k(z')$. They do not seem to have acquired a standard name. They are called *étale neighborhoods* in Milne (1980, p. 36), *strongly étale* in Kurke *et al.* (1975, p. 108) and *strictly étale* by some authors. None of these names became generally accepted and in fact they are used with different meaning by others, so we leave them nameless.

The *Henselization* (Milne, 1980, p. 36) of the local ring $\mathcal{O}_{Z,z}$ is the direct limit of these local rings $\mathcal{O}_{Z',z'}$. Thus the basic characterization of Henselian rings (Milne, 1980, I.4.2) implies the following.

Claim 4.38.1 Let $p \colon W \to Z$ be a proper morphism. Then there is an étale morphism $\pi \colon (z' \in Z') \to (z \in Z)$ such that $k(z) = k(z')$ and taking the fiber induces a one-to-one correspondence between connected components of $W \times_Z Z'$ and connected components of $p^{-1}(z) \simeq p'^{-1}(z')$. $\square$

$\mathbb{P}^1$-linked lc centers

Definition 4.39 ($\mathbb{P}^1$-linking) (Kollár, 2011d) Let $g: (X, \Delta) \to S$ be a crepant, dlt log structure and $Z_1, Z_2 \subset X$ two lc centers. We say that Z_1, Z_2 are *directly* $\mathbb{P}^1$-*linked* if there is an lc center $W \subset X$ containing the Z_i such that $g(W) = g(Z_1) = g(Z_2)$ and $(W, \mathrm{Diff}^*_W \Delta)$ is birational to a standard $\mathbb{P}^1$-link (4.36) with Z_i mapping to D_i. (As in (4.15), we allow $W = X$.)

We say that $Z_1, Z_2 \subset X$ are $\mathbb{P}^1$-*linked* if there is a sequence of lc centers $Z'_1, \dots, Z'_m$ such that $Z'_1 = Z_1$, $Z'_m = Z_2$ and Z'_i is directly $\mathbb{P}^1$-linked to Z'_{i+1} for $i = 1, \dots, m - 1$ (or $Z_1 = Z_2$). Every $\mathbb{P}^1$-linking defines a birational map between $(Z_1, \mathrm{Diff}^*_{Z_1} \Delta)$ and $(Z_2, \mathrm{Diff}^*_{Z_2} \Delta)$.

Note that Z_1 is a minimal lc center if and only if $(Z_1, \mathrm{Diff}^*_{Z_1} \Delta)$ is klt. By (4.35) the latter holds if and only if $(Z_2, \mathrm{Diff}^*_{Z_2} \Delta)$ is klt, hence when Z_2 is also a minimal lc center.

The following theorem, which is a strengthening of Kollár and Kovács (2010, 1.7), plays an important role in the study of lc centers in the non-dlt case.

Theorem 4.40 (Kollár, 2011d) *Let k be a field of characteristic 0 and S essentially of finite type over k. Let $f: (X, \Delta) \to S$ be a proper morphism such that $K_X + \Delta \sim_{\mathbb{Q}, f} 0$ and (X, Δ) is dlt. Let $s \in S$ be a point such that $f^{-1}(s)$ is connected (as a $k(s)$-scheme). Let $Z \subset X$ be minimal (with respect to inclusion) among the lc centers of (X, Δ) such that $s \in f(Z)$. Let $W \subset X$ be an lc center of (X, Δ) such that $s \in f(W)$.*

Then there is an lc center $Z_W \subset W$ such that Z and Z_W are $\mathbb{P}^1$-linked.

In particular, all the minimal (with respect to inclusion) lc centers $Z_i \subset X$ such that $s \in f(Z_i)$ are $\mathbb{P}^1$-linked to each other.

Proof First we show that the conclusions hold over an étale neighborhood $(s' \in S') \to (s \in S)$ such that $k(s) \simeq k(s')$ (4.38). We use induction on $\dim X$ and on $\dim Z$.

If $f^{-1}(s) \cap \lfloor \Delta \rfloor$ is disconnected then by (4.37), after an étale base change there are exactly two lc centers intersecting $f^{-1}(s)$ and they are $\mathbb{P}^1$-linked; thus we are done.

If $f^{-1}(s) \cap \lfloor \Delta \rfloor$ is connected, write $\lfloor \Delta \rfloor = \sum D_i$. By passing to an étale neighborhood of $s \in S$ without changing $k(s)$, we may assume that each D_i has connected fiber over s and every lc center of (X, Δ) intersects $f^{-1}(s)$ (4.38.1).

By suitable indexing we may assume that $Z \subset D_1$, $W \subset D_r$ and $f^{-1}(s) \cap D_i \cap D_{i+1} \neq \emptyset$ for $i = 1, \dots, r - 1$.

By induction, we can apply (4.40) to $D_1 \to S$ with Z as Z and $D_1 \cap D_2$ as W. We get that there is an lc center $Z_2 \subset D_1 \cap D_2$ such that Z and Z_2 are $\mathbb{P}^1$-linked. As we noted in (4.39), Z_2 is also minimal (with respect to inclusion) among the lc centers of (X, Δ) such that $s \in f(Z_2)$. Note that Z_2 is an lc center of $(D_1, \mathrm{Diff}^*_{D_1} \Delta)$. By adjunction, it is an lc center of (X, Δ) and also an lc center of $(D_2, \mathrm{Diff}^*_{D_2} \Delta)$.

Next we apply (4.40) to $D_2 \to S$ with Z_2 as Z and $D_2 \cap D_3$ as W, and so on. At the end we work on $D_r \to S$ with Z_r as Z and W as W to get an lc center $Z_W \subset W$ such that Z and Z_W are $\mathbb{P}^1$-linked.

Finally let us show that the étale base change was not necessary.

Let $\tilde{f} \colon (X, \Delta) \to \tilde{S}$ be the Stein factorization and $\tilde{s} \in \tilde{S}$ the unique preimage of s. Let $Z_i \subset X$ be the minimal lc centers (with respect to inclusion) satisfying $s \in f(Z_i)$. Since lc centers commute with étale base change, we see that there is a unique irreducible subvariety $\tilde{s} \in \tilde{V} \subset \tilde{S}$ such that $\tilde{V} = \tilde{f}(Z_i)$ for every i.

Let $\tilde{v} \in \tilde{V}$ be the generic point and apply (4.40) to $\tilde{f} \colon (X, \Delta) \to \tilde{S}$ and $\tilde{v} \in \tilde{S}$. (The fibers of $\tilde{f}$ are connected; this is why we took the Stein factorization.) We already proved that the conclusions hold after an étale base change $\tilde{\pi} \colon (\tilde{v}' \in \tilde{S}') \to (\tilde{v} \in \tilde{S})$ that induces an isomorphism of the fibers

$$\tilde{\pi} \colon (\tilde{f}')^{-1}(\tilde{v}') \simeq \tilde{f}^{-1}(\tilde{v}).$$

Thus the Z_i canonically lift to $Z_i' \simeq Z_i$ and the $\mathbb{P}^1$-links between the Z_i' descend to $\mathbb{P}^1$-links between the Z_i. $\square$

There are two immediate consequences about lc centers.

Corollary 4.41 $f \colon (X, \Delta_X) \to S$ *be a crepant log structure. Then*

(1) *Every point $s \in S$ is contained in a unique smallest (with respect to inclusion) lc center $W_s \subset S$. This W_s is unibranch (1.44) at s.*

(2) *Any intersection of lc centers is also a union of lc centers.*

(The unibranch property gives us conditions (U) and (HU) in (9.18). We will also prove in Section 5.5 that W_s is in fact normal.)

Proof $W_s = f(Z_i)$ where Z_i is any minimal lc center as in (4.40). Thus W_s is unique. Uniqueness continues to hold in every étale neighborhood of $s \in S$, thus W_s is unibranch at s. If W_1, W_2 are lc centers, let $s \in W_1 \cap W_2$ be any point. Then $W_s \subset W_1 \cap W_2$, proving (2). $\square$

Corollary 4.42   *Let $f \colon (X, \Delta_X) \to Z$ be a dlt, crepant log structure. Let $Y \subset X$ be an lc center. Consider the Stein factorization*

$$f|_Y \colon Y \xrightarrow{f_Y} Z_Y \xrightarrow{\pi} Z$$

and set $\Delta_Y := \mathrm{Diff}^*_Y \Delta_X$. Then

(1) *$f_Y \colon (Y, \Delta_Y) \to Z_Y$ is a dlt, crepant log structure.*

(2) *Let $W_Y \subset Z_Y$ be an lc center of $f_Y \colon (Y, \Delta_Y) \to Z_Y$. Then $\pi(W_Y) \subset Z$ is an lc center of $f \colon (X, \Delta_X) \to Z$.*

(3) *Let $W \subset Z$ be an lc center of $f: (X, \Delta_X) \to Z$. Then every irreducible component of $\pi^{-1}(W)$ is an lc center of $f_Y: (Y, \Delta_Y) \to Z_Y$.*

Proof (1) follows from (4.16.4). To see (2), note that W_Y is dominated by an lc center V_Y of $(Y, \mathrm{Diff}_Y^* \Delta)$. Thus, by (4.9), V_Y is also an lc center of (X, Δ), hence $\pi(W_Y) = f(V_Y)$ is an lc center of Z.

Finally let $W \subset Z$ be an lc center of $f: (X, \Delta_X) \to Z$ and $s \in W$ the generic point. Let $V_X \subset X$ be a minimal lc center that dominates W. By (4.40), there is an lc center $V_Y \subset Y$ that is $\mathbb{P}^1$-linked to V_X. By (4.9), V_Y is also an lc center of $(Y, \mathrm{Diff}_Y^* \Delta)$. Thus $f_Y(V_Y) \subset Z_Y$ is an lc center of $f_Y: (Y, \Delta_Y) \to Z_Y$ and it is also one of the irreducible components of $\pi^{-1}(W)$.

In order to get (3), after replacing Z by an étale neighborhood of s, we may assume that $Y = \cup Y_j$ such that each $f^{-1}(s) \cap Y_j$ is connected. By the previous argument, each Y_j yields an lc center $f_{Y_j}(V_{Y_j}) \subset Z_{Y_j}$ and together these show that every irreducible component of $\pi^{-1}(W)$ is an lc center of $f_Y: (Y, \Delta_Y) \to Z_Y$. $\square$

Note that in (4.40) we do not assume that f has connected fibers; this is useful over non-closed fields. The following example illustrates some of the subtler aspects.

Example 4.43 Set $X = \mathbb{A}^3$ and D_1, D_2, D_3 planes intersecting only at the origin. Let $\pi: B_0 X \to X$ denote the blow-up of the origin with exceptional divisor E. Then $K_{B_0 X} + E + \sum D_i' \sim \pi^*(K_X + \sum D_i)$ where $D_i' := \pi_*^{-1} D_i$. There are three minimal lc centers over 0, given by $p_i := E \cap D_{i-1}' \cap D_{i+1}'$ (with indexing modulo 3).

Assume now that we are over $\mathbb{Q}$, D_1 is defined over $\mathbb{Q}$ and $\mathrm{Gal}(\bar{\mathbb{Q}}/\mathbb{Q})$ interchanges D_2, D_3. Now there are two minimal lc centers over 0. One is p_1 the other is the irreducible $\mathbb{Q}$-scheme $p_2 + p_3$. Thus p_1 and $p_2 + p_3$ cannot be $\mathbb{P}^1$-linked. This is not a contradiction since $(B_0 X, E + \sum D_i')$ is not dlt; the divisor $D_2' + D_3'$ (which is irreducible over $\mathbb{Q}$) is not normal. We get a dlt model by blowing up the curve $D_2' \cap D_3'$. Now there are two minimal lc centers over 0, both isomorphic to $p_2 + p_3$.

Similarly, if $\mathrm{Gal}(\bar{\mathbb{Q}}/\mathbb{Q})$ permutes the three planes D_i, then we need to blow-up all three intersections $D_i' \cap D_j'$ to get a dlt model. Over $\bar{\mathbb{Q}}$, there are six minimal lc centers over 0. Over $\mathbb{Q}$ there is either only one (if $\mathrm{Gal}(\bar{\mathbb{Q}}/\mathbb{Q})$ acts on the planes as the symmetric group S_3) or two, both consisting of three conjugate points and isomorphic as $\mathbb{Q}$-schemes to each other (if $\mathrm{Gal}(\bar{\mathbb{Q}}/\mathbb{Q})$ permutes cyclically).

Finally let us see what happens with lc centers for birational morphisms between dlt pairs.

Lemma 4.44 *Let (X, Δ) be dlt and $f: (Y, \Delta_Y) \to (X, \Delta)$ a log resolution. Let $T \subset X$ be an lc center of (X, Δ) and $S \subset Y$ a minimal lc center of (Y, Δ_Y) dominating T. Then $f: S \to T$ is birational.*

Proof It is enough to prove that $Y \to X$ is dominated by a $Y' \to X$ that satisfies the conclusion. We construct such $Y' \to X$ in two steps.

As we see in (10.45.2), there is a log resolution $g: X' \to X$ of (X, Δ) such that every lc center of (X', Δ') maps birationally to an lc center of (X, Δ). This reduces us to the case when Δ is a snc divisor. Then, by induction, it is sufficient to show that (4.44) holds for one blow-up $p_Z: B_Z X \to X$ where $Z \subset X$ has snc with Δ.

If $T \not\subset Z$ then p_Z is an isomorphism over the generic point of T and the birational transform of T is the only lc center of $(B_Z X, \Delta_{B_Z X})$ that dominates T.

It remains to consider the case when $T \subset Z$. Let $D_1, \ldots, D_r$ be the components of $\lfloor \Delta \rfloor$ such that T is an irreducible component of $D_1 \cap \cdots \cap D_r$. Since Z has snc with Supp Δ, we see that Z is the intersection of some of the D_i, say $Z = D_1 \cap \cdots \cap D_j$.

The exceptional divisor E_Z of p_Z is a $\mathbb{P}^{j-1}$-bundle over Z and it appears in $\Delta_{B_Z X}$ with coefficient 1. There are r different minimal lc centers that dominate Z, obtained by intersecting E_Z with $j - 1$ of the birational transforms of the $D_1, \ldots, D_j$ and with the birational transforms of the $D_{j+1}, \ldots, D_r$. Each of these centers maps isomorphically to Z. (Note that this step could fail if Δ is only nc but not snc.) $\qquad\square$

4.5 Sources and springs of log canonical centers

Let (X, Δ) be an lc pair and $S \subset X$ a codimension 1 lc center, that is, an irreducible component of $\lfloor \Delta \rfloor$. The main assertion of adjunction theory is that the pair $(S, \mathrm{Diff}_S^* \Delta)$ encodes most of the important characteristics of a neighborhood of S in (X, Δ). One would like to find similar results for higher codimension lc centers.

For inductive purposes one should switch from pairs to crepant, dlt, log structures $f: (X, \Delta) \to Y$ and work with lc centers $Z \subset Y$.

When $Y = \mathrm{Spec}\,\mathbb{C}$, we have the theory of Calabi–Yau pairs (4.29). The following theorem is thus a generalization of their properties (4.29.1–3). Note, however, that the Galois property (4.45.5), which is key to many applications, has no analog in the classical setting.

Theorem–Definition 4.45 (Kollár, 2011d) *Let $f: (X, \Delta) \to Y$ be a crepant, dlt, log structure and $Z \subset Y$ an lc center with normalization $n: Z^n \to Z$.*

Let $S \subset X$ be a minimal (with respect to inclusion) lc center of (X, Δ) that dominates Z. Set $\Delta_S := \mathrm{Diff}^*_S \Delta$ and let $f^{\mathrm{n}}_S \colon S \to Z_S \to Z^{\mathrm{n}}$ denote the Stein factorization. Then:

(1) *(Uniqueness of sources)* The crepant-birational equivalence class of (S, Δ_S) does not depend on the choice of S. It is called the source of Z and denoted by $\mathrm{Src}(Z, X, \Delta)$.

(2) *(Uniqueness of springs)* The isomorphism class of Z_S does not depend on the choice of S. It is called the spring of Z and denoted by $\mathrm{Spr}(Z, X, \Delta)$.

(3) *(Crepant log structure)* (S, Δ_S) is dlt, $K_S + \Delta_S \sim_{\mathbb{Q}, f} 0$ and (S, Δ_S) is klt on the generic fiber of $f|_S$.

(4) *(Poincaré residue map)* For $m > 0$ sufficiently divisible, there are well defined isomorphisms

$$\omega^{[m]}_X(m\Delta)|_S \simeq \omega^{[m]}_S(m\Delta_S) \quad \text{and}$$
$$n^*\big(f_*(\omega^{[m]}_X(m\Delta))\big) \simeq \big((f^{\mathrm{n}}_S)_*\omega^{[m]}_S(m\Delta_S)\big)^{\mathrm{inv}}$$

where the exponent inv denotes the invariants under the action of the group of crepant, birational self-maps $\mathrm{Bir}^{\mathrm{c}}_Z(S, \Delta_S)$.

(5) *(Galois property)* The extension $Z_S \to Z$ is Galois and $\mathrm{Bir}^{\mathrm{c}}_Z(S, \Delta_S) \twoheadrightarrow \mathrm{Gal}(Z_S/Z)$ is surjective.

(6) *(Adjunction)* Let $W \subset X$ be an lc center; $f_W \colon W \to W_Y$ and $n_W \colon W_Y \to Y$ the Stein factorization of $f|_W$. Let $Z_W \subset W_Y$ be an irreducible subvariety such that $n_W(Z_W) = Z$. Then Z_W is an lc center of $(W, \mathrm{Diff}^*_W \Delta)$ and

$$\mathrm{Src}(Z_W, W, \mathrm{Diff}^*_W \Delta) \overset{\mathrm{cbir}}{\sim} \mathrm{Src}(Z, X, \Delta) \quad \text{and}$$
$$\mathrm{Spr}(Z_W, W, \mathrm{Diff}^*_W \Delta) \simeq \mathrm{Spr}(Z, X, \Delta).$$

(7) *(Birational invariance)* Let $f_i \colon (X_i, \Delta_i) \to Y$ be birational crepant, dlt, log structures for $i = 1, 2$. Then

$$\mathrm{Src}(Z, X_1, \Delta_1) \overset{\mathrm{cbir}}{\sim} \mathrm{Src}(Z, X_2, \Delta_2) \quad \text{and}$$
$$\mathrm{Spr}(Z, X_1, \Delta_1) \simeq \mathrm{Spr}(Z, X_2, \Delta_2).$$

We emphasize that the source is a pair (up to birational equivalence) but the spring is a variety (up to isomorphism).

Both the source and the spring localize in the obvious way in the Zariski topology but they both vary with base field extensions. In fact, the spring carries rather interesting arithmetic information about the lc center Z.

Proof By (4.40), different choices of S are $\mathbb{P}^1$-linked to each other, hence they are birational, proving (4.45.1). This implies (4.45.2) while (4.45.3) is clear.

Next consider the Poincaré residue map

$$\mathcal{R}^m_{X \to S} \colon \omega^{[m]}_X(m\Delta)|_S \overset{\simeq}{\longrightarrow} \omega^{[m]}_S(m \cdot \mathrm{Diff}^*_S \Delta)$$

as defined in (4.18). Choose $m > 0$ even such that $\omega_X^{[m]}(m\Delta) \sim f^*L$ for some line bundle L on Y. We can then view the Poincaré residue map as

$$\mathcal{R}_{X \to S}^m \colon f^*L|_S \simeq \omega_X^{[m]}(m\Delta)|_S \xrightarrow{\simeq} \omega_S^{[m]}(m \cdot \mathrm{Diff}_S^* \Delta).$$

Let S_1, S_2 be minimal lc centers of (X, Δ) such that $f(S_1) = f(S_2)$. We claim that there is a birational map $\phi \colon S_2 \dashrightarrow S_1$ such that the following diagram commutes

$$
\begin{array}{ccccc}
\omega_X^{[m]}(m\Delta) & \xrightarrow{\simeq} & f^*L & \xrightarrow{\simeq} & \omega_X^{[m]}(m\Delta) \\
{\scriptstyle \mathcal{R}_{X \to S_1}^m} \downarrow & & & & \downarrow {\scriptstyle \mathcal{R}_{X \to S_2}^m} \\
\omega_{S_1}^{[m]}(m\, \mathrm{Diff}_{S_1}^* \Delta) & & \xrightarrow{\phi^*} & & \omega_{S_2}^{[m]}(m\, \mathrm{Diff}_{S_2}^* \Delta)
\end{array}
\qquad (4.45.8)
$$

By (4.40) it is sufficient to prove this in case S_1, S_2 are directly $\mathbb{P}^1$-linked (4.39) in $(W, \mathrm{Diff}_W^* \Delta)$ for some lc center $W \subset X$. By definition we have a map $W \to U$ whose generic fiber is $\mathbb{P}^1$ with S_1, S_2 as sections. Thus projection to U provides a birational isomorphism $\phi \colon S_2 \dashrightarrow S_1$.

Since $\mathcal{R}_{X \to S_i}^m = \mathcal{R}_{W \to S_i}^m \circ \mathcal{R}_{X \to W}^m$, we may assume that $X = W$. The sheaves in (4.45.8) are torsion free, hence it is enough to check commutativity after localizing at the generic point of U. This reduces us to the case when $W = \mathbb{P}_L^1$ with coordinates $(x{:}y)$, $S_1 = (0{:}1)$ and $S_2 = (1{:}0)$. A generator of $H^0(\mathbb{P}^1, \omega_{\mathbb{P}^1}(S_1 + S_2))$ is dx/x which has residue 1 at S_1 and -1 at S_2. Thus (4.45.8) commutes for m even and anti-commutes for m odd.

This gives a Poincaré residue map as stated in (4.45.4) but it is not yet completely canonical. We think of (S, Δ_S) as an element of a birational equivalence class, thus so far $\mathcal{R}^m$ is defined only up to the action of $\mathrm{Bir}_Y^c(S, \Delta_S)$. However, by (10.61), the image of this action is a finite group of rth roots of unity for some r. Thus the $\mathrm{Bir}_Y^c(S, \Delta_S)$ action is trivial on $\omega_S^{[mr]}(mr\Delta_S) = (\omega_S^{[m]}(m\Delta_S))^{\otimes r}$ hence

$$\mathcal{R}^{mr} \colon \omega_X^{[mr]}(mr\Delta)|_S \simeq \omega_S^{[mr]}(mr\Delta_S) \qquad (4.45.9)$$

is completely canonical. Assume next that $\omega_X^{[mr]}(mr\Delta) \sim f^*L$. Let us factor $f|_S \colon S \to f(S)$ using $g \colon S \to W$ and the normalization $n \colon W \to f(S)$. Then we can push-forward (4.45.9) to get an isomorphism

$$n^*L \simeq \left(g_* \omega_S^{[m]}(m\Delta_S) \right)^{\mathrm{inv}}. \qquad (4.45.10)$$

This shows the second isomorphism in (4.45.4).

In order to see (4.45.6), note that Z_W is an lc center by (4.42) and we can actually choose representatives such that

$$\mathrm{Src}(Z_W, W, \mathrm{Diff}_W^* \Delta) = \mathrm{Src}(Z, X, \Delta);$$

the rest then follows from (4.45.1–2).

Applying (4.44) to a common crepant log resolution of the (X_i, Δ_i) gives (4.45.7).

Finally, in order to prove (4.45.5) we may localize at the general point of Z. Thus we may assume that Z is a point and then prove the following more precise result. $\square$

Lemma 4.46 *Let $g: (X, \Delta) \to Y$ be a crepant log structure over a field k and $z \in Y$ an lc center. Assume that (X, Δ) is dlt and X is $\mathbb{Q}$-factorial. Then there is a unique smallest finite field extension $K(z) \supset k(z)$ such that*

(1) *Every lc center of $(X_{\bar{k}}, \Delta_{\bar{k}})$ that intersects $g^{-1}(z)$ is defined over $K(z)$.*
(2) *Let $W_{\bar{z}} \subset X_{\bar{k}}$ be a minimal lc center contained in $g^{-1}(z)$. Then $K(z) = k_{ch}(W_{\bar{z}})$, the field of definition of $W_{\bar{z}}$.*
(3) *$K(z) \supset k(z)$ is a Galois extension.*
(4) *Let W_z be a minimal lc center contained in $g^{-1}(z)$. Then*

$$\mathrm{Bir}^c_{k(z)}(W_z, \mathrm{Diff}^*_{W_z} \Delta) \to \mathrm{Gal}(K(z))/k(z)) \quad \text{is surjective.}$$

Proof It may be helpful to keep in mind the examples in (4.43).

There are only finitely many lc centers and a Galois conjugate of an lc center is also an lc center. Thus the field of definition of any lc center is a finite extension of k. Since $K(z)$ is the composite of some of them, it is finite over $k(z)$.

Let $W_{\bar{z}} \subset X_{\bar{k}}$ be a minimal lc center contained in $g^{-1}(z)$ and $k_{ch}(W_{\bar{z}})$ its field of definition. By (4.17) every lc center of $(X_{\bar{k}}, \Delta_{\bar{k}})$ containing $W_{\bar{z}}$ is also defined over $k_{ch}(W_{\bar{z}})$. Therefore, any lc center that is $\mathbb{P}^1$-linked to $W_{\bar{z}}$ is defined over $k_{ch}(W_{\bar{z}})$. By (4.40) this implies that every lc center of $(X_{\bar{k}}, \Delta_{\bar{k}})$ that intersects $g^{-1}(z)$ is defined over $k_{ch}(W_{\bar{z}})$, hence $k_{ch}(W_{\bar{z}}) \supset K(z)$. By construction, $k_{ch}(W_{\bar{z}}) \subset K(z)$, thus $k_{ch}(W_{\bar{z}}) = K(z)$.

A conjugate of $W_{\bar{z}}$ over $k(z)$ is defined over the corresponding conjugate field of $k_{ch}(W_{\bar{z}})$. By the above, every conjugate of the field $k_{ch}(W_{\bar{z}})$ over $k(z)$ is itself, hence $k_{ch}(W_{\bar{z}}) = K(z)$ is Galois over $k(z)$.

Finally, in order to see (4), fix $\sigma \in \mathrm{Gal}(K(z)/k(z))$ and let $W_{\bar{z}}^\sigma$ be the corresponding conjugate of $W_{\bar{z}}$. By (4.40), $W_{\bar{z}}^\sigma$ and $W_{\bar{z}}$ are $\mathbb{P}^1$-linked over $K(z)$; fix one such $\mathbb{P}^1$-link. The union of the conjugates of this $\mathbb{P}^1$-link over $k(z)$ defines an element of $\mathrm{Bir}^c_{k(z)}(W_z, \mathrm{Diff}^*_{W_z} \Delta)$ which induces σ on $K(z)/k(z)$. (The $\mathbb{P}^1$-link is not unique, hence the lift is not unique. Thus in (4) we only claim surjectivity, not a splitting.) $\square$

Example 4.47 The Galois extension $K(z)/k(z)$ can be arbitrary. To see this, pick a Galois extension $K = k(\alpha)/k$ of degree n. In $\mathbb{A}^n_k$ consider the divisor $(\sum_i \alpha^{i-1} x_i = 0)$ and its n conjugates $D_1, \ldots, D_n$. Then $(\mathbb{A}^n_k, \sum D_i)$ is lc, the origin is an lc center and its spring gives the Galois extension K/k.

From the classification of 2-dimensional lc pairs we see that if $\operatorname{codim}_X Z = 2$ then $\operatorname{Gal}(\operatorname{Spr}(Z, X, \Delta)/Z)$ is cyclic or dihedral.

The examples in Section 3.4 can be used to show that if $\operatorname{codim}_X Z = 3$ then $\operatorname{Gal}(\operatorname{Spr}(Z, X, \Delta)/Z)$ can be arbitrary.

Let us reformulate the most important special case of (4.45.6).

Lemma 4.48 *Let $(Y, D + \Delta)$ be lc, D a reduced divisor and $Z \subset D$ an lc center. Let $n_D \colon D^n \to D$ be the normalization and $Z_D \subset D^n$ an irreducible subvariety such that $n_D(Z_D) = Z$. Then Z_D is an lc center of $(D^n, \operatorname{Diff}_{D^n} \Delta)$ and there is a commutative diagram*

$$
\begin{array}{ccc}
\operatorname{Src}(Z_D, D^n, \operatorname{Diff}_{D^n} \Delta_1) & \overset{\text{cbir}}{\underset{\sim}{-\!-\!\longrightarrow}} & \operatorname{Src}(Z, Y, D + \Delta) \\
\downarrow & & \downarrow \\
Z_D & \overset{n_D}{-\!-\!-\!-\!-\!-\!\longrightarrow} & Z.
\end{array}
$$

$\square$

5

Semi-log canonical pairs

It was discovered by Kollár and Shepherd-Barron (1988) that in order to compactify the moduli theory of higher dimensional varieties we need stable pairs. That is, pairs (X, Δ) with semi-log canonical singularities and ample log canonical class $K_X + \Delta$. The aim of this chapter is to study these semi-log canonical singularities.

In general X is neither normal nor irreducible. Such varieties can be studied either using semi-log resolutions or by focusing on their normalization. Both of these approaches have difficulties.

A stable curve C has ordinary nodes, and we can encode C by giving a triple $(\bar{C}, \bar{D}, \tau)$ where $\bar{C}$ is the normalization of C, $\bar{D} \subset \bar{C}$ is the preimage of the nodes and $\tau \colon \bar{D} \to \bar{D}$ is a fixed point free involution which tells us which point pairs are identified in C.

Correspondingly, a higher dimensional stable variety has ordinary self-intersection in codimension 1, and we encode (X, Δ) by a quadruple $(\bar{X}, \bar{D}, \bar{\Delta}, \tau)$ where $\pi \colon \bar{X} \to X$ is the normalization, $\bar{D}$ the preimage of the double normal crossing locus of X, $\bar{\Delta}$ the preimage of Δ and the involution τ tells us which point pairs are identified in X. Since X can have complicated self-intersections in higher codimensions, τ is an actual involution only on the normalization $\bar{D}^n$ of $\bar{D}$ and it is generically fixed point free, that is, its fixed point set does not contain any irreducible components of $\bar{D}^n$.

It is easy to see that $(\bar{X}, \bar{D}, \bar{\Delta}, \tau)$ uniquely determines (X, Δ). Our principal aim is to understand which quadruples come from an slc pair (X, Δ).

Semi-log canonical pairs are not normal, but normality fails in the mildest possible way. This leads to the class of "demi-normal" schemes, to be introduced in Section 5.1.

Section 5.2 gives the precise definition of semi-log canonical and states our main theorems. We work out the complete theory for surfaces in Section 5.3 and for semi-dlt pairs in Section 5.4.

The log canonical stratification of lc and slc pairs is studied in Section 5.5. Our inductive methods make it easier to investigate the normal and non-normal cases together.

A structure theorem on gluing relations is proved in Section 5.6; this leads to the main finiteness property. These rely on the sources of log canonical centers introduced in Section 4.5.

Even if (X, Δ) exists, $K_X + \Delta$ may not be $\mathbb{Q}$-Cartier. It is easy to see that if $K_X + \Delta$ is $\mathbb{Q}$-Cartier then the Poincaré residue map $\omega_{\bar{X}}(\bar{D}) \to \omega_{\bar{D}}$ has to be compatible with the involution τ (5.12). If $(\bar{X}, \bar{D} + \bar{\Delta})$ is lc, we show in Section 5.7 that the converse also holds (5.38). Again the key ingredient is the study of sources in Section 4.5.

The treatment of Chapter 5 is very much intertwined with the topics in Chapter 9. Our aim is to treat questions about abstract stratified schemes in Chapter 9 and the geometric properties and applications in Chapter 5. This separation is not always clear cut and it leads to some repetitions.

Assumptions   In Sections 5.1–5.2 we work with excellent schemes. Starting with Section 5.5 we work with schemes (or algebraic spaces) that are of finite type over a base scheme S that is essentially of finite type over a field of characteristic 0. See (5.24) for further clarification.

5.1 Demi-normal schemes

In this section we introduce demi-normal schemes and study their basic properties.

Definition 5.1   Recall that, by Serre's criterion, a scheme X is normal if and only if it is S_2 and regular at all codimension 1 points. As a weakening of normality, it is natural to consider schemes that are S_2 and whose codimension 1 points are either regular points or nodes (1.41). Such schemes will be called *demi-normal*. The initial "d" is supposed to remind us that they have double normal crossings in codimension 1. (We really would like to call these schemes "seminormal," but that name is already taken.)

Using the classification of seminormal curve singularities (5.9.3) one sees that X is demi-normal if and only if it is S_2 and its codimension 1 points are seminormal with locally free canonical sheaf.

The *demi-normalization* of a scheme is usually not defined. (What should the demi-normalization of $(x^n = y^n) \subset \mathbb{A}^2$ be for $n \geq 3$?) However, if $j : U \hookrightarrow X$ is an open subscheme with only regular points and nodes such that $X \setminus U$ has codimension ≥ 2, then $\mathrm{Spec}_X j_* \mathscr{O}_U$ is the smallest demi-normal scheme dominating X. It is called the demi-normalization of X.

Roughly speaking, the concept of semi-log canonical is obtained by replacing "normal" with "demi-normal" in the definition of log canonical (2.8), but some basic definitions and foundational results need to be in place first.

5.2 (Normalization of demi-normal schemes) Let X be a reduced scheme and $\pi\colon \bar{X} \to X$ its normalization. The *conductor ideal*

$$cond_X := \mathcal{H}om_X(\pi_*\mathcal{O}_{\bar{X}}, \mathcal{O}_X) \subset \mathcal{O}_X \qquad (5.2.1)$$

is the largest ideal sheaf on X that is also an ideal sheaf on $\bar{X}$. We write it as $cond_{\bar{X}}$ when we view the conductor as an ideal sheaf on $\bar{X}$. The *conductor subschemes* are defined as

$$D := \operatorname{Spec}_X(\mathcal{O}_X/cond_X) \quad \text{and} \quad \bar{D} := \operatorname{Spec}_{\bar{X}}(\mathcal{O}_{\bar{X}}/cond_{\bar{X}}). \qquad (5.2.2)$$

If $\mathcal{O}_X$ is S_2, so is the conductor ideal (2.62) and so $D \subset X$ and $\bar{D} \subset \bar{X}$ are both of pure codimension 1 and without embedded points. If X has only nodes at its codimension 1 points then D and $\bar{D}$ are generically reduced.

Assume that X is demi-normal. Then D and $\bar{D}$ are both reduced subschemes of pure codimension 1. Let $x_i \in D$ be a generic point. Then $\mathcal{O}_{X,x_i}$ is a node, thus, if char $k(x_i) \neq 2$, $\pi\colon \bar{D} \to D$ is an étale double cover in a neighborhood of x_i. (We discuss the general case in (5.9).)

In general, $\bar{D} \to D$ is not everywhere étale and not even flat, but the map between the normalizations $n\colon \bar{D}^n \to D^n$ has degree 2 over every irreducible component. Thus it defines a Galois involution $\tau\colon \bar{D}^n \to \bar{D}^n$. Note that τ is *generically fixed point free*, that is, its fixed point set does not contain any irreducible components of $\bar{D}^n$.

Note that in general, τ does not define an involution of $\bar{D}$, not even set-theoretically. As a simple example, consider $X := (xyz = 0) \subset \mathbb{A}^3$. Here $D \subset \mathbb{A}^3$ is the three coordinate axes with a triple point at the origin. $\bar{X}$ is the disjoint union of three planes each containing a pair of intersecting lines $\bar{D}_i$ and $\bar{D}$ is their disjoint union. The origin $0 \in D$ has three preimages in $\bar{D}$ and they would all have to be in the same τ-orbit.

Proposition 5.3 *Let X be a demi-normal scheme such that $\frac{1}{2} \in \mathcal{O}_X$. Then the triple $(\bar{X}, \bar{D}^n, \tau)$ defined in (5.2) uniquely determines X.*

Proof Note that $\pi\colon \bar{X} \to X$ is a finite surjection and $\pi \circ n\colon \bar{D}^n \to X$ is τ-invariant. Assume that $\pi'\colon \bar{X} \to X'$ is another finite surjection such that $\pi' \circ n\colon \bar{D}^n \to X$ is τ-invariant. We prove that there is a unique $g\colon X \to X'$ such that $\pi' = g \circ \pi$; this gives a characterization of X.

Let $X^* \subset X \times X'$ be the image of (π, π'). Let $x \in X$ be a codimension 1 point. Then either X is regular at x, hence $\bar{X} \to X^* \to X$ are isomorphisms near x, or X has a node at x with preimage $\bar{x} \in \bar{X}$. By assumption $\bar{x} \to X^*$

factors through $x \to X^*$, hence again $X^* \to X$ is an isomorphism near x. Since X is S_2, this implies that the first projection $X^* \to X$ is an isomorphism. Thus the second projection $X^* \to X'$ gives the required g. $\qquad\qquad\qquad\square$

Note that (5.3) asserts uniqueness, but it does not claim existence nor does it describe the points of X explicitly in terms of $(\bar{X}, \bar{D}^n, \tau)$. The latter will be accomplished in Section 9.1. The question of existence turns out to be quite difficult, and we have to deal with two main problems.

Question 5.4 Let $\tilde{X}$ be a normal scheme, $\tilde{D} \subset \tilde{X}$ a reduced divisor with normalization $\tilde{n}: \tilde{D}^n \to \tilde{D}$ and $\tilde{\tau}: \tilde{D}^n \to \tilde{D}^n$ an involution. Under what conditions does there exist a demi-normal variety X with normalization $(\bar{X}, \bar{D}, \tau)$ as in (5.2) such that $(\tilde{X}, \tilde{D}, \tilde{\tau}) = (\bar{X}, \bar{D}, \tau)$?

Question 5.5 Assume that X is demi-normal and $K_{\bar{X}} + \bar{D}$ is $\mathbb{Q}$-Cartier. Under what conditions is K_X also $\mathbb{Q}$-Cartier?

In order to answer (5.4), first we aim to describe the closed fibers of the putative $\pi: \tilde{X} \to X$. Since $\tilde{D}^n \to X$ is $\tilde{\tau}$-invariant, we see that for any closed point $q \in \tilde{D}^n$, the points $\tilde{n}(q) \in \tilde{X}$ and $\tilde{n}(\tilde{\tau}(q)) \in \tilde{X}$ must be in the same fiber of π. The relation $\tilde{n}(q) \sim \tilde{n}(\tilde{\tau}(q))$ generates an equivalence relation $R(\tau)$ on the closed points of $\tilde{X}$, called the *gluing relation*. A necessary condition for the existence of X is that the gluing relation should have finite equivalence classes.

Even assuming finiteness, the first question seems rather intractable in general, as shown by the examples in Kollár (20012c, section 2). Thus we consider the case when $(\tilde{X}, \tilde{D})$ is assumed lc. The main result of Section 5.5 gives a positive answer to (5.4) when $\tilde{\tau}$ is compatible with the lc structure in a weak sense (5.33).

The second question (5.5) may at first seem puzzling in view of the formula $\pi^* K_X \sim_{\mathbb{Q}} K_{\bar{X}} + \bar{D}$ (5.7.5). However, as the examples (5.16) show, in general K_X need not be $\mathbb{Q}$-Cartier, not even if $K_{\bar{X}} + \bar{D}$ is Cartier. We show in (5.12) that a necessary condition is that the different (5.11) be τ-invariant. We prove in Section 5.7 that the converse holds if $(\tilde{X}, \tilde{D})$ is lc (5.38), but not in general (5.17). An explicit study of the surface case is in (5.18).

5.6 (Divisors and divisorial sheaves on demi-normal schemes) Let X be demi-normal. $\mathbb{Z}$-divisors whose support does not contain any irreducible component of the conductor $D \subset X$ (5.2.2) form a subgroup

$$\text{Weil}^*(X) \subset \text{Weil}(X). \qquad\qquad (5.6.1)$$

A rank 1 reflexive sheaf that is locally free at the generic points of D_X is called a *divisorial sheaf* on X. Divisorial sheaves form a subgroup

$$\text{Cl}^*(X) \subset \text{Cl}(X). \qquad\qquad (5.6.2)$$

As in (1.2), in this group the product of two divisorial sheaves L_1, L_2 is given by $L_1 \hat{\otimes} L_2 := (L_1 \otimes L_2)^{**}$, the double dual or reflexive hull of the usual tensor product. For powers we use the notation $L^{[m]} := (L^{\otimes m})^{**}$.

Let B be a $\mathbb{Z}$-divisor whose support does not contain any irreducible component of the conductor. Then there is a closed subset $Z \subset X$ of codimension ≥ 2 such that $X^0 := X \setminus Z$ has only regular and double nc points and $B^0 := B|_{X^0}$ is regular. Thus B^0 is a Cartier divisor on X^0 and $\mathscr{O}_{X^0}(B^0)$ is an invertible sheaf. Let $j: X^0 \hookrightarrow X$ denote the natural injection and set

$$\mathscr{O}_X(B) := j_* \mathscr{O}_{X^0}(B^0). \tag{5.6.3}$$

This establishes a surjective homomorphism $\mathrm{Weil}^*(X) \to \mathrm{Cl}^*(X)$ whose kernel consists of divisors linearly equivalent to 0.

Similarly, K_{X^0} is a Cartier divisor on X^0 and $\omega_{X^0} \simeq \mathscr{O}_{X^0}(K_{X^0})$ is an invertible sheaf. For every $m \in \mathbb{Z}$, we get the rank 1 reflexive sheaves

$$\omega_X := j_* \omega_{X^0} \quad \text{and} \quad \omega_X^{[m]}(B) := j_* \big(\omega_{X^0}^m(B^0)\big). \tag{5.6.4}$$

Thus it makes sense to talk about K_X or B being Cartier or $\mathbb{Q}$-Cartier. (Even on a nodal curve it is ill advised to view a node p as a Weil divisor such that $2[p]$ is Cartier. Fortunately, in the slc case, we mostly need to deal with divisors in $\mathrm{Weil}^*(X)$.)

Let $\pi: \bar{X} \to X$ be the normalization. For any B in $\mathrm{Weil}^*(X)$, let $\bar{B}$ denote the divisorial part of $\pi^{-1}(B)$, as a divisor on $\bar{X}$. This establishes a one-to-one correspondence between $\mathbb{Z}$-divisors (resp. $\mathbb{Q}$-divisors) on X whose support does not contain any irreducible component of the conductor $D_X \subset X$ and $\mathbb{Z}$-divisors (resp. $\mathbb{Q}$-divisors) on $\bar{X}$ whose support does not contain any irreducible component of $\bar{D}_X \subset \bar{X}$.

5.7 Let Y be a scheme and $\pi: \bar{Y} \to Y$ its normalization. Assume that Y is S_2, ω_Y is locally free and $\bar{Y}$ is regular. From

$$\pi_* \omega_{\bar{Y}} = \mathscr{H}om_Y(\pi_* \mathscr{O}_{\bar{Y}}, \omega_Y)$$

we conclude that $\pi_* \omega_{\bar{Y}} = \omega_Y(-D)$. (Note that D is not a Cartier divisor on Y.) Since the conductor ideals $\mathscr{O}_Y(-D) = \mathscr{O}_{\bar{Y}}(-\bar{D})$ agree, the latter is equivalent to $\omega_{\bar{Y}} = (\pi^* \omega_Y)(-\bar{D})$. Since $\bar{D}$ is a Cartier divisor, we can take it to the other side to obtain the natural isomorphism

$$\pi^* \omega_Y = \omega_{\bar{Y}}(\bar{D}). \tag{5.7.1}$$

If X is an arbitrary demi-normal scheme, we can apply the above consideration to an open subset $X^0 \subset X$ such that $X \setminus X^0$ has codimension ≥ 2. By pushing forward from X^0 (resp. $\bar{X}^0$) to X (resp. $\bar{X}$) we obtain that

$$\pi_* \omega_{\bar{X}} = \omega_X(-D) \quad \text{and} \quad (\pi^* \omega_X)^{**} = \omega_{\bar{X}}(\bar{D}), \tag{5.7.2}$$

where the double dual is necessary in general since the pull-back of an S_2 sheaf need not be S_2. Similarly, for any $\mathbb{Z}$-divisor B and integer m we obtain a natural isomorphism

$$\left(\pi^* \omega_X^{[m]}(B)\right)^{**} \simeq \omega_{\bar{X}}^{[m]}(m\bar{D} + \bar{B}). \tag{5.7.3}$$

If Δ is a $\mathbb{Q}$-divisor, $m\Delta$ is integral and $m(K_X + \Delta)$ is Cartier, this simplifies to

$$\pi^*\left(\omega_X^{[m]}(m\Delta)\right) \simeq \omega_{\bar{X}}^{[m]}(m\bar{D} + m\bar{\Delta}), \tag{5.7.4}$$

which we frequently abbreviate as

$$\pi^*(K_X + \Delta) \sim_{\mathbb{Q}} K_{\bar{X}} + \bar{D} + \bar{\Delta}. \tag{5.7.5}$$

It is more interesting to study which sections of $\omega_{\bar{X}}^{[m]}(m\bar{D} + m\bar{\Delta})$ descend to a section of $\omega_X^{[m]}(m\Delta)$. The only question is at the generic points of D, hence we can work on X^0 and ignore Δ.

Assume first that we are over a perfect field whose characteristic is $\neq 2$. We give an answer in terms of the Poincaré residue map (4.1)

$$\mathcal{R}_{\bar{X}^0 \to \bar{D}^0} : \omega_{\bar{X}^0}(\bar{D}^0) \to (\omega_{\bar{X}^0}(\bar{D}^0))|_{\bar{D}^0} = \omega_{\bar{D}^0}.$$

By taking tensor powers, we get

$$\mathcal{R}_{\bar{X}^0 \to \bar{D}^0}^{\otimes m} : (\omega_{\bar{X}^0}(\bar{D}^0))^{\otimes m} \to \omega_{\bar{D}^0}^m.$$

As an étale local model, we can take $X := (xy = 0) \subset \mathbb{A}^2$. A generator of ω_X is given by $\sigma := \mathcal{R}_{\mathbb{A}^2 \to X}((xy)^{-1}dx \wedge dy)$. Note that

$$\mathcal{R}_{\mathbb{A}^2 \to X}\left(\frac{dx \wedge dy}{xy}\right)\bigg|_{(y=0)} = \frac{dx}{x} \quad \text{and} \quad \mathcal{R}_{\mathbb{A}^2 \to X}\left(\frac{dx \wedge dy}{xy}\right)\bigg|_{(x=0)} = -\frac{dy}{y}.$$

Note that dx/x has residue 1 at the origin and $-dy/y$ has residue -1. The two residues differ by a minus sign, thus we obtain the following.

Proposition 5.8 *Let X be a demi-normal variety over a perfect field whose characteristic is $\neq 2$. A section ϕ of $\omega_{\bar{X}}^{[m]}(m\bar{D} + m\bar{\Delta})$ descends to a section of $\omega_X^{[m]}(m\Delta)$ if and only if $\mathcal{R}_{\bar{X} \to \bar{D}}^{\otimes m}(\phi)$ is τ-invariant if m is even and τ-anti-invariant if m is odd.* $\square$

5.9 (Demi-normal schemes in characteristic 2) In general, let X be a demi-normal scheme, $\pi : \bar{X} \to X$ its normalization, $D \subset X$ and $\bar{D} \subset \bar{X}$ the conductors.

If $k(D)$ has characteristic 2 then there is no Galois involution τ in general. Note that if τ exists then $D^{\mathrm{n}} = \bar{D}^{\mathrm{n}}/(\tau)$, thus instead of using τ we can encode the normalization as

$$(\bar{X}, \sigma : \bar{D}^{\mathrm{n}} \to D^{\mathrm{n}}) \tag{5.9.1}$$

where σ has degree 2 over every irreducible component of D^n. If 2 is invertible, this carries the same information as $(\bar{X}, \bar{D}, \tau)$.

The results of this section all work in general, only (5.8) needs some thought. Assume that X is CM, seminormal and $\bar{X}$ is regular. (As in (5.7), these conditions hold outside a codimension ≥ 2 subscheme of any demi-normal scheme.) We have an exact sequence

$$0 \to \mathcal{O}_X(-D) \to \mathcal{O}_X \to \mathcal{O}_D \to 0.$$

Mapping it to ω_X we get

$$0 \to \omega_X \to \mathcal{H}om_X(\mathcal{O}_X(-D), \omega_X) \to \mathcal{E}xt^1_X(\mathcal{O}_D, \omega_X) \simeq \omega_D \to 0.$$

Note that

$$\mathcal{H}om_X(\mathcal{O}_X(-D), \omega_X) \simeq \mathcal{H}om_{\bar{X}}(\mathcal{O}_X(-D), \mathcal{H}om_X(\mathcal{O}_{\bar{X}}, \omega_X))$$
$$\simeq \mathcal{H}om_{\bar{X}}(\mathcal{O}_{\bar{X}}(-\bar{D}), \omega_{\bar{X}}) \simeq \omega_{\bar{X}}(\bar{D}).$$

Thus we get an exact sequence

$$0 \to \omega_X \to \pi_*\omega_{\bar{X}}(\bar{D}) \overset{\partial}{\to} \omega_D \to 0 \qquad (5.9.1)$$

which tells us which sections of $\omega_{\bar{X}}(\bar{D})$ descend to X. We can also obtain ∂ as the composition

$$\omega_{\bar{X}}(\bar{D}) \overset{\mathcal{R}}{\longrightarrow} \omega_{\bar{D}} \overset{\mathrm{tr}(\sigma)}{\longrightarrow} \omega_D \qquad (5.9.2)$$

of the residue map $\mathcal{R}$ with the trace map $\mathrm{tr}(\sigma)$.

Aside 5.9.3 It is instructive to compute the dualizing sheaf of a seminormal curve singularity $p \in C$. Let $\bar{p} \subset \bar{C}$ be the (reduced) preimage of p in the normalization. Then (5.9.1–2) show that

$$\omega_C|_p \simeq \ker[\omega_{\bar{p}} \to \omega_p] \simeq \ker[k(\bar{p}) \overset{\mathrm{tr}}{\to} k(p)]$$

where tr is the trace map of the field extension $k(\bar{p}) \supset k(p)$. In particular, we see that ω_C is locally free at $p \Leftrightarrow k(\bar{p}) \supset k(p)$ has degree 2 $\Leftrightarrow C$ has a node at p.

5.2 Statement of the main theorems

In this section we define semi-log canonical pairs (X, Δ) and state three main theorems describing their properties. The rest of the book is essentially devoted to their proofs.

Definition–Lemma 5.10 Let X be a demi-normal scheme with normalization $\pi \colon \bar{X} \to X$ and conductors $D \subset X$ and $\bar{D} \subset \bar{X}$ (5.2.2). Let Δ be an effective $\mathbb{Q}$-divisor whose support does not contain any irreducible component of D and $\bar{\Delta}$ the divisorial part of $\pi^{-1}(\Delta)$,

The pair (X, Δ) is called *semi-log canonical* or *slc* if

(1) $K_X + \Delta$ is $\mathbb{Q}$-Cartier, and
(2) one of the following equivalent conditions holds:
  (a) $(\bar{X}, \bar{D} + \bar{\Delta})$ is lc, or
  (b) $a(E, X, \Delta) \geq -1$ for every exceptional divisor E of a birational morphism $f: Y \to X$ such that Y is regular at the generic point of E.

Note that (2.b) is the exact analog of the definition of log canonical given in (2.8).

In order to see that the conditions (2.a) and (2.b) are equivalent, pick any $f: Y \to X$ and let $\bar{Y} \to Y$ be the normalization. Then we have a commutative diagram

$$
\begin{array}{ccc}
\bar{Y} & \xrightarrow{\ \pi_Y\ } & Y \\
\bar{f} \downarrow & & \downarrow f \\
\bar{X} & \xrightarrow{\ \pi\ } & X
\end{array}
$$

and $\pi^*(K_X + \Delta) \sim_{\mathbb{Q}} K_{\bar{X}} + \bar{D} + \bar{\Delta}$ by (5.7.5).

Since Y is regular at the generic points of $\mathrm{Ex}(f)$, we see that π_Y is an isomorphism over the generic points of $\mathrm{Ex}(f)$. Thus

$$a(E, X, \Delta) = a(E, \bar{X}, \bar{D} + \bar{\Delta}) \qquad (5.10.3)$$

for every exceptional divisor E. Thus (2.a) $\Rightarrow$ (2.b) and, using (2.13), the converse also follows from (1.13.4); see also (10.56). $\square$

The discrepancy $a(E, X, \Delta)$ is not defined if $K_X + \Delta$ is not $\mathbb{Q}$-Cartier, thus (5.10.2.b) does not make sense unless (5.10.1) holds. By contrast, (5.10.2.a) makes sense if $K_{\bar{X}} + \bar{D} + \bar{\Delta}$ is $\mathbb{Q}$-Cartier, even if $K_X + \Delta$ is not. The point of Question (5.5) is to understand the difference between these two versions.

The answer to Question (5.5) is given in terms of the different (4.2), which we recall next.

5.11 Let $(Y, D + \Delta)$ be a pair where Y is normal, D a reduced divisor and Δ a $\mathbb{Q}$-divisor whose support does not contain any irreducible component of D. Assume that $\frac{1}{2} \in \mathscr{O}_Y$. Let $n: D^n \to D$ be the normalization. Assume that $m\Delta$ is an integral divisor and $m(K_Y + D + \Delta)$ is a Cartier divisor. By (4.2) there is a unique $\mathbb{Q}$-divisor $\mathrm{Diff}_{D^n} \Delta$ on D^n such that

(1) $m \cdot \mathrm{Diff}_{D^n} \Delta$ is integral and $m(K_{D^n} + \mathrm{Diff}_{D^n} \Delta)$ is Cartier, and
(2) for m even, the mth tensor power of the Poincaré residue map (4.1) extends to a natural isomorphism

$$n^*\big(\omega_Y^{[m]}(mD + m\Delta)\big) \simeq \omega_{D^n}^{[m]}(m \cdot \mathrm{Diff}_{D^n} \Delta).$$

Note that the Poincaré residue isomorphism is defined over the snc locus of $(Y, D + \Delta)$ and the different is then chosen as the unique $\mathbb{Q}$-divisor for which the extension is an isomorphism.

Let us now apply the above to $(\bar{X}, \bar{D}, \bar{\Delta})$ obtained as the normalization of a pair (X, Δ). Using (5.7.5), for m sufficiently divisible, we have isomorphisms

$$\sigma^*\pi^*\big(\omega_X^{[m]}(m\Delta)\big) \simeq \sigma^*\big(\omega_{\bar{X}}^{[m]}(m\bar{D} + m\bar{\Delta})\big) \simeq \omega_{\bar{D}^n}^{[m]}(m \cdot \operatorname{Diff}_{\bar{D}^n} \bar{\Delta}). \quad (5.11.3)$$

Note that the composite $\bar{D}^n \to \bar{X} \to X$ is τ-invariant, hence the composite isomorphism in (5.11.3) is also τ-invariant. As noted above, the isomorphism

$$\sigma^*\pi^*\big(\omega_X^{[m]}(m\Delta)\big) \simeq \omega_{\bar{D}^n}^{[m]}(m \cdot \operatorname{Diff}_{\bar{D}^n} \bar{\Delta}).$$

uniquely determines the different $\operatorname{Diff}_{\bar{D}^n} \bar{\Delta}$. Thus we have proved the following:

Proposition 5.12 *Let X be a demi-normal scheme such that $\frac{1}{2} \in \mathcal{O}_X$ and Δ a $\mathbb{Q}$-divisor whose support does not contain any irreducible component of the conductor $D \subset X$. Let $(\bar{X}, \bar{D}, \bar{\Delta})$ and $\tau\colon \bar{D} \to \bar{D}$ be as in (5.2) and (5.6). If $K_X + \Delta$ is $\mathbb{Q}$-Cartier then $\operatorname{Diff}_{\bar{D}^n} \bar{\Delta}$ is τ-invariant.* □

Note that the τ-invariance of $\operatorname{Diff}_{\bar{D}^n} \bar{\Delta}$ depends only on the codimension 2 points of X and we prove in (5.18) that the converse of (5.12) holds outside a codimension ≥ 3 subset of X. Thus the key question is whether there are further conditions at higher codimension points or not.

The following theorem gives a complete answer in the projective, lc case. The examples (5.17) show that there are further conditions if $(\bar{X}, \bar{D}, \bar{\Delta})$ is not lc.

Theorem 5.13 (Kollár, 2011d) *Over a field of characteristic 0, normalization gives a one-to-one correspondence:*

$$\left\{ \begin{array}{c} \textit{Proper slc pairs} \\ (X, \Delta) \textit{ such that} \\ K_X + \Delta \textit{ is ample.} \end{array} \right\} \longleftrightarrow \left\{ \begin{array}{c} \textit{Proper lc pairs } (\bar{X}, \bar{D} + \bar{\Delta}) \textit{ plus} \\ \textit{a generically fixed point free} \\ \textit{involution } \tau \textit{ of } (\bar{D}^n, \operatorname{Diff}_{\bar{D}^n} \bar{\Delta}) \\ \textit{such that } K_{\bar{X}} + \bar{D} + \bar{\Delta} \textit{ is ample.} \end{array} \right\}$$

Here the arrow $\to$ is quite easy but establishing $\leftarrow$ is surprisingly difficult. We obtain it as a combination of (5.32), (5.36) and (5.38). Most of Chapters 5 and 9 are devoted to proving it.

As we noted in Section 4.3, lc centers play a key role in understanding lc pairs and the same holds for slc pairs. The main structure theorems also hold in general.

Theorem 5.14 *Let (X, Δ) be an slc pair that is essentially of finite type over a field of characteristic 0. Then:*

(1) *(Intersection) Any intersection of lc centers is also a union of lc centers.*
(2) *(Seminormality) Every union of lc centers is seminormal.*
(3) *(Normality) Every minimal (with respect to inclusion) lc center is normal.*
(4) *(Du Bois) Every union of lc centers is Du Bois (6.10).*

If X is normal, (5.14.1–3) are due to Ambro (2003, 2011) and Fujino (2009b). We prove the normal and non-normal versions together as special cases of (9.26). If X is normal, (5.14.4) is due to Kollár and Kovács (2010). The general case is established in (6.32).

The above result investigates centers of divisors whose discrepancy equals -1, but it is also of interest to study those divisors whose discrepancy is negative. As in (2.9), we use mld $:= 1 +$ discrepancy in the statement. Thus $W \subset X$ is a union of lc centers if and only if $\mathrm{mld}(W) = 0$.

Theorem 5.15 *Let (X, Δ) be an slc pair that is essentially of finite type over a field of characteristic 0. Let W, W_1, W_2 be closed, reduced subschemes. Then*

(1) *(Intersection)* $\mathrm{mld}(W_1 \cap W_2, X, \Delta) \leq \mathrm{mld}(W_1, X, \Delta) + \mathrm{mld}(W_2, X, \Delta)$
    provided $\mathrm{mld}(W_1, X, \Delta) < 1$ *for* $i = 1, 2$.
(2) *(Seminormality) If* $\mathrm{mld}(W, X, \Delta) < \frac{1}{6}$ *then W is seminormal.*
(3) *(Normality) Let* $W^{\mathrm{norm}} \subset W$ *denote the set of normal points of W. Then*
    $\mathrm{mld}(W \setminus W^{\mathrm{norm}}, X, \Delta) \leq 2 \cdot \mathrm{mld}(W, X, \Delta)$ *if* $\mathrm{mld}(W, X, \Delta) < 1$.
(4) $W_1 \cap W_2$ *is reduced if* $\mathrm{mld}(W_1, X, \Delta) + \mathrm{mld}(W_2, X, \Delta) < \frac{1}{2}$.

If X is normal, these results are proved in Kollár (2011c). The general versions are all special cases of (7.18) and of (7.19). The assertions (5.15.1–3) are generalizations of (5.14.1–3). There should be a corresponding variant of the Du Bois version (5.14.4) as well, but the bound on $\mathrm{mld}(W, X, \Delta)$ depends on the dimension and converges to 0 very rapidly; see (8.6) or Kollár (2011c) for details. If $W_1 \cup W_2$ is seminormal, then $W_1 \cap W_2$ is reduced by (10.21), thus (5.15.2) implies (5.15.4) if $\mathrm{mld}(W_i, X, \Delta) < \frac{1}{6}$ for $i = 1, 2$. For the applications in Section 8.8 the most interesting case is when $\mathrm{mld}(W_1, X, \Delta) = 0$ and $\mathrm{mld}(W_2, X, \Delta) < \frac{1}{2}$.

5.3 Semi-log canonical surfaces

In this section we work out by hand (5.13) for surfaces, but first we consider a series of examples that seem quite close to being slc yet (5.13) fails for them. The results of this section are superseded by the general methods of Chapter 9, but it is useful to get a good feeling for how the proof works in some concrete cases.

Example 5.16 In $\mathbb{A}^4$ consider the surface S that consists of three planes, $P_{xy} := (z = t = 0)$, $P_{yz} := (x = t = 0)$, $P_{zt} := (x = y = 0)$. Its normalization is the disjoint union $\bar{S} = P_{xy} \sqcup P_{yz} \sqcup P_{zt}$ and, correspondingly, the conductor D has three pieces $L_1 := (x = 0) \subset P_{xy}$, $L_1' + L_2' := (yz = 0) \subset P_{yz}$ and $L_2 := (t = 0) \subset P_{zt}$. Its normalization $\bar{D}^n$ is the disjoint union of the 4 lines L_i, L_i'. Thus $(\bar{S}, \bar{D})$ is dlt and both $K_{\bar{S}}$ and $\bar{D}$ are Cartier.

We see that the origin appears with coefficient 0 in the different on L_1 and L_2 but with coefficient 1 on L_1' and L_2'. The involution τ interchanges L_1 with L_1' and L_2 with L_2'. Thus $\mathrm{Diff}_{\bar{D}}(0)$ is not τ-invariant, hence ω_S is not Cartier and not even $\mathbb{Q}$-Cartier.

Note that S is a cone over a curve $C \subset \mathbb{P}^3$ which is a chain of three lines, ω_C has degree -1 on the two ends and 0 on the middle line. Thus ω_C is not $\mathbb{Q}$-linearly equivalent to a rational multiple of the hyperplane class and (3.14) also implies that ω_S is not $\mathbb{Q}$-Cartier.

The next example shows that in the non-lc case there is no numerical condition that decides whether a demi-normal surface has $\mathbb{Q}$-Cartier canonical class or not.

Example 5.17 We describe a flat family of demi-normal surfaces parametrized by $\mathbb{C}^* \times \mathbb{C}^*$ such that the canonical class of the fibers is $\mathbb{Q}$-Cartier for a Zariski dense set of pairs $(\lambda, \mu) \in \mathbb{C}^* \times \mathbb{C}^*$ and not $\mathbb{Q}$-Cartier for another Zariski dense set of pairs.

Start with a cone S over a hyperelliptic curve and two rulings $C_x, C_y \subset S$. Take two copies of S and glue them together by the isomorphisms $C_x^1 \to C_x^2$ and $C_y^1 \to C_y^2$ which are multiplication by $\lambda \in \mathbb{C}^*$ (resp. $\mu \in \mathbb{C}^*$) to get a non-normal surface $T(\lambda, \mu)$. We show that its canonical class is $\mathbb{Q}$-Cartier if and only if λ/μ is a root of unity.

To get concrete examples, fix an integer $a \geq 0$ and set

$$S := (z^2 = xy(x^{2a} + y^{2a})) \subset \mathbb{A}^3 \quad \text{and} \quad C := C_x + C_y$$

where $C_x = (y = z = 0)$ and $C_y = (x = z = 0)$. Note that C is not Cartier but $2C = (xy = 0)$ is. Furthermore, ω_S is locally free with generator $z^{-1} dx \wedge dy$ and so $\omega_S^2(2C)$ is locally free with generator

$$\frac{1}{xyz^2}(dx \wedge dy)^{\otimes 2} = \frac{1}{x^2 y^2 (x^{2a} + y^{2a})}(dx \wedge dy)^{\otimes 2}.$$

The restriction of $\omega_S^2(2C)$ to C_x is thus locally free with generator

$$\mathcal{R}_{S \to C_x}\left(\frac{1}{x^2(x^{2a} + y^{2a})}\left(dx \wedge \frac{dy}{y} \right)^{\otimes 2} \right) = \frac{1}{x^{2+2a}}(dx)^{\otimes 2}.$$

Hence the different on C_x is the origin with coefficient $1 + a$. Similarly, the restriction of $\omega_S^2(2C)$ to C_y is locally free with generator $y^{-2-2a}(dy)^{\otimes 2}$.

Take now two copies S_i with coordinates (x_i, y_i, z_i) for $i \in \{1, 2\}$. Let $\tau(\lambda, \mu): C_1 \to C_2$ be an isomorphism such that $\tau(\lambda, \mu)^* x_2 = \lambda x_1$ and $\tau(\lambda, \mu)^* y_2 = \mu y_1$. Let $T(\lambda, \mu)$ be obtained by gluing $C_1 \subset S_1$ to $C_2 \subset S_2$ using $\tau(\lambda, \mu)$.

Assume that $\omega_{T(\lambda,\mu)}^{2m}$ is locally free with generator σ. Then the restriction of σ to S_i is of the form

$$\sigma|_{S_i} = \frac{1}{x_i^{2m} y_i^{2m} (x_i^{2a} + y_i^{2a})^m} (dx_i \wedge dy_i)^{\otimes 2m} \cdot f_i(x_i, y_i, z_i)$$

for some f_i such that $f_i(0, 0, 0) \neq 0$. Furthermore,

$$\tau^*(\sigma|_{S_2})|_{C_2} = (\sigma|_{S_1})|_{C_1}.$$

Further restricting to the x-axis, this gives

$$\frac{1}{(\lambda x_1)^{2m+2am}} (\lambda dx_1)^{\otimes 2m} f_2(\lambda x_1, 0, 0) = \frac{1}{x_1^{2m+2am}} (dx_1)^{\otimes 2m} f_1(x_1, 0, 0).$$

which implies that $f_2(0, 0, 0) = \lambda^{2am} f_1(0, 0, 0)$. Similarly, computing on the y-axis we obtain that $f_2(0, 0, 0) = \mu^{2am} f_1(0, 0, 0)$. If $\lambda^{2am} \neq \mu^{2am}$, these imply that $f_1(0, 0, 0) = f_2(0, 0, 0) = 0$, hence $\omega_{T(\lambda,\mu)}^{[2m]}$ is not locally free. If $\lambda^{2am} = \mu^{2am}$ then $f_1(x_1, y_1, z_1) \equiv 1$ and $f_2(x_2, y_2, z_2) \equiv \lambda^{2am}$ give a global generator of $\omega_{T(\lambda,\mu)}^{[2m]}$.

For $a \geq 1$, we have our required examples. As λ, μ vary in $\mathbb{C}^* \times \mathbb{C}^*$, we get a flat family of demi-normal surfaces $T(\lambda, \mu)$. The set of pairs (λ, μ) such that λ/μ is a root of unity is a Zariski dense subset of $\mathbb{C}^* \times \mathbb{C}^*$ whose complement is also Zariski dense.

Note, however, that for $a = 0$, $\omega_{T(\lambda,\mu)}^{[2]}$ is locally free for every λ, μ. In this case, $S := (z^2 = xy) \subset \mathbb{A}^3$ is a quadric cone and $T(\lambda, \mu)$ is slc. (In fact $T(\lambda, \mu)$ is isomorphic to the reducible quartic cone $(x^2 + y^2 + z^2 + t^2 = xy = 0) \subset \mathbb{A}^4$ for every λ, μ.)

We are now ready to prove (5.13) for surfaces.

Proposition 5.18 *Let X be a demi-normal scheme such that $\frac{1}{2} \in \mathcal{O}_X$ and Δ a $\mathbb{Q}$-divisor whose support does not contain any irreducible component of the conductor $D \subset X$. Let $(\bar{X}, \bar{D}, \bar{\Delta})$ and $\tau: \bar{D}^n \to \bar{D}^n$ be as in (5.2) and (5.6). The following are equivalent.*

(1) *$\mathrm{Diff}_{\bar{D}^n} \bar{\Delta}$ is τ-invariant.*
(2) *There is a codimension 3 set $W \subset X$ such that $(X \setminus W, \Delta|_{X \setminus W})$ is slc.*

Proof We already saw in (5.12) that (2) $\Rightarrow$ (1).

The converse is étale local near codimension 2 points of X. We can thus localize at such a point $p \in X$ and assume from now on that X is 2-dimensional and local. By passing to a suitable étale neighborhood of p we may assume that the irreducible components of the conductor $\bar{D} \subset \bar{X}$ are regular. (This will make book-keeping easier.)

It is easiest to use case analysis, relying on some of the classification results in Section 2.2, but we use only (2.31).

(5.18.3) *Plt normalization case*   Assume that there is an irreducible component $\bar{X}_1 \subset \bar{X}$ such that $(\bar{X}_1, \bar{D}_1 + \bar{\Delta}_1)$ is plt. By (2.31), $\bar{D}_1$ is a regular curve, $\lfloor \bar{\Delta}_1 \rfloor = 0$, and $(\bar{D}_1, \mathrm{Diff}_{\bar{D}_1} \bar{\Delta}_1)$ is klt by adjunction (2.35).

There are 2 cases:

(a) If τ is an involution of $\bar{D}_1$ then $\bar{X} = \bar{X}_1$ is the only component and $X = X_1$ is not normal.
(b) If τ maps $\bar{D}_1$ to another double curve $\bar{D}_2$, then, by the τ-invariance of the different (5.18.1), $\mathrm{Diff}_{\bar{D}_1} \bar{\Delta}_1 = \mathrm{Diff}_{\bar{D}_2} \bar{\Delta}_2$ and so $(\bar{X}_2, \bar{D}_2 + \bar{\Delta}_2)$ is also plt by inversion of adjunction (2.35). Thus $\bar{X} = \bar{X}_1 + \bar{X}_2$ has 2 irreducible components, both plt and the $\bar{X}_i$ are also irreducible components of X.

In the first case, choose $m \in \mathbb{N}$ such that $m \, \mathrm{Diff}_{\bar{D}_1} \bar{\Delta}_1$ is a $\mathbb{Z}$-divisor and let σ be a τ-invariant generator of $\omega_{\bar{D}_1}^{2m}(2m \cdot \mathrm{Diff}_{\bar{D}_1} \bar{\Delta}_1)$. Since

$$H^0\big(\bar{X}, \omega_{\bar{X}}^{[2m]}(2m\bar{D}_1 + 2m\bar{\Delta})\big) \twoheadrightarrow H^0\big(\bar{D}_1, \omega_{\bar{D}_1}^{2m}(2m \cdot \mathrm{Diff}_{\bar{D}_1} \bar{\Delta}_1)\big)$$

is surjective, we can lift σ to a generator $\phi \in H^0(\bar{X}, \omega_{\bar{X}}^{[2m]}(2m\bar{D}_1 + 2m\bar{\Delta}))$, and by (5.8), ϕ descends to a nowhere zero section

$$\Phi_0 \in H^0\big(X \setminus p, \omega_X^{[2m]}(2m\Delta)|_{X \setminus p}\big)$$

which then extends to a local generator $\Phi \in H^0(X, \omega_X^{[2m]}(2m\Delta))$.

In the second case, for $i = 1, 2$, pick local generators

$$\sigma_i \in H^0\big(\bar{D}_i, \omega_{\bar{D}_i}^{2m}(2m \, \mathrm{Diff}_{\bar{D}_i} \bar{\Delta}_i)\big)$$

that are interchanged by τ and lift them to sections

$$\phi_i \in H^0\big(\bar{X}_i, \omega_{\bar{X}_i}^{[2m]}(2m\bar{D}_i + 2m\bar{\Delta}_i)\big).$$

As before, the pair (ϕ_1, ϕ_2) descends to a section

$$\Phi_0 \in H^0\big(X \setminus p, \omega_X^{[2m]}(2m\Delta)|_{X \setminus p}\big)$$

which then extends to a section $\Phi \in H^0(X, \omega_X^{[2m]}(2m\Delta))$. Then Φ is a local generator of $\omega_X^{[2m]}(2m\Delta)$, thus $2m(K_X + \Delta)$ is Cartier at P.

(5.18.4) *Key point of the proof*

Let $(p_j \in \bar{D}_j)$ be the irreducible components of $\bar{D}^n$. In the remaining cases $(\bar{X}, \bar{D} + \bar{\Delta})$ is lc but not plt at any of the preimages of p, thus we know that $\mathrm{Diff}_{\bar{D}_j}(\bar{D} - \bar{D}_j + \bar{\Delta}) = 1 \cdot [p_j]$ for every j. Therefore

$$H^0\big(\bar{D}_j, \omega_{\bar{D}_j}^{2m}(2m \, \mathrm{Diff}_{\bar{D}_j}(\bar{D} - \bar{D}_j + \bar{\Delta}))\big) = H^0\big(\bar{D}_j, \omega_{\bar{D}_j}^{2m}(2m[p_j])\big)$$

and the Poincaré residue of a section σ_j lies in

$$\mathcal{R}(\sigma_j) \in \omega_{p_j}^{\otimes 2m} = k(p_j)^{\otimes 2m} = k(p_j) \qquad (5.18.5)$$

where the $=$ signs are *canonical* isomorphisms. Thus it makes sense to ask if a section has residue $= 1$ or not.

(5.18.6) *Non-plt normalization case*

Pick local generators

$$\sigma_j \in H^0\big(\bar{D}_j, \omega_{\bar{D}_j}^{2m}(2m\,\mathrm{Diff}_{\bar{D}_j}(\bar{D} - \bar{D}_j + \bar{\Delta}))\big)$$

that have residue 1 at p_j (under the identifications (5.18.5)) and such that together they give a τ-invariant section of $\omega_{\bar{D}^n}^{2m}(2m\,\mathrm{Diff}_{\bar{D}^n}\bar{\Delta})$. Since the σ_j all have the same residue, by (5.8), they descend to a section

$$\sigma \in H^0\big(\bar{D}, \omega_{\bar{D}}^{2m}(2m\,\mathrm{Diff}_{\bar{D}}\bar{\Delta})\big).$$

As before, σ lifts back to $\phi \in H^0(\bar{X}, \omega_{\bar{X}}^{[2m]}(2m\bar{D} + 2m\bar{\Delta}))$ and then descends to

$$\Phi \in H^0\big(X, \omega_X^{[2m]}(2m\Delta)\big). \qquad \square$$

(5.18.7) *Comments on higher dimensions*

A difficulty in higher dimensions is that the residues at a point $p \in X$ lie in a 1-dimensional $k(p)$-vector space that cannot be identified with $k(p)$.

For instance, if X is a cone with vertex v over an Abelian variety A then there is a natural isomorphism

$$\omega_X^{[m]}|_v = \omega_X^m|_v \simeq H^0(A, \omega_A)^{\otimes m}$$

and the latter does not have a canonical isomorphism with $\mathbb{C}$. (Indeed, as A moves in the moduli space of Abelian varieties, the $H^0(A, \omega_A)^{\otimes m}$ are fibers of an ample line bundle on the moduli space.) We return to this in Section 5.7.

5.4 Semi-divisorial log terminal pairs

In this section we establish (5.13) for semi-dlt pairs (5.19). Dlt pairs play a crucial role in our treatment of lc pairs, mostly through the use of (1.34). The analog of the latter is not known for slc pairs; see (5.22).

Definition 5.19   An slc pair (X, Δ) is *semi-divisorial log terminal* or *semi-dlt* if $a(E, X, \Delta) > -1$ for every exceptional divisor E over X such that (X, Δ) is not snc (1.10) at the generic point of center$_X$ E.

By (5.10.3), this implies that $(\bar{X}, \bar{D} + \bar{\Delta})$ is dlt. The converse is not quite true, for instance $S := (xy = zt = 0) \subset \mathbb{A}^4$ is not semi-dlt but its normalization is dlt. We see, however, in (5.20) that this difference appears only in codimension 2.

If $(Y, B + \Delta)$ is dlt and B is reduced, then $(B, \mathrm{Diff}_B \Delta)$ is semi-dlt, and this is the main reason for our definition. However, if $(B, \mathrm{Diff}_B \Delta)$ is semi-dlt then $(Y, B + \Delta)$ need not be dlt. For instance, take $Y := (x_1 \cdots x_n + x_{n+1}^m = 0) \subset \mathbb{A}^{n+1}$ and $B := (x_{n+1} = 0)$. Then B is simple normal crossing but (Y, B) is not, hence (Y, B) is not dlt.

It might be useful to develop a variant of dlt/semi-dlt that is compatible both with adjunction and inversion of adjunction. There are, however, enough flavors of "log terminal" floating around, so we will not do this.

In the dlt case, we have the following positive answer to (5.5).

Proposition 5.20 *Let (X, Δ) be a demi-normal pair over a field of characteristic 0. Assume that the normalization $(\bar{X}, \bar{D}, \bar{\Delta})$ is dlt and there is a codimension 3 set $W \subset X$ such that $(X \setminus W, \Delta|_{X \setminus W})$ is semi-dlt. Then*

(1) *the irreducible components of X are normal,*
(2) *$K_X + \Delta$ is $\mathbb{Q}$-Cartier and*
(3) *(X, Δ) is semi-dlt.*

Proof We may assume that X is affine. At a codimension 2 point $p \in X$, the pair (X, Δ) is either snc, and hence the irreducible components of X are regular near p, or $(\bar{X}, \bar{D}, \bar{\Delta})$ is plt above p and we are in case (5.18.3b). Thus the irreducible components of X are normal in codimension 2.

Let $X_1, \ldots, X_n$ be the irreducible components of X with normalization $\bar{X}_j \to X_j$. Let $\bar{W} \subset \bar{X}$ be the preimage of W.

Set $B_j := X_j \cap (X_1 \cup \cdots \cup X_{j-1})$, as a divisor on X_j. Note that $X_j \setminus W \simeq \bar{X}_j \setminus \bar{W}$ and under this isomorphism B_j is identfied with a sum of certain irreducible components of $\bar{D}|_{\bar{X}_j}$. Since $(\bar{X}, \bar{D}, \bar{\Delta})$ is dlt, (2.88) implies that $\mathscr{O}_{\bar{X}_j}(-\bar{B}_j)$ is a CM sheaf. In particular, $\mathrm{depth}_{\bar{X}_j \cap \bar{W}} \mathscr{O}_{\bar{X}_j}(-\bar{B}_j) \geq 3$. By (2.58), this implies that

$$H^1(X_j \setminus W, \mathscr{O}_{X_j}(-B_j)|_{X_j \setminus W}) = H^1(\bar{X}_j \setminus \bar{W}, \mathscr{O}_{\bar{X}_j}(-\bar{B}_j)|_{\bar{X}_j \setminus \bar{W}}) = 0.$$

Hence, by (5.21.2), each X_j is S_2 and hence normal. Thus X and $\bar{X}$ have the same irreducible components.

There is an $m > 0$ such that $m(K_X + \Delta)|_{X_j}$ is locally free for every j. We can now apply (5.21.3) to $L := \omega_X^m(m\Delta)|_{X \setminus W}$ to conclude that $K_X + \Delta$ is $\mathbb{Q}$-Cartier.

Let E be an exceptional divisor over X such that $a(E, X, \Delta) = -1$. We need to prove that (X, Δ) is snc at the generic point of $\mathrm{center}_X E$. By localizing, we may assume that center$_X$ $E =: p \in X$ is a closed point and (X, Δ) is dlt outside p. By assumption, we are done if codim$_X$ $p \le 2$.

Thus assume that dim $X \ge 3$ and let $(X_i, D_i + \Delta_i)$ denote the irreducible components of $(\bar{X}, \bar{D} + \bar{\Delta})$. By permuting the indices, we may assume that E is an exceptional divisor over X_1. Then $(X_1, D_1 + \Delta_1)$ is snc at p. If X_i is any other irreducible component such that $\dim_p(X_1 \cap X_i) \ge \dim X - 1$, then adjunction and inversion of adjunction shows that there is an exceptional divisor E_i over X_i with discrepancy -1 whose center is $p \in X_i$. Thus $(X_i, D_i + \Delta_i)$ is also snc at p. Since X is S_2, the complement of any codimension ≥ 2 subset is still connected, thus every $(X_j, D_j + \Delta_j)$ is snc at p.

We claim that $\dim T_p X = \dim X + 1$. Note that $\dim T_p X_1 = \dim X$ and for any $i \ne 1$, $\dim T_p(X_1 + X_i) = \dim X + 1$. Thus we are done if $T_p(X_1 + X_i) = T_p(X_1 + X_j)$ for every $i \ne 1 \ne j$. For this it is enough to find a tangent vector

$$v \in (T_p(X_1 + X_i) \cap T_p(X_1 + X_j)) \setminus T_p X_1.$$

Note that $(X_1 \cap X_i)$ and $(X_1 \cap X_j)$ are divisors in X_1, hence their intersection has dimension $\ge \dim X - 2$. Since $\dim X \ge 3$, we conclude that $X_i \cap X_j$ is strictly larger than p. Thus $\dim_p(X_i \cap X_j) \ge \dim X - 1$. In particular, $X_i \cap X_j$ has a tangent vector v which is not a tangent vector to $X_1 \cap X_i \cap X_j$.

Thus X has embedding dimension $\dim X + 1$ and so it is snc. $\qquad\qquad$ $\square$

Proposition 5.21 *Let X be affine, pure dimensional and $X_1, \ldots, X_m$ the irreducible components of X. Let $W \subset X$ be a closed subset of codimension ≥ 3. Let F be a coherent sheaf on X and set*

$$I_j := \ker[F|_{X_1 \cup \cdots \cup X_j} \to F|_{X_1 \cup \cdots \cup X_{j-1}}].$$

Assume that $H^1(X_j \setminus W, I_j|_{X_j \setminus W}) = 0$ for $j \ge 2$. Then

(1) *The restriction maps*

$$H^0(X \setminus W, F|_{X \setminus W}) \to H^0(X_1 \cup \cdots \cup X_j \setminus W, F|_{X_1 \cup \cdots \cup X_j \setminus W})$$

    are surjective.

(2) *If depth$_W$ $F \ge 2$ then depth$_W$ $F|_{X_1 \cup \cdots \cup X_j} \ge 2$ for every j.*

(3) *If $F|_{X_j \setminus W} \simeq \mathcal{O}_{X_j \setminus W}$ for every j then $F \simeq \mathcal{O}_X$.*

Proof The first claim follows from the cohomology sequence of

$$0 \to I_{j+1} \to F|_{X_1 \cup \cdots \cup X_{j+1}} \to F|_{X_1 \cup \cdots \cup X_j} \to 0$$

and induction on j. If depth$_W$ $F|_{X_1 \cup \cdots \cup X_j} < 2$ then $F|_{X_1 \cup \cdots \cup X_j \setminus W}$ has a section ϕ that does not extend to a section of $F|_{X_1 \cup \cdots \cup X_j}$. By lifting ϕ back to a section of $F|_{X \setminus W}$, we would get a contradiction. This proves (2).

Finally, we prove by induction on j that, under the assumptions of (3), $F|_{X_1 \cup \cdots \cup X_j \setminus W}$ has a nowhere zero section. For $j = 1$ we have assumed this. Next we lift the section, going from $j - 1$ to j.

Since $H^1(X_j \setminus W, I_j|_{X_j \setminus W}) = 0$, we have a surjection

$$H^0(X_1 \cup \cdots \cup X_j \setminus W, F|_{X_1 \cup \cdots \cup X_j \setminus W})$$
$$\twoheadrightarrow H^0(X_1 \cup \cdots \cup X_{j-1} \setminus W, F|_{X_1 \cup \cdots \cup X_{j-1} \setminus W}).$$

Thus $F|_{X_1 \cup \cdots \cup X_j \setminus W}$ has a section σ_j which is nowhere zero on $X_1 \cup \cdots \cup X_{j-1} \setminus W$. Note that $\sigma_j|_{X_j \setminus W}$ is the section of a trivial line bundle. Thus, if it vanishes at all, then it vanishes along a Cartier divisor D_j on X_j.

Since $\operatorname{depth}_W F|_{X_1 \cup \cdots \cup X_j} \geq 2$, $X_j \cap (X_1 \cup \cdots \cup X_{j-1})$ has pure codimension 1 in X_j. Thus, if $D_j \neq 0$ then $D_j \cap (X_1 \cup \cdots \cup X_{j-1})$ is a nonempty codimension 2 set of $X_1 \cup \cdots \cup X_{j-1}$. On the other hand, W has codimension 3 and σ_j does not vanish on $X_1 \cup \cdots \cup X_{j-1} \setminus W$.

This implies that $D_j = 0$ and so σ_j is nowhere zero on $X_1 \cup \cdots \cup X_j \setminus W$. $\qquad\square$

5.22 (Semi-dlt models)   In the study of lc pairs (X, Δ) it is very useful that there is a dlt model, that is, a projective, birational morphism $f \colon (X', \Delta') \to (X, \Delta)$ such that (X', Δ') is dlt, $K_{X'} + \Delta' \sim_{\mathbb{Q}} f^*(K_X + \Delta)$ and every f-exceptional divisor has discrepancy -1 (1.36).

It would be convenient to have a similar result for slc pairs. An obvious obstruction is given by codimension 1 self-intersections of the irreducible components of X. Indeed, this is not allowed on a semi-dlt pair but a semi-resolution cannot remove such self-intersections.

Every demi-normal scheme has a natural double cover that removes such codimension 1 self-intersections (5.23), thus it is of interest to ask for semi-dlt models assuming that every irreducible component of X is normal in codimension 1.

Question 5.22.1   Let (X, Δ) be slc such that every irreducible component of X is normal in codimension 1. Is there a projective, birational morphism $f \colon (X', \Delta') \to (X, \Delta)$ such that (X', Δ') is semi-dlt, f maps $\operatorname{Sing} X'$ birationally on the double normal crossing locus $D \subset \operatorname{Sing} X$, $K_{X'} + \Delta' \sim_{\mathbb{Q}} f^*(K_X + \Delta)$ and every f-exceptional divisor has discrepancy -1?

5.23 (A natural double cover)   Every demi-normal scheme has a natural double cover, constructed as follows. (For a normal scheme X, one just gets two disjoint copies of X.)

Let X^0 be a scheme whose singularities are double nc points only.

First let us work over $\mathbb{C}$. Let $\gamma \colon S^1 \to X^0(\mathbb{C})$ be a path that intersects the singular locus only finitely many times. Let $c(\gamma) \in \mathbb{Z}/2\mathbb{Z}$ be the number of those intersection points where γ moves from one local branch to another. It is easy to see that $c \colon \pi_1(X^0, p) \to \mathbb{Z}/2\mathbb{Z}$ is a well defined group homomorphism. Let $\pi^0 \colon \tilde{X}^0 \to X^0$ be the corresponding double cover.

For instance, if C_m is a cycle of smooth rational curves of length m (as in the diagram (3.27.3)), then $\tilde{C}_m$ is a cycle C_{2m} of length $2m$ if m is odd and the disjoint union of two copies of C_m if m is even.

Let now X be a demi-normal scheme and $j \colon X^0 \hookrightarrow X$ an open subset with double nc points only and such that $X \setminus X^0$ has codimenson ≥ 2. Let $\pi^0 \colon \tilde{X}^0 \to X^0$ be as above. Then $j_* \pi^0_* \mathscr{O}_{\tilde{X}^0}$ is a coherent sheaf of algebras on X. Set

$$\tilde{X} := \operatorname{Spec}_X j_* \pi^0_* \mathscr{O}_{\tilde{X}^0}$$

with projection $\pi \colon \tilde{X} \to X$.

By construction, $\tilde{X}$ is S_2, π is étale in codimension 1 and the normalization of $\tilde{X}$ is a disjoint union of two copies of the normalization of X. Furthermore, the irreducible components of $\tilde{X}$ are smooth in codimension 1.

Note that even if X is nc, usually $\tilde{X}$ is not nc. For instance, take $X_3 := (xyz = 0) \subset \mathbb{A}^3$. It is also a cone over the length 3 cycle of smooth rational curves C_3. We see that $\tilde{X}_3$ is a cone over C_6, the length 6 cycle of smooth rational curves.

As shown by the examples (9.57), (9.58) and (9.59), in general the irreducible components of $\tilde{X}$ need not be normal.

For a general scheme, we can construct $\tilde{X}^0$ as follows. Let $\pi^0 \colon \bar{X}^0 \to X^0$ denote its normalization with conductors $D^0 \subset X^0$, $\bar{D}^0 \subset \bar{X}^0$ and Galois involution $\tau \colon \bar{D}^0 \to \bar{D}^0$. Take two copies $\bar{X}^0_1 \sqcup \bar{X}^0_2$ and on $\bar{D}^0_1 \sqcup \bar{D}^0_2$ consider the involution

$$\varrho(p, q) = (\tau(q), \tau(p)).$$

Note that $(\bar{D}^0_1 \sqcup \bar{D}^0_2)/\varrho \simeq \bar{D}^0$ but the isomorphism is non-canonical. Let $\tilde{X}^0$ be the universal push-out (9.30) of

$$\left(\bar{D}^0_1 \sqcup \bar{D}^0_2 \right)/\varrho \leftarrow \left(\bar{D}^0_1 \sqcup \bar{D}^0_2 \right) \hookrightarrow \left(\bar{X}^0_1 \sqcup \bar{X}^0_2 \right).$$

Then $\pi^0 \colon \tilde{X}^0 \to X^0$ is an étale double cover and the irreducible components of $\tilde{X}^0$ are smooth. The normalization of $\tilde{X}^0$ is a disjoint union of two copies of the normalization of X^0.

Another way to construct $\pi^0 \colon \tilde{X}^0 \to X^0$ is the following. There is a natural quotient map $q \colon \pi_* \mathscr{O}_{\tilde{X}^0} \to \pi_* \mathscr{O}_{\bar{D}^0}$ and τ decomposes the latter as the τ-invariant part $\mathscr{O}_D$ and the τ-anti-invariant part, call it L_D. Then $q^{-1} \mathscr{O}_D \subset \pi_* \mathscr{O}_{\tilde{X}^0}$ is naturally $\mathscr{O}_X$ and $L_X := q^{-1}(L_D) \subset \pi_* \mathscr{O}_{\tilde{X}^0}$ is also an invertible sheaf. Its tensor square is $\mathscr{O}_X$, since $L_D \cdot L_D = \mathscr{O}_D$ (multiplication as in $\pi_* \mathscr{O}_{\bar{D}^0}$). Thus L_X is 2-torsion in $\operatorname{Pic}(X^0)$ and $\tilde{X}^0 = \operatorname{Spec}_{X^0}(\mathscr{O}_X + L_X)$.

5.5 Log canonical stratifications

In this section we answer Question 5.4 for semi-log canonical pairs. Or, rather, here we establish fundamental properties of log canonical centers of lc pairs and then refer to the abstract theorems of Chapter 9 on geometric quotients by finite set-theoretic equivalence relations to complete the proofs. First we need some discussion on algebraic spaces.

5.24 *Warning on schemes versus algebraic spaces* Recall that from now on we work with schemes (or algebraic spaces) that are of finite type over a base scheme S that is essentially of finite type over a field of characteristic 0.

Given such an X, let $\pi\colon \bar{X} \to X$ be its normalization. If X is quasi-projective over S then so is $\bar{X}$ and if X is a scheme then so is $\bar{X}$. However, it can happen that $\bar{X}$ is quasi-projective but X is not or that $\bar{X}$ is a scheme but X is only an algebraic space. There are two ways to go around these problems.

The most natural is to work with algebraic spaces. Thus we will talk about "stratified spaces" instead of "stratified schemes."

Those who wish to stay with schemes can work with schemes X such that every finite subscheme is contained in an open affine subscheme. We call this the *Chevalley–Kleiman property*. These include all quasi-projective schemes. We prove in (9.28–9.31) that all our constructions stay within this class.

In the main application (5.13), both X and $\bar{X}$ are projective, so these subtleties do not matter much.

Definition 5.25 (Log canonical stratification) Let $f\colon (X, \Delta) \to Z$ be a crepant log structure. Let $S_i^*(Z, X, \Delta) \subset Z$ be the union of all $\leq i$-dimensional lc centers (4.15, 4.28) of $f\colon (X, \Delta) \to Z$ and set

$$S_i(Z) := S_i^*(Z, X, \Delta) \setminus S_{i-1}^*(Z, X, \Delta).$$

Thus $S_i(Z)$ is a locally closed subspace of Z of pure dimension i and Z is the disjoint union of the $S_i(Z)$. We call $S_{\dim Z}(Z)$ the *open stratum*.

The collection of all the $S_i(Z)$ is called the *log canonical stratification* or *lc stratification* of Z. It is denoted by (Z, S_*).

The *boundary* of (Z, S_*) is the closed subspace

$$B(Z, S_*) := Z \setminus S_{\dim Z}(Z) = \cup_{i < \dim Z} S_i(Z).$$

In Section 9.1 we study abstract stratified spaces (X, S_*), where the $S_i(X) \subset X$ are locally closed of pure dimension i. We always assume that $\cup_{i \leq j} S_i(X)$ is closed for every j. The notion of open stratum and boundary are as above. In analogy with lc centers, an *lc center* of (X, S_*) is the closure of an irreducible component of some stratum $S_i(X)$.

The next two properties of log canonical stratifications are especially important for us. The first, summarizing (4.41.1) and (4.32), describes the local
structure of (Z, S_*). The second, which is a restatement of (4.42), shows the
compatibility of log canonical stratifications with the different.

Lemma 5.26 *Let $f: (X, \Delta) \to Z$ be a crepant log structure and (Z, S_*) the
corresponding log canonical stratification. Then*

(1) *the $S_i(Z)$ are unibranch (1.44) for every i and*
(2) *the boundary $B(Z, S_*)$ is seminormal.*   □

Lemma 5.27 *Let $f: (X, \Delta) \to Z$ be a dlt, crepant log structure and (Z, S_*)
the corresponding log canonical stratification. For an lc center $Y \subset X$ con-
sider the Stein factorization $(Y, \Delta_Y := \mathrm{Diff}_Y^* \Delta) \xrightarrow{f_Y} W \xrightarrow{\pi} Z$. Then f_Y:
$(Y, \Delta_Y) \to W$ is a crepant log structure; let (W, S_*) be the corresponding
log canonical stratification.*
Then $S_i(W) = \pi^{-1}(S_i(Z))$ for every i.   □

Formalizing the conclusion in (5.27) we get:

Definition 5.28 Let (Z_j, S_*^j) be two stratified spaces. A finite morphism
$\pi: Z_1 \to Z_2$ is *stratified* if the following equivalent conditions hold.

(1) $S_i^1(Z_1) = \pi^{-1} S_i^2(Z_2)$ for every i,
(2) for every irreducible component $W_{ij} \subset S_i^1(Z_1)$, its image $\pi(W_{ij})$ is an
    irreducible component of $S_i^2(Z_2)$.

Definition 5.29 We say that a seminormal stratified space (Y, S_*) is *of log
canonical origin* if

(1) the $S_i(Y)$ are all unibranch,
(2) there are crepant log structures $f_j: (X_j, \Delta_j) \to Z_j$ with log canoni-
    cal stratifications (Z_j, S_*^j) and a finite, surjective, stratified morphism
    $\pi: \amalg_j (Z_j, S_*^j) \to (Y, S_*)$.

The identity map shows that every log canonical stratification is of log canonical
origin. More importantly, let $f: (X, \Delta) \to W$ be a crepant log structure and
$Y \subset W$ any union of lc centers. Set $S_i(Y) := Y \cap S_i(W)$. We can apply (5.27)
to each lc center separately to conclude that the seminormalization of (Y, S_*)
is of log canonical origin. (We prove in (9.26) that Y is in fact seminormal.)

5.29.3   If Y itself is normal, then condition (1) follows from (2). Indeed,
there is a Z_j such that $\pi_j: Z_j \to Y$ is surjective. By assumption π_j is stratified,
hence $\pi_j^{-1}(S_i(Y)) = S_i(Z_j)$. Since Y is normal, π_j is universally open, hence its
restriction $S_i(Z_j) \to S_i(Y)$ is also universally open. Since $S_i(Z_j)$ is unibranch
(5.26), so is $S_i(Y)$ (1.44).

Similarly, since the boundary $B(Z_j, S_*)$ is seminormal by (5.26), the boundary $B(Y, S_*)$ is seminormal by (10.26). The latter also holds if Y is not normal, but the proof is more complicated (9.26).

Definition 5.30 Let (Y, S_*) be a stratified space of log canonical origin, shown by the maps $f_j: (X_j, \Delta_j) \to Z_j$ and $\pi_j: Z_j \to Y$. Using Noetherian induction, we define the *hereditary lc centers* of (Y, S_*) as follows.

(1) Let $W \subset Y$ be an lc center with normalization $\tau: W^n \to Y$. Set $S_i(W^n) = \tau^{-1}(S_i(Y))$. Then (W^n, S_*) is a hereditary lc center of (Y, S_*).
    For some j there is an lc center $Y \subset X_j$ such that $\pi_j \circ f_j(Y) = W$. Thus $\pi_j \circ f_j$ lifts to $g_Y: (Y, \mathrm{Diff}_Y^* \Delta_j) \to W^n$. Together with (5.29.3) this shows that (W^n, S_*) is also of log canonical origin.

(2) If $W \ne Y$ then every hereditary lc center of (W^n, S_*) is also a hereditary lc center of (Y, S_*).

Each hereditary lc center (W, S_*) is a normal, stratified space of log canonical origin and comes with a finite stratified morphism $\tau: (W, S_*) \to (Y, S_*)$. As noted in (5.29.3), the strata $S_i(W)$ are unibranch (5.26) and the boundary $B(W, S_*)$ is seminormal.

The last two properties are crucial to the approach in Section 9.1.

We saw in Section 5.2 that a demi-normal pair (X, Δ) can be described by a quadruple $(\bar{X}, \bar{D} + \bar{\Delta}, \tau)$. Furthermore, if (X, Δ) is slc, then $(\bar{X}, \bar{D} + \bar{\Delta})$ is lc and τ is an involution of $(\bar{D}^n, \mathrm{Diff}_{\bar{D}^n} \bar{\Delta})$ (5.12). Note that each irreducible component of $\bar{D}$ is a log canonical center of $(\bar{X}, \bar{D} + \bar{\Delta})$, thus each irreducible component of $\bar{D}^n$ is a hereditary lc center of $(\bar{X}, \bar{D} + \bar{\Delta})$.

The next definition generalizes this setting by working with arbitrary hereditary lc centers (not just $\bar{D}^n$) and with any isomorphisms of hereditary lc centers (not just involutions). Also, although τ is an automorphism of $(\bar{D}^n, \mathrm{Diff}_{\bar{D}^n} \bar{\Delta})$, for now we only need that it is an automorphism of the lc stratification. (The invariance of the full $\mathrm{Diff}_{\bar{D}^n} \bar{\Delta}$ will be important in Section 5.7.) To simplify notation, we drop the bars.

Definition 5.31 (Gluing relation) Let (Y, S_*) be a stratified space of log canonical origin and $\pi_j: (W_j, S_*) \to (Y, S_*)$ its hereditary lc centers. Let

$$\tau := \{\tau_{ijk} : (W_i, S_*) \to (W_j, S_*)\}$$

be a collection of stratified isomorphisms. The corresponding *gluing relation* is the equivalence relation $R(\tau) \rightrightarrows Y$ on the geometric points of Y generated by the relations

$$\pi_i(\bar{w}) \sim \pi_j(\tau_{ijk}(\bar{w})): \quad \forall i, j, k, \quad \forall \bar{w} \to \amalg_i W_i.$$

The *geometric quotient* of (Y, S_*) by $R(\tau)$ is a seminormal stratified space (X, S_*) together with a finite, surjective, stratified morphism $q\colon (Y, S_*) \to (X, S_*)$ such that the geometric fibers of q are exactly the $R(\tau)$-equivalence classes. (For more details see Section 9.1.) We frequently call $X := Y/R(\tau)$ the geometric quotient.

In Section 9.1 we develop an abstract theory of stratified relations of stratified spaces. For now we need the following special case of (9.21).

Theorem 5.32 *Assumptions as in (5.24). Let (Y, S_*) be a stratified space of log canonical origin and $R(\tau)$ a gluing relation on (Y, S_*). Assume that the $R(\tau)$-equivalence classes are all finite. Then the geometric quotient*

$$q\colon (Y, S_*) \to (Y/R(\tau), q_* S_*)$$

exists and $(Y/R(\tau), q_* S_*)$ is also a stratified space of log canonical origin. $\square$

Now we are ready to answer question (5.4) for slc pairs.

Corollary 5.33 *Assumptions as in (5.24). Let $\tilde{X}$ be normal, $\tilde{D} \subset \tilde{X}$ a reduced divisor, $\tilde{\Delta}$ an effective $\mathbb{Q}$-divisor on $\tilde{X}$ and $\tilde{\tau}\colon \tilde{D}^n \to \tilde{D}^n$ an involution on the normalization $\tilde{n}\colon \tilde{D}^n \to \tilde{D}$. Assume that*

(1) $(\tilde{X}, \tilde{D} + \tilde{\Delta})$ *is lc,*
(2) $\tilde{\tau}$ *maps lc centers of $(\tilde{D}^n, \mathrm{Diff}_{\tilde{D}^n} \tilde{\Delta})$ to lc centers and*
(3) *the gluing relation $R(\tilde{\tau}) \rightrightarrows \tilde{X}$ has finite equivalence classes.*

Then there is a demi-normal pair (X, Δ) with normalization $(\tilde{X}, \tilde{D} + \tilde{\Delta}, \tau)$ as in (5.2) such that

(4) $(\tilde{X}, \tilde{D} + \tilde{\Delta}, \tilde{\tau}) = (\tilde{X}, \tilde{D} + \tilde{\Delta}, \tau)$ *and*
(5) $R(\tilde{\tau}) = \mathrm{red}(\tilde{X} \times_X \tilde{X})$.

As noted after (5.5), assumption (3) is obviously necessary. The theorem can fail without (1); in the examples of Holmann (1963, p. 342) and Kollár (20012c, 10), $\tilde{X}$ is a smooth 3-fold, $\tilde{D}$ has cusps along a curve, (2) and (3) both hold yet X does not exist. Here $(\tilde{X}, \tilde{D})$ is not lc but it is not far from it; $(\tilde{X}, \frac{5}{6}\tilde{D})$ is lc.

Proof   The assumptions of (5.32) are satisfied and so the geometric quotient $\tilde{X}/\tilde{R}$ exists. Set $X := \tilde{X}/\tilde{R}$. Then $\tilde{X} \to X$ is the normalization and the equality $(\tilde{X}, \tilde{D}, \tilde{\tau}) = (\tilde{X}, \tilde{D}, \tau)$ holds if and only if it holds in codimension 1.

Outside a codimension 2 set $\tilde{W} \subset \tilde{X}$, we realize $\tilde{\tau}$ as an involution on $\tilde{D} \setminus \tilde{W}$, and the universal push-out of

$$(\tilde{D} \setminus \tilde{W})/\tilde{\tau} \leftarrow (\tilde{D} \setminus \tilde{W}) \hookrightarrow (\tilde{X} \setminus \tilde{W})$$

gives the appropriate open chart of X. Thus $(\tilde{X}, \tilde{D}, \tilde{\tau})$ and $(\tilde{X}, \tilde{D}, \tau)$ agree outside $\tilde{W}$, hence everywhere. Finally set Δ to be the image of $\tilde{\Delta}$. $\square$

5.6 Gluing relations and sources

We formulate and prove a structure theorem for log canonical gluing relations: they are groupoids on the springs of every stratum of the log canonical stratification (5.36). If the log canonical class is ample, these groupoids are finite, giving a basic finiteness result (5.37).

It seems that (5.37) is a rather special property of log canonical gluing relations. A pre-relation that is compatible with an ample line bundle does not always generate a finite set-theoretic equivalence relation. Bogomolov and Tschinkel (2009) give examples of étale pre-relations $R_C \rightrightarrows C$ on smooth curves C of genus ≥ 2 that generate non-finite equivalence relations. These are even compatible with the ample canonical line bundle. For a more concrete example, see (9.61).

The examples (9.56–9.60) show that if $K_X + \Delta$ is not ample, non-finiteness can occur in high codimension, even for dlt pairs. In all of these examples, the restriction of the gluing relation to any stratum is given by a group action.

In order to know where the groups come from, we need the concepts of source and spring of an lc center (4.45).

Definition 5.34 (Source-automorphism group) Let (X, Δ) be lc and $Z_i \subset X$ lc centers with springs $\mathrm{Spr}(Z_i, X, \Delta)$. An isomorphism $\phi \colon \mathrm{Spr}(Z_1, X, \Delta) \simeq \mathrm{Spr}(Z_2, X, \Delta)$ is called a *source-isomorphism* if it can be lifted to a crepant, birational map of the sources $\tilde{\phi} \colon \mathrm{Src}(Z_1, X, \Delta) \overset{\mathrm{cbir}}{\sim} \mathrm{Src}(Z_2, X, \Delta)$.

Correspondingly, we define its source-automorphism group as

$$\mathrm{Aut}^s \mathrm{Spr}(Z, X, \Delta) := \mathrm{im}[\mathrm{Bir}^c \mathrm{Src}(Z, X, \Delta) \to \mathrm{Aut}\,\mathrm{Spr}(Z, X, \Delta)].$$

Note the dual nature of this definition. For us the true invariant is the source, which is a crepant-birational equivalence class. The easy-to-describe part is the spring, which is an isomorphism class. Thus we study birational maps of the source but we see then as automorphisms of the spring. Usually $\mathrm{Aut}\,\mathrm{Spr}(Z, X, \Delta)$ is much bigger than $\mathrm{Aut}^s \mathrm{Spr}(Z, X, \Delta)$.

Recall that $\mathrm{Gal}(\mathrm{Spr}(Z, X, \Delta)/Z) \subset \mathrm{Aut}^s \mathrm{Spr}(Z, X, \Delta)$ by (4.45.5).

Definition 5.35 (Open springs) Let (X, Δ) be lc and $Z \subset X$ an lc center. Set $Z^0 := Z \setminus$ (lower dimensional lc centers), or, equivalently, $Z^0 := Z \cap S_{\dim Z}(X, \Delta)$ (5.25). Z has a spring $p_Z \colon \mathrm{Spr}(Z, X, \Delta) \to Z$ (4.45). The *open spring* of Z is

$$\mathrm{Spr}^0(Z, X, \Delta) := p_Z^{-1}(Z^0) \subset \mathrm{Spr}(Z, X, \Delta).$$

Their disjoint union

$$\mathrm{Spr}^0(X, \Delta) := \amalg_Z \mathrm{Spr}^0(Z, X, \Delta),$$

where Z runs through all lc centers, is called the *open spring* of (X, Δ). Note that the natural projection $p \colon \mathrm{Spr}^0(X, \Delta) \to X$ is surjective and quasi-finite, but not finite. However, p factors as

$$p : \mathrm{Spr}^0(X, \Delta) \xrightarrow{q} \amalg_i S_i(X, \Delta) \to X$$

and q is finite and universally open.

Theorem 5.36 *Assumptions as in (5.24). Let $(X, D + \Delta)$ be lc, τ an involution of $(D^n, \mathrm{Diff}_{D^n} \Delta)$ and $R(\tau) \subset X \times X$ the corresponding gluing relation (5.31). Let p: $\mathrm{Spr}^0(X, D + \Delta) \to X$ be the open spring of $(X, D + \Delta)$. Then*

(1) *every irreducible component of $(p \times p)^* R$ is the graph of an isomorphism, that is, $(p \times p)^* R$ is a groupoid (9.32) and*
(2) *for every log canonical center $Z \subset X$, the stabilizer of its open spring $\mathrm{Spr}^0(Z, X, D + \Delta)$ is a subgroup of $\mathrm{Aut}^s \mathrm{Spr}(Z, X, \Delta)$.*

Proof   We need to describe how the generators of $R(\tau)$ pull-back to the open spring $\mathrm{Spr}^0(X, D + \Delta)$.

First, for any lc center $Z_j \subset X$, the preimage of the diagonal of $Z_j^0 \times Z_j^0$ is a union of the graphs of the natural $G_j := \mathrm{Gal}(\mathrm{Spr}(Z_j, X, D + \Delta)/Z_j)$-action and G_j is a subgroup of $\mathrm{Aut}^s \mathrm{Spr}(Z_j, X, D + \Delta)$ by (5.34).

Second, let $Z_{jk} \subset D^n$ be an irreducible component of the preimage of Z_j. Then Z_{jk} is an lc center of $(D^n, \mathrm{Diff}_{D^n} \Delta)$ and

$$\mathrm{Src}(Z_{jk}, D^n, \mathrm{Diff}_{D^n} \Delta) \overset{\mathrm{cbir}}{\sim} \mathrm{Src}(Z_j, X, D + \Delta)$$

by (4.48). Thus, for each jk, the isomorphism τ: $D^n \simeq D^n$ lifts to crepant, birational maps

$$\tilde{\tau}_{jkl} : \mathrm{Src}\left(Z_j^0, X, D + \Delta\right) \overset{\mathrm{cbir}}{\sim} \mathrm{Src}\left(Z_l^0, X, D + \Delta\right).$$

Given jk, the value of l is determined by $Z_l := n(\tau(Z_{jk}))$, but the lifting is defined only up to left and right multiplication by crepant-birational self-maps of $\mathrm{Src}(Z_j^0, X, D + \Delta) \to Z_j^0$ and $\mathrm{Src}(Z_l^0, X, D + \Delta) \to Z_l^0$. Thus the $\tilde{\tau}_{jkl}$ descend to source-isomorphisms

$$\tau_{jkl} : \mathrm{Spr}\left(Z_j^0, X, D + \Delta\right) \simeq \mathrm{Spr}\left(Z_l^0, X, D + \Delta\right)$$

defined up to left and right multiplication by G_j and G_l.

Thus $(p \times p)^* R$ is the groupoid generated by the G_j and the τ_{jkl}, hence the stabilizer of $\mathrm{Spr}(Z_j^0, X, D + \Delta)$ is generated by the groups $\tau_{jkl}^{-1} G_l \tau_{jkl}$. The latter are all subgroups of $\mathrm{Aut}^s \mathrm{Spr}(Z_j, X, D + \Delta)$. $\square$

Corollary 5.37 *Let S be a scheme that is essentially of finite type over a field of characteristic 0. Let $(\bar{X}, \bar{D} + \bar{\Delta})$ be an lc pair such that X is proper over S and $K_{\bar{X}} + \bar{D} + \bar{\Delta}$ is ample on the generic fiber of $\bar{Z} \to S$ for every lc center $\bar{Z} \subset \bar{X}$. Let τ be an involution of $(\bar{D}^n, \mathrm{Diff}_{\bar{D}^n} \bar{\Delta})$.*
Then the gluing relation $R(\tau)$, defined in (5.31), is finite.

Proof We apply (5.36) to $(\bar{X}, \bar{D} + \bar{\Delta})$.

Since $\mathrm{Spr}^0(\bar{X}, \bar{D} + \bar{\Delta})$ has finitely many irreducible components, the groupoid is finite if and only if the stabilizer of $\mathrm{Spr}^0(\bar{Z}, \bar{X}, \bar{D} + \bar{\Delta})$ is finite for every lc center $\bar{Z} \subset \bar{X}$. By (5.36) this holds if the groups $\mathrm{Aut}^s_S \mathrm{Spr}(\bar{Z}, \bar{X}, \bar{D} + \bar{\Delta})$ of source-automorphisms over S are finite. We have a sequence of proper morphisms

$$\mathrm{Src}(\bar{Z}, \bar{X}, \bar{D} + \bar{\Delta}) \xrightarrow{\sigma} \mathrm{Spr}(\bar{Z}, \bar{X}, \bar{D} + \bar{\Delta}) \to S \qquad (5.37.1)$$

where σ is a crepant log structure. From now on we forget that (5.37.1) arose as the source of $\bar{Z}$. Thus we have a crepant log structure

$$(Y, \Delta_Y) \xrightarrow{\sigma} W \xrightarrow{g} S \qquad (5.37.2)$$

where g is proper and $K_Y + \Delta_Y \sim_{\mathbb{Q}} \sigma^* L$ for some $\mathbb{Q}$-divisor L that is ample on the generic fiber of g. We need to prove that the image of

$$\sigma_*\colon \mathrm{Bir}^c_S(Y, \Delta_Y) \to \mathrm{Aut}_S W \quad \text{is finite.} \qquad (5.37.3)$$

The automorphism group of a variety W over a base scheme S injects into the automorphism group of the generic fiber W. We can thus replace S with the image of the generic point of W.

Finally we have a crepant log structure $\sigma\colon (Y_k, \Delta_{Y_k}) \to W_k$ where W_k is proper over a field k and $K_{Y_k} + \Delta_{Y_k} \sim_{\mathbb{Q}} \sigma^* L_k$ where L_k is ample on W_k. Then (10.69) implies that $\mathrm{im}[\mathrm{Bir}^c_k(Y_k, \Delta_{Y_k}) \to \mathrm{Aut}_k W_k]$ is finite. $\square$

5.7 Descending the canonical bundle

In this Section we answer Question 5.5 for slc pairs.

Theorem 5.38 *Assumptions as in (5.24). Let X be demi-normal and Δ a $\mathbb{Q}$-divisor on X. As in (5.2), let $(\bar{X}, \bar{D}, \bar{\Delta})$ be the normalization of (X, Δ) and $\tau\colon \bar{D}^n \to \bar{D}^n$ the corresponding involution. The following are equivalent:*

(1) (X, Δ) *is slc.*
(2) $(\bar{X}, \bar{D} + \bar{\Delta})$ *is lc and* $\mathrm{Diff}_{\bar{D}^n} \bar{\Delta}$ *is τ-invariant.*

Proof We saw in (5.12) that (1) $\Rightarrow$ (2).

By (5.10), for the converse we only need to prove that $K_X + \Delta$ is $\mathbb{Q}$-Cartier. The set of points where $K_X + \Delta$ is $\mathbb{Q}$-Cartier is open. After localizing at a generic point of the locus where $K_X + \Delta$ is not $\mathbb{Q}$-Cartier, we may assume that X is local with closed point $x \in X$, $k(x)$ is algebraically closed and $K_X + \Delta$ is $\mathbb{Q}$-Cartier on $X^0 := X \setminus \{x\}$.

Choose $m > 0$ even such that $m\Delta$ is a $\mathbb{Z}$-divisor, $\omega_{\bar{X}}^{[m]}(m\bar{D} + m\bar{\Delta})$ is locally free and $\omega_X^{[m]}(m\Delta)$ is locally free on X^0. Consider the $\mathbb{A}^1$-bundle

$$\bar{X}_L := \operatorname{Spec}_{\bar{X}} \sum_{r \geq 0} \omega_{\bar{X}}^{[rm]}(rm\bar{D} + rm\bar{\Delta}) \quad \text{with projection } p \colon \bar{X}_L \to \bar{X}.$$

Set $\bar{D}_L := p^{-1}\bar{D}$ and $\bar{\Delta}_L := p^{-1}\bar{\Delta}$. Then $(\bar{X}_L, \bar{D}_L + \bar{\Delta}_L)$ is lc. The normalization $\bar{D}_L^n$ of $\bar{D}_L$ can be obtained either as the fiber product $\bar{D}^n \times_{\bar{D}} \bar{D}_L$ or as $\operatorname{Spec}_{\bar{D}} \sum_{r \geq 0} \omega_{\bar{D}^n}^{[rm]}(rm\operatorname{Diff}_{\bar{D}^n} \bar{\Delta})$. In particular, $\operatorname{Diff}_{\bar{D}_L^n} \bar{\Delta}_L = p^* \operatorname{Diff}_{\bar{D}^n} \bar{\Delta}$ and the lc centers of $(\bar{D}_L^n, \operatorname{Diff}_{\bar{D}_L^n} \bar{\Delta}_L)$ are the preimages of the lc centers of $(\bar{D}^n, \operatorname{Diff}_{\bar{D}^n} \bar{\Delta})$.

The τ-invariance of $\operatorname{Diff}_{\bar{D}^n} \bar{\Delta}$ is equivalent to saying that τ lifts to an involution $\tau_L \colon \bar{D}_L^n \to \bar{D}_L^n$ and $\operatorname{Diff}_{\bar{D}_L^n} \bar{\Delta}_L$ is τ_L-invariant.

Thus we have an lc pair $(\bar{X}_L, \bar{D}_L + \bar{\Delta}_L)$ and an involution $\tau_L \colon \bar{D}_L^n \to \bar{D}_L^n$ that maps log canonical centers of $(\bar{D}_L^n, \operatorname{Diff}_{\bar{D}_L^n} \bar{\Delta}_L)$ to log canonical centers. We are in a situation considered in (5.33). We have just established that (5.33.1–2) both hold.

5.38.3 In order to apply (5.33), we need to check the assumption (5.33.3). That is, we need to prove that $(n_L, \tau_L \circ n_L) \colon \bar{D}_L^n \to \bar{X}_L \times \bar{X}_L$ generates a *finite* set theoretic equivalence relation $\bar{R}_L \rightrightarrows \bar{X}_L$.

Note that finiteness holds over X^0. Indeed, we assumed that $\omega_X^{[m]}(m\Delta)$ is locally free on X^0; let $X_L^0 \to X^0$ denote the total space of $\omega_X^{[m]}(m\Delta|_{X^0})$. Set $\bar{X}_L^0 := p^{-1}(X^0) \subset \bar{X}_L$. There is a natural finite morphism $\bar{X}_L^0 \to X_L^0$ and in fact $X_L^0 = \bar{X}_L^0/\bar{R}_L^0$ where $\bar{R}_L^0$ denotes the restriction of $\bar{R}_L$ to $\bar{X}_L^0$. Thus $\bar{R}_L^0 \rightrightarrows \bar{X}_L^0$ is finite and therefore $\bar{R}_L \rightrightarrows \bar{X}_L$ is finite if and only if it is finite over $\bar{X}_L \setminus \bar{X}_L^0$.

In order to study the latter, let $\bar{x}_1, \dots, \bar{x}_r \in \bar{X}$ be the preimages of x. If none of the $\bar{x}_i$ are lc centers of $(\bar{X}, \bar{D} + \bar{\Delta})$ then every lc center of $(\bar{X}_L, \bar{D}_L + \bar{\Delta}_L)$ intersects $\bar{X}_L^0$, hence $\bar{R}_L \rightrightarrows \bar{X}_L$ is finite by the easy (9.55).

Thus we are left with the case when at least one of the $\bar{x}_i$ is an lc center. Then, by (5.38.2) and adjunction (4.8), all the $\bar{x}_i$ are lc centers of $(\bar{X}, \bar{D} + \bar{\Delta})$.

The fiber of $p \colon \bar{X}_L \to \bar{X}$ over $\bar{x}_i$ is the 1-dimensional $k(x)$-vectorspace

$$V_i := \omega_{\bar{X}}^{[m]}(m\bar{D} + m\bar{\Delta}) \otimes_{\bar{X}} k(\bar{x}_i).$$

Thus $\bar{X}_L \setminus \bar{X}_L^0 = V_1 \cup \cdots \cup V_r$ and τ_L gives a collection of isomorphisms $\tau_{ijk} \colon V_i \to V_j$. (A given $\bar{x}_i$ can have several preimages in $\bar{D}^n$ and each of these gives an isomorphism of V_i to some V_j.) As in (5.36), the τ_{ijk} generate a groupoid and $\bar{R}_L \rightrightarrows \bar{X}_L$ is a finite set theoretic equivalence relation if and only if $\operatorname{stab}(V_i) \subset \operatorname{Aut}(V_i) = k(x)^*$, the group of all possible composites $\tau_{ij_2k_2} \circ \tau_{j_2j_3k_3} \circ \cdots \circ \tau_{j_nik_n} \colon V_i \to V_i$ is a finite for every i.

In the surface case (5.18.4) the Poincaré residue map gives canonical isomorphisms

$$\mathcal{R}^m \colon V_i \simeq H^0\big(\bar{x}_i, \omega_{\bar{x}_i}^m\big) = k(\bar{x}_i).$$

With these choices, τ_{ijk} is the canonical isomorphism $\tau_{ijk}\colon k(\bar{x}_i) = k(x) = k(\bar{x}_j)$, thus stab$(V_i)$ is the identity.

As noted in (5.18.7), in the higher dimensional case the V_i are not canonically isomorphic to $k(\bar{x}_i)$.

A replacement for this isomorphism is provided by the source (4.45)

$$(S_i, \Delta_i) := \mathrm{Src}(\bar{x}_i, \bar{X}, \bar{D} + \bar{\Delta}).$$

We are in the simplest, Calabi–Yau, case when S_i is proper over $k(\bar{x}_i)$ and $K_{S_i} + \Delta_i \sim_{\mathbb{Q}} 0$ (4.29). We proved in (4.45.4) and (4.45.8) that:

5.38.4 The Poincaré residue map gives a canonical isomorphism

$$\mathcal{R}^m : V_i \simeq H^0\big(S_i, \omega_{S_i}^{[m]}(m\Delta_i)\big) \quad \text{and}$$

5.38.5 Each $\tau_{ijk}\colon V_i \to V_j$ becomes the pull-back by a crepant, birational map $\phi_{jik}\colon (S_j, \Delta_j) \dashrightarrow (S_i, \Delta_i)$.

From these we conclude that

$$\mathrm{stab}(V_i) \subset \mathrm{im}\big[\,\mathrm{Bir}^c(S_i, \Delta_i) \to \mathrm{Aut}_{k(x)}\, H^0(S_i, \omega_{S_i}^{[m]}(m\Delta_i))\big]. \qquad (5.38.6)$$

We prove in (10.61) that, although $\mathrm{Bir}^c(S_i, \Delta_i)$ itself can be infinite, the right hand side of (5.38.6) is finite, hence the stabilizers $\mathrm{stab}(V_i) \subset \mathrm{Aut}(V_i) = k(x)^*$ are finite.

Thus $\bar{R}_L \rightrightarrows \bar{X}_L$ is a finite set theoretic equivalence relation and the geometric quotient $\bar{X}_L/\bar{R}_L$ exists by (5.33).

We prove in (9.48) that the complement of the zero section $\bar{X}_S/\bar{R}_S \subset \bar{X}_L/\bar{R}_L$ is a Seifert bundle over $X = \bar{X}/\bar{R}$ (9.50). By (9.53) this implies that, for sufficiently divisible m, $\omega_X^{[m]}(m\Delta)$ is a line bundle over X. $\qquad\square$

Corollary 5.39 *Let* (X, Δ) *be slc with normalization* $\pi\colon \bar{X} \to X$. *Let B be a $\mathbb{Q}$-divisor on X such that B is $\mathbb{Q}$-Cartier at the generic point of every lc center of* (X, Δ). *Then B is $\mathbb{Q}$-Cartier if and only if $\bar{B} := \pi^*B$ is.*

Proof It is clear that if B is $\mathbb{Q}$-Cartier then so is $\bar{B}$. The converse is a local question, hence we may assume that X is affine. We are allowed to replace B by any other divisor $\mathbb{Q}$-linearly equivalent to it. Thus we may assume that B is effective and it does not contain the generic point of any lc center. Then $(\bar{X}, \bar{D} + \bar{\Delta} + \varepsilon\bar{B})$ is lc for $0 < \varepsilon \ll 1$. By (5.38) $K_X + \Delta + \varepsilon B$ is also $\mathbb{Q}$-Cartier hence so is B. $\qquad\square$

A more elementary proof can be obtained using (7.20) and the method of (5.20).

6

Du Bois property

Hodge theory of smooth projective varieties is a very powerful tool. For many applications the most important piece is the surjectivity of the natural map

$$H^i(X^{\mathrm{an}}, \mathbb{C}) \twoheadrightarrow H^i\left(X^{\mathrm{an}}, \mathscr{O}_X^{\mathrm{an}}\right).$$

This surjectivity continues to hold for certain mildly singular varieties. Steenbrink identified the correct class and dubbed them *Du Bois singularities.*

One should think of Du Bois as a weakening of rationality. If a pair (X, Δ) is dlt then X has rational singularities (2.88). If (X, Δ) is lc or slc, in general it does not have rational singularities, but we prove in (6.32) that X is Du Bois.

A crucial advantage of the concept of Du Bois singularities is that, on the one hand, a normal Du Bois singularity is quite close to being rational or CM, and, on the other hand, Du Bois makes sense even for non-equidimensional schemes. This makes it possible to use induction on the union of log canonical centers, which is usually not equidimensional.

A disadvantage of Du Bois singularities is that their definition is rather complicated, relying on a technical generalization of the De Rham complex, the Deligne–Du Bois complex (6.4). However, once these technical difficulties are settled, the theory is very powerful.

The definition and the basic properties of Du Bois singularities are discussed in Section 6.1. The construction of the Deligne–Du Bois complex relies on cubic hyperresolutions; their theory is recalled in Section 10.6. The applications to lc and slc pairs are proved in Section 6.2. The main result of Kollár and Kovács (2010) shows that lc singularities are Du Bois. Here we extend this to show that slc singularities are Du Bois. The inductive proof works best with stratified schemes of log canonical origin (5.29).

Assumptions   In this chapter we work over a field of characteristic 0. It is not even known how to define Du Bois singularities in positive characteristic.

6.1 Du Bois singularities

The starting point is Du Bois's construction, following Deligne's ideas, of the generalized de Rham complex, which we call the *Deligne Du Bois complex*. Recall that if X is a smooth complex algebraic variety of dimension n, then the sheaves of differential p-forms with the usual exterior differentiation give a resolution of the constant sheaf $\mathbb{C}_X$. That is, one has a complex of sheaves,

$$\mathcal{O}_X \xrightarrow{d} \Omega^1_X \xrightarrow{d} \Omega^2_X \xrightarrow{d} \Omega^3_X \xrightarrow{d} \cdots \xrightarrow{d} \Omega^n_X \simeq \omega_X,$$

which is quasi-isomorphic to the constant sheaf $\mathbb{C}_X$ in the euclidean topology via the natural map $\mathbb{C}_X \to \mathcal{O}_X$ given by considering constants as holomorphic functions on X. The sheaves in the complex are quasi-coherent, but the maps between them are not $\mathcal{O}_X$-module homomorphisms. (As $\mathbb{C}_X$ is not quasi-coherent, it can not have a quasi-coherent resolution.)

The Deligne–Du Bois complex is a generalization of the de Rham complex to singular varieties. It is a complex of sheaves on X that is quasi-isomorphic to the constant sheaf $\mathbb{C}_X$. The terms of this complex are harder to describe but its properties, especially cohomological properties, are very similar to the de Rham complex of smooth varieties. In fact, for a smooth variety the Deligne–Du Bois complex is quasi-isomorphic to the de Rham complex, so it is indeed a direct generalization.

The original construction of this complex is based on simplicial hyperresolutions. The reader interested in the details is referred to the original article Du Bois (1981). Note also that simplified constructions were later obtained in Carlson (1985) and Guillén *et al.* (1988) via the general theory of polyhedral and cubic hyperresolutions. A brief review of cubic hyperresolutions is included in (10.6). Other useful references are Steenbrink (1985), Peters and Steenbrink (2008) and Kovács and Schwede (2011b). For alternative definitions of Du Bois singularities see Schwede (2007) and Kovács (2012a).

The word "hyperresolution" will refer to a cubic hyperresolution, although we will not make any use of its special properties. Formally, the construction is the same regardless of the type of hyperresolution used.

Definition 6.1 Let X be a a scheme of finite type over $\mathbb{C}$ of dimension n. Let $D_{\text{filt}}(X)$ denote the derived category of filtered complexes of $\mathcal{O}_X$-modules and $D_{\text{filt,coh}}(X)$ the subcategory of $D_{\text{filt}}(X)$ of complexes $K^{\bullet}$ such that for all i, the cohomology sheaves of $Gr^i_{\text{filt}}K^{\bullet}$ are coherent; compare Du Bois (1981) and Guillén *et al.* (1988). Let $D(X)$ and $D_{\text{coh}}(X)$ denote the derived categories with the same definition except that the complexes are assumed to have the trivial filtration. The superscripts $+, -, b$ carry the usual meaning (bounded below, bounded above, bounded). Isomorphism in these categories is denoted by $\simeq_{qis}$. A sheaf $\mathscr{F}$ is also considered as a complex $\mathscr{F}^{\bullet}$ with $\mathscr{F}^0 = \mathscr{F}$ and $\mathscr{F}^i = 0$ for $i \neq 0$. If $K^{\bullet}$ is a complex in any of the above categories, then $h^i(K^{\bullet})$ denotes the i-th cohomology sheaf of $K^\bullet$. We will also use simply K to denote $K^\bullet$ if confusion is unlikely.

The (total) right derived functor of a left exact functor $F\colon \mathsf{A} \to \mathsf{B}$ between abelian categories, if it exists, is denoted by $\mathcal{R}F\colon D(\mathsf{A}) \to D(\mathsf{B})$ and $R^i F$ is short for $h^i \circ \mathcal{R}F$. Furthermore, $\mathbb{H}^i$, $\mathbb{H}^i_c$, $\mathbb{H}^i_Z$, and $\mathcal{H}^i_Z$ will denote $R^i\Gamma$, $R^i\Gamma_c$, $R^i\Gamma_Z$, and $R^i\mathcal{H}_Z$ respectively, where Γ is the functor of global sections, Γ_c is the functor of global sections with proper support, Γ_Z is the functor of global sections with support in the closed subset Z and $\mathcal{H}_Z$ is the functor of the sheaf of local sections with support in the closed subset Z. Note that according to this terminology, if $\phi\colon Y \to X$ is a morphism and $\mathcal{F}$ is a coherent sheaf on Y, then $\mathcal{R}\phi_*\mathcal{F}$ is the complex whose cohomology sheaves give rise to the usual higher direct images of $\mathcal{F}$.

We will often use the notion that a morphism $f\colon A \to B$ in a derived category *has a left inverse*. This means that there exists a morphism $f^\ell\colon B \to A$ in the same derived category such that $f^\ell \circ f\colon A \to A$ is the identity morphism of A. That is, f^ℓ is a *left inverse* of f.

Finally, we will also make the following simplification in notation. First observe that if $\iota\colon \Sigma \hookrightarrow X$ is a closed embedding of schemes then ι_* is exact and hence $\mathcal{R}\iota_* = \iota_*$. This allows one to make the following harmless abuse of notation: If $A \in \mathrm{Ob}\, D(\Sigma)$, then, as usual for sheaves, we will drop ι_* from the notation of the object $\iota_* A$. In other words, we will, without further warning, consider A an object in $D(X)$.

The Deligne–Du Bois complex of a pair

Definition 6.2 A *reduced pair* is a pair (X, Σ) where X is a reduced scheme of finite type over $\mathbb{C}$ and Σ is a reduced, closed subscheme of X. Neither X nor Σ is assumed pure dimensional.

A hyperresolution $(X_\bullet, \Sigma_\bullet) \to (X, \Sigma)$ is a *good hyperresolution* if $X_\bullet \to X$ is a hyperresolution (10.79), and using the notation $U_\bullet = X_\bullet \times_X (X \setminus \Sigma)$ and $\Sigma_\bullet = X_\bullet \setminus U_\bullet$, we have that for all α, either Σ_α is a divisor with simple normal crossings on X_α or $\Sigma_\alpha = X_\alpha$. Notice that it is possible that $X_\bullet$ has components that map into Σ, and then these components are contained in $\Sigma_\bullet$. For more details and the existence of such hyperresolutions see Du Bois (1981, 6.2) and Guillén *et al.* (1988, IV.1.21, IV.1.25, IV.2.1). For a primer on hyperresolutions see (10.6).

Notation 6.3 Let X be a smooth complex variety and Σ an snc divisor on X. We will use the following notation:

$$\Omega^p_{X,\Sigma} := \Omega^p_X(\log \Sigma)(-\Sigma) \simeq \Omega^p_X(\log \Sigma) \otimes \mathscr{I}_{\Sigma \subseteq X}.$$

Using the differentials $d^p \colon \Omega^p_X(\log \Sigma) \to \Omega^{p+1}_X(\log \Sigma)$ and the identity map on $\mathscr{I}_{\Sigma \subseteq X}$ these sheaves form a complex which will be denoted by $\Omega^{\bullet}_{X,\Sigma}$.

Let $j \colon X \setminus \Sigma \hookrightarrow X$ denote the natural embedding and recall that $j_! \mathbb{C}_{X \setminus \Sigma}$ denotes the *extension of* $\mathbb{C}_{X \setminus \Sigma}$ *by zero outside of* $X \setminus \Sigma$. It is part of the short exact sequence

$$0 \longrightarrow j_! \mathbb{C}_{X \setminus \Sigma} \longrightarrow \mathbb{C}_X \longrightarrow \mathbb{C}_\Sigma \longrightarrow 0.$$

A straightforward local computation shows that the natural morphism mapping constants to regular functions induces a quasi-isomorphism.

$$j_! \mathbb{C}_{X \setminus \Sigma} \xrightarrow{\simeq} \Omega^{\bullet}_{X,\Sigma}.$$

Using the usual de Rham resolution of the constant sheaf one obtains the filtered distinguished triangle, cf. Friedman (1983)

$$\Omega^{\bullet}_{X,\Sigma} \longrightarrow \Omega^{\bullet}_X \longrightarrow \Omega^{\bullet}_\Sigma \xrightarrow{+1} . \tag{6.3.1}$$

Next, we will extend this construction to arbitrary pairs.

Definition 6.4 (Steenbrink, 1985, section 3) Let (X, Σ) be a reduced pair of finite type over $\mathbb{C}$. Consider an embedded hyperresolution of $\Sigma \subseteq X$, cf. (10.79):

$$
\begin{array}{ccc}
\Sigma_{\bullet} & \xrightarrow{\varrho_{\bullet}} & X_{\bullet} \\
{\scriptstyle \varepsilon_{\bullet}} \downarrow & & \downarrow {\scriptstyle \varepsilon_{\bullet}} \\
\Sigma & \xrightarrow{\varrho} & X.
\end{array}
$$

Then the *Deligne–Du Bois complex of the reduced pair* (X, Σ) is defined as

$$\underline{\Omega}^{\bullet}_{X,\Sigma} := \mathcal{R}\varepsilon_{\bullet *} \Omega^{\bullet}_{X_{\bullet}, \Sigma_{\bullet}}, \tag{6.4.1}$$

an object in $D_{\text{filt}}(X)$ cf. (10.86). This definition is independent of the hyperresolution chosen by (6.5.2). The associated graded quotients of $\underline{\Omega}^{\bullet}_{X,\Sigma}$ will be denoted by

$$\underline{\Omega}^p_{X,\Sigma} := Gr^p_{\text{filt}} \underline{\Omega}^{\bullet}_{X,\Sigma}[p]. \tag{6.4.2}$$

It follows that

$$\underline{\Omega}^p_{X,\Sigma} \simeq_{qis} \mathcal{R}\varepsilon_{\bullet *} \Omega^p_{X_{\bullet}, \Sigma_{\bullet}}. \tag{6.4.3}$$

Taking $\Sigma = \emptyset$ in (6.4.1) one obtains the definition of the *Deligne–Du Bois complex of the scheme* X:

$$\underline{\Omega}^{\bullet}_X := \underline{\Omega}^{\bullet}_{X,\emptyset} \tag{6.4.4}$$

and similarly

$$\underline{\Omega}_X^p := Gr_{\text{filt}}^p \underline{\Omega}_X^{\bullet}[p] \simeq_{qis} \mathcal{R}\varepsilon_{\bullet *} \Omega_{X_{\bullet}}^p. \qquad (6.4.5)$$

Theorem 6.5 (Du Bois, 1981; Kovács, 2011a) *Let* (X, Σ) *be a reduced pair of finite type over* $\mathbb{C}$. *Then* $\underline{\Omega}_{X,\Sigma}^{\bullet} \in \text{Ob } D_{\text{filt}}(X)$ *satisfies the following properties*

(1)

$$\underline{\Omega}_{X,\Sigma}^{\bullet} \simeq_{qis} j_! \mathbb{C}_{X \setminus \Sigma}.$$

(2) *If* $\varepsilon_{\bullet} \colon X_{\bullet} \to X$ *is any embedded hyperresolution of* $\Sigma \subseteq X$, *then*

$$\underline{\Omega}_{X,\Sigma}^{\bullet} \simeq_{qis} \mathcal{R}\varepsilon_{\bullet *} \Omega_{X_{\bullet}, \Sigma_{\bullet}}^{\bullet}.$$

In particular, $h^i(\underline{\Omega}_{X,\Sigma}^p) = 0$ *for* $i < 0$.

(3) *If* X *is smooth and* Σ *is a normal crossing divisor, then*

$$\underline{\Omega}_{X,\Sigma}^{\bullet} \simeq_{qis} \Omega_{X,\Sigma}^{\bullet},$$

and hence,

$$\underline{\Omega}_{X,\Sigma}^p \simeq_{qis} \Omega_{X,\Sigma}^p.$$

(4) $\underline{\Omega}_{(\_),(\_)}^{\bullet}$ *is functorial, that is, if* $\phi \colon (Y, \Gamma) \to (X, \Sigma)$ *is a morphism of reduced pairs of finite type, then there exists a natural map* ϕ^* *of filtered complexes*

$$\phi^* \colon \underline{\Omega}_{X,\Sigma}^{\bullet} \to \mathcal{R}\phi_* \underline{\Omega}_{Y,\Gamma}^{\bullet}.$$

Furthermore, $\underline{\Omega}_{X,\Sigma}^{\bullet} \in \text{Ob}(D_{\text{filt,coh}}^b(X))$, *in particular, for any* p, *the cohomology sheaves* $h^i(\underline{\Omega}_{X,\Sigma}^p)$ *are coherent, and if* ϕ *is proper, then* ϕ^* *is a morphism in* $D_{\text{filt,coh}}^b(X)$.

(5) *Let* $U \subseteq X$ *be an open subscheme of* X. *Then*

$$\underline{\Omega}_{X,\Sigma}^{\bullet}\big|_U \simeq_{qis} \underline{\Omega}_{U, U \cap \Sigma}^{\bullet}.$$

In particular,

$$\underline{\Omega}_{X,\Sigma}^{\bullet}\big|_{X \setminus \Sigma} \simeq_{qis} \underline{\Omega}_{X \setminus \Sigma}^{\bullet}.$$

(6) *Let* (X, Σ) *be a reduced pair and let* $H \subset X$ *be a general member of a base point free linear system. Then*

$$\underline{\Omega}_{H, H \cap \Sigma}^{\bullet} \simeq_{qis} \underline{\Omega}_{X,\Sigma}^{\bullet} \otimes_L \mathcal{O}_H.$$

(7) *There exists a distinguished triangle,*

$$\underline{\Omega}_{X,\Sigma}^{\bullet} \longrightarrow \underline{\Omega}_X^{\bullet} \longrightarrow \underline{\Omega}_{\Sigma}^{\bullet} \xrightarrow{+1},$$

and for each p there exists a distinguished triangle,

$$\underline{\Omega}^p_{X,\Sigma} \longrightarrow \underline{\Omega}^p_X \longrightarrow \underline{\Omega}^p_\Sigma \xrightarrow{+1} .$$

(8) *There exists a natural map $\mathcal{I}_{\Sigma \subseteq X} \to \underline{\Omega}^0_{X,\Sigma}$ compatible with (4) and (5).*

(9) *If X is proper, then there exists a spectral sequence degenerating at E_1 and abutting to the singular cohomology with compact support of $X \setminus \Sigma$:*

$$E_1^{pq} = \mathbb{H}^q\left(X, \underline{\Omega}^p_{X,\Sigma}\right) \Rightarrow H_c^{p+q}(X \setminus \Sigma, \mathbb{C}).$$

(10) *Let $\pi \colon \widetilde{X} \to X$ be a projective morphism and $\Sigma \subseteq X$ a reduced closed subscheme such that π is an isomorphism outside of Σ. Set $E = \pi^{-1}(\Sigma)_{\mathrm{red}}$. Then there exists a distinguished triangle,*

$$\underline{\Omega}^\bullet_X \xrightarrow{\alpha} \underline{\Omega}^\bullet_\Sigma \oplus R\pi_*\underline{\Omega}^\bullet_{\widetilde{X}} \xrightarrow{\beta} R\pi_*\underline{\Omega}^\bullet_E \xrightarrow{+1} ,$$

where α is the sum and β is the difference of the natural maps from (4). Similarly, for each p there exists a distinguished triangle,

$$\underline{\Omega}^p_X \xrightarrow{\alpha_p} \underline{\Omega}^p_\Sigma \oplus R\pi_*\underline{\Omega}^p_{\widetilde{X}} \xrightarrow{\beta_p} R\pi_*\underline{\Omega}^p_E \xrightarrow{+1} .$$

(11) *Suppose $X = Y \cup Z$ is the union of two closed subschemes and denote their intersection by $W := Y \cap Z$. Then there exists a distinguished triangle,*

$$\underline{\Omega}^\bullet_X \longrightarrow \underline{\Omega}^\bullet_Y \oplus \underline{\Omega}^\bullet_Z \xrightarrow{-} \underline{\Omega}^\bullet_W \xrightarrow{+1} ,$$

and for each p there exists a distinguished triangle,

$$\underline{\Omega}^p_X \longrightarrow \underline{\Omega}^p_Y \oplus \underline{\Omega}^p_Z \xrightarrow{-} \underline{\Omega}^p_W \xrightarrow{+1} .$$

Proof Here we only indicate how to obtain the stated results for pairs assuming the same in the absolute case, which was already included in Du Bois (1981). In particular, (5) is simply Du Bois (1981, 3.9) and (6) is Du Bois (1981, 3.8). Next, consider an embedded hyperresolution of $\Sigma \subseteq X$:

$$\begin{array}{ccc} \Sigma_\bullet & \xrightarrow{\varrho_\bullet} & X_\bullet \\ {\scriptstyle \varepsilon_\bullet}\downarrow & & \downarrow{\scriptstyle \varepsilon_\bullet} \\ \Sigma & \xrightarrow{\varrho} & X \end{array}$$

By (6.3.1) there exists a distinguished triangle for all i

$$\Omega^\bullet_{X_i,\Sigma_i} \longrightarrow \Omega^\bullet_{X_i} \longrightarrow \Omega^\bullet_{\Sigma_i} \xrightarrow{+1} ,$$

and these are compatible with respect to the morphisms induced by $\varepsilon_\bullet$. Applying $R\varepsilon_{\bullet *}$ and Du Bois (1981, 3.17) implies (7). Then (7) along with Du Bois (1981,

4.5, 3.17, 3.6, 3.10) imply (1), (2), (4), and (5), and then (3) follows from (2). By Du Bois (1981, 2.11) and (2) (again) there exists a commutative diagram

$$
\begin{array}{ccccc}
\mathscr{I}_{\Sigma \subseteq X} & \longrightarrow & \mathcal{O}_X & \longrightarrow & \mathcal{O}_\Sigma \xrightarrow{+1} \\
{\scriptstyle \alpha} \downarrow & & \downarrow & & \downarrow \\
\underline{\Omega}^0_{X,\Sigma} & \longrightarrow & \underline{\Omega}^0_X & \longrightarrow & \underline{\Omega}^0_\Sigma \xrightarrow{+1} ,
\end{array}
$$

which implies the existence of the morphism α and proving (8). To prove (6) observe that since H is general, the fiber product $X. \times_X H \to H$ provides a hyperresolution of H. Then the special case of the statement with $\Sigma = \emptyset$, that is, $\underline{\Omega}^\bullet_H \simeq_{qis} \underline{\Omega}^\bullet_X \otimes_L \mathcal{O}_H$, follows easily from (2) and then (6) follows from (7) applying the special case to both X and Σ individually.

Finally, (9) is essentially a direct consequence of Deligne (1974, 8.1, 8.2, 9.3). See (6.8) for more details on this. □

Vanishing of the top cohomology

Proposition 6.6 (cf. Guillén *et al.*, 1988, III.1.17; Greb *et al.*, 2011a, 13.7 and Kovács *et al.*, 2010, 4.9) *Let X be a positive dimensional reduced scheme. Then the ith cohomology sheaf of $\underline{\Omega}^p_X$ vanishes for all $i \geq \dim X$, i.e., $h^i(\underline{\Omega}^p_X) = 0$ for all p and for all $i \geq \dim X$.*

Proof Let $\varepsilon_\bullet \colon X_\bullet \to X$ be a cubic hyperresolution of X such that $\dim X_\alpha \leq \dim X - |\alpha| + 1, \forall \alpha \in \square_n$ cf. (10.80).

It follows from the definition that there exists a Grothendieck spectral sequence

$$
E_1^{r,s} = \bigoplus_{|\alpha|=r+1} R^s \varepsilon_{\alpha*} \underline{\Omega}^\bullet_{X_\alpha} \Rightarrow h^{r+s}(\underline{\Omega}^p_X)
$$

and for $|\alpha| = r + 1$ we have $\dim \operatorname{Supp} R^s \varepsilon_{\bullet *} \underline{\Omega}^\bullet_{X_\alpha} \leq \dim X - r - s$, so it follows that

$$
\dim \operatorname{Supp} h^{r+s}(\underline{\Omega}^p_X) \leq \dim X - (r + s).
$$

Hence the statement follows for $i > \dim X$ or $p > 0$, so only the case of $p = 0$ and $i = n := \dim X$ is left to be proven. Let $S := \operatorname{Sing} X$ and $\pi \colon \widetilde{X} \to X$ a log resolution with exceptional divisor E such that $\widetilde{X} \setminus E \simeq X \setminus S$. Consider the distinguished triangle given by (6.5(10)),

$$
\underline{\Omega}^0_X \longrightarrow \underline{\Omega}^0_S \oplus R\pi_*\underline{\Omega}^0_{\widetilde{X}} \longrightarrow R\pi_*\underline{\Omega}^0_E \xrightarrow{+1} . \tag{6.6.1}
$$

Notice that $\widetilde{X}$ is smooth and E is an snc divisor, so they are both DB and hence $\underline{\Omega}^0_{\widetilde{X}} \simeq \mathcal{O}_{\widetilde{X}}$ and $\underline{\Omega}^0_E \simeq \mathcal{O}_E$. Next, consider the long exact sequence of cohomology sheaves induced by the above distinguished triangle,

$$\cdots \longrightarrow h^{n-1}(\underline{\Omega}_S^0) \oplus R^{n-1}\pi_*\mathscr{O}_{\widetilde{X}} \xrightarrow{\varsigma} R^{n-1}\pi_*\mathscr{O}_E$$

$$\longrightarrow h^n(\underline{\Omega}_X^0) \longrightarrow h^n(\underline{\Omega}_S^0) \oplus R^n\pi_*\mathscr{O}_{\widetilde{X}}.$$

Since $\dim S < n$, the case we have already proven implies that $h^n(\underline{\Omega}_S^0) = 0$. Furthermore, as π is birational, the dimension of any fiber of π is at most $n-1$ and hence $R^n\pi_*\mathscr{O}_{\widetilde{X}} = 0$. This implies that $h^n(\underline{\Omega}_X^0) \simeq \operatorname{coker}\varsigma$.

On the other hand, it follows from the construction of the distinguished triangle (6.6.1) that $\varsigma(0,\underline{\quad})$ is the map $R^{n-1}\pi_*\mathscr{O}_{\widetilde{X}} \to R^{n-1}\pi_*\mathscr{O}_E$ induced by the short exact sequence

$$0 \to \mathscr{O}_{\widetilde{X}}(-E) \to \mathscr{O}_{\widetilde{X}} \to \mathscr{O}_E \to 0.$$

Again, the same dimension bound as above implies that $R^n\pi_*\mathscr{O}_{\widetilde{X}}(-E) = 0$, so we obtain that $\varsigma(0,\underline{\quad})$ is surjective, and then so is ς. Therefore, as desired, $h^n(\underline{\Omega}_X^0) \simeq \operatorname{coker}\varsigma = 0$. $\qquad\square$

It follows easily from the definition and (6.6) that we have the following bounds on the nonzero cohomology sheaves of $\underline{\Omega}_{X,\Sigma}^p$.

Proposition 6.7 *Let X be a positive dimensional variety. Then the ith cohomology sheaf of $\underline{\Omega}_{X,\Sigma}^p$ vanishes for all $i \geq \dim X$, that is, $h^i(\underline{\Omega}_{X,\Sigma}^p) = 0$ for all p and for all $i \geq \dim X$.*

Proof This follows directly from (6.6) using the long exact cohomology sequence associated to (6.5.7). $\qquad\square$

Cohomology with compact support

Let (X, Σ) be a pair of finite type over $\mathbb{C}$. Deligne's Hodge theory applied in this situation gives the following theorem:

Theorem 6.8 (Deligne, 1974) *Let (X, Σ) be a pair of finite type over $\mathbb{C}$ and set $U = X \setminus \Sigma$. Then*

(1) *The natural composition map $j_!\mathbb{C}_U \to \mathscr{I}_{\Sigma \subseteq X} \to \underline{\Omega}_{X,\Sigma}^{\bullet}$ is a quasi-isomorphism, that is, $\underline{\Omega}_{X,\Sigma}^{\bullet}$ is a resolution of the sheaf $j_!\mathbb{C}_U$.*
(2) *The natural map $H_c^{\bullet}(U, \mathbb{C}) \to \mathbb{H}^{\bullet}(X, \underline{\Omega}_{X,\Sigma}^{\bullet})$ is an isomorphism.*
(3) *If in addition X is proper, then the spectral sequence,*

$$E_1^{p,q} = \mathbb{H}^q(X, \underline{\Omega}_{X,\Sigma}^p) \Rightarrow H_c^{p+q}(U, \mathbb{C})$$

degenerates at E_1 and abuts to the Hodge filtration of Deligne's mixed Hodge structure.

Proof Consider an embedded hyperresolution of $\Sigma \subseteq X$:

$$
\begin{array}{ccc}
\Sigma_\bullet & \xrightarrow{\varrho_\bullet} & X_\bullet \\
{\scriptstyle\varepsilon_\bullet}\downarrow & & \downarrow{\scriptstyle\varepsilon_\bullet} \\
\Sigma & \xrightarrow{\varrho} & X
\end{array}
$$

Then by (6.5(2)) $\underline{\Omega}^\bullet_{X,\Sigma} \simeq_{qis} \mathcal{R}\varepsilon_{\bullet *}\underline{\Omega}^\bullet_{X_\bullet,\Sigma_\bullet}$. The statements follow from Deligne (1974, 8.1, 8.2, 9.3). See also Guillén *et al.* (1988, IV.4). □

Corollary 6.9 *Let X be a proper complex scheme of finite type, $\iota\colon \Sigma \hookrightarrow X$ a closed subscheme. Then the natural map*

$$
H^i(X, \mathscr{I}_{\Sigma \subseteq X}) \to \mathbb{H}^i\left(X, \underline{\Omega}^0_{X,\Sigma}\right)
$$

is surjective for all $i \in \mathbb{N}$.

Proof Let $j\colon U := X \setminus \Sigma \hookrightarrow X$ be the natural embedding. By (6.8(3)) the natural composition map

$$
H^i_c(U, \mathbb{C}) \to H^i(X, \mathscr{I}_{\Sigma \subseteq X}) \to \mathbb{H}^i\left(X, \underline{\Omega}^0_{X,\Sigma}\right)
$$

is surjective. This implies the statement. □

DB pairs and the DB defect

Definition 6.10 (Kovács, 2011a) A reduced pair (X, Σ) is called a *DB pair* at $x \in X$ if the natural morphism $(\mathscr{I}_{\Sigma \subseteq X})_x \to (\underline{\Omega}^0_{X,\Sigma})_x$ of stalks induced by the morphism in (6.5.8) is a quasi-isomorphism. We define the *non-DB*, or *non-Du Bois locus* as the set containing those points where this condition fails.

The pair (X, Σ) is called a *DB pair* if the above quasi-isomorphism holds for every $x \in X$, or equivalently the non-DB locus is empty.

In the special case when $\Sigma = \emptyset$, this reduces the original definition of Du Bois singularities: a reduced complex scheme of finite type X is said to have *Du Bois* or *DB singularities* if the pair $(X, \emptyset)$ is a DB pair, that is, if the natural morphism $\mathscr{O}_X \to \underline{\Omega}^0_X$ is a quasi-isomorphism. Note that just like the notion of a rational pair, the notion of a *DB pair* describes the "singularity" of the relationship between X and Σ.

As one expects from a good notion of singularity, smooth points are DB. For pairs, being smooth is replaced by being snc.

Example 6.11 Let (X, Δ) be an snc pair or more generally let X be smooth and Σ a normal crossing divisor. Then (X, Σ) is a DB pair by (6.5.3).

We also introduce the notion of the DB defect of the pair (X, Σ):

Definition 6.12 (Kovács, 2011a)   The *DB defect* of the pair (X, Σ) is the mapping cone of the morphism $\mathscr{I}_{\Sigma \subseteq X} \to \underline{\Omega}^0_{X, \Sigma}$. It is denoted by $\underline{\Omega}^\times_{X, \Sigma}$. By definition one has a distinguished triangle,

$$\mathscr{I}_{\Sigma \subseteq X} \longrightarrow \underline{\Omega}^0_{X, \Sigma} \longrightarrow \underline{\Omega}^\times_{X, \Sigma} \xrightarrow{+1} \qquad (6.12.1)$$

and then another one using (6.5.7):

$$\underline{\Omega}^\times_{X, \Sigma} \longrightarrow \underline{\Omega}^\times_X \longrightarrow \underline{\Omega}^\times_\Sigma \xrightarrow{\cdot +1} . \qquad (6.12.2)$$

It follows that

$$h^0\big(\underline{\Omega}^\times_{X, \Sigma}\big) \simeq h^0\big(\underline{\Omega}^0_{X, \Sigma}\big)/\mathscr{I}_{\Sigma \subseteq X}, \quad \text{and} \qquad (6.12.3)$$

$$h^i\big(\underline{\Omega}^\times_{X, \Sigma}\big) \simeq h^i\big(\underline{\Omega}^0_{X, \Sigma}\big) \quad \text{for } i > 0. \qquad (6.12.4)$$

Remark 6.13   By definition, the non-Du Bois locus of (X, Σ) is exactly the support of $\underline{\Omega}^\times_{X, \Sigma}$, that is, the union of the supports of its cohomology sheaves. By (6.5.2), (6.6), and (6.5.4) this set consists of finitely many coherent sheaves and hence the non-Du Bois locus is Zariski closed.

Lemma 6.14   *Let (X, Σ) be a reduced pair. Then the following are equivalent:*

(1) *The pair (X, Σ) is DB.*
(2) *The DB defect of (X, Σ) is acyclic, that is, $\underline{\Omega}^\times_{X, \Sigma} \simeq_{qis} 0$.*
(3) *The induced natural morphism $\underline{\Omega}^\times_X \to \underline{\Omega}^\times_\Sigma$ is a quasi-isomorphism.*
(4) *The induced natural morphism $h^i(\underline{\Omega}^\times_X) \to h^i(\underline{\Omega}^\times_\Sigma)$ is an isomorphism for all $0 \le i \le \dim X$.*
(5) *The induced natural morphism $h^i(\underline{\Omega}^0_X) \to h^i(\underline{\Omega}^0_\Sigma)$ is an isomorphism for all $0 < i \le \dim X$ and a surjection with kernel isomorphic to $\mathscr{I}_{\Sigma \subseteq X}$ for $i = 0$.*

Proof   The equivalence of (1) and (2) follows from (6.12.1), the equivalence of (2) and (3) follows from (6.12.2), the equivalence of (3) and (4) follows from the definition of quasi-isomorphism, and the equivalence of (4) and (5) follows from the definition of the DB defect $\underline{\Omega}^\times_{X, \Sigma}$ (6.12) and the distinguished triangle (6.12.1). $\qquad \qquad \square$

Proposition 6.15   *Let (X, Σ) be a reduced pair. If two of $\{X, \Sigma, (X, \Sigma)\}$ are DB then so is the third.*

Proof   Consider the exact triangle (6.12.2):

$$\underline{\Omega}^\times_{X, \Sigma} \longrightarrow \underline{\Omega}^\times_X \longrightarrow \underline{\Omega}^\times_\Sigma \xrightarrow{+1} .$$

Clearly, if one of the objects in this triangle is acyclic, then it is equivalent that the other two are acyclic. Then the statement follows by (6.14.2) $\qquad \square$

Corollary 6.16 *If X has snc singularities, then it is DB.*

Proof Follows from (6.11) and (6.15). □

We also have the following excision type statement.

Proposition 6.17 *Let $X = (Y \cup Z)_{\text{red}}$ be a union of closed reduced subschemes with intersection $W = (Y \cap Z)_{\text{red}}$. Then the DB defects of the pairs (X, Y) and (Z, W) are quasi-isomorphic. That is,*

$$\underline{\Omega}^{\times}_{X,Y} \simeq_{qis} \underline{\Omega}^{\times}_{Z,W}.$$

Proof Consider the following diagram of distinguished triangles,

$$
\begin{array}{ccccccc}
\underline{\Omega}^{\times}_{X,Y} & \longrightarrow & \underline{\Omega}^{\times}_{X} & \longrightarrow & \underline{\Omega}^{\times}_{Y} & \xrightarrow{+1} & \\
\downarrow{\alpha} & & \downarrow{\beta} & & \downarrow{\gamma} & & \\
\underline{\Omega}^{\times}_{Z,W} & \longrightarrow & \underline{\Omega}^{\times}_{Z} & \longrightarrow & \underline{\Omega}^{\times}_{W} & \xrightarrow{+1} & ,
\end{array}
$$

where β and γ are the natural restriction morphisms and α is the morphism induced by β and γ on the mapping cones. Then by Kollár and Kovács (2010, 2.1) there exists a distinguished triangle

$$\mathbf{Q} \longrightarrow \underline{\Omega}^{\times}_{Y} \oplus \underline{\Omega}^{\times}_{Z} \longrightarrow \underline{\Omega}^{\times}_{W} \xrightarrow{+1}$$

and a map $\sigma \colon \underline{\Omega}^{\times}_{X} \to \mathbf{Q}$ compatible with the above diagram such that α is an isomorphism if and only if σ is one. On the other hand, σ is indeed an isomorphism by (6.5.6) and so the statement follows. □

General hyperplane cuts and taking roots

Lemma 6.18 *Let (Z, Λ) be a reduced pair of finite type and $\phi_{\bullet} \colon Z_{\bullet} \to Z$ an embedded hyperresolution of $\Lambda \subseteq Z$. Let $\pi \colon (W, \Theta) \to (Z, \Lambda)$ be a morphism of pairs such that $\psi_{\bullet} \colon W_{\bullet} := W \times_Z Z_{\bullet} \to W$ is an embedded hyperresolution of $\Theta \subseteq W$. Let $\pi_i \colon W_i \to Z_i$ be the morphisms induced by π and assume that the natural transformation $L\pi^* R\phi_{\bullet *} \to R\psi_{\bullet *} L\pi_{\bullet}^*$ induces an isomorphism*

$$L\pi^* R\phi_{\bullet *} \mathscr{I}_{\Lambda_{\bullet} \subseteq Z_{\bullet}} \simeq R\psi_{\bullet *} L\pi_{\bullet}^* \mathscr{I}_{\Lambda_{\bullet} \subseteq Z_{\bullet}}.$$

Then

$$L\pi^* \underline{\Omega}^0_{Z,\Lambda} \simeq \underline{\Omega}^0_{W,\Theta}.$$

In particular, if (Z, Λ) is a DB pair then (W, Θ) is a DB pair.

Proof The hyperresolutions $\phi_\bullet$ and $\psi_\bullet$ fit into the commutative diagram:

$$
\begin{array}{ccc}
Z_\bullet & \xleftarrow{\;\pi_\bullet\;} & W_\bullet \\
{\scriptstyle\phi_\bullet}\downarrow & & \downarrow{\scriptstyle\psi_\bullet} \\
Z & \xleftarrow[\;\pi\;]{} & W,
\end{array}
$$

and we obtain the following representations of the Deligne–Du Bois complexes of (Z, Λ) and (W, Θ):

$$\underline{\Omega}^0_{Z,\Lambda} \simeq \mathcal{R}\phi_{\bullet *}\mathscr{I}_{\Lambda_\bullet \subseteq Z_\bullet} \quad \text{and} \quad \underline{\Omega}^0_{W,\Theta} \simeq \mathcal{R}\psi_{\bullet *}\mathscr{I}_{\Theta_\bullet \subseteq W_\bullet}.$$

Then by assumption

$$
\begin{aligned}
L\pi^*\underline{\Omega}^0_{Z,\Lambda} &\simeq L\pi^*\mathcal{R}\phi_{\bullet *}\mathscr{I}_{\Lambda_\bullet \subseteq Z_\bullet} \simeq \mathcal{R}\psi_{\bullet *}L\pi_\bullet^*\mathscr{I}_{\Lambda_\bullet \subseteq Z_\bullet} \\
&\simeq \mathcal{R}\psi_{\bullet *}\mathscr{I}_{\Theta_\bullet \subseteq W_\bullet} \simeq \underline{\Omega}^0_{W,\Theta}. \qquad \square
\end{aligned}
$$

Proposition 6.19 *Let (Z, Λ) be a DB pair and $\pi\colon W \to Z$ a smooth morphism. Let $\Theta := \pi^*\Lambda$. Then (W, Θ) is a DB pair.*

Proof Let $\phi_\bullet\colon Z_\bullet \to Z$ be an embedded hyperresolution of $\Lambda \subseteq Z$. Since π is smooth, it follows that $\psi_\bullet\colon W_\bullet := W \times_Z Z_\bullet \to W$ is an embedded hyperresolution of $\Theta \subseteq W$. Since π is flat, the functors $L\pi^*\mathcal{R}\phi_{\bullet *}$ and $\mathcal{R}\psi_{\bullet *}L\pi_\bullet^*$ are naturally isomorphic, so the statement follows by (6.18). $\square$

It easily follows that DB defects are invariant under restriction to a general hyperplane section the same way as the Deligne–Du Bois complex is:

Proposition 6.20 *Let (X, Σ) be a reduced pair and let $H \subset X$ be a general member of a base point free linear system. Then*

$$\underline{\Omega}^\times_{H,H\cap\Sigma} \simeq_{qis} \underline{\Omega}^\times_{X,\Sigma} \otimes_L \mathscr{O}_H.$$

Thus (X, Σ) is a DB pair in a (Zariski) neighborhood of H if and only if $(H, H \cap \Sigma)$ is a DB pair.

Proof This follows easily from (6.5.6), (6.12.1), and (6.13). $\square$

Corollary 6.21 *Let (Z, Λ) be a DB pair and $\mathscr{M}$ a semi-ample line bundle on Z. Let $\pi\colon W \to Z$ be the cyclic cover associated to a general section of $\mathscr{M}^m$ for some $m \gg 0$, compare Kollár and Mori (1998, 2.50) and $\Theta = \pi^*\Lambda$. Then (W, Θ) is a DB pair.*

Proof One can easily prove that π satisfies the conditions of (6.18) or argue as follows: Let M denote the total space of $\mathscr{M}$ and $\mu\colon M \to Z$ the associated natural map. Then $(M, \mu^*\Lambda)$ is a DB pair by (6.19) and then the statement follows by (6.20) applied to the divisor $W \subseteq M$, compare Kollár (1995, 9.4). $\square$

A DB criterion

Du Bois singularities play an important role in dealing with deformations of stable varieties. The main result of this section is a sufficient condition for Du Bois singularities. This was first proven in Kollár and Kovács (2010, 1.6). The proof here is a somewhat streamlined version using DB pairs compare Kovács (2012c, 3.3).

We start with a meta-theorem which is a key ingredient in various criteria that guarantee that a scheme or a pair have DB singularities. The main point of this statement is to relay the surjectivity we get for projective schemes from Hodge theory (cf. (6.9)) to quasi-projective ones. In order to avoid losing important information this is done by using local cohomology.

Theorem 6.22 (Kovács, 2000b, 1.4; Kovács, 2012c, 2.4) *Let X be a scheme of finite type over $\mathbb{C}$ and Σ and $P \subset X$ two closed subschemes of X. Assume that P is proper and that $\operatorname{Supp} \underline{\Omega}^\times_{X,\Sigma} \subseteq P$. Then $\mathbb{H}^i_P(X, \mathscr{I}_{\Sigma \subseteq X}) \twoheadrightarrow \mathbb{H}^i_P(X, \underline{\Omega}^0_{X,\Sigma})$ is surjective for all i.*

Proof Let $\overline{X}$ be a proper scheme that contains X as an open subscheme. Further let $Q = \overline{X} \setminus X$, $Z = P \overset{.}{\cup} Q$, and observe that $U := X \setminus P = \overline{X} \setminus Z$. Consider the distinguished triangle of functors,

$$\mathbb{H}^{\boldsymbol{\cdot}}_Z(\overline{X}, \underline{\quad}) \longrightarrow \mathbb{H}^{\boldsymbol{\cdot}}(\overline{X}, \underline{\quad}) \longrightarrow \mathbb{H}^{\boldsymbol{\cdot}}(U, \underline{\quad}) \overset{+1}{\longrightarrow} \qquad (6.22.1)$$

and apply it to the morphism $\mathscr{I}_{\Sigma \subseteq \overline{X}} \to \underline{\Omega}^0_{\overline{X},\Sigma}$. One obtains a morphism of two long exact sequences:

$$
\begin{array}{ccccccc}
\mathbb{H}^{i-1}(U, \mathscr{I}_{\Sigma \subseteq X}) & \longrightarrow & \mathbb{H}^i_Z(\overline{X}, \mathscr{I}_{\Sigma \subseteq \overline{X}}) & \longrightarrow & \mathbb{H}^i(\overline{X}, \mathscr{I}_{\Sigma \subseteq \overline{X}}) & \longrightarrow & \mathbb{H}^i(U, \mathscr{I}_{\Sigma \subseteq X}) \\
\downarrow{\scriptstyle \alpha_{i-1}} & & \downarrow{\scriptstyle \beta_i} & & \downarrow{\scriptstyle \gamma_i} & & \downarrow{\scriptstyle \alpha_i} \\
\mathbb{H}^{i-1}(U, \underline{\Omega}^0_{X,\Sigma}) & \longrightarrow & \mathbb{H}^i_Z(\overline{X}, \underline{\Omega}^0_{\overline{X},\Sigma}) & \longrightarrow & \mathbb{H}^i(\overline{X}, \underline{\Omega}^0_{\overline{X},\Sigma}) & \longrightarrow & \mathbb{H}^i(U, \underline{\Omega}^0_{X,\Sigma}).
\end{array}
$$

By assumption, α_{i-1}, α_i are isomorphisms and by (6.9) γ_i is surjective for all i. Then by the 5-lemma, β_i is also surjective for all i.

By construction $P \cap Q = \emptyset$ and hence

$$\mathbb{H}^i_Z(\overline{X}, \mathscr{I}_{\Sigma \subseteq \overline{X}}) \simeq \mathbb{H}^i_P(\overline{X}, \mathscr{I}_{\Sigma \subseteq \overline{X}}) \oplus \mathbb{H}^i_Q(\overline{X}, \mathscr{I}_{\Sigma \subseteq \overline{X}})$$

$$\mathbb{H}^i_Z(\overline{X}, \underline{\Omega}^0_{\overline{X},\Sigma}) \simeq \mathbb{H}^i_P(\overline{X}, \underline{\Omega}^0_{\overline{X},\Sigma}) \oplus \mathbb{H}^i_Q(\overline{X}, \underline{\Omega}^0_{\overline{X},\Sigma}).$$

It follows that the natural map (which is also the restriction of β_i),

$$\mathbb{H}^i_P(\overline{X}, \mathscr{I}_{\Sigma \subseteq \overline{X}}) \to \mathbb{H}^i_P(\overline{X}, \underline{\Omega}^0_{\overline{X},\Sigma})$$

is surjective for all i. Since $P \subseteq X$, by excision on local cohomology one has that

$$\mathbb{H}^i_P(\overline{X}, \mathscr{I}_{\overline{\Sigma} \subseteq \overline{X}}) \simeq \mathbb{H}^i_P(X, \mathscr{I}_{\Sigma \subseteq X}) \quad \text{and} \quad \mathbb{H}^i_P(\overline{X}, \underline{\Omega}^0_{\overline{X}, \overline{\Sigma}}) \simeq \mathbb{H}^i_P(X, \underline{\Omega}^0_{X, \Sigma}),$$

and so the desired statement follows. $\qquad\square$

Theorem 6.23 (Kovács, 2012c, 2.5)   *Let (X, Σ) be a reduced pair. If for any general complete intersection $H \subseteq X$ and any closed subscheme $Z \subseteq H \setminus U$,*

$$\mathbb{H}^i_Z(H, \mathscr{I}_{\Sigma \subseteq X} \otimes_L \mathscr{O}_H) \hookrightarrow \mathbb{H}^i_Z(H, \underline{\Omega}^0_{X, \Sigma} \otimes_L \mathscr{O}_H) \tag{6.23.1}$$

is injective for all i, then $\mathscr{I}_{\Sigma \subseteq X} \xrightarrow{\simeq} \underline{\Omega}^0_{X, \Sigma}$ is a quasi-isomorphism.

Proof   We need to prove that $\underline{\Omega}^\times_{X, \Sigma} \simeq_{qis} 0$. Let $P = \operatorname{Supp} \underline{\Omega}^\times_{X, \Sigma}$. By assumption X is reduced and hence $P \neq X$. It follows that if $H \subset X$ is a general complete intersection of X of the appropriate codimension, then $Z := X \cap P$ is a finite closed subset and if $P \neq \emptyset$, then Z is *nonempty*. Then $\underline{\Omega}^\times_{H, H \cap \Sigma} \simeq_{qis} \underline{\Omega}^\times_{X, \Sigma} \otimes_L \mathscr{O}_H$ by (6.5.6), so $Z = \operatorname{Supp} \underline{\Omega}^\times_{H, H \cap \Sigma}$. This implies that by replacing X by H we may assume that P is a finite set. Then we may also assume that X is affine and hence that it is contained as an open subset $X \subseteq \overline{X}$ in a projective scheme $\overline{X}$. Again by P being a finite set, it is proper and then it follows by (6.22) and the assumption in (6.23.1) that

$$\mathbb{H}^i_P(X, \mathscr{I}_{\Sigma \subseteq X}) \xrightarrow{\simeq} \mathbb{H}^i_P(X, \underline{\Omega}^0_{X, \Sigma}) \tag{6.23.2}$$

is an isomorphism for all i. In turn, it follows that $\mathbb{H}^i_P(X, \underline{\Omega}^\times_{X, \Sigma}) = 0$ for all i. Since $\operatorname{Supp} \underline{\Omega}^\times_{X, \Sigma} = P$ it also follows that $\mathbb{H}^i(X \setminus P, \underline{\Omega}^\times_{X, \Sigma}) = 0$ for all i as well, and therefore

$$\mathbb{H}^i(X, \underline{\Omega}^\times_{X, \Sigma}) = 0 \tag{6.23.3}$$

for all i by the long exact sequence induced by (6.22.1) and applied with $\overline{X} \leftrightarrow X$.

Since X is affine, the spectral sequence that computes hypercohomology from the cohomology of the cohomology sheaves of the complex $\underline{\Omega}^\times_{X, \Sigma}$ degenerates and gives that $\mathbb{H}^i(X, \underline{\Omega}^\times_{X, \Sigma}) = H^0(X, h^i(\underline{\Omega}^\times_{X, \Sigma}))$ for all i. It follows by (6.23.3) that $h^i(\underline{\Omega}^\times_{X, \Sigma}) = 0$ for all i, and hence $\underline{\Omega}^\times_{X, \Sigma}$ is acyclic which implies the desired statement. $\qquad\square$

Theorem 6.23 implies a number of DB criteria that have been proven earlier, some with somewhat different methods.

Corollary 6.24 (Kovács, 1999, 2.3; Kovács, 2011a, 5.4)   *Let (X, Σ) be a reduced pair. Assume that the natural morphism $\mathscr{I}_{\Sigma \subseteq X} \to \underline{\Omega}^0_{X, \Sigma}$ has a left inverse. Then (X, Σ) is a DB pair. In particular, if the natural map $\mathscr{O}_X \to \underline{\Omega}^0_X$ admits a left inverse, then X has DB singularities.*

Proof If $\mathscr{I}_{\Sigma \subseteq X} \to \underline{\Omega}^0_{X,\Sigma}$ has a left inverse, then the condition in (6.23.1) is easily satisfied and hence the statement follows.   □

Corollary 6.25 (Kovács, 1999, 2.6; Kovács, 2011a, 5.6)   *Let (X, Σ) be a rational pair. Then (X, Σ) is a DB pair. In particular, if X has rational singularities, then it has DB singularities.*

Proof   Let $\pi: (Y, \Gamma) \to (X, \Sigma)$ be a log resolution. Consider the commutative diagram induced by the natural morphisms (cf. (Kovács, 2011a, 3.15)):

$$
\begin{array}{ccc}
\mathscr{I}_{\Sigma \subseteq X} & \xrightarrow{\;\beta\;} & \mathcal{R}\pi_* \mathscr{I}_{\Gamma \subseteq Y} \\
{\scriptstyle \alpha} \downarrow & & \downarrow {\scriptstyle \gamma} \\
\underline{\Omega}^0_{X,\Sigma} & \xrightarrow{\;\delta\;} & \mathcal{R}\pi_* \underline{\Omega}^0_{Y,\Gamma}.
\end{array}
$$

Since (X, Σ) is a rational pair, β is a quasi-isomorphism. Since (Y, Γ) is an snc pair, γ is a quasi-isomorphism. Therefore, $\beta^{-1} \circ \gamma^{-1} \circ \delta$ is a left inverse to α, and hence the statement follows by (6.24)   □

Corollary 6.26 (Kovács, 1999, 2.4)   *Let $\sigma: (Y, \Gamma) \to (X, \Sigma)$ be a finite dominant morphism of normal pairs and assume that $\sigma|_\Gamma$ dominates Σ. If (Y, Γ) is a DB pair, then so is (X, Σ). In particular, if $\sigma: Y \to X$ is a finite dominant morphism of normal varieties and Y has DB singularities, then so does X.*

Proof   Consider the natural morphisms,

$$
\mathscr{I}_{\Sigma \subseteq X} \to \underline{\Omega}^0_{X,\Sigma} \to \mathcal{R}\sigma_* \mathscr{I}_{\Gamma \subseteq Y} \cong_{qis} \sigma_* \mathscr{I}_{\Gamma \subseteq Y}
$$

and observe that the normalized trace map (10.27) splits the composition.   □

The following is a slight improvement of Kollár and Kovács (2010, 1.6).

Theorem 6.27 (Kovács, 2012c, 3.3)   *Let $f: Y \to X$ be a proper surjective morphism between reduced schemes of finite type over $\mathbb{C}$, $W \subseteq X$ an arbitrary subscheme and $F := f^{-1}(W)$ equipped with the reduced subscheme structure. Assume that the natural map ϱ*

$$
\begin{array}{c}
\overset{\varrho'}{\underset{}{\nearrow}} \qquad \nwarrow \\
\mathscr{I}_{W \subseteq X} \xrightarrow{\quad \varrho \quad} \mathcal{R}f_* \mathscr{I}_{F \subseteq Y}
\end{array}
$$

admits a left inverse ϱ'. Then if (Y, F) is a DB pair, then so is (X, W). In particular, if (Y, F) is a DB pair, then X is DB if and only if W is DB.

Proof By functoriality one obtains a commutative diagram

$$
\begin{array}{ccc}
\mathscr{I}_{W \subseteq X} & \xrightarrow{\varrho} & \mathcal{R}f_* \mathscr{I}_{F \subseteq Y} \\
{\scriptstyle \alpha} \downarrow & & {\scriptstyle \gamma} \downarrow {\scriptstyle \approx_{qis}} \\
\underline{\Omega}^0_{W \subseteq X} & \xrightarrow{\beta} & \mathcal{R}f_* \underline{\Omega}^0_{Y,F}.
\end{array}
$$

Since (Y, F) is assumed to be a DB pair, it follows that γ is a quasi-isomorphism and hence $\varrho' \circ \gamma^{-1} \circ \beta$ is a left inverse to α. Then the statement follows by (6.24). $\qquad \square$

Corollary 6.28 (Kollár and Kovács, 2010, 1.6) *Let* $f \colon Y \to X$ *be a proper morphism between reduced schemes of finite type over* $\mathbb{C}$, $W \subseteq X$ *an arbitrary subscheme, and* $F := f^{-1}(W)$, *equipped with the reduced subscheme structure. Assume that the natural map* ϱ

$$
\begin{array}{ccc}
& \xrightarrow{\varrho'} & \\
\mathscr{I}_{W \subseteq X} & \xrightarrow{\varrho} & \mathcal{R}f_* \mathscr{I}_{F \subseteq Y}
\end{array}
$$

admits a left inverse ϱ', *that is,* $\varrho' \circ \varrho = \mathrm{id}_{\mathscr{I}_{W \subseteq X}}$. *Then if* Y, F *and* W *all have DB singularities, then so does* X.

As a straightforward consequence of (6.29) we get that Du Bois singularities are invariant under (small) deformation.

Theorem 6.29 (Kovács and Schwede, 2011a) *Let* X *be a scheme of finite type over* $\mathbb{C}$ *and* H *a reduced effective Cartier divisor. If* H *has Du Bois singularities, then* X *has Du Bois singularities near* H.

Thus if $f \colon X \to S$ *is a proper flat family of varieties over a smooth curve* S *and the fiber* X_s *has Du Bois singularities for some* $s \in S$, *then so do all the fibers near* s *(in a Zariski open set).*

Remark 6.30 The second statement of the theorem follows from the first by (6.20).

6.2 Semi-log canonical singularities are Du Bois

The main theorem of Kollár and Kovács (2010) says that log canonical singularities are Du Bois. Here we prove this, and extend it to the semi-log canonical case at the same time.

Theorem 6.31 *Let* (X, S_*) *be a stratified scheme of log canonical origin (5.29). Then* X *is Du Bois.*

Corollary 6.32 *Let* (X, Δ) *be slc. Then any union of lc centers of* X *is Du Bois. In particular,* X *is Du Bois.*

Proof Let $W \subset X$ be a union of lc centers of X. Then W is a stratified subscheme of the log canonical stratification (5.25), hence it is of log canonical origin by (5.29) and (9.26). Thus (6.31) implies (6.32). $\square$

Proof of (6.31) We use induction on the dimension.

Let $\pi \colon (X^n, S^n_*) \to (X, S_*)$ denote the normalization. Let $B(X) \subset X$ and $B(X^n) \subset X^n$ denote the union of all lc centers (5.25). By (9.15.1), we have a universal push-out diagram

$$
\begin{array}{ccc}
B(X^n) & \hookrightarrow & X^n \\
\downarrow & & \downarrow{\scriptstyle \pi} \\
B(X) & \hookrightarrow & X.
\end{array}
$$

Here $B(X)$ and $B(X^n)$ are of log canonical origin by (5.29) and (9.26), hence Du Bois by induction.

Since π is finite it follows that $R\pi_* \mathscr{I}_{B(X^n) \subseteq X^n} = \pi_* \mathscr{I}_{B(X^n) \subseteq X^n}$. Furthermore, $\pi_* \mathscr{I}_{B(X^n) \subseteq X^n} = \pi_* \mathscr{I}_{B(X) \subseteq X}$ is a basic property of universal push-out diagrams (9.30). Thus (6.27) applies and therefore X is Du Bois if X^n is.

By assumption, for each irreducible component $X^n_i \subset X^n$ there is a crepant log structure $f_i \colon (Y_i, \Delta_i) \to Z_i$ and a finite surjection $Z_i \to X^n_i$. By (6.26), if Z_i is Du Bois then so is X^n_i.

Finally, let $B(Z_i) \subset Z_i$ be the boundary of the log canonical stratification of Z_i. Then $B(Z_i)$ is of log canonical origin by (5.29) and (9.26), hence Du Bois by induction. By (6.33) the pair $(Z_i, B(Z_i))$ is DB, this implies that Z_i is Du Bois. $\square$

The following theorem shows that the DB criterion (6.27) applies to crepant log structures.

Theorem 6.33 *Let* $f \colon (Y, \Delta) \to X$ *be a crepant log structure with klt general fiber and* $W \subset X$ *the union of all of its lc centers. Then* (X, W) *is a DB pair.*

Proof Let $\pi \colon \tilde{Y} \to Y$ be a proper birational morphism and $F := \tilde{f}^{-1}(W)_{\mathrm{red}}$, where $\tilde{f} := f \circ \pi$. We claim that the natural map

$$
\varrho \colon \mathscr{I}_{W \subseteq X} \simeq \tilde{f}_* \mathscr{I}_{F \subseteq \tilde{Y}} \to R\tilde{f}_* \mathscr{I}_{F \subseteq \tilde{Y}} \quad \text{has a left inverse.} \tag{6.33.1}
$$

If we choose $\pi \colon \tilde{Y} \to Y$ such that $\tilde{Y}$ is smooth and $F \subset \tilde{Y}$ is a snc divisor, then $(\tilde{Y}, F)$ is a DB pair. Thus (6.27) implies that (X, W) is a DB pair.

Now to the proof of (6.33.1). First, observe that if $\tau \colon \hat{Y} \to \tilde{Y}$ is a log resolution of $(\tilde{Y}, F)$ that factors through π, then it is enough to prove the statement for $\sigma = \pi \circ \tau$ instead of π. Indeed let $\widehat{F} = \tau^{-1}(F)_{\text{red}}$, an snc divisor, and $\widehat{f} = f \circ \sigma$. Suppose that the natural map

$$\widehat{\varrho} : \mathscr{I}_{W \subseteq X} \simeq \widehat{f}_* \mathscr{O}_{\widehat{Y}}(-\widehat{F}) \to \mathcal{R}\widehat{f}_* \mathscr{O}_{\widehat{Y}}(-\widehat{F})$$

has a left inverse, $\widehat{\delta} : \mathcal{R}\widehat{f}_* \mathscr{O}_{\widehat{Y}}(-\widehat{F}) \to \mathscr{I}_{W \subseteq X}$ such that $\widehat{\delta} \circ \widehat{\varrho} = \mathrm{id}_{\mathscr{I}_{W \subseteq X}}$. Then, as $\widehat{f} = \widetilde{f} \circ \tau$, one has that $\mathcal{R}\widehat{f}_* \mathscr{O}_{\widehat{Y}}(-\widehat{F}) \simeq \mathcal{R}\widetilde{f}_* \mathcal{R}\tau_* \mathscr{O}_{\widehat{Y}}(-\widehat{F})$ and applying the functor $\mathcal{R}\widetilde{f}_*$ to the natural map $\overline{\varrho} \colon \mathscr{I}_{F \subseteq \widetilde{Y}} \simeq \tau_* \mathscr{O}_{\widehat{Y}}(-\widehat{F}) \to \mathcal{R}\tau_* \mathscr{O}_{\widehat{Y}}(-\widehat{F})$ shows that $\widehat{\varrho} = \mathcal{R}\widetilde{f}_*(\overline{\varrho}) \circ \varrho$:

$$\widehat{\varrho} \colon \mathscr{I}_{W \subseteq X} \simeq \widetilde{f}_* \mathscr{I}_{F \subseteq \widetilde{Y}} \xrightarrow{\ \varrho\ } \mathcal{R}\widetilde{f}_* \mathscr{I}_{F \subseteq \widetilde{Y}} \xrightarrow{\ \mathcal{R}\widetilde{f}_*(\overline{\varrho})\ } \mathcal{R}\widehat{f}_* \mathscr{O}_{\widehat{Y}}(-\widehat{F}).$$

Therefore, $\delta = \widehat{\delta} \circ \mathcal{R}\widetilde{f}_*(\overline{\varrho})$ is a left inverse to ϱ showing that it is indeed enough to prove the statement for σ. In particular, we may assume that F is an snc divisor and may replace π with its composition with any further blow-up. Next write

$$\pi^*(K_Y + \Delta) \sim_{\mathbb{Q}} K_{\widetilde{Y}} + E + \widetilde{\Delta} - B,$$

where E is the sum of all (not necessarily exceptional) divisors with discrepancy -1, B is an effective exceptional integral divisor and $\lfloor \widetilde{\Delta} \rfloor = 0$. By the argument above we may also assume that $\widetilde{f}^{-1}\widetilde{f}(E)$ is an snc divisor. Since $B - E \geq -F$, we have natural maps

$$\widetilde{f}_* \mathscr{O}_{\widehat{Y}}(-F) \to \mathcal{R}\widetilde{f}_* \mathscr{O}_{\widehat{Y}}(-F) \to \mathcal{R}\widetilde{f}_* \mathscr{O}_{\widehat{Y}}(B - E).$$

Note that $B - E \sim_{\mathbb{Q}, \widetilde{f}} K_{\widetilde{Y}} + \widetilde{\Delta}$, hence by (10.41)

$$\mathcal{R}\widetilde{f}_* \mathscr{O}_{\widehat{Y}}(B - E) \simeq_{qis} \sum_i R^i \widetilde{f}_* \mathscr{O}_{\widehat{Y}}(B - E)[-i].$$

Thus we get a morphism

$$\widetilde{f}_* \mathscr{O}_{\widehat{Y}}(-F) \to \mathcal{R}\widetilde{f}_* \mathscr{O}_{\widehat{Y}}(-F) \to \mathcal{R}\widetilde{f}_* \mathscr{O}_{\widehat{Y}}(B - E) \to \widetilde{f}_* \mathscr{O}_{\widehat{Y}}(B - E).$$

Note that $\pi_* \mathscr{O}_{\widehat{Y}}(B - E) = \mathscr{I}_{B(Y)}$, the ideal of the union of all lc centers of (Y, Δ). Furthermore, for any $U \subseteq X$ open subset with preimage $U_Y := f^{-1}(U)$, a global section of $\mathscr{O}_{U_Y}$ vanishes along a fiber of f if and only if it vanishes at one point of that fiber. Thus

$$\widetilde{f}_* \mathscr{O}_{\widehat{Y}}(-F) = f_* \mathscr{I}_{B(Y)} = \widetilde{f}_* \mathscr{O}_{\widehat{Y}}(B - E),$$

and so the statement follows. $\qquad\square$

7

Log centers and depth

In this chapter we study two topics that have important applications to flips and to moduli questions.

In Section 4.1 we studied the log canonical centers of an lc pair (X, Δ); these are centers of divisors of discrepancy -1.

Here we study a larger class of interesting subvarieties called *log centers*, which are centers of divisors of negative discrepancy. As a general principle, the closer the discrepancy is to -1, the more a log center behaves like a log canonical center. Thus log canonical centers are the most special among the log centers.

The case when X is normal is treated in Section 7.1 and the general semi-log canonical version is derived from it in Section 7.2.

The depth of the structure sheaf and of the dualizing sheaf of a semi-log canonical pair is studied in Section 7.3.

Assumptions In this chapter we work with schemes (or algebraic spaces) that are of finite type over a base scheme S that is essentially of finite type over a field of characteristic 0.

7.1 Log centers

Definition 7.1 Let $f \colon (X, \Delta) \to Z$ be a weak crepant log structure and $W \subset Z$ an irreducible subvariety. The *minimal log discrepancy* of (or over) W is defined as the infimum of the numbers $1 + a(E, X, \Delta)$ where E runs through all divisors over X such that $f(\mathrm{center}_X(E)) = W$. It is denoted by

$$\mathrm{mld}(W, X, \Delta) \quad \text{or by} \quad \mathrm{mld}(W) \tag{7.1.1}$$

if the choice of $f \colon (X, \Delta) \to Z$ is clear. We set $\mathrm{mld}(Z, X, \Delta) = 0$. If f is birational, then $(Z, f_*\Delta)$ is lc and $\mathrm{mld}(W, Z, f_*\Delta) = \mathrm{mld}(W, X, \Delta)$ (2.9).

We say that a divisor E over X *computes* $\mathrm{mld}(W, X, \Delta)$ if $f(\mathrm{center}_X(E)) = W$ and $\mathrm{mld}(W, X, \Delta) = 1 + a(E, X, \Delta)$. We see below that for crepant log structures the infimum is a minimum, hence each $\mathrm{mld}(W, X, \Delta)$ is computed by some divisor.

We can also write $\mathrm{mld}(W, X, \Delta)$ as the infimum of all $\mathrm{mld}(W_X, X, \Delta)$ where W_X runs through all irreducible subvarieties of X such that $f(W_X) = W$.

If (X, Δ) is lc, $a(E, X, \Delta) \geq -1$, hence $\mathrm{mld}(W, X, \Delta) \geq 0$. Furthermore, $W \subset Z$ is a lc center of (X, Δ) if and only if $\mathrm{mld}(W, X, \Delta) = 0$.

If $W \subset Z$ is a closed subscheme with irreducible components W_i then set

$$\mathrm{mld}(W, X, \Delta) := \max_i \{\mathrm{mld}(W_i, X, \Delta)\}.$$

By (2.23), birational weak crepant log structures give the same minimal log discrepancies. Thus in order to compute them, we may use a weak crepant log structure $h: (Y, \Delta_Y) \to Z$ where (Y, Δ_Y) is snc. If (Y, Δ_Y) is lc, then, by (2.10), for an irreducible subvariety $W_Y \subset Y$,

$$\mathrm{mld}(W_Y, Y, \Delta_Y) = \mathrm{codim}_Y W_Y - \sum_{i: D_i \supset W_Y} \mathrm{coeff}_{D_i} \Delta_Y.$$

Thus the infimum in the definition of mld is a minimum.

Definition 7.2 Let $f: (X, \Delta) \to Z$ be a crepant log structure. An irreducible subvariety $W \subset Z$ is a *log center* of (X, Δ) if $\mathrm{mld}(W, X, \Delta) < 1$. By (2.23), birational crepant log structures have the same log centers.

Every log center on Z is the image of a log center of (X, Δ), thus a crepant log structure has only finitely many log centers. To focus on these further, we set

$$\overline{\mathrm{mld}}(W, X, \Delta) := \begin{cases} \mathrm{mld}(W, X, \Delta) & \text{if} \quad \mathrm{mld}(W, X, \Delta) < 1, \quad \text{and} \\ +\infty & \text{otherwise.} \end{cases}$$

The following is a very powerful result which allows us to reduce many questions about log centers first to the divisorial case and then to dimension 2. After a remark, we derive several consequences, and then prove (7.3) at the end of the section.

Theorem 7.3 *Let $f: (X, \Delta) \to Z$ be a crepant log structure and $W_i \subset Z$ log centers, none of which is contained in another. Set $W := \cup_i W_i$. Then $f: (X, \Delta) \to Z$ is crepant birational to another crepant log structure $f^m: (X^m, \Delta^m) \to Z$ such that*

(1) *X^m is $\mathbb{Q}$-factorial,*
(2) *there are divisors D_i^m such that $f^m(D_i^m) = W_i$ and $\mathrm{mld}(D_i^m, X^m, \Delta^m) = \mathrm{mld}(W_i, X, \Delta)$ for every i,*
(3) *$(X^m, \Delta^m - \eta D^m)$ is dlt for every $\eta > 0$ where $D^m := \sum_i D_i^m$,*
(4) *$-D^m$ is f^m-nef,*

(5) Supp $D^m = \text{Supp}(f^m)^{-1}(W)$,

(6) $f_*^m \mathcal{O}_{D^m} = \mathcal{O}_W$ and

(7) $X \dashrightarrow X^m$ is an isomorphism over $Z \setminus W$.

7.4 (Note on vicious circles) Many of the results about log *canonical* centers proved in Section 4.1 are special cases of the results in this section. There are three reasons why we chose not to treat them together.

Log *canonical* centers are essential building blocks of log canonical and semi-log canonical pairs, whereas log centers are merely useful additional structures. Thus it makes sense to treat them separately.

Most arguments in this section rely on (1.30.8) whose proof uses various properties of log *canonical* centers. Thus the two stage treatment is logically necessary.

We also use (10.39) which we have not proved.

The next result can be viewed as a generalization of inversion of adjunction.

Theorem 7.5 (Kollár, 2011c) *Let* $f: (X, \Delta) \to Z$ *be a crepant log structure and* W_1, W_2 *log centers on* Z. *Then*

$$\text{mld}(W_1 \cap W_2, X, \Delta) \leq \text{mld}(W_1, X, \Delta) + \text{mld}(W_2, X, \Delta).$$

Proof Let us consider first the very special case when $X = Z$ and the $W_i := D_i$ are reduced $\mathbb{Q}$-Cartier divisors. Then $\text{mld}(W_i, X, \Delta) = 1 - \max_j \{\text{coeff}_{D_{ij}} \Delta\}$ where D_{ij} are the irreducible components of D_i.

Note that every irreducible component of $D_1 \cap D_2$ has codimension 2. Thus, by localizing at a generic point of $D_1 \cap D_2$, we are reduced to the case when $\dim X = 2$. The latter is proved in (2.38).

In the general case we apply (7.3) to get $f^m: (X^m, \Delta^m) \to Z$. Note that by (7.3.5) and the connectedness of the fibers, every irreducible component $V_Z \subset W_1 \cap W_2$ is dominated by an irreducible component $V_Y \subset D_1^m \cap D_2^m$. Thus

$$\begin{aligned}
\text{mld}(V_Z, X, \Delta) &\leq \text{mld}(V_Y, X^m, \Delta^m) \\
&\leq \text{mld}(D_1^m, X^m, \Delta^m) + \text{mld}(D_2^m, X^m, \Delta^m) \\
&= \text{mld}(W_1, X, \Delta) + \text{mld}(W_2, X, \Delta),
\end{aligned}$$

where the second $\leq$ follows from the already established divisorial case. □

The next three results of Kollár (2011c) investigate seminormality of log centers.

Theorem 7.6 *Let* $f: (X, \Delta) \to Z$ *be a crepant log structure and* $W \subset Z$ *a reduced subscheme. If* $\text{mld}(W, X, \Delta) < \frac{1}{6}$ *then* W *is seminormal.*

Proof We again start with the special case when (X, Δ) is dlt, $\mathbb{Q}$-factorial and W is a divisor. Then X is CM and so is any divisor on X by (2.88). Thus, by (10.14), W is seminormal if and only if it is seminormal at its codimension 1 points. By localization at codimension 1 points we are reduced to the case when $\dim X = 2$. This was settled in (3.44).

In the general case we apply (7.3) with $W = \cup_i W_i$ to get $f^m \colon (X^m, \Delta^m) \to Z$. By (7.3.6), $f_*^m \mathcal{O}_{D^m} = \mathcal{O}_W$ and D^m is seminormal by the above divisorial case. Thus W is seminormal by (10.15). $\qquad\square$

A slight variant of the proof yields a normality condition.

Proposition 7.7 *Let $f \colon (X, \Delta) \to Z$ be a crepant log structure. Let $W \subset Z$ be a reduced subscheme and $W^{\mathrm{norm}} \subset W$ the open subset of normal points. If $\mathrm{mld}(W, X, \Delta) < 1$ then $\mathrm{mld}(W \setminus W^{\mathrm{norm}}, X, \Delta) \le 2 \cdot \mathrm{mld}(W, X, \Delta)$.*

Proof The reduction to the case $\dim X = 2$ goes as before. Next use (2.43) and (2.53.1) to reduce to the case when X is smooth and D is a singular curve. Blowing up the singular point gives the exceptional divisor with discrepancy $\le -1 + 2 \cdot (1 - (\text{coefficient of } D))$. $\qquad\square$

Theorem 7.8 *Let $f \colon (X, \Delta) \to Z$ be a crepant log structure and $W_1, W_2 \subset Z$ reduced subschemes. If $\mathrm{mld}(W_1, X, \Delta) + \mathrm{mld}(W_2, X, \Delta) < \frac{1}{2}$ then $W_1 \cap W_2$ is reduced. Equivalently, $W_1 \cup W_2$ is seminormal in $W_1 \amalg W_2$ (10.20).*

Proof As before, we start with the special case when (X, Δ) is dlt, $\mathbb{Q}$-factorial and W_1, W_2 are divisors. Then X is CM and so is $\mathcal{O}_{W_1 \cup W_2}$. So, by (10.14), $W_1 \cup W_2$ is seminormal in $W_1 \amalg W_2$ if and only if this holds at all codimension 1 points. By localization at codimension 1 points we are reduced to the case when $\dim X = 2$. This was settled in (3.44).

In the general case we apply (7.3) to get $f^m \colon (X^m, \Delta^m) \to Z$. Write $D^m = D_1^m \cup D_2^m$ where $f^m(D_i^m) = W_i$. We already know that $D_1^m \cup D_2^m$ is seminormal in $D_1^m \amalg D_2^m$. Since $f_*^m \mathcal{O}_{D^m} = \mathcal{O}_W$, by (10.15) this implies that $W_1 \cup W_2$ is seminormal in $W_1 \amalg W_2$. $\qquad\square$

Example 7.9 The following examples show that the numerical conditions of (7.6) and (7.8) are sharp.

(1) $(\mathbb{A}^2, \frac{5}{6}(x^2 = y^3))$ is lc, the minimal log discrepancy of the curve $(x^2 = y^3)$ is $\frac{1}{6}$ but it is not seminormal.
(2) Consider $(\mathbb{A}^3, \frac{11}{12}(z - x^2 - y^3) + \frac{11}{12}(z + x^2 + y^3))$. One can check that this is lc. The irreducible components of the boundary are smooth, but their intersection is a cuspidal curve, hence not seminormal. It again has minimal log discrepancy $\frac{1}{6}$.
(3) Let $Z \subset \mathbb{C}^4$ be the image of $\mathbb{C}^2_{uv}$ by the map $x = u, y = v^3, z = v^2, t = uv$. Z is defined by the equations $xy - zt = yt - xz^2 = y^2 - z^3 = t^2 - x^2 z = 0$. Set $X := (xy = zt) \subset \mathbb{C}^4$ and let $D \subset X$ be the divisor defined

by

$$(y^2 - z^3) - (t^2 - x^2 z) = (y^2 - t^2) + z(x^2 - z^2) = 0.$$

From this we see that D has three irreducible components: $D_1 := Z$, $D_2 := (y - t = x - z = 0)$ and $D_3 := (y + t = x + z = 0)$.

We claim that $(X, \frac{5}{6}D_1 + \frac{5}{6}D_2 + \frac{5}{6}D_3)$ is lc. Here D_1 has minimal log discrepancy $\frac{1}{6}$ but seminormality fails in codimension 3 on X. In order to check the claim, blow-up the ideal (x, z). On $\mathbb{C}^2_{uv}$ this corresponds to blowing up the ideal (u, v^2).

On one of the charts we have coordinates $x_1 := x/z, y, z$ and the birational transform D_1' of D_1 is given by $(y^2 = z^3)$. On the other chart we have coordinates $x, z_1 := z/x, t$ and D_1' is given by $(z_1 x^3 = t^2)$. Thus we see that $(B_{(x,z)}X, \frac{5}{6}D_1')$ is lc. The linear system $|(y = z = 0)|$ becomes base point free on the blow-up, hence $(B_{(x,z)}X, \frac{5}{6}D_1' + \frac{5}{6}D_2' + \frac{5}{6}D_3')$ is lc and so is $(X, \frac{5}{6}D_1 + \frac{5}{6}D_2 + \frac{5}{6}D_3)$.

(4) $(\mathbb{A}^2, (\frac{3}{4} - c)(x - y^2 = 0) + (\frac{3}{4} + c)(x + y^2 = 0))$ is lc for any $0 \le c \le \frac{1}{4}$, the sum of the minimal log discrepancies is $\frac{1}{2}$ but $(x - y^2 = 0) \cup (x + y^2 = 0)$ is not seminormal in $(x - y^2 = 0) \amalg (x + y^2 = 0)$.

(5) $(\mathbb{A}^2, (x = 0) + \frac{1}{2}(y^2 = x^2 + x^3))$ is lc, the sum of the minimal log discrepancies is $\frac{1}{2}$ but $(x = 0) \cup (y^2 = x^2 + x^3)$ is not seminormal in $(x = 0) \amalg (y^2 = x^2 + x^3)$.

(6) Let X be a cone over $Z \times \mathbb{P}^1$ where Z is smooth and $K_Z \sim 0$. Pick points $p_1, p_2, p_3 \in \mathbb{P}^1$. Let $D_1 \subset X$ be the cone over $Z \times \{p_1\}$ and $D_2 \subset X$ the cone over $Z \times \{p_2, p_3\}$. Then $(X, D_1 + \frac{1}{2}D_2)$ is lc, the sum of the minimal log discrepancies is $\frac{1}{2}$ but $D_1 \cup D_2$ is not seminormal in $D_1 \amalg D_2$. Here seminormality fails only at the vertex of the cone, thus in high codimension.

The following result is another generalization of (4.42).

Theorem 7.10 (Precise inversion of adjunction) *Let* $f : (X, \Delta) \to Z$ *be a crepant log structure on* Z *with* (X, Δ) *dlt. Let* $Y \subset X$ *be an lc center,* $f|_Y : Y \xrightarrow{g} Z_Y \xrightarrow{\pi} Z$ *the Stein factorization and* $V \subset Z_Y$ *an irreducible subvariety. Then* $g : (Y, \Delta_Y) \to Z_Y$ *is a crepant log structure on* Z_Y *and*

$$\overline{\mathrm{mld}}(V, Y, \mathrm{Diff}_Y^* \Delta) = \overline{\mathrm{mld}}(\pi(V), X, \Delta). \qquad (7.10.1)$$

Proof By induction and (4.16) it is enough to prove this when Y is a divisor on X.

Set $\Delta_Y := \mathrm{Diff}_Y^*(\Delta)$. Let E_Y be a divisor over Y with $a(E_Y, Y, \Delta_Y) < 0$. By (4.8), there is a divisor E over X such that $\mathrm{center}_X E = \mathrm{center}_Y(E_Y)$ and $a(E, X, \Delta) = a(E_Y, Y, \Delta_Y)$. This shows that $\overline{\mathrm{mld}}(V, Y, \Delta_Y) \ge \overline{\mathrm{mld}}(\pi(V), X, \Delta)$.

The reverse inequality is interesting only if $\mathrm{mld}(\pi(V), X, \Delta) < 1$. Let E be a divisor over X that computes $\mathrm{mld}(\pi(V), X, \Delta)$. By (1.38) we may assume that E is a $\mathbb{Q}$-Cartier divisor on X. If $V = Z_Y$ then $\operatorname{mld}(\pi(Z_Y), X, \Delta) = 0$ is computed by Y. By definition, $\operatorname{mld}(Z_Y, Y, \Delta_Y) = 0$ hence the equality (7.10.1) holds.

Thus assume that $V \neq Z_Y$, hence $f(E)$ does not contain $\pi(Z_Y)$. Apply (7.3) with $W = f(E)$. We get $f^m : (X^m, \Delta^m) \to Z$ such that $D^m := (f^m)^{-1}(f(E))$ is in irreducible divisor and $\operatorname{coeff}_{D^m} \Delta^m \leq \operatorname{coeff}_E \Delta$. (In fact, we can choose D^m to be the birational transform of E, so equality holds.)

By the usual adjunction (4.9), each irreducible component V_j of $Y^m \cap D^m$ is dominated by a divisor F_j such that

$$a(F_j, Y^m, \operatorname{Diff}^*_{Y^m}(\Delta^m)) \leq a(D^m, X^m, \Delta^m) = a(E, X, \Delta).$$

Since $(Y^m, \operatorname{Diff}^*_{Y^m}(\Delta^m))$ is crepant-birational to $(Y, \operatorname{Diff}^*_Y \Delta)$ (4.6), these divisors show that $\operatorname{mld}(V, Y, \Delta_Y) \leq \operatorname{mld}(\pi(V), X, \Delta)$. $\qquad \square$

Proof of (7.3) By (1.38) we may assume that X is $\mathbb{Q}$-factorial and there are divisors D_i such that $f(D_i) = W_i$ and $\operatorname{mld}(D_i, X, \Delta) = \operatorname{mld}(W_i, X, \Delta)$ for every i. Fix $0 < c < \min_i \{\operatorname{mld}(W_i, X, \Delta)\}$ and set $D := \sum_i D_i$, $\Delta' := \Delta - cD$, $\Delta'' := cD$ and run an (X, Δ')-MMP over Z. The assertions (7.3.1–5) are checked in (7.11) and (7.3.6) is proved in (7.12). $\qquad \square$

Lemma 7.11 *Let $f : (X, \Delta) \to Z$ be a crepant log structure on Z. Write Δ as a sum of effective divisors $\Delta = \Delta' + \Delta''$ where Δ'' is $\mathbb{Q}$-Cartier and run an (X, Δ')-MMP over Z.*

(1) *Every step of the MMP is an isomorphism over $Z \setminus f(\Delta'')$.*
(2) *If Δ'' dominates Z then the MMP terminates with a Fano contraction $\phi_i : (X_i, \Delta'_i) \to X_{i+1}$.*
(3) *If Δ'' does not dominate Z then the MMP terminates with a minimal model $f_m : (X_m, \Delta'_m) \to Z$. Furthermore,*
   (a) *$f(\Delta'') = f_m(\Delta''_m)$ and*
   (b) *$\operatorname{Supp} \Delta''_m = \operatorname{Supp} f_m^{-1}\big(f_m(\Delta''_m)\big) = \operatorname{Supp} f_m^{-1}(f(\Delta''))$.* $\qquad \square$

Proof As we run an (X, Δ')-MMP, at each step $K_{X_i} + \Delta'_i \sim_{f_i, \mathbb{Q}} -\Delta''_i$. Thus any curve that is contained in a fiber of f_i and has negative intersection with $K_{X_i} + \Delta'_i$ is contained in $f_i^{-1}(f_i(\Delta''_i))$, proving (1).

Conversely, if $\operatorname{Supp} \Delta''_i \neq \operatorname{Supp} f_i^{-1}(f_i(\Delta''_i))$ then there is such a curve, hence $K_{X_i} + \Delta'_i$ is not f_i-nef. The termination is assured by (1.30.8). Thus we have (2) and the first equality in (3.b).

Finally we claim that at every intermediate step of the MMP we have $f(\Delta'') = f_i(\Delta''_i)$. This is clear for $i = 0$. As we go from i to $i + 1$, the image $f_i(\Delta''_i)$ is unchanged if $X^i \dashrightarrow X^{i+1}$ is a flip. Thus let $\phi_i : X_i \to X_{i+1}$ be a divisorial contraction with exceptional divisor E_i and let $F_i \subset E_i$ be a general fiber of $E_i \to X$. It is clear that

$$f_{i+1}(\Delta''_{i+1}) \subset f_i(\Delta''_i)$$

and equality fails only if E_i is an irreducible component of Δ''_i but no other irreducible component of Δ''_i intersects F_i. Since ϕ_i contracts a $(K_{X_i} + \Delta'_i)$-negative extremal ray, $-\Delta''_i \sim_{\mathbb{Q},f_i} K_{X_i} + \Delta'_i$ shows that Δ''_i is ϕ_i-nef. However, an exceptional divisor has negative intersection with some contracted curve; a contradiction. This proves (3.a). $\qquad\square$

Lemma 7.12 *Let Y, Z be normal varieties and $g \colon Y \to Z$ a proper morphism such that $g_* \mathcal{O}_Y = \mathcal{O}_Z$. Let D be a reduced divisor on Y and Δ' an effective $\mathbb{Q}$-divisor on Y. Fix some $0 < c \le 1$. Assume that*

(1) (Y, Δ') *is dlt,*
(2) $K_Y + cD + \Delta' \sim_{\mathbb{Q},g} 0$,
(3) $(Y, \eta D + \Delta')$ *is lc for some $\eta > 0$ and*
(4) $-D$ *is g-nef and hence* $\operatorname{Supp} D = \operatorname{Supp} g^{-1}(g(D))$.

Then $g_* \mathcal{O}_D = \mathcal{O}_{g(D)}$.

Proof By pushing forward the exact sequence

$$0 \to \mathcal{O}_Y(-D) \to \mathcal{O}_Y \to \mathcal{O}_D \to 0$$

we obtain

$$\mathcal{O}_Z = g_* \mathcal{O}_Y \to g_* \mathcal{O}_D \to R^1 g_* \mathcal{O}_Y(-D).$$

Note that

$$-D \sim_{\mathbb{Q},g} K_Y + \Delta' + (1 - c)(-D)$$

and the right hand side is of the form $K + \Delta' + (g\text{-nef})$. Let $W \subset Y$ be an lc center of (Y, Δ'). Then W is not contained in D since then $(Y, \eta D + \Delta')$ would not be lc along W. In particular, D is disjoint from the general fiber of $W \to X$ by (4). Thus from (10.39) we conclude that none of the associated primes of $R^1 g_* \mathcal{O}_Y(-D)$ are contained in $g(D)$. On the other hand, $g_* \mathcal{O}_D$ is supported on $g(D)$, hence $g_* \mathcal{O}_D \to R^1 g_* \mathcal{O}_Y(-D)$ is the zero map.

This implies that $\mathcal{O}_Z \to g_* \mathcal{O}_D$ is surjective. This map factors through $\mathcal{O}_{g(D)}$, hence $g_* \mathcal{O}_D = \mathcal{O}_{g(D)}$. $\qquad\square$

7.13 (The conjectures of Shokurov and Ambro) So far we have studied small values of the minimal log discrepancy. In defining $\overline{\mathrm{mld}}$ we chose to ignore all values that were at least 1. However, there are two interesting conjectures about larger minimal log discrepancies.

First, one can ask about the largest possible value of mld(p, X, Δ). If p is a smooth point not contained in Supp Δ then mld(p, X, Δ) = dim X and a conjecture of Shokurov (1988a) says that mld(p, X, Δ) $\leq$ dim X should always hold. This was extended by Ambro (1999a,b) to the conjecture that $p \mapsto$ mld(p, X, Δ) is a lower semi-continuous function.

Both of these are open already in dimension 4.

7.2 Minimal log discrepancy functions

In Section 7.1 we studied log centers of lc pairs and of crepant log structures. Here we extend these results to log centers of semi-log canonical pairs. Since we are especially interested in the minimal log discrepancy values < 1, we use the truncated minimal log discrepancy function $\overline{\mathrm{mld}}(W)$ defined in (7.2).

Definition 7.14 A *minimal log discrepancy function* – abbreviated as an $\overline{\mathrm{mld}}$-*function* – on a scheme X is a function

$$\overline{\mathrm{mld}}_X: \{\text{irreducible subvarieties of } X\} \longrightarrow [0, 1) \cup \{\infty\}. \tag{7.14.1}$$

Alternatively, we can think of $\overline{\mathrm{mld}}_X$ as a function from the set of (not necessarily closed) points of X to $[0, 1) \cup \{\infty\}$. If $W \subset X$ is reducible with irreducible components W_i then set $\overline{\mathrm{mld}}_X(W) := \max_i\{\overline{\mathrm{mld}}_X(W_i)\}$.

An irreducible subvariety $W \subset X$ is called a *log center* if $\overline{\mathrm{mld}}_X(W) < 1$ and a *log canonical center* (or an *lc center*) if $\overline{\mathrm{mld}}_X(W) = 0$. We are only interested in the cases when there are only finitely many log centers.

As in (5.25), we define the *log canonical stratification* (or *lc stratification*) of $(X, \overline{\mathrm{mld}}_X)$ as follows. Let $S_i^* \subset X$ be the union of all $\leq i$-dimensional lc centers and set $S_i := S_i^* \setminus S_{i-1}^*$. Note that the definition of the lc stratification involves only the log canonical centers, but not the log centers.

As our main example, any crepant log structure $f: (Y, \Delta) \to X$ defines a minimal log discrepancy function $\overline{\mathrm{mld}}_{(Y,\Delta)}(*) := \overline{\mathrm{mld}}(*, Y, \Delta)$ on X as in (7.1–7.2). It is convenient to set $\overline{\mathrm{mld}}_{(Y,\Delta)}(X) = 0$. As noted in (7.1), there are only finitely many log centers. The above definitions give the same log canonical centers and log canonical stratification as defined earlier (4.15, 5.25).

Definition 7.15 (Maps and $\overline{\mathrm{mld}}$-functions) Let $(X_i, \overline{\mathrm{mld}}_i)$ be two schemes with $\overline{\mathrm{mld}}$-functions. A finite morphism $g: X_1 \to X_2$ is said to be *compatible* with the $\overline{\mathrm{mld}}$-functions if $\overline{\mathrm{mld}}_1(W_1) = \overline{\mathrm{mld}}_2(g(W_1))$ for every subvariety $W_1 \subset X_1$.

Since lc centers are exactly the subvarieties with $\overline{\mathrm{mld}} = 0$, such a g is also a stratified morphism $g: (X_1, S_*^1) \to (X_2, S_*^2)$.

Precise adjunction, as stated in (7.10), gives our first main example of maps compatible with $\overline{\mathrm{mld}}$-structures.

Proposition 7.16 *Let* $f: (Y, \Delta) \to X$ *be a dlt, crepant log structure. For an lc center* $Z \subset Y$ *consider the Stein factorization* $(Z, \Delta_Z := \mathrm{Diff}_Z^* \Delta) \xrightarrow{f_Z} W \xrightarrow{\pi} X$. *Then* $f_Z: (Z, \Delta_Z) \to W$ *is a crepant log structure and* $\pi: (W, \overline{\mathrm{mld}}_{(Z, \Delta_Z)}) \to (X, \overline{\mathrm{mld}}_{(Y, \Delta)})$ *is compatible with the* $\overline{\mathrm{mld}}$*-functions.* □

In close parallel with (5.29) we make the following definition.

Definition 7.17 We say that a seminormal scheme Y with a function $\overline{\mathrm{mld}}_Y$ is *of log canonical origin* if

(1) the $S_i(Y)$ are all unibranch,
(2) there are crepant log structures $f_j: (X_j, \Delta_j) \to Z_j$ and a finite, surjective morphism

$$\pi : \amalg_j (Z_j, \overline{\mathrm{mld}}_{(X_j, \Delta_j)}) \to (Y, \overline{\mathrm{mld}}_Y)$$

that is compatible with the $\overline{\mathrm{mld}}$-functions.

As in (5.29), the key example is the following.

Let $f: (X, \Delta) \to W$ be a crepant log structure and $V \subset W$ any union of lc centers. We can apply (7.16) to each lc center separately to conclude that $(V, \overline{\mathrm{mld}}_{(X, \Delta)}|_V)$ is of log canonical origin. More generally, if $(Y, \overline{\mathrm{mld}}_Y)$ is of log canonical origin and $V \subset Y$ is any union of lc centers then $(V, \overline{\mathrm{mld}}_Y|_V)$ is also of log canonical origin.

Iterating this procedure, we obtain that if $(Y, \overline{\mathrm{mld}}_Y)$ is of log canonical origin and $\tau: W \to Y$ is a hereditary lc center (5.30) then W also has a natural $\overline{\mathrm{mld}}$-function and τ is compatible with the $\overline{\mathrm{mld}}$-functions.

Let $R(\tau)$ be a gluing relation (5.31) on a pair $(Y, \overline{\mathrm{mld}}_Y)$ that is of log canonical origin such that all the defining isomorphisms τ_{ijk} are compatible with $\overline{\mathrm{mld}}_Y$. It is clear that in this case there is a unique minimal log discrepancy function $\overline{\mathrm{mld}}_{Y/R(\tau)}$ on $Y/R(\tau)$ such that the quotient map $q: Y \to Y/R(\tau)$ is compatible with $\overline{\mathrm{mld}}_Y$ and $Y/R(\tau)$. Thus $(Y/R(\tau), \overline{\mathrm{mld}}_{Y/R(\tau)})$ is also of log canonical origin.

The three main properties (7.5–7.8) of minimal log discrepancies of lc pairs (X, Δ) continue to hold for $(Y, \overline{\mathrm{mld}}_Y)$ of log canonical origin. We are mainly interested in the case when (Y, Δ) is slc and $\overline{\mathrm{mld}} := \overline{\mathrm{mld}}_{(Y, \Delta)}$, but the inductive proof uses general pairs of log canonical origin.

Theorem 7.18 *Let* Y *be a seminormal scheme and* $\overline{\mathrm{mld}}$ *a minimal log discrepancy function such that* $(Y, \overline{\mathrm{mld}})$ *is of log canonical origin. Let* W, W_1, W_2 *be closed, reduced subschemes. Then*

(1) $\overline{\mathrm{mld}}(W_1 \cap W_2) \leq \overline{\mathrm{mld}}(W_1) + \overline{\mathrm{mld}}(W_2)$.
(2) *If* $\overline{\mathrm{mld}}(W) < \frac{1}{6}$ *then* W *is seminormal.*

(3) *If* $\overline{\mathrm{mld}}(W_1) + \overline{\mathrm{mld}}(W_2) < \frac{1}{2}$ *then* $W_1 \cap W_2$ *is reduced, equivalently,* $W_1 \cup W_2$ *is seminormal in* $W_1 \amalg W_2$ *(10.20).*

Proof Let $V \subset W_1 \cap W_2$ be an irreducible component. We need to prove that $\overline{\mathrm{mld}}_Y(V) \leq \overline{\mathrm{mld}}_Y(W_1) + \overline{\mathrm{mld}}_Y(W_2)$. We use induction on the dimension of Y.

By assumption there is a crepant log structure $f: (X, \Delta) \to Z$ and a finite morphism $\pi: Z \to Y$ that is compatible with the $\overline{\mathrm{mld}}$-functions, where, to simplify notation, we allow Z to be disconnected.

Assume first that V is not contained in the boundary of the induced lc stratification of Y (7.14). Then π is universally open over the generic point of V (1.44), hence there are irreducible components $W_i^Z \subset \pi^{-1}(W_i)$ and $V^Z \subset W_1^Z \cap W_2^Z$ such that V^Z dominates V. Thus, using (7.5) on Z, we get that

$$\overline{\mathrm{mld}}_Y(V) = \overline{\mathrm{mld}}_Z(V^Z) \leq \overline{\mathrm{mld}}_Z(W_1^Z) + \overline{\mathrm{mld}}_Z(W_2^Z)$$
$$= \overline{\mathrm{mld}}_Y(W_1) + \overline{\mathrm{mld}}_Y(W_2).$$

We are left with the case when $V \subset B(Y)$. In this case V is also an irreducible component of

$$(B(Y) \cap W_1) \cap (B(Y) \cap W_2).$$

Since π is surjective, there are subvarieties $W_i^Z \subset Z$ such that $\pi: W_i^Z \to W_i$ is surjective. Since $B(Z) = \pi^{-1} B(Y)$ and $\overline{\mathrm{mld}}_Z(B(Z)) = 0$, we see that

$$\overline{\mathrm{mld}}_Y(B(Y) \cap W_i) = \overline{\mathrm{mld}}_Z(B(Z) \cap W_i^Z) \leq \overline{\mathrm{mld}}_Z(W_i^Z) = \overline{\mathrm{mld}}_Y(W_i).$$

Since $\dim B(Y) < \dim Y$, we can use induction to see that

$$\overline{\mathrm{mld}}_Y(V) \leq \overline{\mathrm{mld}}_Y(B(Y) \cap W_1) + \overline{\mathrm{mld}}_Y(B(Y) \cap W_2)$$
$$\leq \overline{\mathrm{mld}}_Y(W_1) + \overline{\mathrm{mld}}_Y(W_2).$$

This proves (1). Most of the proof of (2–3) will be done in Section 9.2; here we note two special instances.

7.18.4 (Normal case) Assume that Y is normal. Then there is a crepant log structure $f: (X, \Delta) \to Z$ and a finite morphism $\pi: Z \to Y$ that is compatible with the $\overline{\mathrm{mld}}$-functions.

Let $W \subset Y$ be a log center with $\overline{\mathrm{mld}}_Y(W) < \frac{1}{6}$. Then $\overline{\mathrm{mld}}_{(X,\Delta)}(\pi^{-1}(W)) < \frac{1}{6}$, hence $\pi^{-1}(W)$ is seminormal by (7.6). Thus W is seminormal by (10.26). This shows (2) while (3) is obtained similarly using (10.30).

7.18.5 (Hereditary case) Let $\tau: Z \to Y$ be any hereditary lc center. For a log center $W \subset Y$, set $W_Z := \mathrm{red}\,\tau^{-1}(W)$. As we noted in (7.17), $\overline{\mathrm{mld}}_Z(W_Z) = \overline{\mathrm{mld}}_Y(W)$. Since the boundary has $\overline{\mathrm{mld}} = 0$, we conclude that

$$\overline{\mathrm{mld}}_Z(W_Z \cup B(Z)) = \overline{\mathrm{mld}}_Y(W).$$

By the already established normal case we conclude the following.

Claim 7.18.6   With the above notation, let $W \subset Y$ be a log center with $\overline{\mathrm{mld}}_Y(W) < \frac{1}{6}$ and $\tau \colon Z \to Y$ be any hereditary lc center. Then $\mathrm{red}\, \tau^{-1}(W) \cup B(Z)$ is seminormal.   □

We prove in (9.38–9.40) that the conclusion of (7.18.6) implies that W is seminormal. This will complete the proof of (2).   □

Similarly, using (7.8), we get the following.

Claim 7.18.7   With the above notation, let $W_1, W_2 \subset Y$ be log centers with $\overline{\mathrm{mld}}_Y(W_1) + \overline{\mathrm{mld}}_Y(W_2) < \frac{1}{2}$ and $\tau \colon Z \to Y$ be any hereditary lc center. Then $\mathrm{red}\, \tau^{-1}(W_1) \cup \mathrm{red}\, \tau^{-1}(W_2) \cup B(Z)$ is seminormal in $\mathrm{red}\, \tau^{-1}(W_1) \amalg \mathrm{red}\, \tau^{-1}(W_2) \amalg B(Z)$.   □

We prove in (9.43–9.45) that the conclusion of (7.18.7) implies that $W_1 \cup W_2$ is seminormal in $W_1 \amalg W_2$. By (10.21) this will complete the proof of (3).   □

Remark 7.19   The analog of (7.7) also holds but the original proof covers all cases. Indeed, assume that $(Y, \overline{\mathrm{mld}})$ is of log canonical origin and let $W \subset Y$ be a closed, reduced subscheme with $\overline{\mathrm{mld}}(W) < 1$. We are allowed to discard log centers $V \subset W$ with $\overline{\mathrm{mld}}(V) \leq 2 \cdot \overline{\mathrm{mld}}(W)$ without changing the conclusion.

If $W_i \subset W$ are the irreducible components then $\overline{\mathrm{mld}}(W_i \cap W_j) \leq 2 \cdot \overline{\mathrm{mld}}(W)$ by (7.18) thus it is enough to consider one W_i at a time. Let Y_i be the smallest lc center that contains W_i. If $Y_i' \subsetneq Y_i$ is another lc center, then $\overline{\mathrm{mld}}(W_i \cap Y_i') \leq \overline{\mathrm{mld}}(W_i)$. Thus we are reduced to the case when Y_i is normal. Then we conclude by (7.7) and (10.26).

7.3 Depth of sheaves on slc pairs

If (X, Δ) is dlt, then $\mathcal{O}_X$ is a CM sheaf (2.88), but if (X, Δ) is lc or slc, then in general $\mathcal{O}_X$ is not even S_3 (3.6). However, it was noticed by Alexeev (2008), that $\mathcal{O}_X$ fails to be S_3 only at the lc centers. Here we generalize this to certain divisorial sheaves on slc pairs.

While going from S_2 to S_3 seems like a small step, these results are very important in the deformation theory of slc pairs.

Theorem 7.20 (Kollár, 2011a)   *Let (X, Δ) be slc and $x \in X$ a point that is not an lc center. Let D be a $\mathbb{Z}$-divisor such that none of the irreducible components of D are contained in $\mathrm{Sing}\, X$. Assume that there is an effective $\mathbb{Q}$-divisor $\Delta' \leq \Delta$ such that $D \sim_{\mathbb{Q}} \Delta'$. Then*

$$\mathrm{depth}_x\, \mathcal{O}_X(-D) \geq \min\{3, \mathrm{codim}_X x\}.$$

Note that $\mathcal{O}_X(-D)$ is reflexive, hence S_2, for any D. Thus the only interesting part is the claim that $\mathrm{depth}_x\, \mathcal{O}_X(-D) \geq 3$ if $\mathrm{codim}_X x \geq 3$.

Before giving a proof, let us derive several consequences. First, the $D = 0$ and $\Delta' = 0$ case gives the above mentioned result of Alexeev (2008) and Fujino (2009b). (For stronger results, see Alexeev and Hacon, 2011.)

Corollary 7.21 *Let (X, Δ) be slc and $x \in X$ a point that is not an lc center. Then* $\operatorname{depth}_x \mathscr{O}_X \geq \min\{3, \operatorname{codim}_X x\}$. $\qquad\square$

As another application, take $D = -K_X$. By passing to an open neighborhood of $x \in X$ we may assume that $K_X + \Delta \sim_{\mathbb{Q}} 0$, hence $-K_X \sim_{\mathbb{Q}} \Delta$. Since $\mathscr{O}_X(-(-K_X)) \simeq \omega_X$, we get the following.

Corollary 7.22 *Let (X, Δ) be slc and $x \in X$ a point that is not an lc center. Then* $\operatorname{depth}_x \omega_X \geq \min\{3, \operatorname{codim}_X x\}$. $\qquad\square$

Note that while $\mathscr{O}_X$ is CM if and only if ω_X is CM (see Kollár and Mori (1998, 5.70)), it can happen that $\mathscr{O}_X$ is S_3 but ω_X is not; see Patakfalvi (2010). Thus (7.22) is not a formal consequence of (7.21). These two corollaries imply that restrictions of $\mathscr{O}_X$ and of ω_X to certain Cartier divisors are well behaved.

Corollary 7.23 *Let (X, Δ) be slc and $S \subset X$ a Cartier divisor that does not contain any lc centers. Then $\mathscr{O}_S$ is S_2 and $\omega_S \simeq \omega_X(S)|_S$.* $\qquad\square$

Corollary 7.24 *Let (X, Δ) be slc and $x \in X$ a point that is not an lc center. Assume that $\Delta = \sum(1 - \frac{1}{m_i})D_i$ with $m_i \in \mathbb{N} \cup \{\infty\}$. Then, for any $r \geq 0$,*

$$\operatorname{depth}_x \mathscr{O}_X(rK_X + \lfloor r\Delta \rfloor) \geq \min\{3, \operatorname{codim}_X x\}.$$

Proof As before, we may assume that $K_X + \Delta \sim_{\mathbb{Q}} 0$, thus

$$-rK_X - \lfloor r\Delta \rfloor \sim_{\mathbb{Q}} -r(K_X + \Delta) + (r\Delta - \lfloor r\Delta \rfloor) \sim_{\mathbb{Q}} r\Delta - \lfloor r\Delta \rfloor.$$

Note that $r\Delta - \lfloor r\Delta \rfloor = \sum_i \frac{c_i}{m_i} D_i$ for some $c_i \in \mathbb{N}$ where $0 \leq c_i < m_i$ for every i. Thus $c_i \leq m_i - 1$ for every i, that is, $r\Delta - \lfloor r\Delta \rfloor \leq \Delta$. Thus, we can take $\Delta' := r\Delta - \lfloor r\Delta \rfloor$ and (7.20) applies. $\qquad\square$

The assumption in (7.24) that the coefficients of $\Delta = \sum a_i D_i$ be of the form $a_i = 1 - \frac{1}{m_i}$ seems artificial. It is not clear what the optimal form of (7.24) should be, but the following examples show that some strong numerical assumptions are necessary.

Example 7.25 Set $X := (xy - uv = 0) \subset \mathbb{A}^4$ and let $B_1, B_2 \in |B|$ be two members of one family of planes on X and $A_1, A_2, A_3 \in |A|$ three members of the other family.

(1) Fix a natural number $m \equiv -1 \mod 3$ and consider the klt pair

$$\left(X, \Delta := \left(\tfrac{2}{3} - \tfrac{1}{m}\right)(A_1 + A_2 + A_3) + \left(1 - \tfrac{3}{2m}\right)(B_1 + B_2)\right).$$

Then $\lfloor 2m\Delta \rfloor \sim (4m - 8)A + (4m - 6)B$. Thus

$$\mathscr{O}_X(2mK_X + \lfloor 2m\Delta \rfloor) \simeq \mathscr{O}_X(-2A)$$

has only depth 2 at the origin by (3.15.2).

(2) Fix $m \geq 3$ and consider the klt pair

$$\left(X, \Delta := \left(1 - \tfrac{1}{m}\right)(A_1 + A_2) + \tfrac{2}{m(m+1)}A_3 + \left(1 - \tfrac{1}{m+1}\right)(B_1 + B_2) \right).$$

Then $\lfloor (m + 1)\Delta \rfloor = (m - 1)A_1 + (m - 1)A_2 + mB_1 + mB_2$ involves only those divisors with coefficients of the form $1 - \tfrac{1}{m_i}$, but $\mathscr{O}_X((m + 1)K_X + \lfloor (m + 1)\Delta \rfloor) \simeq \mathscr{O}_X(-2A)$ is not S_3.

7.26 *Proof of (7.20)*   As we noted, the only interesting part of (7.20) is the assertion that $\mathrm{codim}_X x \geq 3$ implies $\mathrm{depth}_x \mathscr{O}_X(-D) \geq 3$. By (2.58), the latter is equivalent to $H_x^i(X, \mathscr{O}_X(-D)) = 0$ for $i \leq 2$. Since $\mathscr{O}_X(-D)$ is reflexive, the $i = 0, 1$ cases are clear. Thus we only need to prove that $H_x^2(X, \mathscr{O}_X(-D)) = 0$.

To show this, we use the method of two spectral sequences introduced in Kollár and Mori (1998, 5.22). In our case, we need to use local cohomology and exploit a special property of the second cohomology group; these variants were developed in Alexeev (2008) and Fujino (2009b).

7.27 (The method of two spectral sequences)   Let $f : Y \to X$ be a proper morphism, $V \subset X$ a closed subscheme and $W := f^{-1}V \subset Y$. Let F be a coherent sheaf on Y. Our aim is to prove the vanishing of certain cohomology groups $H_V^i(X, f_*F)$ using some information about Y and F, where H_V^i denotes cohomology with supports in V; see Grothendieck (1968, chapter 1).

The basic tool is the Leray spectral sequence

$$H_V^i(X, R^j f_*F) \Longrightarrow H_W^{i+j}(Y, F). \qquad (7.27.1)$$

If $R^j f_*F = 0$ for every $j > 0$ then the spectral sequence degenerates and we get isomorphisms $H_V^i(X, f_*F) \simeq H_W^i(Y, F)$. Thus if the $R^j f_*F$ and $H_W^i(Y, F)$ both vanish for $i, j \geq 1$, then $H_V^i(X, f_*F) = 0$ for all $i \geq 1$. However, in most applications, one cannot achieve both vanishings simultaneously. We get around this problem by arranging for the $H_W^i(Y, F)$ to vanish for F and the $R^j f_*F'$ to vanish for some other sheaf F'.

Given a map of sheaves $F \to F'$ we get, for each i, a commutative diagram

$$
\begin{array}{ccc}
H_V^i(X, f_*F') & \xrightarrow{\ \alpha_i'\ } & H_W^i(Y, F') \\
\uparrow & & \uparrow \\
H_V^i(X, f_*F) & \xrightarrow{\ \alpha_i\ } & H_W^i(Y, F).
\end{array}
\qquad (7.27.2)
$$

Assume that

i) $f_* F = f_* F'$,
ii) $H_W^i(Y, F) = 0$ for $i > 0$ and
iii) $R^j f_* F' = 0$ for $i > 0$.

Then α_i' is an isomorphism and, by easy diagram chasing, $H_V^i(X, f_* F) = 0$ for $i > 0$.

For the cases $i = 1, 2$, weaker hypotheses suffice. $H_V^1(X, f_* F) \hookrightarrow H_W^1(Y, F)$ is injective, thus $H_W^1(Y, F) = 0$ alone yields $H_V^1(X, f_* F) = 0$.

The $i = 2$ case is more interesting. If $H_W^2(Y, F) = 0$ then $\alpha_2 = 0$. Together with $f_* F = f_* F'$ this implies that $\alpha_2' = 0$. On the other hand, α_2' sits in the exact sequence

$$H_V^0(X, R^1 f_* F') \to H_V^2(X, f_* F') \xrightarrow{\alpha_2'} H_W^2(Y, F'),$$

hence we get the following.

Proposition 7.28 *With the above notation, assume that*

(1) $f_* F = f_* F'$,
(2) $H_W^2(Y, F) = 0$ *and*
(3) $H_V^0(X, R^1 f_* F') = 0$.

Then $H_V^2(X, f_* F) = 0$. $\qquad\qquad\square$

7.29 *Proof of (7.20), continued* We need to prove that $H_x^2(X, \mathscr{O}_X(-D)) = 0$ if $\operatorname{codim}_X x \geq 3$.

First consider the case when X is normal. We may assume that X is affine and $\lfloor \Delta \rfloor = 0$ (3.71). Let $f \colon Y \to X$ be a log resolution of (X, Δ) and write

$$K_Y + f_*^{-1}\Delta \sim_{\mathbb{Q}} f^*(K_X + \Delta) + B - A - E, \qquad (7.29.1)$$

where A, B, E are effective, f-exceptional, B, E are $\mathbb{Z}$-divisors, E is reduced and $\lfloor A \rfloor = 0$. Write

$$f^*(D - \Delta') = f_*^{-1}D - f_*^{-1}\Delta' - F, \qquad (7.29.2)$$

where F is f-exceptional. Set $D_Y := f_*^{-1}D - \lfloor F \rfloor$. Then

$$D_Y = f_*^{-1}\Delta' + \{F\} + f^*(D - \Delta') \sim_{\mathbb{Q},f} f_*^{-1}\Delta' + \{F\}. \qquad (7.29.3)$$

We apply (7.28) with $F := \mathscr{O}_Y(-D_Y)$ and $F' := \mathscr{O}_Y(B_Y - D_Y)$ where B_Y is an effective and f-exceptional divisor to be specified later.

First, (7.30) applies to D_Y (we can take $D_h = 0$ and $D_v = f_*^{-1}\Delta' + \{F\}$) hence $\mathscr{O}_X(-D) = f_* \mathscr{O}_Y(-D_Y) = f_* \mathscr{O}_Y(B_Y - D_Y)$. This shows (7.28.1).

Next apply (10.44) with $M = \mathscr{O}_Y(K_Y + D_Y)$. By (7.29.3) and (10.32), $R^{n-i} f_* M = 0$ for $i < \dim X$, hence also $H_W^i(Y, \mathscr{O}_Y(-D_Y)) = 0$ for $i < \dim X$, giving (7.28.2).

It remains to check (7.28.3), that is, the vanishing of $H_x^0(X, R^1 f_* \mathscr{O}_Y(B_Y - D_Y))$. To this end choose B_Y, A_Y, E_Y such that

$$B_Y - A_Y - E_Y = B - A - E + \{F\},$$

where A_Y, B_Y, E_Y are effective, f-exceptional, B_Y, E_Y are $\mathbb{Z}$-divisors, $E_Y \leq E$ is reduced and $\lfloor A_Y \rfloor = 0$. Set $\Delta'' := \Delta - \Delta'$. With these choices,

$$
\begin{aligned}
B_Y - D_Y &\sim_{\mathbb{Q}} B - A - E + \{F\} + A_Y + E_Y - f_*^{-1}\Delta' - \{F\} - f^*(D - \Delta') \\
&\sim_{\mathbb{Q}} K_Y + f_*^{-1}\Delta + A_Y + E_Y - f_*^{-1}\Delta' - f^*(K_X + \Delta + D - \Delta') \\
&\sim_{\mathbb{Q}} K_Y + f_*^{-1}\Delta'' + A_Y + E_Y - f^*(K_X + \Delta + D - \Delta').
\end{aligned}
$$

$$(7.29.4)$$

Since $E_Y \leq E$, every lc center W of $(Y, f_*^{-1}\Delta'' + A_Y + E_Y)$ is also an lc center of (Y, E), hence $f(W)$ is an lc center of (X, Δ). Thus $x \in X$ is not the f-image of any lc center of $(Y, f_*^{-1}\Delta'' + A_Y + E_Y)$. We can apply (10.39) to conclude that x is not an associated prime of $R^i f_* \mathscr{O}_Y(B_Y - D_Y)$. Equivalently, $H_x^0(X, R^i f_* \mathscr{O}_Y(B_Y - D_Y)) = 0$, thus (7.28.3) also holds and so $H_x^2(X, \mathscr{O}_X(-D)) = 0$.

The above proof works with little change if X is not normal. By (5.23) there is a double cover $\pi \colon (\tilde{X}, \tilde{\Delta}) \to (X, \Delta)$ that is étale in codimension 1 such that every irreducible component of $(\tilde{X}, \tilde{\Delta})$ is smooth in codimension 1. Set $\tilde{D} := \pi^{-1}(D)$. Then $\mathscr{O}_X(-D)$ is a direct summand of $\pi_* \mathscr{O}_{\tilde{X}}(-\tilde{D})$, hence it is enough to prove the depth bounds for $\mathscr{O}_{\tilde{X}}(-\tilde{D})$.

By (10.56) there is a semi-resolution $\tilde{f} \colon (\tilde{Y}, \tilde{\Delta}_Y) \to (\tilde{X}, \tilde{\Delta})$. The rest of the proof works as before.   □

During the proof we have used the following.

Lemma 7.30 (Fujita, 1985) *Let $f \colon Y \to X$ be a proper, birational morphism. Let D be a $\mathbb{Z}$-divisor on X and assume that $D \sim_{\mathbb{Q},f} D_h + D_v$ where D_v is f-exceptional, $\lfloor D_v \rfloor = 0$ and D_h is effective without exceptional components. Let B be any effective, f-exceptional divisor. Assume that*

(1) *either X and Y are normal,*
(2) *or X and Y are S_2, f is an isomorphism outside a codimension 2 subscheme of X and Y is normal at the generic point of every exceptional divisor.*

Then $\mathscr{O}_X(-f_*D) = f_* \mathscr{O}_Y(-D) = f_* \mathscr{O}_Y(B - D)$.

Proof Note that in general $f_* \mathscr{O}_Y(-D) \subset \mathscr{O}_X(-f_*D)$ and there is an effective, f-exceptional divisor B' such that $\mathscr{O}_X(-f_*D) \subset f_* \mathscr{O}_Y(B' - D)$. Thus it is enough to prove that $f_* \mathscr{O}_Y(-D) = f_* \mathscr{O}_Y(B - D)$.

For some B_1, fix a section $s \in H^0(X, f_* \mathscr{O}_Y(B_1 - D))$ and then take $0 \leq B \leq B_1$ as small as possible such that $s \in H^0(X, f_* \mathscr{O}_Y(B - D))$ still holds. We need to prove that $B = 0$.

By assumption, there is a $\mathbb{Q}$-Cartier $\mathbb{Q}$-divisor M on X such that $D + f^*M = D_h + D_v$. Choose $n \in \mathbb{N}$ such that nD_h, nD_v, $f^*(nM)$ are all $\mathbb{Z}$-divisors. Then

$$
\begin{aligned}
(f^*s)^n &\in H^0(Y, \mathscr{O}_Y(nB - nD)) \\
&= H^0(Y, \mathscr{O}_Y(nB - nD_h - nD_v + f^*(nM))) \\
&\subset H^0(Y, \mathscr{O}_Y(nB - nD_v + f^*(nM))).
\end{aligned}
$$

Adding exceptional divisors to a pull-back never creates new sections. By assumption, every irreducible component of B appears in $nB - nD_v$ with positive coefficient. Thus $(f^*s)^n$ vanishes along every irreducible component of B and so does f^*s, contradicting the minimality of B. Thus $B = 0$ and so s is a section of $f_*\mathscr{O}_Y(-D)$. $\qquad\square$

Let us also note that (7.20) becomes much stronger in the dlt case.

Theorem 7.31 (Kollár, 2011a) *Let* (X, Δ) *be semi-dlt,* D *a (not necessarily effective)* $\mathbb{Z}$-*divisor and* $\Delta' \leq \Delta$ *an effective* $\mathbb{Q}$-*divisor on* X *such that* $D \sim_{\mathbb{Q}} \Delta'$. *Then* $\mathscr{O}_X(-D)$ *is CM.*

The proof is actually simpler than (7.29). The main difference is that in (7.29.4) $E_Y = 0$, hence $R^i f_*\mathscr{O}_Y(B_Y - D_Y) = 0$ for $i > 0$ by (10.32). See Kollár (2011a) for details.

Remark 7.32 Note that while the assumptions of (7.20) and (7.31) depend only on the $\mathbb{Q}$-linear equivalence class of D, being CM is not preserved by $\mathbb{Q}$-linear equivalence in general. For instance, let X be a cone over an Abelian variety A of dimension ≥ 2. Let D_A be a $\mathbb{Z}$-divisor on A such that $mD_A \sim 0$ for some $m > 1$ but $D_A \not\sim 0$. Let D_X be the cone over D_A. Then $D_X \sim_{\mathbb{Q}} 0$, $\mathscr{O}_X(D_X)$ is CM by (3.15) but $\mathscr{O}_X$ is not CM by (3.11).

8

Survey of further results and applications

The aim of this chapter is to give a short summary of further results about the singularities we have been studying, without going into the details of the definitions and results. Some of the topics are covered in recent books or surveys; for these our treatment is especially cursory.

The main topics are

- Linear systems and ideal sheaves.
- Connections with complex analysis.
- Log canonical thresholds.
- The ACC conjecture.
- Arc spaces of lc singularities.
- F-regular and F-pure singularites.
- Differential forms on lc singularities.
- The topology of lc singularities.
- Applications to the abundance conjecture and to moduli problems.
- Unexpected applications of lc pairs.

Assumptions With the exception of Section 8.4, the results in this chapter have been worked out for varieties over fields of characteristic 0. Most of them should hold in more general settings.

8.1 Ideal sheaves and plurisubharmonic funtions

In Section 2.1 we defined discrepancies of divisors and related concepts for pairs (X, Δ) where Δ is a $\mathbb{Q}$-divisor. It is straightforward and useful to define these notions when divisors are replaced by linear systems or ideal sheaves.

Definition 8.1 (Discrepancies for linear systems) Let X be a normal scheme, $|M_i|$ linear systems and $c_i \in \mathbb{Q}$. Assume that K_X and the M_i are $\mathbb{Q}$-Cartier.

Let $f: Y \to X$ be a proper morphism with exceptional divisors E_j. Write $f^*|M_i| = |F_i| + B_i$ where $|F_i|$ has no fixed divisors and B_i is the base locus of $f^*|M_i|$. Thus we can formally write

$$f^*(K_X + \sum_i c_i |M_i|) \sim_{\mathbb{Q}} K_Y + \sum_i c_i |F_i| + (\sum_i c_i B_i + \sum_j a(E_j, X, 0) E_j).$$

The coefficient of a divisor E in the last term is called its *discrepancy*. It is denoted by

$$a(E, X, \sum_i c_i |M_i|) := a(E, X, 0) + \sum_i c_i \operatorname{coeff}_E B_i.$$

If the $|M_i|$ are 0 dimensional, that is, when they can be identified with divisors, this is the same as the usual discrepancy.

If X is affine, one can more or less identify linear subsystems of $|\mathcal{O}_X|$ with ideal sheaves. This leads to the following variant.

Definition 8.2 (Discrepancies for ideal sheaves) Let X be a normal scheme, $I_i \subset \mathcal{O}_X$ ideal sheaves and $c_i \in \mathbb{Q}$. Assume that K_X is $\mathbb{Q}$-Cartier. Let $f: Y \to X$ be a proper morphism and $f^{-1} I_i \cdot \mathcal{O}_Y$ the inverse image ideal sheaves. If $E \subset Y$ is a divisor, we define its *discrepancy* by the formula

$$a\left(E, X, \prod I_i^{c_i}\right) := a(E, X, 0) + \sum c_i \operatorname{mult}_E(f^{-1} I_i \cdot \mathcal{O}_Y).$$

(Here $\prod I_i^{c_i}$ is a formal symbol. Adding divisors corresponds to multiplying their ideals, hence the change to multiplicative notation.)

For principal ideals we again recover the usual divisor version.

While both of these notions are more general than (2.4), in most theorems and proofs it is quite easy to go between the different versions. For instance, let X be a variety over $\mathbb{C}$ and E_j countably many divisors over X. If $D_i \in |M_i|$ are very general then

$$a(E_j, X, \sum c_i |M_i|) = a(E_j, X, \sum c_i D_i) \quad \forall E_j.$$

However, equality always fails for some divisors if at least one of the $|M_i|$ is positive dimensional. Indeed, if D is an irreducible component of $D_i \in |M_i|$ that is not in the base locus then

$$a(D, X, \sum c_i |M_i|) = 0 \quad \text{but} \quad a(D, X, \sum c_i D_i) = -c_i.$$

As a rule, one can easily switch between the linear system/ideal sheaf and the divisor versions as long as the c_i are small. As in (3.71), this can frequently be achieved by the simple trick of replacing $c|M|$ by $\frac{c}{m}|mM|$ for some $m \in \mathbb{N}$. However, in some questions the ideal sheaf version is a more convenient language and many formulas end up more natural, see, for example, (8.4.3–4).

The ideal sheaf version is especially well suited to asymptotic constructions. See Lazarsfeld (2004, chapters 9–11) for an excellent treatment.

Connections with complex analysis

This is a vast subject. Here we explain how the klt condition can be translated into an analytic statement and show the first connections between singular Hermitian metrics defined using *plurisubharmonic funtions* and singularities of algebraic pairs. For detailed introductions to these topics see Demailly (2001) and McNeal and Mustaţă (2010).

Proposition 8.3 *Let f be a holomorphic function on an open set $U \subset \mathbb{C}^n$ and $D := (f = 0)$. Then (U, cD) is klt if and only if $|f|^{-c}$ is locally L^2.*

Proof Set $\omega = dz_1 \wedge \cdots \wedge dz_n$. Then $|f|^{-c}$ is locally L^2 if and only if, on any compact $K \subset U$, the integral

$$\int_K (f\bar{f})^{-c} \, \omega \wedge \bar{\omega} \quad \text{is finite.} \tag{8.3.1}$$

(We can ignore the power of $\sqrt{-1}$ that makes this integral real.) Let $\pi: Y \to U$ be a proper bimeromorphic morphism. We can rewrite the above integral as

$$\int_K (f\bar{f})^{-c} \, \omega \wedge \bar{\omega} = \int_{\pi^{-1}(K)} ((f \circ \pi)\overline{(f \circ \pi)})^{-c} \, \pi^*\omega \wedge \pi^*\bar{\omega}. \tag{8.3.2}$$

If π is a log resolution of (U, D) then at any point $q \in X$ we can choose local coordinates $x_1, \ldots, x_n$ such that

$$f \circ \pi = (\text{invertible}) \prod_i x_i^{m(i,q)} \quad \text{and}$$

$$\pi^*\omega = (\text{invertible}) \prod_i x_i^{a(i,q)} \cdot dx_1 \wedge \cdots \wedge dx_n.$$

Thus the integral (8.3.2) is finite near $q \in X$ if and only if

$$\int_{\pi^{-1}(K)} \prod_i (x_i \bar{x}_i)^{a(i,q) - c \cdot m(i,q)} \, dV \quad \text{is finite.}$$

As in (10.68.2) this holds if and only if $a(i, q) - c \cdot m(i, q) > -1$ for every i. Let E_i be the exceptional divisor locally defined by $(x_i = 0)$. Then $m(i, q) = \text{mult}_{E_i} f^*D$ and $a(i, q) = a(E_i, U, 0)$, thus by (2.5), $a(i, q) - c \cdot m(i, q) = a(E_i, U, cD)$. $\qquad\square$

In complex analysis it is customary to view (8.3) as a statement about the singularities of plurisubharmonic functions. In the notation of (8.3), $\phi := \log|f|$ is plurisubharmonic and we ask whether $e^{-c\phi}$ is locally L^2.

Conversely, if ψ is an arbitrary plurisubharmonic funtion, then one can reasonably well approximate it with functions of the form $\sum_{i=1}^m c_i \log|f_i|$

where the f_i are holomorphic functions or even polynomials (Demailly and Kollár, 2001). This allows one to go between questions about singular Hermitian metrics and singularities of algebraic pairs.

Note, however, that the analytic approach allows certain operations that are difficult to perform algebraically. For instance, one can work with infinite sums $\psi := \sum_{i=1}^{\infty} c_i \log |f_i|$ and then $e^{-\psi}$ carries information about infinitely many divisors ($f_i = 0$). Since polynomial rings are Noetherian, the algebraic variants (8.1–8.2) capture information only about finitely many divisors or polynomials at a time.

8.2 Log canonical thresholds and the ACC conjecture

Definition 8.4 Let f be a holomorphic function and $D := (f = 0)$ its divisor of zeros. The *log canonical threshold* or *complex singularity exponent* of f or of D at a point $p \in \mathbb{C}^n$ is the number denoted by $c_p(f)$ or by $\mathrm{lcth}_p(D)$ such that the following equivalent conditions hold.

(1) $|f|^{-s}$ is L^2 in a neighborhood of p if and only if $s < c_p(f)$.
(2) $(\mathbb{C}^n, sD)$ is klt near p if and only if $s < c_p(f)$.

The second variant defines the log canonical threshold for any effective $\mathbb{Q}$-Cartier divisor on a scheme. For instance, one can compute Kollár (1997, 8.15) that

$$\mathrm{lcth}\left(\mathbb{A}^n, \left(x_1^{a_1} + \cdots + x_n^{a_n} = 0\right)\right) = \min\{1, \sum \tfrac{1}{a_i}\}. \qquad (8.4.3)$$

One can also define the log canonical thresholds for ideal sheaves in the natural way. As an example we see that

$$\mathrm{lcth}\left(\mathbb{A}^n, \left(x_1^{a_1}, \ldots, x_n^{a_n}\right)\right) = \sum \tfrac{1}{a_i}. \qquad (8.4.4)$$

This concept first appears in connection with the "division problem for distributions," see Schwartz (1950), Hörmander (1958), Łojasiewicz (1958) and Gel'fand and Šilov (1958). The general question is considered in Atiyah (1970) and Bernšteĭn (1971). Applications to singularity theory are summarized in Arnol'd *et al.* (1985).

The log canonical threshold is a very interesting function on the space of power series that vanish at the origin. It is lower semicontinuous (Varčenko, 1976; Demailly and Kollár, 2001) and it is a metric, that is, $c_0(f + g) \le c_0(f) + c_0(g)$ (Demailly and Kollár (2001) theorem 2.9). Applying this to the Taylor polynomials $t_m(f)$ of f and to $f - t_m(f)$ gives the following uniform approximation result.

Corollary 8.5   *Let $f \in \mathbb{C}[[x_1, \ldots, x_n]]$ be a power series and $t_m(f)$ its degree m Taylor polynomial. Then*

$$|c_0(f) - c_0(t_m(f))| \leq \tfrac{n}{m+1}.$$

An open problem about log canonical thresholds with applications to (8.32) is the following.

8.6 (Gap conjecture (Kollár, 1997, 8.16))   Consider the sequence defined recursively by $c_{k+1} = c_1 \cdots c_k + 1$ starting with $c_1 = 2$. (It is called Euclid's or Sylvester's sequence, see Graham *et al.* (1989, section 4.3) or Sloane (2003, A00058).) It starts as

$$2, 3, 7, 43, 1807, 3263443, 10650056950807, \ldots$$

The gap conjecture asserts that $c_0(z_1^{c_1} + \cdots + z_n^{c_n}) = 1 - \frac{1}{c_1 \cdots c_n}$ is an extremal example. That is, if $f(z_1, \ldots, z_n)$ is holomorphic near the origin then $c_0(f)$ can not lie in the open interval $(1 - \frac{1}{c_1 \cdots c_n}, 1)$.

This example is also expected to be extremal for the Du Bois property. That is, we conjecture the following. Let (X, Δ) be an n-dimensional slc pair and $W \subset X$ a closed, reduced subscheme such that $\mathrm{mld}(W, X, \Delta) < \frac{1}{c_1 \cdots c_n}$. Then W is Du Bois.

The $n = 1, 2$ cases are easy to check by hand; for $n = 3$ this is proved in Kollár (1994, 5.5.7).

Next we consider a far reaching generalization of the gap conjecture.

Shokurov realized that while every rational number is the log canonical threshold of some polynomial, for each fixed dimension n the set of log canonical thresholds of all n-variable functions has remarkable properties (Shokurov, 1988b). Let us start with the most general case.

Definition 8.7   Let $\mathcal{S}_n$(local) be the set of all sequences $(b_1, \ldots, b_m)$ of positive real numbers such that there is an n-dimensional $\mathbb{Q}$-factorial log canonical pair $\left(X, \sum_{i=1}^m b_i D_i\right)$ that has an lc center contained in $\cap_{i=1}^m D_i$. (Note that these are precisely the pairs which are *maximally lc*, that is, increasing any one of the b_i results in a non-lc pair.)

Define a partial ordering of these sequences by $(b_1, \ldots, b_{m_1}) \geq (b'_1, \ldots, b'_{m_2})$ if and only if either $m_1 < m_2$ or $m_1 = m_2$ and $b_i \geq b'_i$ for every i.

We can now state two of the most general conjectures, due to Shokurov (1988b) and Kollár (1992, 1997), as follows.

Conjecture 8.8 (ACC conjecture)   *For each n, the set $\mathcal{S}_n$(local) does not contain any infinite increasing subsequences.*

Conjecture 8.9 (Accumulation conjecture)   *For each n, the limit points of the set $\mathcal{S}_n$(local) are exactly $\mathcal{S}_{n-1}$(local) \ {1}.*[1]

Even very special cases of this are quite interesting and difficult. For instance, fix n and let f be a holomorphic function that vanishes at the origin. Then $(\mathbb{C}^n, c_0(f)(f = 0))$ is maximally lc, hence the length 1 sequence $(c_0(f))$ is in $\mathcal{S}_n$(local). In particular, we see that the set

$$\{c_0(f) : f \in \mathbb{C}[[z_1, \ldots, z_n]]\} \subset [0, 1] \cap \mathbb{Q}$$

does not contain any infinite increasing subsequences. This was proved in the series of papers of de Fernex and Mustaţă (2009), Kollár (2008b) and de Fernex *et al.* (2010, 2011); see Totaro (2010) for a survey.

There are many interesting ways of associating numerical sequences to an lc pair and there is a whole series of conjectures connecting these and their properties, see Shokurov (1988b) and Kollár (1992, section 18).

8.3 Arc spaces of log canonical singularities

Let X be a variety over a field k. For $m \geq 0$, let $X_m := \mathrm{Mor}(\mathrm{Spec}\, k[t]/(t^{m+1}), X)$ be the space of mth order jets on X as a k-scheme of finite type. There are natural projection maps $X_{m+1} \to X_m$ and passing to the inverse limit we get the *space of arcs* $X_\infty := \mathrm{Mor}(\mathrm{Spec}\, k[[t]], X)$ which is a non-noetherian k-scheme.

For example a morphism $\mathrm{Spec}\, k[t]/(t^{m+1}) \to \mathbb{A}^n$ is given by n polynomials $\phi_i(t) = \sum_{j=0}^m a_{ij} t^j$, thus

$$\mathbb{A}_m^n := \mathrm{Mor}(\mathrm{Spec}\, k[t]/(t^{m+1}), \mathbb{A}^n) \simeq \mathbb{A}^{n(m+1)}.$$

More generally, if X is smooth then the projection $X_m \to X_0 = X$ is a bundle with fibers $\mathbb{A}^{nm}$. (It is, however, *not* a vector bundle since the transition functions between charts are not linear.) Thus interesting local behavior of X_m comes from the singularities of X.

Assume that $X \subset \mathbb{A}^n$ is defined by the equations $f_\ell(x_1, \ldots, x_n) = 0$. Then $X_m \subset \mathbb{A}_m^n$ is defined by the equations

$$f_\ell(\textstyle\sum_{j=0}^m a_{1j} t^j, \ldots, \sum_{j=0}^m a_{nj} t^j) \equiv 0 \quad \mathrm{mod}\ (t^{m+1}).$$

In general these give rather complicated systems of equations, but in some simple examples their structure is transparent. The reader may enjoy verifying (8.10) by hand in the examples $X := (x_1^d + \cdots + x_n^d = 0)$.

---

[1] (8.8) and (8.9) were recently proven in Hacon *et al.* (2012).

It was discovered by Mustață (2002) that, at least for local complete intersection varieties, there is a close relationship between the properties of the jet spaces and discrep(X, 0). The main cases can be summarized as follows.

Theorem 8.10 (Mustață, 2002; Ein and Mustață, 2004; Ein *et al.*, 2003)   *Let X be a normal, local complete intersection variety over a field of characteristic 0.*

(1) X_m *has pure dimension* $(m + 1)$ dim X *for every m if and only if X is log canonical.*
(2) X_m *is irreducible for every m if and only if X is canonical.*
(3) X_m *is normal for every m if and only if X is terminal.*

One can also prove inversion of adjunction (again for local complete intersection varieties) using arc spaces (Ein and Mustață, 2004). For a summary of further results see Ein and Mustață (2009).

8.4 *F*-regular and *F*-pure singularites

There is an interesting and somewhat surprising and deep connection between singularities of the minimal model program and singularities defined via the behavior of the Frobenius morphism in characteristic $p > 0$.

This section is a brief overview of this growing subject.

Definition 8.11   A reduced ring of characteristic $p > 0$ is called *F-finite* if $R^{1/p}$ is a finite R-module. Every ring essentially of finite type over a perfect field is *F*-finite. A scheme over a field of characteristic $p > 0$ is *F-finite* if all of its local rings are *F*-finite.

Definition 8.12   Let R be a reduced F-finite ring of characteristic $p > 0$ and $I \subseteq R$ an ideal. Then $I^{[p^e]}$ denotes the ideal generated by the p^eth powers of elements of I. It is called the eth *Frobenius power of* I. The *tight closure* of I is defined as follows:

$$I^* := \{z \in R | \text{there exists a nonzero divisor } c \in R, \text{ so that } cz^{p^e} \in I^{[p^e]}, \forall e \geq 0\}.$$

The ideal I is called *tightly closed* if $I = I^*$.

Definition 8.13   Let R be a local ring which is essentially of finite type over a perfect field of characteristic $p > 0$ with maximal ideal $\mathfrak{m} \subseteq R$. Then R is called

weakly F-regular	if all ideals in R are tightly closed,
F-regular	if all localizations of R are weakly F-regular,
F-rational	if all ideals generated by a full system of parameters are tightly closed,
F-pure	if for every R-module M, the map $M \to M \otimes_R R^{1/p}$ is injective (if R is of finite type over a perfect field, then this is equivalent to the condition that $R \to R^{1/p}$ splits as an R-module homomorphism), and
F-injective	if $H^i_{\mathfrak{m}}(R) \to H^i_{\mathfrak{m}}(R^{1/p})$ is injective for all $i \geq 0$.

Let Z be an irreducible, reduced essentially of finite type scheme over a perfect field of characteristic $p > 0$. Then Z is called *F-regular, F-rational, F-pure*, respectively *F-injective* if all of its local rings satisfy the corresponding notion for rings.

Let X be an irreducible, reduced essentially of finite type scheme over a field of characteristic 0. Then X is called *(open) F-regular type, (open) F-rational type, (open) F-pure type*, respectively *(open) F-injective type* if the reduction of X mod p (cf. Kollár and Mori (1998, p. 14); Kollár (1996, II.5.10)) satisfies the corresponding condition for an open and dense set of primes. Similarly, X is called *dense F-regular type, dense F-rational type, dense F-pure type*, respectively *dense F-injective type* if X_p satisfies the corresponding condition for a dense set of primes. It is relatively easy to see that these definitions are independent of the way one reduces X mod p.

The connection between singularities of the minimal model program and these F-singularities uses F-$(\dots)$ type singularities. The following diagram shows the currently known implications.

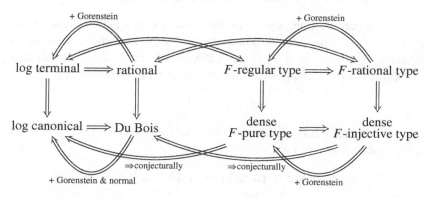

Most of these implications are highly nontrivial. For details see Elkik (1981), Kovács (1999), Kovács (2000a), Kovács (2000b), Saito (2000), Kollár and Kovács (2010), Fedder and Watanabe (1989), Mehta and Srinivas (1991), Smith

(1997), Mehta and Srinivas (1997), Hara (1998a), Hara (1998b), Smith (2000), Hara and Watanabe (2002), Hara and Yoshida (2003), Takagi and Watanabe (2004), Takagi (2004), Hara (2005), Takagi (2008), Schwede (2009), Shibuta and Takagi (2009), Mustaţă and Srinivas (2010), Mustaţă (2010), Hernández (2011) and Takagi (2011).

The connection between these sets of seemingly unrelated classes of singularities runs deep. Many notions defined in one of these theories have been identified in the other.

An important notion in higher dimensional geometry is the *multiplier ideal* (Lazarsfeld, 2004, II.9.3.G) a similarly important notion in tight closure theory is the *test ideal* (Hochster and Huneke, 1990). It turns out that these two notions are not entirely equivalent, but morally they should be considered so: Blickle *et al.* (2011) gave a characteristic free description of an ideal that in characteristic 0 gives the multiplier ideal while in characteristic $p > 0$ gives the test ideal. Interestingly, one may define the multiplier ideal in characteristic $p > 0$ which will be different from the test ideal (and hence the above remark that the multiplier and test ideals are not entirely equivalent), but the mentioned result suggests that in characteristic $p > 0$ the test ideal may be a better object to consider than the multiplier ideal and it may end up playing a role there that is similar to the role the multiplier ideal plays in characteristic 0.

There are many more interesting results and open conjectures in this intriguing connection. For more on this and related issues the reader is invited to consult the recent surveys of Schwede and Tucker (2012) and Blickle and Schwede (2012).

8.5 Differential forms on log canonical pairs

Let X be a normal, affine variety of dimension n over a field a characteristic 0, $X^{ns} \subset X$ the smooth locus and $\pi \colon \widetilde{X} \to X$ a log resolution. Since ω_X is reflexive, $H^0(X, \omega_X) = H^0(X^{ns}, \omega_{X^{ns}})$. If X has log terminal singularities then $H^0(\widetilde{X}, \omega_{\widetilde{X}}) = H^0(X, \omega_X)$ by (2.3.3) hence the natural restriction map

$$\text{rest}_n \colon H^0(\widetilde{X}, \omega_{\widetilde{X}}) \to H^0(X^{ns}, \omega_{X^{ns}})$$

is an isomorphism. It is interesting to consider the analogous restriction maps for other differential forms

$$\text{rest}_p \colon H^0(\widetilde{X}, \Omega^p_{\widetilde{X}}) \to H^0(X^{ns}, \Omega^p_{X^{ns}}).$$

The connecting role of ω_X is played not by the sheaves Ω^p_X – which can be rather complicated – but by the push-forward sheaves $\pi_* \Omega^p_{\widetilde{X}}$.

The main results are the following.

Theorem 8.14 (Extension theorem for differential forms) (Greb *et al.*, 2011a, 1.4) *Let X be a variety with log terminal singularities over a field of characteristic 0 and $\pi\colon \widetilde{X} \to X$ a log resolution. Then, for all $p \le \dim X$,*

(1) $H^0(\widetilde{X}, \Omega^p_{\widetilde{X}}) \xrightarrow{\simeq} H^0(X^{\mathrm{ns}}, \Omega^p_{X^{\mathrm{ns}}})$ *and*
(2) $\pi_* \Omega^p_{\widetilde{X}}$ *is reflexive.*

Theorem 8.15 (Extension theorem on lc pairs) (Greb *et al.*, 2011a, 1.5) *Let (X, D) be a log canonical pair over a field a characteristic 0 and $\pi\colon \widetilde{X} \to X$ a log resolution. Set $\widetilde{D} := \operatorname{Supp} \pi^{-1}(\text{non-klt}(X, D))$. Then, for all $p \le \dim X$,*

(1) $H^0(\widetilde{X}, \Omega^p_{\widetilde{X}}(\log \widetilde{D})) \xrightarrow{\simeq} H^0(X^{\mathrm{ns}}, \Omega^p_{X^{\mathrm{ns}}}(\log \lfloor D \rfloor))$ *and*
(2) $\pi_* \Omega^p_{\widetilde{X}}(\log \widetilde{D})$ *is reflexive.*

The extension problem has been studied in the literature in many special cases. For a variety X with only isolated singularities, reflexivity of $\pi_* \Omega^p_{\widetilde{X}}$ was shown by van Straten and Steenbrink (1985, theorem 1.3) for $p \le \dim X - 2$ without any further assumption on the singularities. This was extended to normal varieties by Flenner (1988), assuming $p \le \operatorname{codim}_X \operatorname{Sing} X - 2$. The $p \in \{1, 2\}$ cases were treated in Namikawa (2001, theorem 4) for canonical Gorenstein singularities. For quotient singularities similar results were obtained in de Jong and Starr (2004). For a log canonical pair with reduced boundary divisor, the cases $p \in \{1, \dim X - 1, \dim X\}$ were settled in Greb *et al.* (2010, theorem 1.1).

A related setup where the pair (X, D) is snc, and where $\pi\colon \widetilde{X} \to X$ is the composition of a finite Galois covering and a subsequent resolution of singularities has been studied in Esnault and Viehweg (1982) with additional results on the vanishing of higher direct image sheaves.

For applications to rational connectedness, the Lipman–Zariski conjecture and Bogomolov–Sommese type results see Greb *et al.* (2011a) and Greb *et al.* (2011b). The main theorem (8.15) is also an important ingredient in the proof of several Shafarevich-type hyperbolicity theorems (Kebekus and Kovács, 2010).

The following examples illustrate (8.15) and show that its statement is sharp.

Example 8.16 (Non-log canonical singularities) This example shows that the log canonical assumption on (X, D) is necessary to obtain any extension result allowing no worse than log poles along the exceptional divisor.

Let X be the affine cone over a smooth curve C of degree 4 in $\mathbb{P}^2$. Observe that X is a normal hypersurface singularity, in particular, K_X is Cartier. Let $\widetilde{X}$ be the total space of the line bundle $\mathscr{O}_C(-1)$. Then, the contraction of the zero section E of $\widetilde{X}$ yields a log resolution $\pi\colon \widetilde{X} \to X$. An elementary computation shows that the discrepancy of E with respect to X is equal to -2 cf. (Reid, 1987, p. 351, exercise (1)). Hence, X has worse than log canonical singularities. If τ is a local generator of the locally free sheaf $\Omega^{[2]}_X$ near the vertex $P \in X$, the discrepancy computation implies that τ acquires poles of order 2 when pulled back to $\widetilde{X}$. By abusing notation we denote the rational form obtained on $\widetilde{X}$ by $\pi^*\tau$.

Next, let ξ be the vector field induced by the natural $\mathbb{C}^*$-action on $\widetilde{X}$ coming from the cone structure. By contracting $\pi^*\tau$ by ξ we obtain a regular 1-form on $\widetilde{X} \setminus E$ that does not extend to an element of $H^0(\widetilde{X}, \Omega^1_{\widetilde{X}}(\log E))$. For further details see Greb *et al.* (2010, 6.3) and Greb *et al.* (2011a, 3.1).

Example 8.17 (Non-klt locus and discrepancies (Greb *et al.*, 2011a, 3.2))　It is easy to see that for, $p = \dim X$, one can replace $\widetilde{D}$ in (8.15) by the smaller divisor $\widetilde{D}_{(-1)}$ which is the sum of all divisors whose diecrepancy equals -1. The next example shows that this does not work in general for $p < n$.

Let $X = \{uw - v^2\} \subset \mathbb{C}^3_{u,v,w}$ be the quadric cone, and let $D = \{v = 0\} \cap X$ be the union of two rays through the vertex. The pair (X, D) is log canonical. Let $\widetilde{X} \subset \mathrm{Bl}_{(0,0,0)}(\mathbb{C}^3) \subset \mathbb{C}^3_{u,v,w} \times \mathbb{P}^2_{[y_1:y_2:y_3]}$ be the strict transform of X in the blow-up of $\mathbb{C}^3$ at $(0, 0, 0)$ and $\pi : \widetilde{X} \to X$ the corresponding resolution. The intersection U of $\widetilde{X}$ with $\{y_1 \neq 0\}$ is isomorphic to $\mathbb{C}^2$ and choosing coordinates x, z on this $\mathbb{C}^2$, the blow-up is given by $\varphi : (x, z) \mapsto (z, xz, x^2z)$. In these coordinates the exceptional divisor E is defined by the equation $\{z = 0\}$. The form $d \log v := \frac{1}{v} dv$ defines an element in $H^0(X, \Omega^{[1]}_X(\log D))$. Pulling back we obtain

$$\varphi^*(d \log v) = d \log x + d \log z,$$

which has log-poles along the exceptional divisor. If $f : \widetilde{X}' \to \widetilde{X}$ is the blow-up at a point $p \in E \setminus \pi_*^{-1}(D)$, we obtain a further resolution $\pi' = \pi \circ f$ of X. This resolution has an additional exceptional divisor $E' \subset \widetilde{X}'$ with discrepancy 0. Note however that the pull-back of $d \log v$ via π' has logarithmic poles along E'. To be explicit we compute on $f^{-1}(U)$: we have

$$f^*\varphi^*(d \log v) = d \log(f^*x) + d \log(f^*z),$$

and we note that f^*z vanishes along E' since we have blown up a point in $E = \{z = 0\}$.

Example 8.18 (Other tensor powers (Greb *et al.*, 2010, exercise 3.1.3))　This example shows that the statement of (8.15) fails for other reflexive tensor powers of Ω^1_X, even when X is canonical.

Let $\widetilde{X}$ be the total space of $\mathcal{O}_{\mathbb{P}^1}(-2)$, and E the zero-section. Then E contracts to a Du Val singularity $\pi : \widetilde{X} \to X$. It is reasonably easy to write down a section

$$\sigma \in H^0\big(\widetilde{X}, \mathrm{Sym}^2 \, \Omega^1_{\widetilde{X}}(\log E)\big) \setminus H^0\big(\widetilde{X}, \mathrm{Sym}^2 \, \Omega^1_{\widetilde{X}}\big),$$

which implies that $\pi_* \, \mathrm{Sym}^2 \, \Omega^1_X(\log E)$ is not reflexive.

For an explicit example for σ, consider the standard coordinate cover of $\widetilde{X}$ with open sets $U_1, U_2 \simeq \mathbb{A}^2$, where U_i carries coordinates x_i, y_i and coordinate change is given as

$$\phi_{1,2}: (x_1, y_1) \mapsto (x_2, y_2) = \left(x_1^{-1}, x_1^2 y_1\right).$$

In these coordinates the bundle map $U_i \to \mathbb{P}^1$ is given by $(x_i, y_i) \to x_i$, and the zero-section E is given as $E \cap U_i = \{y_i = 0\}$. Now take

$$\sigma_2 := y_2^{-1}(dy_2)^2 \in H^0\left(U_2, \text{Sym}^2\left(\Omega_{\bar{X}}^1(\log E)\right)\right)$$

and observe that $\phi_{1,2}^*(\sigma_2)$ extends to a section in $H^0(U_1, \text{Sym}^2(\Omega_{\bar{X}}^1(\log E)))$.

8.6 The topology of log canonical singularities

Question 8.19 Let (X, Δ) be a dlt or lc pair over $\mathbb{C}$. What can be said about the local topological properties of $X(\mathbb{C})$?

Let X be an algebraic variety over $\mathbb{C}$ and $x \in X$ a closed point. Choose an affine neighborhood $x \in X^0$ and an embedding $X^0 \subset \mathbb{C}^N$. Let $\mathbb{S}_x^{2N-1}(\varepsilon)$ be the $(2N - 1)$-sphere of radius ε centered at x and $\mathbb{B}_x^{2N}(\varepsilon)$ the closed $2N$-ball of radius ε centered at x. Up to homeomorphism, the intersection $X^0 \cap \mathbb{S}_x^{2N-1}(\varepsilon)$ does not depend on $0 < \varepsilon \ll 1$ and on the embedding. It is called the *link* of $x \in X$ and is denoted by $\text{link}(x \in X)$. The intersection $X \cap \mathbb{B}_x^{2N}(\varepsilon)$ is homeomorphic to the cone over $\text{link}(x \in X)$, thus the link carries all the information about the local topology of X near x. See Goresky and MacPherson (1988, p. 41) for details. Thus we can reformulate (8.19) as follows.

Question 8.20 Let (X, Δ) be a dlt or lc pair over $\mathbb{C}$. What can be said about the topological properties of its links?

In some ways links, especially links of isolated singularities, are quite similar to smooth projective varieties. For instance, the Lefschetz hyperplane theorems all hold for links, see Goresky and MacPherson (1988). It was realized only recently that links are quite different from quasi-projective varieties in some respects. The most striking difference is in their fundamental groups.

8.21 There are at least four natural ways to attach a fundamental group to a normal singularity $(x \in X)$. Let $p: Y \to X$ be a resolution such that $E_x := p^{-1}(x)$ is an snc divisor. For $0 < \varepsilon \ll 1$ set $Y(\varepsilon) := p^{-1}(X \cap \mathbb{B}_x^{2N}(\varepsilon))$. The following 4 groups are all independent of the resolution and of ε.

(1) $\pi_1(\text{link}(x \in X))$.
(2) $\pi_1(Y(\varepsilon) \setminus E_x) = \pi_1(p^{-1} \text{link}(x \in X))$.
(3) $\pi_1(\mathcal{R}(x \in X)) := \pi_1(Y(\varepsilon)) = \pi_1(E_x)$. These groups were first studied in Kollár (1993b) and Takayama (2003).

(4) $\pi_1(\mathcal{DR}(x \in X)) := \pi_1(D(E_x))$ where $D(E_x)$ denotes the dual complex of E_x (3.62). The homotopy type $\mathcal{DR}(x \in X)$ of $D(E_x)$, and hence the fundamental group, are independent of the resolution, see Thuillier (2007), Stepanov (2008) and Arapura *et al.* (2011).

There are natural surjections between these groups:

$$\pi_1(\text{link}(x \in X)) \twoheadleftarrow \pi_1(Y(\varepsilon) \setminus E_x) \twoheadrightarrow \pi_1(\mathcal{R}(x \in X)) \twoheadrightarrow \pi_1(\mathcal{DR}(x \in X)).$$
$$(8.21.5)$$

Even for isolated 3-fold singularities, these groups can be arbitrary.

Theorem 8.22 (Kapovich and Kollár, 2011) *Let Γ be a finitely presented group. Then there is a 3-dimensional isolated singularity $(x \in X_\Gamma)$ such that the 4 groups in (8.21) are all isomorphic to Γ.* $\square$

Usually none of the maps in (8.21.5) is an isomorphism but in the dlt case, two of these groups are trivial.

Theorem 8.23 (Kollár, 1993b; Takayama, 2003) *If (X, Δ) is dlt then*

$$\pi_1(\mathcal{R}(x \in X)) = \pi_1(\mathcal{DR}(x \in X)) = 1.$$

This implies that if (X, Δ) is dlt and $p: Y \to X$ is a resolution of singularities then for any connected open subset $U \subset X(\mathbb{C})$ the induced map

$$\pi_1(p^{-1}(U)) \to \pi_1(U) \quad \text{is an isomorphism.}$$

In particular, $\pi_1(\text{link}(x \in X)) \simeq \pi_1(Y(\varepsilon) \setminus E)$, so the only remaining open question about fundamental groups associated to dlt singularities is the following.

Conjecture 8.24   *If (X, Δ) is dlt then $\pi_1(\text{link}(x \in X))$ is finite.*[2]

If $\dim X = 2$ then (8.24) holds, but the answer is not known already when $\dim X = 3$. The kernel of $\pi_1(\text{link}(x \in X)) \twoheadrightarrow \pi_1(\mathcal{R}(x \in X))$ is generated by loops around the irreducible components of E_x, but the relations between these loops are not well understood; see Mumford (1961) for some computations in the 2-dimensional case.

We saw in (3.47) that the four groups in (8.21) can be very large if (X, Δ) is lc, but the following is wide open.

Question 8.25   What can one say about the above four groups if (X, Δ) is lc?

---

[2] The algebraic version was recently proven in Xu (2012).

Even the homotopy type of the dual complex is quite well understood for general singularities.

Theorem 8.26 (Kollár, 2012a)   *Let T be a finite, connected cell complex.*

(1) *There is a normal singularity $(0 \in X)$ whose dual complex $\mathcal{DR}(0 \in X)$ is homotopy equivalent to T.*

(2) *There is a rational singularity $(0 \in X)$ whose dual complex $\mathcal{DR}(0 \in X)$ is homotopy equivalent to T if and only if $H^i(T, \mathbb{Q}) = 0$ for $i > 0$.* $\square$

Thus, if (X, Δ) is dlt then $\mathcal{DR}(0 \in X)$ is simply connected and its cohomology groups with $\mathbb{Q}$-coefficient are zero. If they are zero with $\mathbb{Z}$-coefficient, then $\mathcal{DR}(0 \in X)$ is contractible.

Question 8.27   Let (X, Δ) be dlt. Is it true that $\mathcal{DR}(0 \in X)$ is contractible?[3]

More speculatively, it is possible that if (X, Δ) is dlt then $\mathcal{DR}(0 \in X)$ has a natural contraction to a "specific stratum" of the exceptional divisor described in some way by algebraic geometry. The results of Kollár (2007a) and Hogadi and Xu (2009) give some indication that such a special stratum should exist. See also (2.37).

8.7 Abundance conjecture

The abundance conjecture says that if X is a projective variety with terminal singularities and K_X is nef then it is semi-ample, that is, $|mK_X|$ is base point free for some $m > 0$. More generally, this is conjectured to hold for lc pairs (X, Δ): If $K_X + \Delta$ is nef then it is semi-ample. See Kollár (1992, sections 9–15) for a detailed treatment.

An approach, initiated by Miyaoka (1988) and Kawamata (1992), is to find another divisor $D \in |m'K_X|$ for some $m' > 0$ such that $K_X + \Delta + \text{red } D$ is nef and then use restriction to D, adjunction and induction on the dimension to conclude. Finding such D is already quite difficult. Another problem is that usually D is reducible, and one needs a method to piece together a section over $D = \cup_i D_i$ from sections over the irreducible D_i. It is in this context that the problems studied in Chapter 5 first appeared.

These ideas were developed in a series of papers by Kollár (1992), Keel *et al.* (1994), Fujino (2000), Kollár and Kovács (2010), Gongyo (2010), Fujino (2011), Kollár (2011d) and Fujino and Takagi (2011). For further applications see Birkar (2011) and Hacon and Xu (2011b).

---

[3] This was recently proven in de Fernex *et al.* (2013).

8.8 Moduli spaces for varieties

As we already mentioned, slc surfaces were defined by Kollár and Shepherd-Barron (1988) in order to compactify the moduli space of surfaces of general type. Similarly, slc varieties (with ample canonical class) are needed to compactify the moduli space of higher dimensional varieties of general type. Much of the theory in this book was developed with these applications in mind. A thorough treatment will be given in Kollár (2013); see also Kollár (2012b). For now let us just outline three of the main applications.

Existence of stable limits   The stable reduction theorem for curves says that if $\pi^0 \colon S^0 \to B^0$ is a smooth family of genus $g \geq 2$ curves over a curve B^0 then there is a finite surjection $p \colon A^0 \to B^0$ and a smooth compactification $A^0 \subset A$ such that $S_A^0 := S^0 \times_{B^0} A^0 \to A^0$ uniquely extends to a flat family of genus g stable curves $S_A \to A$.

The proof consists of two steps. First we use the semistable reduction theorem of Kempf et al. (1973) to get $\tilde{S}_A \to A$ whose fibers are reduced nodal curves and then run the minimal model program over A to get $S_A \to A$ as the canonical model of $\tilde{S}_A \to A$.

As pointed out in Kollár and Shepherd-Barron (1988), this recipe works in higher dimensions, as long as we start with a family $\pi^0 \colon X^0 \to B^0$ of smooth varieties of general type.

Observe that, already for curves, one should prove stable reduction not just for *smooth* families, but also for *flat* families of *stable* curves. The semistable reduction part works in all dimensions and we get $\tilde{X}_A \to A$ whose fibers are reduced, snc varieties. However, it was discovered in Kollár (2011e) that the minimal model program does not work for snc varieties.

This led to an alternate approach. We normalize $\tilde{X}_A$, run the minimal model program on each irreducible component separately and then try to glue to canonical models together to get $X_A \to A$. The gluing is ensured by (5.13). For details see Kollár (2013).

Constancy of numerical invariants   In a proper, flat family of schemes $f \colon X \to S$, the Euler characteristic of the fibers $\chi(X_s, \mathscr{O}_{X_s})$ is a locally constant function on S, but the individual cohomologies $h^i(X_s, \mathscr{O}_{X_s})$ are usually not. The Du Bois condition was essentially invented to ensure that if $f \colon X \to S$ is a flat, proper morphism and X_s is Du Bois for some $s \in S$ then there is an open neighborhood $s \in S^0 \subset S$ such that

(a) $R^i f_* \mathscr{O}_X$ is locally free and compatible with base change over S^0 and
(b) $s \mapsto h^i(X_s, \mathscr{O}_{X_s})$ is a locally constant function on S^0.

Together with (6.32) we have the following consequences. For proofs see Kollár and Kovács (2010) and Kollár (2013).

Let $f: X \to S$ be a proper and flat morphism with slc fibers over closed points; S connected.

(1) Let L be an f-semi-ample line bundle on X. Then, for all i,
   (a) $R^i f_*(L^{-1})$ is locally free and compatible with base change and
   (b) $h^i(X_s, L_s^{-1})$ is independent of $s \in S$.
(2) If one fiber of f is CM then all fibers of f are CM.
(3) $\omega_{X/S}$ exists and is compatible with base change. Furthermore, for all i,
   (a) $R^i f_*\omega_{X/S}$ is locally free and compatible with base change and
   (b) $h^i(X_s, \omega_{X_s})$ is independent of $s \in S$.

Absence of embedded points In analogy with the moduli of pointed curves, one is interested in the moduli space of slc pairs. Assume that we have a morphism $f: (X, \Delta) \to C$ to a smooth curve. We would like to view this as a family whose fiber over a point $c \in C$ is the pair (X_c, Δ_c) where $\Delta_c = \mathrm{Diff}_{X_c} \Delta$.

If $\Delta = \sum_i a_i D_i$ then $\Delta_c = \sum_i a_i D_{ic}$ where D_{ic} is the divisor-theoretic restriction. That is, the divisor corresponding to the scheme-theoretic intersection $D_i \cap X_c$. Examples due to Hassett (cf. Kollár (2012b, section 6)) show that in general the scheme-theoretic restriction can have embedded points. There are even examples where the log canonical rings of the fibers are not flat over the base.

If $a_i > \frac{1}{2}$ for every i, then $D_i \cap X_c$ is reduced by (7.8, 7.18), thus in this case the divisor-theoretic and scheme-theoretic viewpoints agree with each other and the various proposed definitions of the moduli functor agree with each other. If we allow $a_i \le \frac{1}{2}$, then different definitions do yield different moduli functors and their relationship is not yet fully understood.

8.9 Applications of log canonical pairs

Here we discuss several results whose statement does not even mention singularities but lc pairs appear unexpectedly in their proofs. For more examples see Kollár (1997).

8.29 (Effective base point freeness) Let X be a smooth projective variety of dimension n and H an ample divisor on X. It is interesting to find explicit very ample or free divisors on X. Fujita (1987) conjectured what seems to be the optimal form: $K_X + (n + 1)H$ is free and $K_X + (n + 2)H$ is very ample. This is known in dimension 3 (Ein and Lazarsfeld, 1993) and in dimension 4 (Kawamata (1997)) but is open in higher dimensions.

An approach developed to attack this question relies on finding effective $\mathbb{Q}$-divisors $D \sim_{\mathbb{Q}} mH$ such that (X, D) has an lc center at a given point. The smaller one can make m, the closer we get to the conjecture. After exponential bounds in Demailly (1993) and Kollár (1993a) the method of Tsuji (1994) and Angehrn and Siu (1995) obtained freeness of $K_X + \binom{n+1}{2}H$. This was later improved to roughly $K_X + n^{4/3}H$ by Heier (2002).

8.30 (Theta divisors of Abelian varieties)   Understanding the theta divisor of a principally polarized Abelian variety (A, Θ) has been a long standing problem. For instance, it was conjectured in Arbarello and De Concini (1987) that Θ is normal, save when (A, Θ) is a product of two lower dimensional principally polarized Abelian varieties.

A rather short application of the ideas used in (8.29) allows one to prove that the pair (A, Θ) is lc (Kollár, 1995, 17.13). In particular,

$$\dim\{x \in A : \mathrm{mult}_p\,\Theta \geq k\} \leq \dim X - k.$$

A subtler argument shows that (A, Θ) is plt if and only if it is indecomposable (Ein and Lazarsfeld, 1997); see also Smith and Varley (1996). Thus, if (A, Θ) is indecomposable then Θ has rational singularities, hence it is also normal.

8.31 (Birational maps of Fano varieties)   Let X be a smooth projective Fano variety; that is, $-K_X$ is ample. It is usually quite hard to decide whether X is birational to another Fano variety or not. In our language, a theorem going back to M. Noether and G. Fano says the following.

Theorem 8.31.1   Every birational map $\phi\colon X \dashrightarrow X'$ to another Fano variety is an isomorphism if the following holds:

Let $|H| \subset |-mK_X|$ be any pencil for some $m \geq 1$. Then $(X, \frac{1}{m}H_c)$ is canonical for general $H_c \in |H|$. Equivalently, $(X, \frac{1}{m}|H|)$ is canonical, compare (8.1).

Most theorems on the birational classification of Fano varieties rely on this approach. For an introduction to these questions, see Kollár *et al.* (2004, chapter 5). For further results along this direction see Corti (1995), Corti *et al.* (2000), Pukhlikov (2007, 2010) and Hacon and McKernan (2012).

8.32 (Kähler–Einstein metrics on Fano varieties)   The question of existence of Kähler–Einstein metrics on smooth Fano varieties over $\mathbb{C}$ is a large topic; one part of it has to do with log canonical singularities.

Let X be a smooth Fano variety of dimension n over $\mathbb{C}$. By Nadel (1990), X has a Kähler–Einstein metric if there is an $\varepsilon > 0$ such that for every effective $\mathbb{Q}$-divisor $D \sim_\mathbb{Q} -K_X$ the pair

$$\left(X, \tfrac{n+\varepsilon}{n+1}D\right) \quad \text{is klt.} \tag{8.32.1}$$

In this form, the criterion is very rarely applicable directly. To get a truly useful criterion, one needs to work with orbifolds. These are pairs (X, Δ) that are locally (analytically) of the form $\mathbb{C}^n/G$ for some finite group $G \subset \mathrm{GL}(n, \mathbb{C})$. The pseudo-reflections in G give rise to the boundary Δ as in (2.41.6). By

Demailly and Kollár (2001), (X, Δ) has an orbifold Kähler–Einstein metric if there is an $\varepsilon > 0$ such that for every effective $\mathbb{Q}$-divisor $D \sim_{\mathbb{Q}} -(K_X + \Delta)$ the pair

$$\left(X, \Delta + \tfrac{n+\varepsilon}{n+1}D\right) \quad \text{is klt.} \tag{8.32.2}$$

The latter criterion applies in many cases, see for instance Nadel (1990), Chel′tsov and Shramov (2009), Demailly and Kollár (2001) and Johnson and Kollár (2001a,b).

Note that in spirit these criteria are quite similar to (8.31.1) since $\frac{1}{m}H_c \sim_{\mathbb{Q}}$ $-K_X$. However, in (8.31.1) we want the pair to be canonical, whereas in (8.32.1) (X, D) is allowed to be log canonical (or slightly worse).

8.33 (Einstein metrics on odd dimensional manifolds)   Let L be a line bundle over a complex variety X. Fixing a Hermitian metric on L, we can talk about the corresponding unit circle bundle $L \supset M \to X$. By Kobayashi (1963), if X is Fano then a Kähler–Einstein metric on X lifts to an Einstein metric on M.

We again have many more such cases where X is only an orbifold. It happens frequently that X is an orbifold (of complex dimension n) and we can use a Seifert bundle $L \to X$ (9.50) to get a *smooth* manifold (of real dimension $2n + 1$) as a Seifert $\mathbb{S}^1$-bundle over X. (Seifert originally studied these when $\dim X = 1$.) Furthermore, the criterion (8.32.2) can be used to produce an orbifold Kähler–Einstein metric on X and then lift this to an Einstein metric on M.

This approach was used to produce Einstein metrics with positive Ricci curvature on spheres and exotic spheres (Boyer *et al.*, 2005b,a) and on many 5-manifolds (Kollár, 2005, 2007b, 2009).

8.34 (Ax's conjecture)   The conjecture says that if k is a field and $X \subset \mathbb{P}^n_k$ is a hypersurface of degree $\leq n$ then X contains a subvariety $Z \subset X$ that is geometrically irreducible (Ax, 1968). If X is geometrically irreducible then $Z = X$ works, so this is really a conjecture about hypersurfaces with many irreducible components over $\bar{k}$.

This is proved in characteristic 0 in Kollár (2007a). The proof essentially constructs a terminal singularity from X as in Section 3.4 and then solves a special case of (8.27) to get Z as the image of the "special stratum" of a resolution.

9

Finite equivalence relations

In this chapter we study general questions about geometric quotients by the finite set theoretic equivalence relations that were used in Chapter 5. These results are somewhat technical and probably the best introduction is to study the examples (9.7) and (9.61). While they do not help with the proofs, they show that the questions are quite subtle and that the theorems are likely to be optimal.

Section 9.1 studies the existence question for geometric quotients. Usually they do not exist, but we identify some normality and seminormality properties of stratified spaces that ensure the existence of quotients (9.21).

The finiteness of equivalence relations is studied in Section 9.4.

The seminormality of subvarieties of geometric quotients is investigated in Section 9.2.

In Section 9.3 we give necessary and sufficient conditions for a line bundle to descend to a geometric quotient. This makes it possible to decide when such a quotient is projective.

Assumptions The quotient theorems in this chapter hold for quasi-excellent algebraic spaces in characteristic 0. See (9.5) for remarks about schemes.

We do not study the positive characteristic case for two reasons: there are much stronger quotient theorems (9.6) and the assumptions (HN) and (HSN) are not known to hold. Our method crucially relies on the descent of normality of subschemes by finite, generically étale morphisms (10.26) which fails in positive characteristic (10.28).

9.1 Quotients by finite equivalence relations

In this section we study finite set theoretic equivalence relations and the existence of the corresponding geometric quotients. Here we focus on the abstract theory, but the definitions and theorems are all set up to work well for slc pairs and log centers on them. The resulting theory may seem artificial at first, but it works very well for our purposes. In addition, these conditions seem to define the largest reasonable and flexible class where geometric quotients exist.

The applications to slc pairs and log centers were given in Chapter 5.

Finite equivalence relations

Definition 9.1 Let X and R be S-schemes. A pair of morphisms $\sigma_1, \sigma_2 \colon R \rightrightarrows X$, or equivalently a morphism $\sigma \colon R \to X \times_S X$ is called a *pre-relation*. A pre-relation is called *finite* (resp. *quasi-finite*) if the σ_i are both finite (resp. quasi-finite) and a *relation* if σ is a closed embedding.

To any finite pre-relation $\sigma \colon R \to X \times_S X$ one can associate a finite relation $i \colon \sigma(R) \hookrightarrow X \times_S X$. For the purposes of this section, there is no substantial difference between $\sigma \colon R \to X \times_S X$ and $i \colon \sigma(R) \hookrightarrow X \times_S X$. (By contrast, a key idea of Abramovich and Hassett (2011) is to exploit this difference using stacks.)

Definition 9.2 (Set theoretic equivalence relations) Let X and R be reduced S-schemes. We say that a morphism $\sigma \colon R \to X \times_S X$ is a *set theoretic equivalence relation* on X if, for every geometric point $\operatorname{Spec} K \to S$, we get an equivalence relation on K-points

$$\sigma(K) \colon \operatorname{Mor}_S(\operatorname{Spec} K, R) \hookrightarrow \operatorname{Mor}_S(\operatorname{Spec} K, X) \times \operatorname{Mor}_S(\operatorname{Spec} K, X).$$

Equivalently,

(1) σ is geometrically injective.
(2) (reflexive) R contains the diagonal Δ_X.
(3) (symmetric) There is an involution τ_R on R such that $\tau_{X \times X} \circ \sigma \circ \tau_R = \sigma$ where $\tau_{X \times X}$ denotes the involution which interchanges the two factors of $X \times X$.
(4) (transitive) For $1 \le i < j \le 3$ set $X_i := X$ and let $R_{ij} := R$ when it maps to $X_i \times_S X_j$. Then the coordinate projection of $\operatorname{red}(R_{12} \times_{X_2} R_{23})$ to $X_1 \times_S X_3$ factors through R_{13}:

$$\operatorname{red}(R_{12} \times_{X_2} R_{23}) \to R_{13} \xrightarrow{\pi_{13}} X_1 \times_S X_3.$$

Note that the fiber product need not be reduced, and taking its reduced structure is essential.

9.3 (Equivalence closure) Let $R \hookrightarrow Y \times Y$ be a finite relation, R reduced. There is a smallest set theoretic equivalence relation generated by R which is constructed as follows.

First we have to add the diagonal of $Y \times Y$ to R and make R symmetric with respect to the interchange of the two factors. Then we have $R^1 \hookrightarrow Y \times Y$ which is reflexive and symmetric.

Achieving transitivity may be an infinite process. Assume that we have already constructed $R^i \hookrightarrow Y \times Y$ with projections $\sigma_1^i, \sigma_2^i \colon R^i \to Y$. R^{i+1} is obtained by replacing R^i by the image

$$R^{i+1} := \big(\sigma_1^i \circ \pi_1^i, \sigma_2^i \circ \pi_2^i\big)(R^i \times_Y R^i) \subset Y \times Y, \qquad (9.3.1)$$

where the maps are defined by the following diagram.

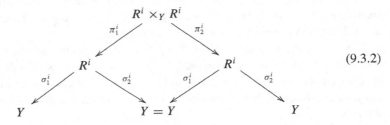

$$(9.3.2)$$

At the end we obtain a countable union of reduced subschemes

$$R \subset R^1 \subset R^2 \subset \cdots \subset Y \times Y$$

and finite projections $\sigma_1^j, \sigma_2^j \colon R^j \rightrightarrows Y$. In general we obtain *pro-finite* set theoretic equivalence relations.

Definition 9.4 (Geometric quotients)   Let $\sigma_1, \sigma_2 \colon R \rightrightarrows X$ be a set theoretic equivalence relation. We say that $q \colon X \to Y$ is a *categorical quotient* of X by R (or the *coequalizer* of $\sigma_1, \sigma_2 \colon R \rightrightarrows X$) if

(1) $q \circ \sigma_1 = q \circ \sigma_2$, and
(2) $q \colon X \to Y$ is universal with this property. That is, given any $q' \colon X \to Y'$ such that $q' \circ \sigma_1 = q' \circ \sigma_2$, there is a unique $\pi \colon Y \to Y'$ such that $q' = \pi \circ q$.

If $\sigma_1, \sigma_2 \colon R \rightrightarrows X$ is finite, we say that $q \colon X \to Y$ is a *geometric quotient* of X by R if, in addition,

(3) $q \colon X \to Y$ is finite and
(4) for every geometric point $\operatorname{Spec} K \to S$, the fibers of $q_K \colon X_K(K) \to Y_K(K)$ are the $\sigma(R_K(K))$-equivalence classes of $X_K(K)$. The last condition is equivalent to the equality $R = \operatorname{red}(X \times_Y X)$.

Somewhat sloppily, we refer to the last property by saying that "the geometric fibers of q are the R-equivalence classes."

The geometric quotient is denoted by X/R.

It is not hard to see (Kollár, 20012c, 17) that the assumptions (1–3) imply (4), but in our applications we check (4) directly.

Even in very simple situations, it can happen that X is projective and X/R is only proper or that X is proper but X/R exists only as an algebraic space (Kollár, 20012c). To handle these complications we need to choose an appropriate category to work in.

Condition 9.5 For the rest of the section, we work in one of the following categories.

(1) Schemes of finite type over a base scheme such that every finite subscheme is contained in an open affine subscheme. We call the latter the Chevalley–Kleiman property (9.28).
(2) Algebraic spaces of finite type over a base scheme.
(3) Quasi-excellent schemes over a base scheme over $\mathbb{Q}$ such that every finite subscheme is contained in an open affine subscheme.
(4) Quasi-excellent algebraic spaces over a base scheme over $\mathbb{Q}$.

For more details on the choice of these classes see (9.28–9.31) or Kollár (20012c).

It turns out that geometric quotients by finite set theoretic equivalence relations rarely exist in characteristic 0. By contrast, geometric quotients usually exist in positive characteristic.

Theorem 9.6 (Kollár, 20012c) *Let S be a Noetherian $\mathbb{F}_p$-scheme and X an algebraic space that is essentially of finite type over S. Let $R \rightrightarrows X$ be a finite, set theoretic equivalence relation. Then the geometric quotient X/R exists.* □

Example 9.7 (cf. Holmann, 1963, p. 342 or Kollár, 20012c, examples 9–11) Set $Y_1 \simeq Y_2 := \mathbb{A}^3_{xyz}$, $Z := \mathbb{A}^2_{uv}$ and let $p_i \colon Z \to Y_i$ be given by $p_1(u, v) = (u, v^2, v^3)$ and $p_2(u, v) = (u + v, v^2, v^3)$. By a direct computation, in characteristic zero, the universal push-out of

$$
\begin{array}{ccc}
Z & \longrightarrow & Y_1 \\
\downarrow & & \\
Y_2 & &
\end{array}
$$

is given by the spectrum of the ring

$$B := k + (y_1, z_1) + (y_2, z_2) \subset k[x_1, y_1, z_1] + k[x_2, y_2, z_2],$$

where (y_i, z_i) denotes the ideal $(y_i, z_i) \subset k[x_i, y_i, z_i]$. The problem is that (y_i, z_i) is *not* finitely generated as an B-ideal. A minimal generating set of B

is given by

$$y_1 x_1^m, z_1 x_1^m, y_2 x_2^m, z_2 x_2^m \ : \ m = 0, 1, 2, \dots$$

In particular, we see that the natural maps $Y_i \to \operatorname{Spec} B$ are not finite; the x-axis in Y_i is mapped to a point in $\operatorname{Spec} B$.

We can construct an equivalence relation on $Y_1 \amalg Y_2$ by letting R be the union of the diagonal with two copies of Z, one of which maps as

$$(p_1, p_2)\colon Z \to Y_1 \times Y_2 \subset (Y_1 \amalg Y_2) \times (Y_1 \amalg Y_2),$$

and the other as its symmetric pair. The categorical quotient $((Y_1 \amalg Y_2)/R)^{cat}$ equals the universal push-out $\operatorname{Spec} B$.

Thus the geometric quotient $(Y_1 \amalg Y_2)/R$ does not exist.

Basic properties

Elementary properties of seminormal schemes (10.16) imply that geometric quotients are even simpler for seminormal spaces.

Lemma 9.8 *Let $R \rightrightarrows X$ be a set theoretic equivalence relation.*

(1) *If X is seminormal then so is X/R.*
(2) *Let $q\colon X \to Y$ be a morphism that satisfies conditions (1), (3) and (4) of (9.4). If Y is seminormal and the characteristic is 0 then it also satisfies (2), hence $Y = X/R$.* $\square$

Definition 9.9 Let $R \rightrightarrows X$ be a finite relation and $g\colon Y \to X$ a finite (resp. quasi-finite) morphism. Then

$$g^* R := R \times_{(X \times X)} (Y \times Y) \rightrightarrows Y$$

defines a finite (resp. quasi-finite) relation on Y. It is called the *pull-back* of $R \rightrightarrows X$. (Strictly speaking, it should be denoted by $(g \times g)^* R$.) When Y is a locally closed subscheme of X, we also call it the *restriction* of R to Y and denote it by $R|_Y$. Usually $R|_Y$ is only quasi-finite but if Y is R-invariant then $R|_Y$ is finite.

Note that if R is a set theoretic equivalence relation then so is $g^* R$ and the $g^* R$-equivalence classes on the geometric points of Y map injectively to the R-equivalence classes on the geometric points of X.

If X/R exists then, by (9.10), $Y/g^* R$ also exists and the natural morphism $j_Y\colon Y/g^* R \to X/R$ is injective on geometric points. (We study in Section 9.2 the case when j_Y is an embedding.) If, in addition, g is surjective then $Y/g^* R \to X/R$ is finite and an isomorphism on geometric points. Thus, if X is seminormal and the characteristic is 0, then $Y/g^* R \simeq X/R$.

Let $h\colon X \to Z$ be a finite morphism and R a finite relation. Then the composite $R \rightrightarrows X \to Z$ defines a finite pre-relation. If, in addition, R is a set theoretic equivalence relation and the geometric fibers of h are subsets of R-equivalence classes, then $R \rightrightarrows X \to Z$ corresponds to a set theoretic equivalence relation

$$h_* R := (h \times h)(R) \subset Z \times Z,$$

called the *push-forward* of $R \rightrightarrows X$. If $Z/h_* R$ exists, then, by (9.10), X/R also exists and the natural morphism $X/R \to Z/h_* R$ is finite and an isomorphism on geometric points.

The next three lemmas describe cases when the construction of the geometric quotient is easy.

Lemma 9.10 *Let $\sigma_1, \sigma_2 \colon R \rightrightarrows X$ be a finite, set theoretic equivalence relation and assume that there is a finite morphism $q' \colon X \to Y'$ such that $q' \circ \sigma_1 = q' \circ \sigma_2$. Set*

$$\mathscr{O}_Y := \ker \left[q'_* \mathscr{O}_X \xrightarrow{\sigma_1^* - \sigma_2^*} (q' \circ \sigma_i)_* \mathscr{O}_R \right].$$

Then $Y = X/R$.

Proof Y clearly satisfies the assumptions (9.4.1–3) and the geometric fibers of $X \to Y$ are finite unions of R-equivalence classes. As we noted above, by Kollár (20012c, 17), Y also satisfies the assumption (9.4.4). $\qquad\square$

Since push-forward commutes with flat base change, we conclude from (9.10) that the geometric quotient commutes with flat base change:

Corollary 9.11 *Let $R \to X \times_S X$ be a finite, set theoretic equivalence relation such that X/R exists. Let $h \colon Y \to X/R$ be a flat morphism. Then*

$$Y \times_{X/R} R \times_{X/R} Y \to (Y \times_{X/R} X) \times_S (X \times_{X/R} Y)$$

is a finite, set theoretic equivalence relation and the corresponding geometric quotient is Y. In particular, the construction of a geometric quotient is local in the Zariski topology. $\qquad\square$

Lemma 9.12 *Let $R \rightrightarrows X$ be a finite, set theoretic equivalence relation with X, R reduced and over a field of characteristic 0. Let $\pi \colon X' \to X$ and $q' \colon X' \to Z$ be finite surjections. Assume that one of the following holds:*

(1) *X, Z are seminormal and the geometric fibers of q' are exactly the preimages of R-equivalence classes, or*

(2) *Z, X are normal, the $\sigma_i \colon R \to X$ are open and, over a dense open subset of Z, the geometric fibers of q' are exactly the preimages of R-equivalence classes.*

Then $Z = X/R$.

Proof Let $X^* \subset Z \times X$ be the image of X' under the diagonal map (q', π).

In the first case, every geometric fiber of π is contained in a geometric fiber of q', thus we see that the projection $X^* \to X$ is one-to-one on geometric points. Since X is seminormal, this implies that $X^* \simeq X$. Thus we get a morphism $q: X \to Z$ whose geometric fibers are exactly the R-equivalence classes.

Therefore, $q \circ \sigma_1$ agrees with $q \circ \sigma_2$ on geometric points. Since R is reduced, this implies that $q \circ \sigma_1 = q \circ \sigma_2$. Define $p: Y \to Z$ as in (9.10). Since X is reduced, so is Y. The geometric fibers of $X \to Y$ are finite unions of R-equivalence classes. On the other hand, every geometric fiber of $X \to Y$ is contained in a geometric fiber of $X \to Z$ which is a single R-equivalence class. Thus $X \to Z$ and $X \to Y$ have the same fibers, hence $Y \to Z$ is an isomorphism on geometric points. Since Z is seminormal, this implies that $Y \simeq Z$.

In the second case, the same argument gives that $X^* \to X$ is birational. Since it is also finite, $X^* \simeq X$ since the latter is normal. We know that $q \circ \sigma_1 = q \circ \sigma_2$ holds over a dense open subset of X, hence over a dense open subset of R. Thus $q \circ \sigma_1 = q \circ \sigma_2$ everywhere. Construct $p: Y \to Z$ as before. Here p is birational and finite, hence an isomorphism since X is normal. $\square$

Corollary 9.13 *Let $R \rightrightarrows X$ be a finite, set theoretic equivalence relation with X, R reduced and $Y \subset X$ a closed, reduced subscheme. Assume that R is the identity on $X \setminus Y$ and the geometric quotient $Y/(R|_Y)$ exists. Then X/R is given by the universal push-out diagram (9.30)*

$$
\begin{array}{ccc}
Y & \lhook\joinrel\longrightarrow & X \\
\downarrow & & \downarrow{\scriptstyle \pi} \\
Y/(R|_Y) & \longrightarrow & X/R
\end{array}
$$

In particular, every irreducible component of the normalization of X/R is either an irreducible component of the normalization of X or of the normalization of $Y/(R|_Y)$.

Furthermore, if X is in one of the classes (9.5) then so is X/R.

Proof Note that the geometric fibers of π are exactly the R equivalence classes. On Y this holds since $Y \to Y/(R|_Y)$ is the geometric quotient and π is an isomorphism on $X \setminus Y$. By the universality of the push-out Y equals X/R. The last claim follows from (9.31). $\square$

Lemma 9.14 *Let X be a normal scheme of pure dimension d. Let $R \rightrightarrows X$ be a finite, set theoretic equivalence relation. Let $R(d) \subset R$ denote the purely d-dimensional part of R.*

(1) *$R(d) \rightrightarrows X$ is a finite, set theoretic equivalence relation (Białynicki-Birula, 2004, 2.7).*

(2) *If X is in one of the classes (9.5) then $X/R(d)$ exists and it is in the same class.*

(3) $X/R(d)$ *is normal.*

Proof Let us prove first that $R(d) \rightrightarrows X$ is a set theoretic equivalence relation. The only question is transitivity. (Easy examples show that transitivity can fail if X is not normal (Kollár, 20012c, 29).) Note that $\sigma_i \colon R(d) \to X$ is finite with normal target. Hence, by (1.44), $R(d) \times_X R(d) \to R(d)$ is open. In particular, $R(d) \times_X R(d)$ has pure dimension d. Thus the image of the finite morphism $R(d) \times_X R(d) \to R$ in (9.2.3) lies in $R(d)$. Therefore $R(d) \rightrightarrows X$ is a set theoretic equivalence relation. It is then necessarily finite.

Next assume that $X/R(d)$ exists and let $Y \to X/R(d)$ be the normalization. Since X is normal, the quotient morphism $X \to X/R(d)$ lifts to $\tau \colon X \to Y$. Thus $\tau \circ \sigma_1 = \tau \circ \sigma_2$ on a dense open set, hence equality holds everywhere. By the universal property of geometric quotients (9.4.2), $X/R(d) = Y$ is normal.

The existence of $X/R(d)$ is shown in (9.33). $\qquad\square$

Stratified equivalence relations

We saw in (9.14) that pure dimensional equivalence relations behave well on normal schemes. Our next aim is to show that $R \rightrightarrows X$ is still well behaved if X and R can be decomposed into normal and pure dimensional pieces and some strong seminormality assumptions hold about the closures of the strata. To do this, we need the concept of a stratification.

Definition 9.15 Let X be a scheme. A *stratification* of X is a decomposition of X into a finite disjoint union of reduced and locally closed subschemes. We will deal with stratifications where the strata are pure dimensional and indexed by the dimension. Then we write $X = \cup_i S_i X$ where $S_i X \subset X$ is the i-dimensional stratum. Such a stratified scheme is denoted by (X, S_*). We also assume that $\cup_{i \le j} S_i X$ is closed for every j.

The *boundary* of (X, S_*) is the closed subscheme

$$BX := \cup_{i < \dim X} S_i X = X \setminus S_{\dim X} X$$

and $S_{\dim X} X$ is the *open stratum*. (Thus, if $X_j \subset X$ is an irreducible component such that $\dim X_j < \dim X$ then $X_j \subset BX$.)

Let (X, S_*) and (Y, S_*) be stratified schemes. We say that $f \colon X \to Y$ is a *stratified morphism* if $f(S_i X) \subset S_i Y$ for every i. Equivalently, if $S_i X = f^{-1}(S_i Y)$ for every i.

Let (Y, S_*) be a stratified scheme and $f \colon X \to Y$ a quasi-finite morphism such that $f^{-1}(S_i Y)$ has pure dimension i for every i. Then $S_i X := f^{-1}(S_i Y)$ defines a stratification of X, denoted by $(X, f^{-1} S_*)$. We say that $f \colon X \to (Y, S_*)$ is *stratifiable*.

Let (X, S_*) be a stratified scheme and $f: X \to Y$ a surjective, quasi-finite morphism such that $f^{-1}(f(S_i X)) = S_i X$ for every i. Then $S_i Y := f(S_i Y)$ defines a stratification of Y, denoted by $(Y, f_*(S_*))$. We say that $f: (X, S_*) \to Y$ is *stratifiable*.

Let (X, S_*) be a stratified scheme such that X is seminormal and $X \setminus B(X)$ is normal. Let $X^n \to X$ be the normalization and $B(X^n) \subset X^n$ the reduced preimage of $B(X)$. We frequently use the following universal push-out diagram

$$\begin{array}{ccc} B(X^n) & \hookrightarrow & X^n \\ \downarrow & & \downarrow \pi \\ B(X) & \hookrightarrow & X. \end{array} \qquad (9.15.1)$$

Definition 9.16 Let (X, S_*) be stratified. A relation $\sigma_i: R \rightrightarrows (X, S_*)$ is called *stratified* if each σ_i is stratifiable and $\sigma_1^{-1} S_* = \sigma_2^{-1} S_*$. Equivalently, there is a stratification $(R, \sigma^{-1} S_*)$ such that

$$r \in \sigma^{-1} S_i R \iff \sigma_1(r) \in S_i X \iff \sigma_2(r) \in S_i X.$$

(Note that this is much stronger than just assuming that $\sigma_1(r) \in S_i X \iff \sigma_2(r) \in S_i X$ for every $r \in R$, as shown by the relation $g^* R_C$ below.)

Let $\sigma_i: R \rightrightarrows (X, S_*)$ be a stratified set theoretic equivalence relation and $f: (X, S_*) \to Y$ a stratifiable morphism. If the geometric fibers of f are subsets of R-equivalence classes then the push-forward (9.9)

$$f_* R \rightrightarrows (Y, f_*(S_*))$$

is also a stratified set theoretic equivalence relation.

By contrast, the pull-back of a stratified relation by a stratified morphism is not always stratified. (Sufficient conditions are given in (9.17).) As an example, let C be a nodal curve with $S_1 = C$, $R_C \rightrightarrows C$ the identity relation and $g: \bar{C} \to C$ the normalization. Then $(g^{-1} S)_1 = \bar{C}$ but $g^* R_C$ has three components. Besides the diagonal, it has 0 dimensional components showing that the two preimages of the node are equivalent.

Lemma 9.17 *Assume that the strata of (Y, S_*) are all unibranch* (1.44).

(1) *Let $f_i: X_i \to (Y, S_*)$ be stratifiable quasi-finite morphisms. Then the induced maps $X_1 \times_Y X_2 \rightrightarrows X_i \rightrightarrows Y$ are all stratifiable.*

(2) *Let $f: X \to (Y, S_*)$ be a stratifiable, finite surjection. Then $R_{Y/X} := X \times_Y X \rightrightarrows X$ is a stratified equivalence relation. If Y is seminormal and characteristic is 0 then $Y = X/R_{Y/X}$.*

(3) *Let $R \rightrightarrows (Y, S_*)$ be a stratified relation. Then the pull-back $g^* R \rightrightarrows (X, g^{-1} S_*)$ by a stratified morphism g is also a stratified relation.*

(4) *Let $R \rightrightarrows (Y, S_*)$ be a stratified relation. Then its equivalence closure* (9.3) *is a stratified pro-finite relation.*

Proof The conditions need to be checked one stratum at a time. Hence we may assume that Y is unibranch and of pure dimension d.

By (1.44), the f_i are universally open and the X_i also have pure dimension d. Thus $X_i \times_Y X_2 \to Y$ is also open hence $X_i \times_Y X_2$ has pure dimension d, proving (1). This implies that $R_{Y/X}$ is a stratified equivalence relation and the rest of (2) follows from (9.8.2).

Similarly, $g^*R := R \times_{(Y \times Y)} (X \times X) \to R$ is also open. Thus g^*R has pure dimension d and so $g^*R \rightrightarrows (X, g^{-1}S_*)$ is stratifiable.

To see (4), we need to show that all the pre-relations R^i constructed in (9.3) are stratified. By induction on i, assume that the maps $\sigma^i_j \colon R^i \rightrightarrows Y$ are stratified. By (1), the fiber products $\tau^i_j \colon R^i \times_Y R^i \rightrightarrows R^i$ are also stratified. Hence all arrows in the diagram (9.3.2) are stratified, and so the composites along the outer edges of the triangle are also stratified. Thus all the $\sigma^{i+1}_1 \colon R^{i+1} \rightrightarrows Y$ are also stratified. $\qquad\square$

The following definitions are the main results of this section. By Ambro (2003), any union of lc centers is seminormal and every lc center is normal away from lower dimensional lc centers. These are axiomatized (more or less) by the conditions (N) and (SN) below. We have proved so far only that the union of all lc centers is seminormal (4.32) and every lc center is unibranch away from lower dimensional lc centers (4.41). The former accounts for the weaker-than-expected formulation of (SN) and the latter for the unibranch version (U) of the condition (N). The proofs seem to go better with the present choices, but it would have been possible to ignore condition (U) and work with a stronger form of (SN).

The hereditary versions (HN) and (HSN) are much more delicate. It seems that they are dictated by the needs of the proofs; at least we do not know other variants that would work.

We eventually prove that (HU) and (HSN) imply (HN), in fact they imply even stronger normality and seminormality results; see (9.26).

As we noted in (4.20), it would have been possible to prove stronger properties of lc centers first and then obtain a variant of (9.21) without using hereditary lc centers.

Definition 9.18 Let X be a scheme and S_* a stratification as in (9.15). We consider three normality conditions on stratifications. We say that

(N) (X, S_*) has *normal strata* if each $S_i X$ is normal,

(U) (X, S_*) has *unibranch strata* if each $S_i X$ is unibranch,

(SN) (X, S_*) has *seminormal boundary* if X and the boundary $BX = \cup_{i < \dim X} S_i X$ are both seminormal.

We often say "(X, S_*) satisfies (N)," etc.

If the hereditary normalization $\mathbf{HN}(X, S_*)$ of (X, S_*) – defined in (9.19) – exists, we say that

(HN) (X, S_*) has *hereditarily normal strata* if $\mathbf{HN}(X, S_*)$ satisfies (N),
(HU) (X, S_*) has *hereditarily unibranch strata* if $\mathbf{HN}(X, S_*)$ satisfies (U),
(HSN) (X, S_*) has *hereditarily seminormal boundary* if $\mathbf{HN}(X, S_*)$ satisfies (SN).

Definition 9.19 (Hereditary normalization) For certain stratified spaces (X, S_*) we define inductively the *hereditary normalization* $\mathbf{HN}(X, S_*)$ as follows.

Assume that the normalization $\pi \colon (X^n, S_*^n) \to (X, S_*)$ is stratifiable and its boundary $B(X^n)$ has a hereditary normalization. Then set

$$\mathbf{HN}(X, S_*) = (X^n, S_*^n) \amalg \mathbf{HN}(B(X^n), B(S_*^n)).$$

(In order to get a correct inductive definition, we should add that $\mathbf{HN}(\emptyset) = \emptyset$.)

In practice it may be quite hard to decide if the normalization $\pi \colon X^n \to X$ is stratifiable or not and we do not know any easy criterion that guarantees the existence of the hereditary normalization. We see in (5.30) that for lc stratifications arising from crepant log structures, the hereditary normalization exists. This is the only case that we will use.

By construction, the hereditary normalization comes with a finite stratified morphism $\pi \colon \mathbf{HN}(X, S_*) \to (X, S_*)$. The image of the $(\le i)$-dimensional irreducible components of $\mathbf{HN}(X, S_*)$ is the union of all $(\le i)$-dimensional strata. The map π is birational over the open stratum but can have high degree over the lower dimensional strata.

The boundary of $\mathbf{HN}(X, S_*)$ is called the *hereditary boundary* and is denoted by $\mathbf{HB}(X, S_*)$.

Let $g \colon X \to Y$ be a finite morphism such that every irreducible component of X dominates an irreducible component of Y. Then g lifts to the normalizations $g^n \colon X^n \to Y^n$. Furthermore, if $g \colon (X, S_*^X) \to (Y, S_*^Y)$ is also stratified, then the above property holds for the boundary $B(g) \colon B(X) \to B(Y)$. Thus, for these maps we get

$$\mathbf{HN}(g) \colon \mathbf{HN}(X, S_*^X) \to \mathbf{HN}(Y, S_*^Y)$$

which is also a surjection.

Remark 9.20 Condition (N) and (HN) are quite reasonable and usually easy to satisfy.

Let X be a reduced scheme and $\sigma_1, \sigma_2 \colon R \rightrightarrows X$ a finite set theoretic equivalence relation on X. There is a unique coarsest stratification S_* that satisfies (HN) such that R is a stratified equivalence relation with respect to S_*.

To construct S_*, let $B(X) \subset X$ be the union of the equivalence class of the non-normal locus of X and of the image by σ_1 of all the irreducible components of R whose dimension is $< \dim X$. The maximal dimensional stratum will be

$X \setminus B(X)$. By induction on the dimension, the required stratification exists on $B(X^n)$. Pushing it forward to $B(X)$ we get our stratification on X.

Nonetheless, (HN) is a quite subtle condition. As an example, take

$$X = (x^2 = y^2(y + z^3)) \subset \mathbb{A}^3 \tag{9.20.1}$$

with $S_1 X = (x = y = 0)$. Then $S_1 X$ and $S_2 X$ are both smooth. The normalization of X is

$$X^n = \left(x_1^2 = y + z^3\right) \subset \mathbb{A}^3$$

where $x_1 = x/y$ and the preimage of $S_1 X$ is $(y = x_1^2 - z^3 = 0)$ which is not seminormal.

Actually, one of the trickiest parts of (HN) is to know when the normalization $\pi \colon X^n \to X$ is stratifiable. For example, let

$$X := (x = y = 0) \cup (z = t = 0) \subset \mathbb{A}^4 \tag{9.20.2}$$

with $S_1 X = (x = y = z = 0)$. As before, $S_1 X$ and $S_2 X$ are both smooth but $\pi \colon X^n \to X$ is not stratifiable since the preimage of $S_1 X$ has a 0-dimensional irreducible component.

By contrast, the conditions (SN) and (HSN) are usually impossible to satisfy since they pose restrictions on the *closures* of strata.

The following is the main theorem of this section.

Theorem 9.21 *Let (X, S_*) be a scheme or algebraic space over a field of characteristic 0 with a stratification as in (9.15). Assume that (X, S_*) satisfies the conditions (HU) and (HSN). Let $R \rightrightarrows X$ be a finite, set theoretic, stratified equivalence relation. Then*

(1) *if X is in one of the classes (9.5) then the geometric quotient X/R exists and it is in the same class,*
(2) *$\pi \colon X \to X/R$ is stratifiable,*
(3) *$S_i(X)/(R|_{S_i(X)}) \to (\pi_* S_i)(X/R)$ is an isomorphism and*
(4) *$(X/R, \pi_* S_*)$ satisfies the conditions (HN) and (HSN).*

(The change from (HU) to (HN) in (4) is not a typo. In particular, the identity relation shows that (HU) and (HSN) imply (HN). Even stronger consequences are proved in (9.26).)

Various special cases of the next Corollary will be established by hand during the proof of (9.21), but it is easiest to derive the general form from (9.21).

Corollary 9.22 *Let (X, S_*) and (Y, S_*) be seminormal stratified spaces with unibranch strata over a field of characteristic 0. Let $f \colon X \to Y$ be a finite, surjective, stratified morphism. If (X, S_*) satisfies the conditions (HN) and (HSN), then so does (Y, S_*).*

Proof By (9.17), $R := X \times_Y X \rightrightarrows X$ is a finite, set theoretic, stratified equivalence relation on (X, S_*). Since Y is seminormal, $Y = X/R$. Thus (9.22) follows from (9.21.)   □

9.23 *Proof of (9.21)* The proof is by induction on $\dim X$. We follow the inductive plan in Kollár (20012c, 30).

For applications it is important that the construction of X/R is obtained by repeated use of the following steps.

(1) Let $(X^n, S^n_*) \to (X, S_*)$ be the normalization of X and $R^n \rightrightarrows X^n$ the pull-back of R. By (9.17), R^n is also a finite, set theoretic, stratified equivalence relation. Then $X/R = X^n/R^n$ by (9.12).
(2) Assume that there are disjoint closed subsets $X_i \subset X$ and finite, set theoretic, equivalence relations $R_i \rightrightarrows X_i$ such that $X = X_1 \sqcup X_2$ and $R = R_1 \sqcup R_2$. Then $X/R = X_1/R_1 \sqcup X_2/R_2$.
(3) Assume that X is normal of pure dimension d and that R also has pure dimension d. Then X/R exists by (9.14). We see in (9.24) that (9.21.3) also holds in this case.
(4) Assume that R is the identity on $X \setminus B(X)$ and $B(X)/B(R)$ exists. By (9.13) X/R is given by the universal push-out diagram

$$
\begin{array}{ccc}
B(X) & \hookrightarrow & X \\
\downarrow & & \downarrow{\scriptstyle \pi} \\
B(X)/B(R) & \longrightarrow & X/R
\end{array}
$$

As we noted in (9.13), the normalization of X/R is an open and closed subscheme of the normalization of $X \sqcup (B(X)/B(R))$. Thus if (X, S_*) and $B(X)/B(R)$ satisfy (HU) and (HSN) then $(X/R, \pi_* S_*)$ also satisfies (HU) and (HSN).

Now to the actual construction.

Let $\pi \colon X^n \to X$ be the normalization. By (9.17.2), $B(X)$ is a geometric quotient of $B(X^n)$. Thus, by induction, $B(X)$ satisfies (HN). The open stratum $X \setminus B(X)$ is unibranch and seminormal, hence normal (1.44). Thus (X, S_*) satisfies (HN).

Using (1) we may replace $R \rightrightarrows X$ by $R^n \rightrightarrows X^n$.

Let $X^n(d) \subset X^n$ (resp. $R^n(d) \subset R^n$) be the union of all $\dim X$-dimensional irreducible components. By (9.14) $R^n(d)$ is a finite, set theoretic, stratified equivalence relation on $X^n(d)$, the geometric quotient $X^n(d)/R^n(d)$ is normal and the quotient map $X^n(d) \to X^n(d)/R^n(d)$ is stratifiable. We check in (9.24) that the push-forward of $R^n|_{X^n(d)}$ to $X^n(d)/R^n(d)$ satisfies the conditions (HN) and (HSN). This establishes (9.21.3).

Let $X^n(<d)$ be the union of all lower dimensional irreducible components of X^n. By a slight abuse of notation, we can view $R^n(d)$ as an equivalence relation on X^n which is the identity on $X^n(<d)$. Thus

$$X^n/R^n(d) = (X^n(d)/R^n(d)) \amalg X^n(<d).$$

Let $q: X^n \to X^n/R^n(d)$ denote the quotient map. Then $q_* S_*^n$ is a stratification which agrees with the push-forward of $R^n|_{X^n(d)}$ on $X^n(d)/R^n(d)$ and with $R^n|_{X^n(<d)}$ on $X^n(<d)$. Thus $(X^n/R^n(d), q_* S_*^n)$ also satisfies the conditions (HN) and (HSN).

Furthermore, R^n descends to a stratified equivalence relation $q_* R^n$ on $X^n/R^n(d)$ which is the identity outside the boundary

$$B(X^n/R^n(d)) = B(X^n(d)/R^n(d)) \amalg X^n(<d).$$

By induction on the dimension, the geometric quotient of $B(X^n/R^n(d))$ by the restriction of $q_* R^n$ exists. Let us denote it by $B(X^n/R^n(d))/q_* R^n$.

We are now in a position to use (4). By (9.30) we get a universal push-out diagram

$$
\begin{array}{ccc}
B(X^n(d)/R^n(d)) \amalg X^n(<d) = B(X^n/R^n(d)) & \hookrightarrow & X^n/R^n(d) \\
\downarrow & & \downarrow \\
B(X^n/R^n(d))/q_* R^n & \longrightarrow & Y.
\end{array}
$$

Observe that $Y = X^n/R^n$ by (9.13) and, as we noted in (9.9), $X/R = X^n/R^n$.

The open stratum of X/R is also the open stratum of $X^n/R^n(d)$ which is normal. The lower dimensional strata of X/R are also strata of $B(X^n/R^n(d))/q_* R^n$. These are normal by induction. Thus $(X/R, \pi_* S_*)$ satisfies condition (N). We have seen that both $Y = X/R$ and its boundary $B(X^n/R^n(d))/q_* R^n$ are seminormal. Thus $(X/R, \pi_* S_*)$ also satisfies condition (SN).

Note that $X^n/R^n(d)$ is normal and $X^n/R^n(d) \to Y = X/R$ is an isomorphism at all d-dimensional generic points. Hence the normalization of X/R is an open and closed subscheme of $X^n/R^n(d)$. We have seen during the proof that $(X^n/R^n(d), q_* S_*^n)$ satisfies the conditions (HN) and (HSN), hence the same holds for the normalization of X/R. Together with the previous comments, these show that X/R satisfies the conditions (HN) and (HSN). $\qquad\square$

Lemma 9.24 *Let (X, S_*) and (Y, S_*) be normal stratified spaces over a field of characteristic 0 and $f: X \to Y$ a finite stratified morphism. If (X, S_*) satisfies one of the conditions (N), (SN), (HN), (HSN) then so does (Y, S_*).*

Proof (N) and (HN) follow from (10.26) and the hereditary cases are obtained by induction using the maps $B(X)^n \to B(Y)^n$. $\qquad\square$

Next we show that if (X, S_*) satisfies (HN) and (HSN) then every stratified subscheme of X is seminormal.

Definition 9.25 Let (X, S_*) be a stratified scheme. Following (9.15) a subscheme $j: Z \hookrightarrow X$ is called *stratified* if j is a stratifiable morphism. Equivalently, if Z is the union of some of the irreducible components of the strata $S_i X$.

We say that (X, S_*) satisfies the *stratified closure property* if for every irreducible component $W \subset S_i X$, the injection of its closure $j: \bar{W} \to X$ is stratified. (The example in (9.20.2) does not have the stratified closure property.)

Theorem 9.26 *Assume that (X, S_*) satisfies (HU) and (HSN). Then (X, S_*) satisfies the stratified closure property and every stratified subscheme is seminormal.*

Proof The proof is by induction on the dimension. By (9.21) we know that (X, S_*) also satisfies (HN).

Let $W \subset X$ be an irreducible component of a stratum. Let $\pi: X^n \to X$ be the normalization and set $W^n := \operatorname{red} \pi^{-1}(W)$.

If $\dim W = \dim X$ then the closure $\bar{W}^n$ of W^n is an irreducible component of X^n, hence a union of irreducible components of strata. Thus $\bar{W} = \pi(\bar{W}^n)$ is also a union of irreducible components of strata.

If $\dim W < \dim X$ then $W^n \subset B(\bar{X}^n)$. By induction on the dimension, $\bar{W}^n$ is a union of irreducible components of strata, and so is its image $\bar{W}$. Thus (X, S_*) satisfies the stratified closure property.

To see the second assertion, let $Z \subset X$ be a closed, reduced subscheme that is a union of irreducible components of strata.

Let $\pi: X^n \to X$ be the normalization with irreducible components X_i^n and set $Z^n := \operatorname{red} \pi^{-1}(Z)$. Then Z^n is also a union of irreducible components of strata. Note that an intersection $X_i^n \cap (Z^n \cup B(X^n))$ either equals $X_i^n \cap B(X^n)$ (which is seminormal by induction on the dimension) or equals X_i^n (which is normal, hence seminormal). Thus $Z^n \cup B(X^n)$ is seminormal.

$Z \cap B(X)$ is also a union of irreducible components of strata, hence seminormal by induction on the dimension.

Using the universal push-out diagram (9.15.1) and (10.29), these two imply that Z is seminormal. $\square$

Corollary 9.27 *Assume that (X, S_*) satisfies (HN) and (HSN) and let $j: Z \hookrightarrow X$ be a stratified subscheme. Then $(Z, j^* S_*)$ also satisfies (HN) and (HSN).*

Proof We use induction on the dimension of X. Let $j^n: Z^n \hookrightarrow X^n$ be the preimage of Z in the normalization X^n of X. Then Z^n is the disjoint union of some of the irreducible components X_i^n of X^n (these satisfy (HN) and (HSN)) and of a stratified subscheme $Z' \subset B(X^n)$ (this satisfies (HN) and (HSN) by induction). Thus $(Z^n, (j^n)^*S^n_*)$ satisfies (HN) and (HSN).

Z is seminormal by (9.26), thus $Z = Z^n/(R^n|_{Z^n})$. Thus (Z, j^*S_*) also satisfies (HN) and (HSN) by (9.22). $\qquad\square$

General quotient theorems

Here we recall some general results about quotients by finite group actions and gluing (or pinching) along subvarieties. An inconvenient feature of these operations is that the usual categories we work with (for instance, quasi-projective varieties or schemes) are not well adapted to these constructions. There are two reasonable ways to deal with this problem.

Aiming at full generality, one can work with algebraic spaces Knutson (1971). The only problem is that many references are missing or are rather vague.

To be conservative, one can work with a good subclass of schemes. One such subclass is given by the following.

Definition 9.28 We say that a scheme (or an algebraic space) X has the *Chevalley–Kleiman property* or that X is *CK* if every finite subscheme is contained in an open affine subscheme. In particular, X is necessarily a scheme.

These schemes appeared in Kleiman (1966) and were later studied in Ferrand (2003) and Kollár (20012c).

9.29 (Quotients by finite group actions)   Quotients by finite group actions are discussed in many places. The quasi-projective case is quite elementary; see, for instance Shafarevich (1974, section I.2.3) or the more advanced Mumford (1970, section 12). For general schemes and algebraic spaces, the quotients are constructed in some unpublished notes of Deligne. See Knutson (1971, IV.1.8) for a detailed discussion of this method.

The reduction to the affine case is especially easy if X is CK. To see this, let $P \subset X$ be a finite subscheme. If a finite group G acts on X, then the orbit $G \cdot P$ is still finite, hence, by assumption, it is contained in an open, affine subscheme $U_1 \subset X$. Then $U := \cap_{g \in G} g(U_1)$ is an open, affine, G-invariant subscheme which contains $G \cdot P$. Thus $U/G = \operatorname{Spec} k[U]^G$ exists and commutes with localization. So X/G exists and it is also CK.

In general, if X is of finite type over S then X/G exists, it is of finite type over S and $X \to X/G$ is finite.

If X is not of finite type, the same results hold as long as $|G|$ is invertible on X. See Kollár (20012c) for a summary of the results, proofs and examples.

About gluing (or pinching) along subvarieties we have the following theorem
of Artin (1970), Ferrand (2003) and Raoult (1974). For elementary proofs see
Kollár (20012c, section 6).

Theorem 9.30 *Let X be an algebraic space, $Z \subset X$ a closed subspace and
$g\colon Z \to V$ a finite morphism. Then there is a universal push-out diagram of
algebraic spaces*

$$
\begin{array}{ccc}
Z & \lhook\joinrel\longrightarrow & X \\
\scriptstyle g \downarrow & & \downarrow \scriptstyle \pi \\
V & \lhook\joinrel\longrightarrow & Y := X/(Z \to V).
\end{array}
$$

*Furthermore, π is finite, $V \to Y$ is a closed embedding, $Z = \pi^{-1}(V)$ and the
natural map between the ideal sheaves $\mathscr{I}_{V \subset Y} \to \pi_* \mathscr{I}_{Z \subset X}$ is an isomorphism.*
                                                                              □

It is also useful to know which properties of X, Z, V descend to Y. More
generally, it is interesting to know which properties are inherited by images of
finite morphisms.

9.31 (Descent for finite morphisms) Let $f\colon X \to Y$ be a finite morphism of
schemes or algebraic spaces. There are many properties **P** such that if Y satisfies
P then so does X. One cannot expect the converse unless f is surjective, and
these implications are usually much harder to prove.

Let $f\colon X \to Y$ be a finite surjection of schemes or algebraic spaces. For any
of the following properties **P**, if X satisfies **P** then so does Y:

(1) finite type,
(2) Noetherian (Eakin–Nagata theorem, (Matsumura, 1986, theorem 3.7)),
(3) Chevalley–Kleiman (Kollár, 20012c, corollary 48),
(4) quasi-excellent (Bellaccini, 1983; Ogoma, 1983).

Some properties that do not descend for finite surjections are:

(5) projective (see, for example, Kollár (20012c, example 14)),
(6) catenary (Grothendieck, 1960, IV.7.8.4),
(7) excellent (this follows from the previous).

However, being excellent (or catenary) descends if the codimension function
has the following property:

(8) $\operatorname{codim}_{x_1} x_2 = \dim x_1 - \dim x_2$ for every $x_1 \in X$ and $x_2 \in \overline{\{x_1\}}$.

Almost group actions

The simplest equivalence relations are given by group actions. Here we consider those equivalence relations that become group actions on a finite cover.

Definition 9.32 Let Y be an irreducible variety and G a countable (discrete) group acting on Y. For $g \in G$, let $\Gamma(g) \subset Y \times Y$ be the graph of g. As a set, $\Gamma(g) = \{(y, g(y)) : y \in Y\}$. Their union $\Gamma(G) := \cup_g \Gamma(g)$ is a pro-finite set-theoretic equivalence relation on Y. Note that $\Gamma(G)$ is finitely generated (that is, it is the equivalence closure of finite relation) if and only if G is a finitely generated group.

Somewhat imprecisely, we say that a pro-finite equivalence relation $R \subset Y \times Y$ is a *group action* if $R = \Gamma(G)$ for some group G.

Let X be an irreducible variety and $R \subset X \times X$ a pro-finite set-theoretic equivalence relation. We say that R is *almost a group action* if there is an irreducible variety Y, a finite surjection $p: Y \to X$ and an (at most) countable group G acting on Y such that $\mathrm{red}(p \times p)^{-1}R = \Gamma(G)$. In this case $\mathrm{red}(p \times p)^{-1}$(diagonal of $X \times X$) $\subset \Gamma(G)$ corresponds to a finite subgroup $H \subset G$ and Y/X is Galois with $\mathrm{Gal}(Y/X) = H$. Thus an almost group action is essentially a group action modulo a finite (non-normal) subgroup.

Similarly, if X, Y are reducible, one can define the notion of R being *almost a groupoid action*. This holds if and only if every irreducible component of $\mathrm{red}(p \times p)^{-1}R$ is the graph of an isomorphism between two irreducible components of Y.

Note that not every pro-finite equivalence relation is almost a group action. First of all, $\Gamma(G)$ is pure dimensional of dimension $\dim Y$, thus if R is almost a group action then R is pure dimensional of dimension $\dim X$. We prove in (9.33) that every finite and pure dimensional equivalence relation is almost a group action. This fails for pro-finite equivalence relations, see (9.35), but the problem seems to be entirely field-theoretic.

9.33 (Finite, pure dimensional equivalence relations) Let X be a scheme of pure dimension d and $R \rightrightarrows X$ a set-theoretic equivalence relation of pure dimension d. Assume that $R \to X$ has the same degree m over every irreducible component.

Consider the m-fold product $X \times \cdots \times X$ with coordinate projections π_i. Let R_{ij} (resp. Δ_{ij}) denote the preimage of R (resp. of the diagonal) under (π_i, π_j). A geometric point of $\cap_{ij} R_{ij}$ is a sequence of geometric points $(x_1, \ldots, x_m)$ such that any 2 are R-equivalent and a geometric point of $\cap_{ij} R_{ij} \setminus \cup_{ij} \Delta_{ij}$ is a sequence $(x_1, \ldots, x_m)$ that constitutes a whole R-equivalence class. Let X' be the closure of $\cap_{ij} R_{ij} \setminus \cup_{ij} \Delta_{ij}$. Note that every $\pi_\ell: \cap_{ij} R_{ij} \to X$ is finite, hence the projections $\pi'_\ell: X' \to X$ are finite.

The symmetric group S_m acts on $X \times \cdots \times X$ by permuting the factors and this gives an S_m-action on X'. By construction, the pull-back of R to X' equals this S_m-action, at least over the dense open subset $\cap_{ij} R_{ij} \setminus \cup_{ij} \Delta_{ij}$. If X is normal then equality holds everywhere by (9.14), hence we have proved the following.

Claim 9.33.1   Let X be a normal scheme of pure dimension d and $R \rightrightarrows X$ a set-theoretic equivalence relation of pure dimension d. Assume that $R \to X$ has the same degree over every irreducible component. Then R is an almost group action. $\square$

As we discussed in (9.29), the geometric quotient X'/S_m exists under mild conditions. Over a dense open subset of X, the S_m-orbits on the geometric points of X' are exactly the R-equivalence classes. If X is normal, then, by (9.12), $X'/S_m \simeq X/R$, at least in characteristic 0. Hence the construction of X/R is reduced to the construction of X'/S_m.

Proposition 9.34   *Let X be an irreducible, normal variety and $R = \cup_{i \in I} R_i$ a set-theoretic equivalence relation of pure dimension $\dim X$. Assume that R is generated by the subrelation $R_J := \cup_{i \in J} R_i$ for some subset $J \subset I$. The following are equivalent.*

(1) *R is almost a group action.*
(2) *There is a field K such that for every $i \in I$ there are embeddings j_i: $k(R_i) \hookrightarrow K$ where the composites*

$$k(X) \xrightarrow{\pi_1^*} k(R_i) \xrightarrow{j_i} K \quad and \quad k(X) \xrightarrow{\pi_2^*} k(R_i) \xrightarrow{j_i} K$$

are finite degree Galois extensions.
(3) *There is a field K such that for every $i \in J$ there are embeddings j_i: $k(R_i) \hookrightarrow K$ as above.*

Proof   Assume that $p: Y \to X$ and the group G show that R is almost a group action. We noted above that $K := k(Y)$ is Galois over $k(X)$. Thus $(1) \Rightarrow (2) \Rightarrow (3)$.

Assume (3) and let Y be the normalization of X in K. Since $K/k(X)$ is Galois, $K \otimes_{k(X)} k(R_i)$ is a direct sum of copies of K for both inclusions $\pi_j^*: k(X) \hookrightarrow k(R_i)$. Thus the irreducible components of $\mathrm{red}(p \times p)^{-1}(R_i)$ have degree 1 over Y for both projections. They are also finite, hence graphs of automorphisms. Thus the equivalence relation they generate is a group action. $\square$

Example 9.35   Let $R \subset \mathbb{P}^1 \times \mathbb{P}^1$ be the equivalence relation generated by the graph of $(x : y) \mapsto (x^2 : y^2)$ and by any curve $C \subset \mathbb{P}^1 \times \mathbb{P}^1$ of geometric genus at least 2.

We claim that R is not almost a group action; it is not even a subrelation of an almost group action.

Assume to the contrary that there is a finite morphism $p\colon D \to \mathbb{P}^1$ and a group G acting on D such that $(p \times p)^{-1}R \subset \Gamma(G)$. Note first that $(x : y) \mapsto (x^2 : y^2)$ generates a pro-algebraic equivalence relation most of whose equivalence classes are infinite. Thus the group G has to be infinite. On the other hand, one of the components of $\Gamma(G)$ dominates C, hence $g(D) \geq g(C) \geq 2$. Thus $\mathrm{Aut}(D)$ is finite and so is G, a contradiction.

9.2 Descending seminormality of subschemes

Assumption 9.36 In this section, all schemes are over a field of characteristic 0. All the results fail in positive characteristic, even for separable morphisms, see (10.28).

Let (X, S_*) be a stratified space with a stratified equivalence relation $R \rightrightarrows X$. Let $j\colon Z \hookrightarrow X$ be a reduced R-invariant subscheme. By (9.10), Z/j^*R exists and the natural map $j_{Z/R}\colon Z/j^*R \to X/R$ is a finite monomorphism. Thus Z/j^*R is the seminormalization of the image $j_{Z/R}(Z/j^*R)$. We would like to understand when $j_{Z/R}$ is a closed embedding, that is, when $j_{Z/R}(Z/j^*R)$ is seminormal.

Example 9.37 Let Y be the reducible surface $(xy = 0) \subset \mathbb{A}^3$. Let $\pi\colon X \to Y$ be its normalization and $R := X \times_Y X$. Note that X is the disjoint union of two affine planes $\mathbb{A}^2_{xz} \amalg \mathbb{A}^2_{yz}$ and $R \rightrightarrows X$ satisfies the conditions (HN) and (HSN).

Let $j\colon Z \hookrightarrow X$ be the disjoint union of the curves $Z_x := (z = x^2) \subset \mathbb{A}^2_{xz}$ and $Z_y := (z = y^2) \subset \mathbb{A}^2_{yz}$. Then j^*R also satisfies the conditions (HN) and (HSN) and Z/j^*Z is the reducible nodal curve obtained by identifying the two origins $(0, 0) \in Z_x$ and $(0, 0) \in Z_y$.

The natural map $j_{Z/R}\colon Z/j^*Z \to X/R = Y$ is not a closed embedding. Its image is the curve $Z_x \cup Z_y \subset \mathbb{A}^3$ whose two branches are tangent to each other at the origin.

It is clear that in this example the problem stems from the tangency between Z and the boundary of X. The next definition, which is modeled on (9.18), takes care of this.

Definition 9.38 Assume that (X, S_*) satisfies the conditions (HN) and (HSN) and let $\pi\colon \mathbf{HN}(X, S_*) \to (X, S_*)$ denote the hereditary normalization (9.19). A closed, reduced subscheme $Z \subset X$ is called *hereditarily seminormal* if $\mathrm{red}\,\pi^{-1}(Z) \cup B(\mathbf{HN}(X, S_*))$ is seminormal.

We prove in (9.41) that this definition answers our question but first we summarize its basic properties.

Lemma 9.39 *Assume that (X, S_*) and (X', S'_*) satisfy (HN) and (HSN). Let $g: (X', S'_*) \to (X, S_*)$ be a finite stratified surjection and $Z \subset X$ a closed, reduced subscheme. If $\operatorname{red} g^{-1}(Z)$ is hereditarily seminormal then so is Z.*

Proof The conditions concern only the normalization of X, hence we may assume that X, X' are both normal. In this case every irreducible component of $\mathbf{HN}(X, S_*)$ is dominated by an irreducible component of $\mathbf{HN}(X', S'_*)$. Thus (9.38) follows from (10.26). $\quad\square$

Lemma 9.40 *Assume that (X, S_*) satisfies (HN) and (HSN). Then a hereditarily seminormal subscheme of X is seminormal.*

Proof Assume first that X is normal and let $Z \subset X$ be a closed, reduced, hereditarily seminormal subscheme. Then $Z \cup B(X)$ is seminormal since (X, S_*) is an irreducible component of $\mathbf{HN}(X, S_*)$. Similarly, $Z \cap B(X)$ is hereditarily seminormal since $\mathbf{HN}(B(X), S_*|_{B(X)}) \subset \mathbf{HN}(X, S_*)$. Hence $Z \cap B(X)$ is seminormal by induction on the dimension. These two imply that Z is seminormal by (10.25).

Next we use the usual push-out diagram (9.15.1). Note that $\pi: B(X^{\mathrm{n}}) \to B(X)$ is a stratified morphism and $\operatorname{red} \pi^{-1}Z \cap B(X^{\mathrm{n}})$ is hereditarily seminormal. Thus $Z \cap B(X) = \pi(\operatorname{red} \pi^{-1}Z \cap B(X^{\mathrm{n}}))$ is hereditarily seminormal by (9.39), hence seminormal by induction on the dimension. Furthermore, $\operatorname{red} \pi^{-1}Z \cup B(X^{\mathrm{n}})$ is seminormal by assumption, hence Z is seminormal by (10.29). $\quad\square$

Proposition 9.41 *Assume that (X, S_*) satisfies (HN) and (HSN). Let $j: Z \hookrightarrow X$ be a closed, reduced R-invariant hereditarily seminormal subscheme. Then $j_{Z/R}: Z/j^*R \to X/R$ is a closed embedding.*

Proof Set $W := \pi(Z) \subset X/R$. Then $Z = \operatorname{red} \pi^{-1}(W)$, hence W is hereditarily seminormal by (9.39) and so seminormal by (9.40). As we noted at the beginning, this implies that $j_{Z/R}: Z/j^*R \to W$ is an isomorphism. $\quad\square$

The following relative variant of the above results is especially important for moduli questions, see (7.8, 7.18), Section 8.8 and Kollár (2011c, corollary 3).

Definition 9.42 Assume that (X, S_*) satisfies the conditions (HN) and (HSN) and let $\pi: \mathbf{HN}(X, S_*) \to (X, S_*)$ denote the hereditary normalization (9.19) with hereditary boundary $\mathbf{HB}(X, S_*)$. We say that $Z_1 \cup \cdots \cup Z_r$ is *hereditarily seminormal* in $Z_1 \amalg \cdots \amalg Z_r$ if, in the terminology of (10.13),

$$\mathbf{HB}(X, S_*) \cup \bigcup_i \operatorname{red} \pi^{-1}(Z_i) \quad \text{is seminormal in}$$

$$\mathbf{HB}(X, S_*) \amalg \coprod_i \operatorname{red} \pi^{-1}(Z_i).$$

The next two lemmas summarize the basic properties.

Lemma 9.43 *Assume that (X, S_*) and (X', S'_*) satisfy (HN) and (HSN). Let g: $(X', S'_*) \to (X, S_*)$ be a finite stratified surjection and $Z_i \subset X$ closed, reduced subschemes. If $\cup_i$ red $g^{-1}(Z_1)$ is hereditarily seminormal in $\amalg_i$ red $g^{-1}(Z_1)$ then $\cup_i Z_i$ is hereditarily seminormal in $\amalg_i Z_i$.*

Proof Arguing as in (9.39), this follows from (10.30). □

Lemma 9.44 *Assume that (X, S_*) satisfies (HN) and (HSN). Let $Z_i \subset X$ be closed, reduced subschemes such that $\cup_i Z_i$ is hereditarily seminormal in $\amalg_i Z_i$. Then $\cup_i Z_i$ is seminormal in $\amalg_i Z_i$.*

Proof We follow the proof of (9.40). Assume first that X is normal. Then $B(X) \cup \bigcup_i Z_i$ is seminormal in $B(X) \amalg (\amalg_i Z_i)$ by assumption and we show by induction that $\cup_i$ red$(Z_i \cap B(X))$ is seminormal in $\amalg_i$ red$(Z_i \cap B(X))$. Thus $\cup_i Z_i$ is seminormal in $\amalg_i Z_i$ by (10.29) (applied with $v =$ identity).

Next we use the usual push-out diagram (9.15.1). Note that $\pi \colon B(X^n) \to B(X)$ is a stratified morphism and $\cup_i$ red$(\pi^{-1}(Z_i) \cap B(X^n))$ is hereditarily seminormal in $\amalg_i$ red$(\pi^{-1}(Z_i) \cap B(X^n))$. Thus $\cup_i$ red$(\pi^{-1}(Z_1) \cap B(X^n))$ is seminormal in $\amalg_i$ red$(\pi^{-1}(Z_1) \cap B(X^n))$ by (9.43). Therefor $\cup_i$ red$(Z_i \cap B(X))$ is seminormal in $\amalg_i$ red$(Z_i \cap B(X))$ by induction on the dimension.

Furthermore, $\cup_i$ red $\pi^{-1}(Z_i))$ is seminormal, hence $Z_1 \cup \cdots \cup Z_r$ is seminormal in $Z_1 \amalg \cdots \amalg Z_r$ by (10.29). □

Proposition 9.45 *Assume that (X, S_*) satisfies (HN) and (HSN). Let $Z_i \subset X$ be a closed, reduced R-invariant subscheme such that $Z_1 \cup \cdots \cup Z_r$ is hereditarily seminormal in $Z_1 \amalg \cdots \amalg Z_r$.*

Let $\pi : X \to X/R$ be the geometric quotient and $W_i := \pi(Z_i)$. Then $W_1 \cup \cdots \cup W_r$ is seminormal in $W_1 \amalg \cdots \amalg W_r$.

Proof $Z_i = $ red $\pi^{-1}(W_i)$, hence $W_1 \cup \cdots \cup W_r$ is hereditarily seminormal in $W_1 \amalg \cdots \amalg W_r$ by (9.43) and so $W_1 \cup \cdots \cup W_r$ is seminormal in $W_1 \amalg \cdots \amalg W_r$ by (9.44). □

9.3 Descending line bundles to geometric quotients

We investigate the following problem in one of the four categories considered in (9.5).

Question 9.46 Let $R \rightrightarrows X$ be a finite, set-theoretic equivalence relation with geometric quotient $q \colon X \to X/R$.

Let L be a line bundle on X. Under what conditions does there exist a line bundle $L_{X/R}$ on X/R such that $L \simeq q^* L_{X/R}$? How to classify these $L_{X/R}$?

The following example shows that descending line bundles to geometric quotients is rather subtle in some cases.

Example 9.47 (Triangular pillows)   Take two pairs

$$(\mathbb{P}_i^2 := \mathbb{P}^2(x_i : y_i : z_i), C_i := (x_i y_i z_i = 0)).$$

Given $c_x, c_y, c_z \in \mathbb{C}^*$ define $\phi(c_x, c_y, c_z): C_1 \simeq C_2$ by $(0 : y_1 : z_1) \mapsto (0 : y_1 : c_z z_1), (x_1 : 0 : z_1) \mapsto (c_x x_1 : 0 : z_1)$ and $(x_1 : y_1 : 0) \mapsto (x_1 : c_y y_1 : 0)$ and glue the $\mathbb{P}_i^2$ using $\phi(c_x, c_y, c_z)$ to get the surface $S(c_x, c_y, c_z)$.

We claim that $S(c_x, c_y, c_z)$ is projective if the product $c_x c_y c_z$ is a root of unity but otherwise the only line bundle on $S(c_x, c_y, c_z)$ is the trivial one.

To see this note that $\text{Pic}^0(C_i) \simeq \mathbb{C}^*$ and $\text{Pic}^r(C_i)$, the set of degree r line bundles, is a principal homogeneous space under $\mathbb{C}^*$ for every $r \in \mathbb{Z}$. We can identify $\text{Pic}^3(C_i)$ with $\mathbb{C}^*$ using the restriction of the ample generator L_i of $\text{Pic}(\mathbb{P}_i^2) \simeq \mathbb{Z}$ as the base point.

The key observation is that $\phi(c_x, c_y, c_z)^*: \text{Pic}^3(C_2) \to \text{Pic}^3(C_1)$ is multiplication by $c_x c_y c_z$. Thus if $c_x c_y c_z$ is an rth root of unity then L_1^r and L_2^r glue together to an ample line bundle but not otherwise.

We give a geometric treatment of (9.46); one can no doubt work out a cohomological version as well.

Let $p: X_L \to X$ denote the total space of the line bundle L. That is, $p_* \mathcal{O}_{X_L} = \sum_{m \geq 0} L^{-m}$. It is more convenient to use instead the complement of the zero section $Y_L \subset X_L$ with projection $p: Y_L \to X$. Note that $p: Y_L \to X$ is a $\mathbb{G}_m$-bundle (9.49) and $p_* \mathcal{O}_{Y_L} = \sum_{m \in \mathbb{Z}} L^{-m}$.

Our main result says that, up to replacing L by a suitable power, there is a one-to-one correspondence between $\mathbb{G}_m$-equivariant finite, set theoretic equivalence relations on Y_L and answers to the descent problem (9.46).

This correspondence becomes more natural if we work with Seifert $\mathbb{G}_m$-bundles instead of $\mathbb{G}_m$-bundles. The relevant definitions and proofs are given at the end of the section.

Proposition 9.48   *Let (X, S_*) be a stratified space satisfying (HN) and (HSN). Let $f: Y \to X$ be a $\mathbb{G}_m$-bundle; then $(Y, f^* S_*)$ also satisfies (HN) and (HSN). Let $R_X \rightrightarrows X$ and $R_Y \rightrightarrows Y$ be finite, set theoretic, stratified equivalence relations. Assume that $R_Y \rightrightarrows Y$ is $\mathbb{G}_m$-equivariant and $R_Y \to R_X$ is surjective. Then*

(1) *The induced map $f/R: Y/R_Y \to X/R_X$ is a Seifert $\mathbb{G}_m$-bundle and*
(2) *for sufficiently divisible m, the mth power $(Y/R_Y)^{(m)} \to X/R_X$ (9.53) is a $\mathbb{G}_m$-bundle whose pull-back to X is $Y^{(m)} \to X$.*

Proof   The geometric quotients X/R_X and Y/R_Y exist by (9.21).

It remains to prove that $f/R\colon Y/R_Y \to X/R_X$ is a Seifert $\mathbb{G}_m$-bundle. Since $Y \to X$ is affine and $X \to X/R_X$ is finite, $Y \to X/R_X$ is also affine. Since $Y \to Y/R_Y$ is finite, $Y/R_Y \to X/R_X$ is affine by Chevalley's theorem. Since $R_Y \rightrightarrows Y$ is $\mathbb{G}_m$-equivariant, the $\mathbb{G}_m$ action descends to Y/R_Y, again by the universal property of geometric quotients.

The only remaining question is about the fibers of $Y/R_Y \to X/R_X$. Pick a point $\bar{x} \in X/R_X$, let $x_i \in X$ be its preimages and $Y_i \subset Y$ the reduced Seifert fiber over x_i. Then

$$\mathrm{red}(f/R)^{-1}(\bar{x}) = (\amalg_i Y_i)/(R_Y|_{\amalg_i Y_i}).$$

Thus $\mathrm{red}(f/R)^{-1}(x)$ is a union of $\mathbb{G}_m$-orbits and it is irreducible if and only if for every i, j, every point of Y_i is R_Y-equivalent to some point of Y_j. Using the $\mathbb{G}_m$-action, this holds if and only if for every i, j, some point of Y_i is R_Y-equivalent to some point of Y_j. The latter holds if and only if $R_Y \to R_X$ is surjective. $\square$

Seifert bundles

Seifert bundles were introduced to algebraic geometry in the works of Orlik and Wagreich (1975), Dolgachev (1975) and Pinkham (1977). The main emphasis has been on the case of smooth varieties and orbifolds. We need to study Seifert bundles over non-normal bases, so we go through the basic definitions.

Notation 9.49 $\mathbb{G}_m$ denotes the multiplicative group scheme GL(1). As a scheme over Spec A, it is $\mathrm{Spec}_A A[t, t^{-1}]$. For any natural number $r > 0$, the rth roots of unity form the subgroup scheme

$$\mu_r := \mathrm{Spec}_A A[t, t^{-1}]/(t^r - 1).$$

These are all the subgroup schemes of $\mathbb{G}_m$. (Note that μ_r is nonreduced when the characteristic divides r.)

Every linear representation $\varrho\colon \mathbb{G}_m \to \mathrm{GL}(W)$ is completely reducible, and the same holds for $\mu_r \subset \mathbb{G}_m$ (see, for instance, (SGA3, 1970, I.4.7.3)). The set of vectors $\{v : \varrho(\lambda)(v) = \lambda^i v\}$ is called the λ^i-*eigenspace*. We use this terminology also for μ_r-actions. In this case i is determined modulo r.

This implies that every quasi-coherent sheaf with a $\mathbb{G}_m$-action is a direct sum of eigensubsheaves.

If a group G acts on a scheme X via $\varrho\colon G \to \mathrm{Aut}(X)$, we get an action on rational functions on X given by $f \mapsto f \circ \varrho(g^{-1})$. (The inverse is needed mostly for noncommutative groups only.)

Thus if $\mathbb{G}_m$ acts on itself by multiplication, we get an induced action on $A[t, t^{-1}]$ where $\lambda \in \mathbb{G}_m(\bar{k})$ acts as $t^i \mapsto \lambda^{-i} t^i$. Thus t^i spans the λ^{-i}-eigenspace.

A $\mathbb{G}_m$-action on an A-algebra R is equivalent to a $\mathbb{Z}$ grading $R = \sum_{i \in \mathbb{Z}} R_i$ where R_i is the λ^{-i}-eigenspace.

The natural $\mathbb{G}_m$-action on $\mathbb{G}_m / \mu_r$ corresponds to $\sum_{i \in r\mathbb{Z}} A \simeq A[t^r, t^{-r}]$.

Definition 9.50 Let X be a seminormal scheme (or algebraic space) over S. A *Seifert bundle* (or a Seifert $\mathbb{G}_m$-bundle) over X is a reduced scheme (or algebraic space) Y together with a morphism $f : Y \to X$ and a $\mathbb{G}_m$-action on Y satisfying the following conditions.

(1) f is affine and $\mathbb{G}_m$-equivariant (with the trivial $\mathbb{G}_m$-action on X).
(2) The natural map $\mathcal{O}_X \to (f_* \mathcal{O}_Y)^{\mathbb{G}_m}$ is an isomorphism.
(3) For every point $x \in X$, the $\mathbb{G}_m$-action on the reduced fiber red Y_x is isomorphic to the natural $\mathbb{G}_m$-action on $\mathbb{G}_m / \mu_{m(x)}$ for some $m(x) \in \mathbb{N}$, called the *multiplicity* of the fiber over x.

There are two points where this definition differs from the standard ones.

(4) If X is normal, one usually assumes that $m(x) = 1$ at the generic point. One can assume this without any loss of generality as long as X is irreducible, but for X reducible this is not a natural condition.
(5) If X is normal, one usually assumes that Y is also normal and one could assume in our case that Y is also seminormal. For applications, this is the natural setting, but it would make (9.51) more complicated.

One can thus view the theory of Seifert bundles as a special chapter of the study of algebraic $\mathbb{G}_m$-actions. The emphasis is, however, quite different.

Theorem 9.51 *Let X be a Noetherian seminormal scheme (or algebraic space). There is a one-to-one correspondence between*

(1) *Seifert $\mathbb{G}_m$-bundles $f : Y \to X$ and*
(2) *graded $\mathcal{O}_X$-algebras $\sum_{i \in \mathbb{Z}} L_i$ such that*
    (a) *each L_i is a torsion free, coherent sheaf on X whose rank is 1 or 0 at the generic points,*
    (b) *$L_i \otimes L_j \to L_{i+j}$ are isomorphisms at the generic points,*
    (c) *L_M is locally free for some $M > 0$ and*
    (d) *$L_i \otimes L_M \to L_{i+M}$ is an isomorphism for every i.*

Proof Let $f : Y \to X$ be a Seifert bundle. Since $f : Y \to X$ is affine, $f_* \mathcal{O}_Y$ is a quasi-coherent sheaf with a $\mathbb{G}_m$-action. Thus it decomposes as a sum of quasi-coherent $\mathbb{G}_m$-eigensubsheaves (9.49)

$$f_* \mathcal{O}_Y = \sum_{j \in \mathbb{Z}} L_j, \tag{9.51.3}$$

where L_j is the λ^{-j} eigensubsheaf, with multiplication maps $m_{ij}\colon L_i \otimes L_j \to L_{i+j}$. Note that $L_0 = \mathscr{O}_X$ by (9.50.2).

Pick any point $x \in X$. By assumption red $Y_x \simeq \mathbb{G}_m / \mu_{m(x)}$, thus $t^{-m(x)}$ on $\mathbb{G}_m$ descends to an invertible function h_x on Y_x which is a $\mathbb{G}_m$-eigenfunction with eigencharacter $m(x)$. There is an affine neighborhood $x \in U \subset X$ such that h_x lifts to an invertible function h_U on $f^{-1}(U)$ which is a $\mathbb{G}_m$-eigenfunction with eigencharacter $m(x)$. This h_U is a generator of $L_{m(x)}$ on U and the multiplication maps $L_i \otimes L_{m(x)} \to L_{i+m(x)}$ are isomorphisms over U for every i.

Setting $M = m(X) := \operatorname{lcm}\{m(x) : x \in X\}$, we see that L_M is locally free on X and the multiplication maps $L_i \otimes L_M \to L_{i+M}$ are isomorphisms for every i.

If x is a generic point, then $L_i \otimes k(x) \simeq k(x)$ if $m(x)$ divides i and $L_i \otimes k(x) = 0$ otherwise.

We still need to prove that the L_i are torsion free and coherent. Any torsion section of L_i is killed by $L_i^{\otimes M} \to L_{Mi}$, hence it would give a nilpotent section of $\mathscr{O}_Y$, a contradiction. Thus every L_i is torsion free. Coherence is a local question, thus assume that X is affine. For a generic point $x_g \in X$, $L_i \otimes k(x_g) \neq 0$ if and only if $m(x_g)|i$ if and only if $L_{-i} \otimes k(x_g) \neq 0$. Thus there is a section $s \in H^0(X, L_{-i})$ that is a generator at all generic points x_g such that $L_i \otimes k(x_g) \neq 0$. Then the composite

$$L_i \simeq L_i \otimes \mathscr{O}_X \xrightarrow{(1,s)} L_i \otimes L_{-i} \to L_0 \simeq \mathscr{O}_X$$

is an isomorphism at the generic points, hence an injection. Thus every L_i is a coherent sheaf on X.

Conversely, if $\sum_{i \in \mathbb{Z}} L_i$ satisfies (9.51.2.a–d) then $\sum_{i \in \mathbb{Z}} L_i$ is generated by the coherent submodule $\sum_{-M \leq i \leq M} L_i$. Thus $Y := \operatorname{Spec}_X \sum_{i \in \mathbb{Z}} L_i$ is affine over X. The grading gives a $\mathbb{G}_m$-action.

Pick any $x \in X$, then the fiber Y_x over x is $\operatorname{Spec}_x(\sum_{i \in \mathbb{Z}} L_i \otimes k(x))$. By (9.52), $L_i \otimes k(x)$ is nilpotent unless L_i is locally free at x and $L_i^{\otimes r} \to L_{ri}$ is an isomorphism near x for every r. Hence the reduced fiber is $\operatorname{Spec}_x \sum_{i \in m(x)\mathbb{Z}} k(x) \simeq k(x)[t, t^{-1}]$ for some $m(x) \in \mathbb{N}$. $\square$

Lemma 9.52 *Let L, M be rank 1 torsion free sheaves and assume that there is a surjective map $h\colon L \otimes M \twoheadrightarrow \mathscr{O}_X$. Then L, M are both locally free.*

Proof Pick $x \in X$. By assumption there is an affine neighborhood $x \in U$ and sections $\alpha \in H^0(U, L)$, $\beta \in H^0(U, M)$ such that $h(\alpha \otimes \beta)$ is invertible.

Let $\gamma \in H^0(U, L)$ be arbitrary. Then $h(\gamma \otimes \beta) = f \cdot h(\alpha \otimes \beta)$ for some $f \in \mathscr{O}_U$, thus $h((\gamma - f\alpha) \otimes \beta) = 0$. Thus $\gamma - f\alpha$ is 0 on the open set where M is locally free, hence it is 0 since L is torsion free. Thus α generates $L|_U$ and so L is locally free. $\square$

9.53 (Power maps) There are two ways to think of the mth power of a line bundle in the context of Seifert bundles.

Let $p\colon Y \to X$ be a Seifert bundle. For any $m \in \mathbb{N}$ the group $\mu_m \subset \mathbb{G}_m$ acts on Y; set $Y^{(m)} := Y/\mu_m$. Algebraically, by looking at the μ_m-action on $p_*\mathscr{O}_Y$, we see that if

$$p_*\mathscr{O}_Y = \sum_{i \in \mathbb{Z}} L_i \quad \text{then} \quad p_*\mathscr{O}_{Y^{(m)}} = \sum_{i \in m\mathbb{Z}} L_i.$$

In particular, if all the fiber multiplicities $m(x)$ divide m then $p^{(m)}\colon Y^{(m)} \to X$ is a $\mathbb{G}_m$-bundle.

This way we can think of every Seifert $\mathbb{G}_m$-bundle as a finite cyclic cover of a $\mathbb{G}_m$-bundle.

9.54 (Descending ampleness) Let $q\colon Y \to X$ be a finite, dominant morphism and L_X a line bundle on X. If X is proper and q^*L_X is ample, then L_X is ample, hence X is projective, compare Hartshorne (1977, exercise III.5.7).

Note, however, that this fails if X is not proper. Kollár (2011e) gives an example of a surface T with normal crossing singularities such that ω_T is not ample, yet its pull-back to the normalization $\bar{T} \to T$ is ample.

9.4 Pro-finite equivalence relations

In general it is quite hard to see when a finite pre-relation $R \rightrightarrows Y$ generates a finite set theoretic equivalence relation.

We start with the following obvious finiteness condition which turns out to be quite useful. Note that its assumptions are satisfied if (X, Δ) is lc, S_* is the stratification by lc centers as in (5.25) and Z does not contain any lc centers.

Then we give a series of examples of non-finite gluing relations.

Lemma 9.55 *Let (X, S_*) be a stratified space and $Z \subset X$ a closed subspace which does not contain any of the irreducible components of the $S_i X$. Let $\sigma_i\colon R \rightrightarrows (X, S_*)$ be a pro-finite, stratified, set theoretic equivalence relation. Assume that $R|_{X \setminus Z}$ is a finite, set theoretic equivalence relation. Then R is also a finite, set theoretic equivalence relation.*

Proof Since R is a union of finite relations, it is enough to check that R has finitely many irreducible components. The latter can be checked one stratum at a time, hence we may assume that X and R are pure dimensional. Then every irreducible component of R dominates an irreducible component of X, hence finiteness over the dense open set $X \setminus Z$ implies finiteness. $\square$

Example 9.56 Fix two points $a, b \in \mathbb{A}^1$ and consider two involutions $\tau_1\colon x \mapsto a - x$ and $\tau_2\colon x \mapsto b - x$. The composite $\tau_1 \circ \tau_2$ is translation by $b - a$. Thus it has infinite order if $a \neq b$ and the characteristic is 0.

Example 9.57 Let $X := \mathbb{A}^3$ with coordinates (x, y, t) and $D_1 := (y = 0)$, $D_2 := (x = 0)$ two hyperplanes. Let $L := (x = y = 0)$ be the t-axis. For $a, b \in \mathbb{C}$ define involutions on D_i by

$$\tau_1(x, 0, t) \mapsto (x, 0, a - t) \quad \text{and} \quad \tau_2(0, y, t) \mapsto (0, y, b - t).$$

Note that $D_1/(\tau_1) = \operatorname{Spec} \mathbb{C}[x, t(a - t)]$ and $D_2/(\tau_2) = \operatorname{Spec} \mathbb{C}[y, t(b - t)]$. Thus on $X \setminus L$ the τ_i generate a finite equivalence relation and we obtain a finite morphism

$$\pi^0 : (X \setminus L) \to (X \setminus L)/(\tau_1, \tau_2).$$

Both involutions act on L. Note that $\tau_2|_L \circ \tau_1|_L$ is translation by $b - a$, hence has infinite order if $a \neq b$ and the characteristic is 0.

This shows that π^0 cannot be extended to a finite morphism on X.

Example 9.58 Pick involutions $r_1, r_2, r_3 \in \operatorname{PGL}(2, \mathbb{C})$ such that any two of them generate a finite subgroup but the three together generate an infinite subgroup.

Consider $X = \mathbb{A}^3 \times \mathbb{P}^1$. Let x_i be the coordinates on $\mathbb{A}^3$ and $D_i := (x_i = 0) \times \mathbb{P}^1$. On D_i consider the involution τ_i which is the identity on D_i and r_i on the $\mathbb{P}^1$-factor. Let $R \rightrightarrows X$ be the set theoretic equivalence relation generated by the $\tau_i : i = 1, 2, 3$. Note that

$$\pi_1: (X \setminus D_1) \times \mathbb{P}^1 \to (X \setminus D_1) \times (\mathbb{P}^1/\langle r_2 r_3 \rangle)$$

is finite, thus $R|_{X \setminus D_1}$ is a finite equivalence relation. Similarly, $(X \setminus D_i)/(R|_{X \setminus D_1})$ exists for $i = 2, 3$. Set $\mathbb{P}^1_0 := \{0\} \times \mathbb{P}^1$. Then the geometric quotient

$$\left(X \setminus \mathbb{P}^1_0\right)/\left(R|_{X \setminus \mathbb{P}^1_0}\right)$$

exists. Note, however, that the restriction of R to $\mathbb{P}^1_0$ is not a finite equivalence relation since the subgroup generated by r_1, r_2, r_3 is infinite. Thus R is not a finite relation and there is no geometric quotient of X by R.

In order to find such r_1, r_2, r_3, it is easier to work with $\operatorname{SO}(3) \simeq \operatorname{SU}(2)/(\pm 1)$. Let $L_i \subset \mathbb{R}^3$ be three lines such that the angles between them are rational multiples of π. Let r_i denote the rotations by π around the lines L_i. By assumption, the angle between any two lines is a rational multiple of π, hence any two rotations generate a finite dihedral group.

The finite subgroups of $G \subset \operatorname{SO}(3, \mathbb{R})$ are all known. If G is not cyclic or dihedral, then any rotation in G has order ≤ 6. Thus, as soon as the denominator of the angle between L_i, L_j is large enough, the subgroup generated by r_1, r_2, r_3 is infinite.

Example 9.59 In $\mathbb{R}^n$ consider the hyperplanes

$$H_0 := (x_1 = 1), \ H_n := (x_n = 0) \quad \text{and} \quad H_i := (x_i = x_{i+1})$$

$$\text{for } i = 1, \ldots, n - 1.$$

Note that any n of these hyperplanes have a common point but the intersection of all $n + 1$ of them is empty.

Let r_i denote the reflection on H_i. Each r_i is defined over $\mathbb{Z}$ and maps $\mathbb{Z}^n$ to itself. Thus any n of the r_i generate a reflection group which has a fixed point and preserves a lattice. These are thus finite groups. By contrast, all $n + 1$ of them generate a reflection group with no fixed point. It is thus an infinite group.

As in (9.58) consider $X = \mathbb{A}^{n+1} \times \mathbb{A}^n$. Let $x_0, \ldots, x_n$ be the coordinates on $\mathbb{A}^{n+1}$ and $D_i := (x_i = 0) \times \mathbb{A}^n$. On D_i consider the involution τ_i which is the identity on D_i and r_i on the $\mathbb{A}^n$-factor. Let $R \rightrightarrows X$ be the set theoretic equivalence relation generated by the $\tau_i : i = 0, \ldots, n$. We see that R is not a finite equivalence relation on X but it restricts to a finite equivalence relation on $(\mathbb{A}^{n+1} \setminus \{0\}) \times \mathbb{A}^n$.

The next examples have normal irreducible components but they are only lc.

Example 9.60 Let X be an affine variety, $p \in X$ a point and $D_1, D_2 \subset X$ divisors such that $D_1 \cap D_2 = p$. Take two copies $(X^i, D_1^i + D_2^i)$ for $i = 1, 2$. Choose an isomorphism $\phi(\lambda, \mu)$

$$\left((D_1^1 \setminus \{p^1\}) \times \mathbb{A}^1 \right) \amalg \left((D_2^1 \setminus \{p^1\}) \times \mathbb{A}^1 \right)$$
$$\rightarrow \left((D_1^2 \setminus \{p^2\}) \times \mathbb{A}^1 \right) \amalg \left((D_2^2 \setminus \{p^2\}) \times \mathbb{A}^1 \right)$$

where $\phi(\lambda, \mu) = (1_{D_1} \times \lambda) \amalg (1_{D_2} \times \mu)$ is the identity on the D_i and multiplication by λ (resp. by μ) on the $\mathbb{A}^1$-factor of D_1 (resp. D_2).

The corresponding geometric quotient $Y^*(\lambda, \mu)$ is a non-normal variety whose irreducible components $(X^i \setminus \{p^i\}) \times \mathbb{A}^1$ intersect along $((D_1^i + D_2^i) \setminus \{p^i\}) \times \mathbb{A}^1$.

When can we extend this to a non-normal variety $Y(\lambda, \mu) \supset Y^*(\lambda, \mu)$ whose irreducible components are $X^1 \times \mathbb{A}^1$ and $X^2 \times \mathbb{A}^1$?

Assume that $f^i = \sum_j f_j^i t^j$ is a function on $X^i \times \mathbb{A}^1$ such that f^1, f^2 glue together to a regular function on $Y^*(\lambda, \mu)$. The compatibility conditions are

$$f_j^1 \big|_{D_1^1} = \lambda^j f_j^2 \big|_{D_1^2} \quad \text{and} \quad f_j^1 \big|_{D_2^1} = \mu^j f_j^2 \big|_{D_2^2}.$$

In particular, we get that

$$f_j^1(p^1) = \lambda^j f_j^2(p^2) \quad \text{and} \quad f_j^1(p^1) = \mu^j f_j^2(p^2).$$

If λ / μ is not a root of unity, this implies that the f^i are constant on $\{p^i\} \times \mathbb{A}^1$. Thus there is no scheme or algebraic space $Y(\lambda, \mu) \supset Y^*(\lambda, \mu)$ whose irreducible components are $X^1 \times \mathbb{A}^1$ and $X^2 \times \mathbb{A}^1$.

Let us see now some log canonical examples satisfying the above assumptions.

9.60.1 Set $X = \mathbb{A}^2$, $D_1 = (x = 0)$ and $D_2 = (y = 0)$. Then $(X, D_1 + D_2)$ is dlt. Thus we see that gluing in codimension 2 is not automatic for dlt pairs.

9.60.2 Set $X = (xy - uv = 0)$, $D_1 = (x = u = 0)$, $D_2 = (y = v = 0)$ and $\Delta = (x = v = 0) + (y = u = 0)$. Here $(X, D_1 + D_2 + \Delta)$ is lc but not dlt. We can replace Δ by some other divisor whose coefficients are < 1, but $(X, D_1 + D_2 + \Delta)$ can never be dlt. Thus we see that gluing in codimension 3 is not automatic for lc pairs.

9.60.3 Similar examples exist in any dimension. Let X be the cone over $\mathbb{P}^1 \times \mathbb{P}^n$, D_i the cone over $(i : 1) \times \mathbb{P}^n$ and Δ the cone over some $\mathbb{P}^1 \times B$ where $B \sim_{\mathbb{Q}} (n + 1)H$ on $\mathbb{P}^n$. These examples can be lc but not dlt.

Thus we see that gluing in any codimension is not automatic for lc pairs.

Example 9.61 We give an example of a pair $(\mathbb{P}^1, \Delta)$ such that

(1) $\deg(K_{\mathbb{P}^1} + \Delta) = 1$,
(2) there is a degree 3 map $f \colon (\mathbb{P}^1, \Delta) \to (C_3, \Delta_3)$ such that $K_{\mathbb{P}^1} + \Delta = f^*(K_{C_3} + \Delta_3)$,
(3) there is a degree 2 map $g \colon (\mathbb{P}^1, \Delta) \to (C_2, \Delta_2)$ such that $K_{\mathbb{P}^1} + \Delta = g^*(K_{C_2} + \Delta_2)$ and
(4) the equivalence relation generated by the fibers of f, g has infinite equivalence classes and so the universal push-out of

$$
\begin{array}{ccc}
\mathbb{P}^1 & \xrightarrow{\ f\ } & C_3 \\
{\scriptstyle g}\downarrow & & \\
C_2 & &
\end{array}
\quad \text{is} \quad
\begin{array}{ccc}
\mathbb{P}^1 & \xrightarrow{\ f\ } & C_3 \\
{\scriptstyle g}\downarrow & & \downarrow \\
C_2 & \longrightarrow & \operatorname{Spec}\mathbb{C}.
\end{array}
$$

The starting point is the map $f \colon \mathbb{C} \to \mathbb{C}$ given by $x \mapsto x^3 - 3x$ which has double ramification at ± 1 and triple ramification at ∞. Thus if we set $C_3 \simeq \mathbb{P}^1$ and

$$\Delta_3 := \tfrac{1}{2}[0] + \tfrac{1}{2}[2] + \tfrac{1}{2}[-2] + \tfrac{5}{6}[\infty]$$

then $f^*(K_{C_3} + \Delta_3) = K_{\mathbb{P}^1} + \Delta$ where

$$\Delta := \tfrac{1}{2}[0] + \tfrac{1}{2}[\sqrt{3}] + \tfrac{1}{2}[-\sqrt{3}] + \tfrac{1}{2}[2] + \tfrac{1}{2}[-2] + \tfrac{1}{2}[\infty].$$

The coincidence we exploit is that Δ is invariant under the involution $x \mapsto 2\sqrt{3}/x$. Thus we can set $g := x + 2\sqrt{3}/x$ and then $K_{\mathbb{P}^1} + \Delta = g^*(K_{C_2} + \Delta_2)$ where $C_2 \simeq \mathbb{P}^1$ and

$$\Delta_2 := \tfrac{1}{2}[\infty] + \tfrac{1}{2}[2 + \sqrt{3}] + \tfrac{1}{2}[-2 - \sqrt{3}] + \tfrac{1}{2}[\sqrt{2\sqrt{3}}] + \tfrac{1}{2}[-\sqrt{2\sqrt{3}}].$$

The equivalence relation is generated by two rules:

$$x \sim \frac{2\sqrt{3}}{x} \quad \text{and} \quad x \sim \frac{-x \pm \sqrt{12 - 3x^2}}{2}.$$

For example, starting with $x = \infty$, we apply these rules repeatedly to obtain that its equivalence class contains

$$\infty, 0, \pm\sqrt{3}, \pm2, \pm1, \pm2\sqrt{3}, \pm\sqrt{3}(1 \pm \sqrt{2}), \pm2(1 \pm \sqrt{2}), \ldots \quad (9.61.5)$$

Assume to the contrary that the universal push-out is not $\operatorname{Spec} \mathbb{C}$. Then it is necessarily a curve D (in fact $D \simeq \mathbb{P}^1$). We claim that

$$\deg(\mathbb{P}^1/D) \leq 42. \quad (9.61.6)$$

If this holds then every equivalence class contains ≤ 42 elements. At ∞ we have at least triple ramification, thus its equivalence class contains $\leq 14 = 42/3$ elements. We have already found 18 in (9.61.5) a contradiction.

In order to see (9.61.6), we use that by (9.34) there is a curve $h\colon C \to \mathbb{P}^1$ such that $f \circ h\colon C \to C_3$ and $g \circ h\colon C \to C_2$ are both Galois. Then $D = C/G$ where $G = \langle \operatorname{Gal}(C/C_3), \operatorname{Gal}(C/C_2) \rangle$.

Write $h^*(K_{\mathbb{P}^1} + \Delta) = K_C + \Delta_C$. Then $G \subset \operatorname{Aut}(C, \Delta_C)$ thus $K_C + \Delta_C$ is the pull-back of $K_D + \Delta_D$ for some Δ_D. Then $K_{\mathbb{P}^1} + \Delta$ is also the pull-back of $K_D + \Delta_D$. The rest follows from (9.62). □

Lemma 9.62 *Let $q\colon (C_1, \Delta_1) \to (C_2, \Delta_2)$ be a morphism of smooth projective curves such that $K_{C_1} + \Delta_1 = q^*(K_{C_2} + \Delta_2)$. Assume that the coefficients of the points in Δ_1 are of the form $1 - \frac{1}{m}$ for some $m \in \mathbb{N}$. Then*

(1) *the coefficients of the points in Δ_2 are also of the form $1 - \frac{1}{m}$ and*
(2) *$\deg q \leq 42 \cdot \deg(K_{C_1} + \Delta_1)$ if the latter is positive.*

Proof   The first claim follows from (2.41.5). Since

$$\deg(K_{C_1} + \Delta_1) = \deg q \cdot \deg(K_{C_2} + \Delta_2),$$

the second claim is equivalent to $\deg(K_{C_2} + \Delta_2) \geq \frac{1}{42}$.

If $g(C_2) \geq 1$ then $\deg(K_{C_2} + \Delta_2) \geq \frac{1}{2}$. If $g(C_2) = 0$ then the claim is equivalent to the following exercise: If $-2 + \sum_i (1 - \frac{1}{m_i}) > 0$ then in fact it is $\geq \frac{1}{42}$ with equality holding only for $-2 + (1 - \frac{1}{2}) + (1 - \frac{1}{3}) + (1 - \frac{1}{7}) = \frac{1}{42}$. □

10

Ancillary results

In this chapter we collect several topics for which good references are either not available or are too scattered in the literature.

In Section 10.1 we prove three well-known results on birational maps of surfaces for excellent 2-dimensional schemes.

General properties of seminormal schemes are studied in Section 10.2. Seminormality plays a key role in the study of lc centers and in many inductive methods involving lc and slc pairs.

In Section 10.3 we gather, mostly without proofs, various vanishing theorems that we use. Section 10.4 contains resolution theorems that are useful for non-normal schemes and in Section 10.5 we study the action of birational maps on differential forms. The basic theory of cubic hyperresolutions is recalled in Section 10.6.

Assumptions In Section 10.1 we work with excellent surfaces and in Section 10.2 with arbitrary schemes. In later sections characteristic 0 is always assumed.

10.1 Birational maps of 2-dimensional schemes

Here we prove the Hodge Index theorem, the Grauert–Riemenschneider vanishing theorem, Castelnuovo's contraction theorem and study rational singularities of surfaces. Instead of the usual setting, we consider these for excellent 2-dimensional schemes.

Theorem 10.1 (Hodge Index theorem) *Let X be a 2-dimensional regular scheme, Y an affine scheme and $f \colon X \to Y$ a proper and generically finite morphism with exceptional curves $\cup C_i$. Then the intersection form $(C_i \cdot C_j)$ is negative-definite.*

Proof   It is enough to consider all the exceptional curves that lie over a given $y \in Y$. Then all the curves considered are proper over the residue field $k(y)$, so the intersection numbers

$$(C_i \cdot C_j) := \deg_{k(y)} \mathscr{O}_X(C_i)|_{C_j}$$

are defined as usual. Note the obvious property that

$$(C_i \cdot C_j) \geq 0 \quad \text{for } i \neq j. \tag{10.1.1}$$

By (10.2), there is an effective f-exceptional Cartier divisor. An easy argument (10.3.4), using only (10.1.1) and bilinear algebra, shows that the intersection form is negative-definite. $\square$

Lemma 10.2   *Let $f: X \to Y$ be as in (10.1). Then f is projective and there is an effective f-exceptional Cartier divisor W on X such that $-W$ is f-ample.*

Proof   For each exceptional curve $C_i \subset X$ let $U_i \subset X$ be an open affine subset such that $C_i \cap U_i \neq \emptyset$. Let $H_i \subset U_i$ be an irreducible divisor that intersects C_i such that $H_i \not\subset \cup C_j$ and let $\bar{H}_i \subset X$ denote its closure. Then $H := \sum \bar{H}_i$ has positive intersection number with every exceptional curve, hence it is f-ample. Its push-forward $f(H) \subset Y$ is an effective Weil divisor on Y; it is thus contained in an effective Cartier divisor D since Y is affine. Write $f^*(D) = H + D' + W$ where W is effective, f-exceptional and D' is effective without f-exceptional irreducible components. Then

$$(W \cdot C_i) = (f^*D \cdot C_i) - (H \cdot C_i) - (D' \cdot C_i) \leq -(H \cdot C_i) < 0,$$

so $-W$ is f-ample. $\square$

10.3 (Remarks on quadratic forms)   Fix a basis $C_1, \ldots, C_n$ in a real vector space V. Let $B(\,,\,)$ be a bilinear form on V. We usually write $(C_i \cdot C_j)$ instead of $B(C_i, C_j)$. We are interested in the cases that look like the intersection form of a collection of curves on a surface. That is, we impose the following:

Assumption 10.3.1 $(C_i \cdot C_j) \geq 0$ if $i \neq j$.

Notation 10.3.2   Given two vectors $A = \sum a_i C_i$ and $A' = \sum a_i' C_i$, write $A \geq A'$ if and only if $a_i \geq a_i'$ for every i and $A \gg A'$ if and only if $a_i > a_i'$ for every i. Thus $\sum a_i C_i \geq \mathbf{0}$ (resp. $\gg \mathbf{0}$) iff $a_i \geq 0$ (resp. $a_i > 0$) for every i. Set $\operatorname{Supp}(\sum a_i C_i) := \{C_i : a_i \neq 0\}$.

We say that $B(\,,\,)$ is *decomposable* if we can write $\{1, \ldots, n\} = I \cup J$ such that $I \neq \emptyset \neq J$ and $(C_i \cdot C_j) = 0$ whenever $i \in I$ and $j \in J$. Decomposable forms correspond to disconnected sets of curves.

Lemma 10.3.3   Let $B(\,,\,)$ be an indecomposable bilinear form satisfying (10.3.1). Assume that there is a vector $W = \sum w_i C_i \gg \mathbf{0}$ such that

$(C_i \cdot W) \leq 0$ for every i with strict inequality for some i. Let Z be any vector such that $Z \not\leq 0$. Then there is a C_j such that $(C_j \cdot Z) < 0$ and C_j has positive coefficient in Z.

Proof Write $Z = \sum z_i C_i$ and choose the maximal $a \in \mathbb{R}^+$ such that $az_i \leq w_i$ for every i. Then $\sum (w_i - az_i)C_i \geq 0$; set $I := \{i : w_i - az_i > 0\}$. If $I = \emptyset$ then $Z = \frac{1}{a}W$ and C_j exists by assumption. Otherwise $I \neq \emptyset$, and since $B(\ ,\ )$ is indecomposable, there is an index j such that $w_j - az_j = 0$ and $(C_j \cdot C_i) > 0$ for some $i \in I$. Then

$$(C_j \cdot \sum(w_i - az_i)C_i) \geq (w_i - az_i)(C_j \cdot C_i) > 0.$$

Thus $a(C_j \cdot Z) = (C_j \cdot W) - (C_j \cdot (W - aZ)) < 0.$ □

Claim 10.3.4   Let $B(\ ,\ )$ be a bilinear form satisfying (10.3.1).

(1) $B(\ ,\ )$ is negative definite if and only if there is a vector $W = \sum w_i C_i \geq \mathbf{0}$ such that $(C_i \cdot W) < 0$ for every i.
(2) If $B(\ ,\ )$ is indecomposable, then it is negative definite if and only if there is a vector $W = \sum w_i C_i \geq \mathbf{0}$ such that $(C_i \cdot W) \leq 0$ for every i and strict inequality holds for some i.

Proof Assume that W exists and consider the function $f(x_1, \ldots, x_n) = (\sum x_i C_i \cdot \sum x_i C_i)$ on any cube $0 \leq x_i \leq N$. Since

$$\tfrac{\partial f}{\partial x_j}(Z) = 2(C_j \cdot Z),$$

we see from (10.3.3) that f strictly increases as we move toward the origin parallel to one of the coordinate axes. Thus f has a strict maximum at the origin and so $(Z \cdot Z) < 0$ for any vector $Z \geq \mathbf{0}$ except for $Z = \mathbf{0}$.

Any Z can be written as $Z^+ - Z^-$ where $Z^+, Z^- \geq \mathbf{0}$ have disjoint supports. Then

$$(Z \cdot Z) = (Z^+ \cdot Z^+) + (Z^- \cdot Z^-) - 2(Z^+ \cdot Z^-) \leq (Z^+ \cdot Z^+) + (Z^- \cdot Z^-),$$

and we have seen that the last two terms are ≤ 0 with equality only if $Z = 0$. Thus $B(\ ,\ )$ is negative definite.

Conversely, assume that $B(\ ,\ )$ is negative definite. Since $B(\ ,\ )$ is nondegenerate, there is a vector W such that $(W \cdot C_i) = -1$ for every i. The next, more general result shows that $W \gg \mathbf{0}$. □

Claim 10.3.5   Let $B(\ ,\ )$ be an indecomposable negative definite bilinear form satisfying (10.3.1). Let $Z = \sum a_i C_i$ be a vector such that $(Z \cdot C_j) \geq 0$ for every j. Then

(1) either $Z = 0$ (and so $(Z \cdot C_j) = 0$ for every j),
(2) or $-Z \gg \mathbf{0}$ (and $(Z \cdot C_j) > 0$ for some j).

Proof Write $Z = Z^+ - Z^-$ where $Z^+, Z^- \geq \mathbf{0}$ have disjoint supports. If $Z^+ \neq 0$ then $(Z^+ \cdot Z^+) < 0$. Hence there is a $C_i \subset \operatorname{Supp} Z^+$ such that $(C_i \cdot Z^+) < 0$. C_i is not in $\operatorname{Supp} Z^-$, so $(C_i \cdot Z^-) \geq 0$ and so $(C_i \cdot Z) < 0$, a contradiction.

Finally, if $\emptyset \neq \operatorname{Supp} Z^- \neq \cup_i C_i$, then there is a C_i such that $C_i \not\subset \operatorname{Supp} Z^-$ but $(C_i \cdot Z^-) > 0$. Then $(C_i \cdot Z) = -(C_i \cdot Z^-) < 0$, again a contradiction. $\square$

Definition and Claim 10.3.6 Let $B(\,,\,)$ be an indecomposable negative definite bilinear form satisfying (10.3.1). Then there is a unique, minimal vector $W = \sum w_i C_i$ with $w_i \in \mathbb{N}$ such that $(W \cdot C_i) \leq 0$ for every i. This W is called the *fundamental vector* of $B(\,,\,)$. If $B(\,,\,)$ is the intersection form of a collection of curves on a surface, then $\sum w_i C_i$ is called the *fundamental cycle*.

Proof There is a vector $V \gg \mathbf{0}$ such that $(V \cdot C_i) < 0$ for every i. By perturbing its coefficients, we may assume that they are rational. Clearing denominators then gives an $\mathbb{N}$-linear combination $V' = \sum v_i' C_i$ such that $(V' \cdot C_i) \leq 0$ for every i.

Let $V^1 = \sum v_i^1 C_i$ be a minimal (with respect to $\leq$) $\mathbb{N}$-linear combination such that $(V^1 \cdot C_i) \leq 0$ for every i. In order to show that V^1 is unique, it is enough to prove that if $V^2 = \sum v_i^2 C_i$ is another such vector then so is $V^3 := \sum \min\{v_i^1, v_i^2\} C_i$. Fix an index j. In order to compute $(V^3 \cdot C_j)$ we may assume that $v_j^1 \leq v_j^2$. Then

$$(V^3 \cdot C_j) = (V^1 \cdot C_j) - \sum_{i \neq j} \left(v_i^1 - \min\left\{ v_i^1, v_i^2 \right\} \right)(C_i \cdot C_j) \leq (V^1 \cdot C_j).$$

$\square$

The following is a strengthening of the Grauert–Riemenschneider vanishing theorem for surfaces, using the method of Lipman (1969).

Theorem 10.4 *Let X be a regular surface and $f\colon X \to Y$ a proper, generically finite morphism with exceptional curves C_i such that $\cup_i C_i$ is connected. Let L be a line bundle on X and assume that there exist $\mathbb{Q}$-divisors N and $\Delta_X = \sum d_i C_i$ such that*

(1) $L \cdot C_i = (K_X + N + \Delta_X) \cdot C_i$ *for every* i,
(2) $N \cdot C_i \geq 0$ *for every* i *and*
(3) Δ_X *satisfies one of the following*
    (a) $0 \leq d_i < 1 \; \forall i$,
    (b) $0 < d_i \leq 1 \; \forall i$ *and* $d_j \neq 1$ *for some* j,
    (c) $0 < d_i \leq 1 \; \forall i$ *and* $N \cdot C_j > 0$ *for some* j.

Then $R^1 f_* L = 0$.

Proof Let $Z = \sum_{i=1}^{s} r_i C_i$ be an effective integral cycle. We prove by induction on $\sum r_i$ that

$$H^1(Z, L \otimes \mathcal{O}_Z) = 0. \tag{10.4.4}$$

Using the theorem on Formal Functions, this implies (10.4).

Let C_i be an irreducible curve contained in Supp Z. Set $Z_i = Z - C_i$ and consider the short exact sequence:

$$0 \to \mathcal{O}_{C_i} \otimes \mathcal{O}_X(-Z_i) \simeq \mathcal{O}_X(-Z_i)/\mathcal{O}_X(-Z) \to \mathcal{O}_Z \to \mathcal{O}_{Z_i} \to 0.$$

Tensoring with L we obtain

$$0 \to \mathcal{O}_{C_i} \otimes L(-Z_i) \to L \otimes \mathcal{O}_Z \to L \otimes \mathcal{O}_{Z_i} \to 0.$$

By induction on $\sum r_i$, $H^1(Z_i, L \otimes \mathcal{O}_{Z_i}) = 0$. Thus it is enough to prove that $H^1(C_i, \mathcal{O}_{C_i} \otimes L(-Z_i)) = 0$ for some i. This in turn would follow from

$$L \cdot C_i - Z_i \cdot C_i > \deg \omega_{C_i} = C_i^2 + K_X \cdot C_i,$$

which is equivalent to

$$N \cdot C_i + (\Delta_X - Z) \cdot C_i > 0.$$

By assumption, $N \cdot C_i \geq 0$ always holds.

If $Z \not\leq \Delta_X$ we can apply (10.3.3) to $Z - \Delta_X$ to obtain $C_i \subset$ Supp Z such that $(\Delta_X - Z) \cdot C_i > 0$. This settles case (3.a).

Thus assume that $Z \leq \Delta_X$. If Supp $Z =$ Supp Δ_X then $Z = \Delta_X$. This can happen only in case (3.c), but then $N \cdot C_i > 0$ for some i and we are done.

Finally, if Supp $Z \neq$ Supp Δ_X then Supp Z and Supp$(\Delta_X - Z)$ intersect in finitely many points and this intersection is nonempty in cases (3.b–c). Thus again there is a $C_i \subset$ Supp Z such that $(\Delta_X - Z) \cdot C_i > 0$. $\qquad \square$

Theorem 10.5 (Castelnuovo's contractibility criterion) (Lipman, 1969, theorem 27.1) *Let X be a regular surface, $f: X \to Y$ a projective morphism and $E \subset X$ an irreducible and reduced curve such that $(E \cdot K_X) < 0$ and $f(E)$ is 0-dimensional.*

Then there is a proper morphism to a regular pointed surface $g: (E \subset X) \to (z \in Z)$ such that $g: X \setminus E \to Z \setminus \{z\}$ is an isomorphism.

Proof Pick an f-very ample divisor H on X such that $R^1 f_* \mathcal{O}_X(nH) = 0$ for $n \geq 1$. Using the sequences

$$0 \to \mathcal{O}_X(nH + iE) \to \mathcal{O}_X(nH + (i+1)E) \to \mathcal{O}_X(nH + (i+1)E)|_E \to 0$$

we see that

(1) $|nH + iE|$ is very ample on $X \setminus E$ for $n \geq 1, i \geq 0$,
(2) $R^1 f_* \mathcal{O}_X(nH + iE) = 0$ if $((nH + iE) \cdot E) \geq (E \cdot E)$,

(3) $f_*\mathscr{O}_X(nH + iE) \twoheadrightarrow H^0(E, \mathscr{O}_E((nH + iE)|_E))$ is onto if $((nH + iE) \cdot E) \geq 0$ and

(4) $|nH + iE|$ is base point free if $((nH + iE) \cdot E) \geq 0$.

By the adjunction formula, $\omega_E \simeq \mathscr{O}_X(E)|_E \otimes \mathscr{O}_X(K_X)|_E$ is the product of two negative line bundles. Thus, by (10.6), $\mathscr{O}_X(E)|_E \simeq \mathscr{O}_X(K_X)|_E$ are negative generators of Pic E. In particular, for any line bundle L, $(L \cdot E)/(E \cdot E)$ is an integer.

Set $m = (H \cdot E)/(E \cdot E)$. Then $((H + mE) \cdot E) = 0$, $((H + (m-1)E) \cdot E) = -(E \cdot E)$ and by (3) there is a curve $D \in |H + (m-1)E|$ that does not contain E. Since $\mathscr{O}_X(D)|_E$ is a generator of Pic E, we conclude that D is regular at $D \cap E$. By (4) we can choose $D' \in |H + mE|$ disjoint from E. From the sequence

$$0 \to \mathscr{O}_X((n-1)(H + mE) + E) \to \mathscr{O}_X(n(H + mE))$$
$$\to \mathscr{O}_D(n(H + mE)|_D) \to 0$$

we conclude that $f_*\mathscr{O}_X(n(H + mE)) \twoheadrightarrow f_*\mathscr{O}_D(n(H + mE)|_D)$ is onto for $n \gg 1$. Moreover, $n(H + mE)|_D$ is very ample for $n \gg 1$.

Let $g \colon X \to Z$ be the morphism given by $|n(H + mE)|$ for some $n \gg 1$. As we saw,

(5) g contracts E to a point z,
(6) $|n(H + mE)|$ is very ample on $X \setminus E$ and
(7) $|n(H + mE)|$ is also very ample on D.

Thus $g(D + (n-1)D') = g(E + D + (n-1)D') \subset Z$ is a curve which is regular at z. It is also a hyperplane section of Z since $E + D + (n-1)D' \in |n(H + mE)|$, thus Z is regular at z. $\square$

Lemma 10.6 *Let k be a field and C a reduced and irreducible k-curve such that $H^1(C, \mathscr{O}_C) = 0$ and ω_C is locally free. Then*

(1) $H^1(C, L) = 0$ *for every line bundle L such that* $\deg L \geq 0$.
(2) $\mathrm{Pic}(C) \simeq \mathbb{Z}[H]$ *where H denotes the positive generator.*
(3) *Furthermore, with $K := H^0(C, \mathscr{O}_C)$,*
  (a) *either $\omega_C \simeq H^{-1}$ and C is isomorphic to a conic in $\mathbb{P}^2_K$,*
  (b) *or $\omega_C \simeq H^{-2}$ and C is isomorphic to $\mathbb{P}^1_K$.*
(4) *C is a line as in (3.b) if and only if it carries a line bundle L such that* $H^0(C, L) = H^1(C, L) = 0$.

Proof Set $r = \deg[K : k]$. Then $\chi(\mathscr{O}_C) = r$ and by Riemann–Roch, $\chi(C, L) = \deg L + r$ for any line bundle L. Thus $h^0(C, L) > 0$ for $\deg L \geq 0$ and from

$$0 \to \mathscr{O}_C \to L \to (\text{torsion sheaf}) \to 0$$

we obtain that $H^1(C, L) = 0$ for $\deg L \geq 0$.

If $\deg L = 0$ then $h^0(C, L) > 0$ implies that $L \simeq \mathscr{O}_C$. Thus $\mathrm{Pic}(C)$ is isomorphic to $\mathbb{Z}$; let H be the positive generator.

Since $H^1(C, \omega_C) \neq 0$, this implies that $\omega_C \simeq H^{-n}$ for some $n \geq 1$ and $-r = \chi(C, \omega_C) = -n \deg H + r$.

Since $H^0(C, H)$ and $H^1(C, H)$ are both K-vector spaces, we see that $\deg H$ is a multiple of r, say $\deg H = dr$. Thus $2r = ndr$ and so $nd = 2$, hence $n \in \{1, 2\}$ as claimed.

If $n = 2$ then $\deg H = r$ and $H^0(C, H^m) \simeq K^{m+1}$. Two K-independent sections of $H^0(C, H)$ give a morphism $p \colon C \to \mathbb{P}^1_K$. By dimension count, the injections $p^* \colon H^0(\mathbb{P}^1_k, \mathscr{O}(m)) \to H^0(C, H^m)$ are isomorphisms, thus $C \simeq \mathbb{P}^1_K$.

If $n = 1$ then $\deg H = 2r$ and $H^0(C, H^m) \simeq K^{2m+1}$. Three K-independent sections of $H^0(C, H)$ give a morphism $p \colon C \to \mathbb{P}^2_K$. By counting dimensions, $p^* \colon H^0(\mathbb{P}^2_k, \mathscr{O}(2)) \to H^0(C, H^2)$ is not injective, hence the image of p is a conic $Q \subset \mathbb{P}^2_K$. Again counting dimensions, the injections $p^* \colon H^0(Q, \mathscr{O}_Q(m)) \to H^0(C, H^m)$ are isomorphisms, thus $C \simeq Q$. $\qquad\square$

Rational singularities of surfaces

These results are from Lipman (1969, section 11) with later strengthenings also due to Lipman.

Definition 10.7 Let Y be a normal, excellent surface. A resolution $f \colon X \to Y$ is called *rational* if $R^1 f_* \mathscr{O}_X = 0$. Since any two resolutions are connected by a sequence of blow-ups and their inverses, if Y has a rational resolution then every resolution is rational. Such a Y is said to have *rational singularities*.

Lemma 10.8 *Let* $(y \in Y)$ *be a rational surface singularity and* $f \colon X \to Y$ *a resolution with exceptional curves* C_i. *Let* C_1 *be an exceptional curve for which* $r_i := \dim_{k(y)} H^0(C_i, \mathscr{O}_{C_i})$ *is minimal. Then there is a sequence of effective cycles* $Z_1 \leq \cdots \leq Z_m$ *such that*

(1) $Z_1 = C_1$ *and* Z_m *is the fundamental cycle* (10.3.6).
(2) Z_{i+1} *is obtained from* Z_i *by adding a curve* C_j *such that* $(Z_i \cdot C_j) > 0$.
(3) $H^0(Z_i, \mathscr{O}_{Z_i}) = H^0(C_1, \mathscr{O}_{C_1})$ *and* $H^1(Z_i, \mathscr{O}_{Z_i}) = 0$ *for every* i.
(4) *For* $i \geq 1$ *the curves* C_j *are lines* (10.6.3.b) *and there are exact sequences*

$$0 \to \mathscr{O}_{C_j}(-1) \to \mathscr{O}_{Z_{i+1}} \to \mathscr{O}_{Z_i} \to 0.$$

Proof For any effective cycle $Z = \sum c_i C_i$ there is an exact sequence $R^1 f_* \mathscr{O}_X \to H^1(Z, \mathscr{O}_Z) \to R^2 f_* \mathscr{O}_X(-Z)$. Here $R^1 f_* \mathscr{O}_X = 0$ since the singularity is rational and $R^2 f_* \mathscr{O}_X(-Z) = 0$ since the fiber dimension of f is ≤ 1. Thus $H^1(Z, \mathscr{O}_Z) = 0$ for every i.

Assume that we already have $Z_1, \ldots, Z_i$ satisfying (1–4). If there is a curve C_j such that $(Z_i \cdot C_j) > 0$ then set $Z_{i+1} := Z_i + C_j$, consider the exact sequence

$$0 \to \mathscr{O}_X(-Z_i)|_{C_j} \to \mathscr{O}_{Z_{i+1}} \to \mathscr{O}_{Z_i} \to 0$$

and its exact cohomology sequence

$$0 \longrightarrow H^0(C_j, \mathscr{O}_X(-Z_i)|_{C_j}) \longrightarrow H^0(Z_{i+1}, \mathscr{O}_{Z_{i+1}}) \longrightarrow H^0(Z_i, \mathscr{O}_{Z_i})$$

$$\overset{\partial}{\longrightarrow} H^1(C_j, \mathscr{O}_X(-Z_i)|_{C_j}) \longrightarrow 0.$$

Note that $H^1(C_j, \mathscr{O}_X(-Z_i)|_{C_j})$ is a vector space over $H^0(C_j, \mathscr{O}_{C_j})$, thus it is either 0 or has $k(y)$-dimension at least r_j. We already know that $H^0(Z_i, \mathscr{O}_{Z_i}) \simeq H^0(C_1, \mathscr{O}_{C_1})$ has $k(y)$-dimension $r_1 \le r_j$, thus ∂ is either an isomorphism or $H^1(C_j, \mathscr{O}_X(-Z_i)|_{C_j}) = 0$. Since the constant sections always lift from $\mathscr{O}_{Z_i}$ to $\mathscr{O}_{Z_{i+1}}$, we conclude that $H^1(C_j, \mathscr{O}_X(-Z_i)|_{C_j}) = 0$.

By our choice $\mathscr{O}_X(-Z_i)|_{C_j}$ is an invertible sheaf of negative degree, therefore $H^0(C_j, \mathscr{O}_X(-Z_i)|_{C_j}) = 0$ which shows that $H^0(Z_{i+1}, \mathscr{O}_{Z_{i+1}}) = H^0(Z_i, \mathscr{O}_{Z_i})$ and C_j is line by (10.6.4).

We claim that $Z_i \le W$ for every i where W denotes the fundamental cycle. By induction write $W = Z_i + W_i$ where W_i is effective. If $(Z_i \cdot C_j) > 0$ then

$$(W_i \cdot C_j) = (W \cdot C_j) - (Z_i \cdot C_j) < (W \cdot C_j) \le 0.$$

This implies that $C_j \subset \mathrm{Supp}\, W_i$ hence $Z_{i+1} = Z_i + C_j \le W$.

Thus the procedure eventually stops with a cycle $Z_m \le W$ such that $(Z_m \cdot C_j) \le 0$ for every C_j. By definition, Z_m is the fundamental cycle. $\square$

Proposition 10.9 *Let* $(y \in Y)$ *be a rational surface singularity,* Y *affine and* $f: X \to Y$ *a resolution. Let* $C = \cup_i C_i$ *be the reduced exceptional set.*

(1) *Let* M *be a line bundle on* X *such that* $\deg_{C_i} M \ge 0$ *for every* i. *Then* M *is generated by global sections and* $R^1 f_* M = 0$.

(2) *Let* L *be a line bundle on* X *such that* $\deg_{C_i} L = 0$ *for every* i. *Then* $L \simeq \mathscr{O}_X$ *in a neighborhood of* C.

(3) Y *is* $\mathbb{Q}$-*factorial at* y. *More precisely,* $\det(\Gamma) \cdot D$ *is Cartier for any Weil* $\mathbb{Z}$-*divisor* D *on* Y, *where* Γ *is the dual graph of* C.

(4) *Let* C_1 *be a curve for which* $\dim_{k(y)} H^0(C_i, \mathscr{O}_{C_i})$ *is minimal. Then* $k(y) = H^0(C, \mathscr{O}_C) = H^0(C_1, \mathscr{O}_{C_1})$, *hence* C_1 *is geometrically reduced and irreducible.*

Proof Let W denote the fundamental cycle (10.3.6). Working through the exact sequences (10.8.4) shows by induction that $H^1(W, M|_W) = 0$. Since $M(-nW)$ also has non-negative degree on every curve, we similarly obtain that $H^1(W, M(-nW)|_W) = 0$. Now the exact sequences

$$H^1(W, M(-nW)|_W) \to H^1((n+1)W, M|_{(n+1)W}) \to H^1(nW, M|_{nW})$$

and induction show that $H^1(nW, M|_{nW}) = 0$ for $n \geq 1$, hence $R^1 f_* M = 0$.

The exact sequences (10.8.4) also show that $H^0(Z_{i+1}, M|_{Z_{i+1}}) \twoheadrightarrow$ $H^0(Z_i, M|_{Z_i})$ for every i. We can choose the sequence $Z_1, \ldots$ such that, as long as possible, we add new curves to its support. Thus we may assume that one of the intermediate curves is $Z_n = C$. We prove by induction that $M|_{Z_i}$ is generated by global sections for $i \leq n$. Consider the sequence

$$0 \to M \otimes \mathscr{O}_{C_j}(-1) \to M|_{Z_{i+1}} \to M|_{Z_i} \to 0.$$

As we noted, C_j is line as in (10.6.3.b) and $p_j := C_j \cap Z_i$ is the spectrum of $H^0(C_j, \mathscr{O}_{C_j})$.

If $\deg M|_{C_j} > 0$ then $\deg M \otimes \mathscr{O}_{C_j}(-1) \geq 0$ hence $M \otimes \mathscr{O}_{C_j}(-1)$ is generated by global sections and so is $M|_{Z_{i+1}}$. If $\deg M|_{C_j} = 0$ then $M|_{C_j} \simeq \mathscr{O}_{C_j}$ and $H^0(C_j, M|_{C_j}) \simeq M \otimes k(p_j)$. Thus every section of $M|_{Z_i}$ that is nonzero at p_j lifts to a section of $M|_{Z_{i+1}}$ that is nowhere zero on C_j. So again $M|_{Z_{i+1}}$ is generated by global sections.

All the global sections of $M|_C = M_{Z_n}$ now lift to global sections of $M|_W$. We already saw that $R^1 f_* M(-W) = 0$. Since Y is affine, this implies that $H^1(X, M(-W)) = 0$ hence global sections of $M|_W$ lift to global sections of M. This completes (1).

Assume next that L is a line bundle such that $\deg_{C_i} L = 0$ for every i. By (1), $L|_C$ is generated by global sections. Since it has degree 0, we see that $L|_C \simeq \mathscr{O}_C$. The constant section lifts to a global section of L that is nowhere zero in a neighborhood of C, proving (2). (In characteristic 0 a more conceptual proof is in Kollár and Mori (1992, 12.1.4).)

Let D be any Weil divisor on Y and $f_*^{-1} D$ its birational transform on X. By (10.1) and by Cramer's rule, the system of equations

$$\sum_i a_i (C_i \cdot C_j) = (f_*^{-1} D \cdot C_j) \quad \forall j$$

has a unique solution and $\det(\Gamma) \cdot a_i \in \mathbb{Z}$. Apply (2) to

$$L := \mathscr{O}_X(\det(\Gamma) \cdot f_*^{-1} D - \sum \det(\Gamma) \cdot a_i \cdot C_i)$$

to get that $\mathscr{O}_Y(\det(\Gamma) \cdot D)$ is locally free, hence $\det(\Gamma) \cdot D$ is Cartier.

Finally, applying (2) to $L = \mathscr{O}_X$ gives a surjection $H^0(Y, \mathscr{O}_Y) \twoheadrightarrow H^0(C, \mathscr{O}_C)$ which factors through $k(y)$ and $H^0(C, \mathscr{O}_C) \simeq H^0(C_1, \mathscr{O}_{C_1})$ by (10.8.3). A surjection of fields is an isomorphism, thus $k(y) = H^0(C, \mathscr{O}_C) = H^0(C_1, \mathscr{O}_{C_1})$. $\qquad\square$

The following result holds for geometrically normal singularities, hence for every normal singularity with a perfect residue field. In general, it seems a quite

basic property of log canonical surface singularities and it solves the problem
that we had to deal with at the end of the proof of (2.31).

Corollary 10.10 *Let $(y \in Y, B)$ be a numerically log canonical surface singularity and $f: X \to Y$ a log minimal resolution with exceptional curves C.
Then $k(y) \simeq H^0(C, \mathcal{O}_C)$ and if $(y \in Y)$ is a rational singularity then there is
an irreducible component $C_i \subset C$ such that $k(y) \simeq H^0(C_i, \mathcal{O}_{C_i})$.*

Proof If $(y \in Y)$ is a rational singularity then this follows from (10.9.4). Otherwise, by (2.28), $\Delta = C$ hence $K_X \sim -C$ and $R^1 f_* \mathcal{O}_X(-C) = 0$ by (10.4).
Pushing forward the exact sequence

$$0 \to \mathcal{O}_X(-C) \to \mathcal{O}_X \to \mathcal{O}_C \to 0$$

gives a surjection $f_* \mathcal{O}_X = \mathcal{O}_Y \twoheadrightarrow H^0(C, \mathcal{O}_C)$. Since C is reduced, $H^0(C, \mathcal{O}_C)$
is a field and we can factor this surjection through $k(y) \twoheadrightarrow H^0(C, \mathcal{O}_C)$. A
surjection between fields is an isomorphism, thus $k(y) \simeq H^0(C, \mathcal{O}_C)$. □

10.2 Seminormality

In this section we gather various definitions and results on seminormality and
seminormalization that are needed in the study of log canonical and semi-log
canonical pairs.

By the Riemann extension theorem, an affine variety X over $\mathbb{C}$ is normal
if and only if every bounded rational function on X is regular. Based on this,
Andreotti and Norguet (1967), Andreotti and Bombieri (1969) and Traverso
(1970) defined X to be seminormal if and only if every continuous rational
function on X is regular.

An ordinary node $(xy = 0)$ is seminormal. More generally, the union of the n
coordinate axes in $\mathbb{C}^n$ is seminormal. These examples exhaust all seminormal
curve singularities, but in higher dimensions the situation gets much more
complicated.

Basic definitions

Definition 10.11 (Seminormalization) A finite morphism of schemes $g:
X' \to X$ is called a *partial seminormalization* if

(1) X' is reduced and
(2) for every point $x \in X$, the induced morphism $g^*: k(x) \to k(\text{red } g^{-1}(x))$ is
     an isomorphism. Equivalently, each point $x \in X$ has exactly one preimage
     $x' := g^{-1}(x)$ and $g^*: k(x) \to k(x')$ is an isomorphism.

(In characteristic 0, it is enough to assume (2) for closed points only. Over perfect fields of positive characteristic, the Frobenius morphism satisfies (2) for closed points but not for the generic points.)

Using (2) for the generic points of X we conclude that a partial seminormalization is birational (1.11). Thus if $n\colon X^n \to X$ is the normalization of red X then

$$g_*\mathscr{O}_{X'} \quad \text{is a subsheaf of} \quad n_*\mathscr{O}_{X^n}. \tag{10.11.3}$$

It is clear that the composite of two partial seminormalizations is also a partial seminormalization. Furthermore, if $g\colon X' \to X$ is a partial seminormalization then so is any fiber product

$$g \times_S 1_Z\colon \operatorname{red}(X' \times_S Z) \to X \times_S Z. \tag{10.11.4}$$

(Being reduced is not preserved by fiber products, so we have to ensure that property by hand.) In particular, if $g_i\colon X_i \to X$ are partial seminormalizations then

$$g_1 \times g_2\colon \operatorname{red}(X_1 \times_X X_2) \to X \tag{10.11.5}$$

is also a partial seminormalization which dominates both of the X_i.

Using (10.11.3), we see that if the normalization $\tau\colon X^n \to X$ is a finite morphism (which holds for all excellent schemes) then there is a unique largest partial seminormalization $\pi\colon X^{sn} \to X$ which dominates every other partial seminormalization of X. It is called the *seminormalization* of X.

To be more explicit, $\pi_*\mathscr{O}_{X^{sn}}$ is the subsheaf of $\tau_*\mathscr{O}_{X^n}$ consisting of those sections ϕ such that for every point $x \in X$ with preimage $\bar{x} := \operatorname{red} \tau^{-1}(x)$ we have

$$\phi|_{\bar{x}} \in \operatorname{im}[\tau^* : k(x) \to k(\bar{x})]. \tag{10.11.6}$$

A scheme X is called *seminormal* if its seminormalization $\pi\colon X^{sn} \to X$ is an isomorphism. Equivalently, if every partial seminormalization $X' \to X$ is an isomorphism. Since red $X \to X$ is always a partial seminormalization, a seminormal scheme is reduced.

It is easy to see that an open subscheme of a seminormal scheme is also seminormal. Furthermore, being seminormal is a local property. That is, X is seminormal $\Leftrightarrow$ it is covered by seminormal open subschemes $\Leftrightarrow$ every local ring of X is seminormal. For more details, see Kollár (1996, section I.7.2).

Example 10.12 The node $(xy = 0)$ is seminormal but the cusp $(x^2 = y^3)$ and the ordinary triple point $(x^3 = y^3)$ are not seminormal. The seminormalization of the cusp is $\mathbb{A}^1$ and the seminormalization of $(x^3 = y^3)$ is the union of the three coordinate axes in $\mathbb{A}^3$.

More generally, one easily sees that, over an algebraically closed field, a curve singularity $(0 \in C)$ is seminormal if and only if it is analytically isomorphic to the union of the n coordinate axes in $\mathbb{A}^n$. That is, if and only if

$$\hat{\mathcal{O}}_{0,C} \simeq k[[x_1, \ldots, x_n]]/(x_i x_j : i \neq j).$$

In higher dimensions, any union of coordinate subspaces in $\mathbb{A}^n$ is seminormal, even if it has components of different dimension.

A problem with seminormality in higher dimensions is that an irreducible component of a seminormal scheme need not be seminormal. A concrete example is given in (7.9.3). More generally, let $Z \subset \mathbb{C}^n$ be a reduced subvariety of pure dimension d such that $\dim \mathrm{Sing}\, Z \leq d - 2$. Let $X \supset Z$ be a general d-dimensional complete intersection. Then X is CM, seminormal and Z is a union of some of the irreducible components of X.

Let $X_i \subset X$ be the irreducible components. Then the target of the natural map $\coprod X_i \to X$ is seminormal, the fibers are reduced points (thus seminormal, even smooth) yet $\coprod X_i$ is not seminormal. However, if $f : X \to Y$ is flat, Y and the geometric fibers are seminormal then X is seminormal by Kollár (2010a, 4.1).

Definition 10.13 (Relative seminormalization)   Let X be a scheme and $\tau : V \to X$ a morphism. (The main examples we have in mind are open embeddings and finite morphisms.) A morphism $g : X' \to X$ is called a *partial seminormalization relative to V* if g is a partial seminormalization and one can factor τ through g. As in (10.11), for excellent schemes X there is a unique largest partial seminormalization relative to V; it is called the *seminormalization relative to V*. If τ is a proper surjection, we also call this the *seminormalization in V*.

We say that X is *seminormal relative to V* if it is isomorphic to its own seminormalization relative to V. Equivalently, if every partial seminormalization relative to V is an isomorphism. If τ is a proper surjection, we also say that X is *seminormal in V*.

(We will use this notion mainly when τ is birational on every irreducible component. The definition makes perfect sense in general, but it does not harmonize well with the usual concept of normalization of X in a field extension of $k(X)$.)

If X is seminormal, then it is also seminormal relative to any $\tau : V \to X$. As a partial converse, if V is seminormal and X is seminormal relative to V then X is seminormal. This follows from (10.16).

Similarly, a morphism $g : X' \to X$ is called a *partial normalization relative to V* if X' is reduced, g is finite, birational and τ factors through g. The concepts of *normalization relative to V* and *normal relative to V* are defined the obvious way. If τ is a proper surjection, it is usual to call this the *normalization in V*.

Let $\tau: U \hookrightarrow X$ be an open embedding. If $\mathcal{O}_X$ has depth ≥ 2 at every point of $X \setminus U$ then every finite, birational morphism $g: X' \to X$ that is an isomorphism over U is an isomorphism everywhere. Thus, in this case, X is seminormal relative to U. In particular, we obtain the following.

Lemma 10.14 *Let X be a scheme that satisfies Serre's condition S_2. A partial seminormalization $g: X' \to X$ is an isomorphism if and only if g is an isomorphism at all codimension 1 points $x \in X$.*

In particular, X is seminormal if and only if the local rings $\mathcal{O}_{X,x}$ are seminormal for all codimension 1 points $x \in X$. □

The next elementary property turns out to be surprisingly useful.

Lemma 10.15 *Let $g: Y \to X$ be a proper morphism of reduced schemes such that $g_* \mathcal{O}_Y = \mathcal{O}_X$. If Y is normal (resp. seminormal) than so is X.*

More generally, if $\tau: V \to X$ is a morphism and Y is normal (resp. seminormal) relative to $g^* \tau$: red$(Y \times_X V) \to Y$ then X is normal (resp. seminormal) relative to V.

Proof We prove the seminormal case; the proof of the normal case is then obtained by deleting every occurrence of the adverb "semi."

Assume that we have $\tau: V \xrightarrow{\tau'} X' \xrightarrow{\pi} X$ where π is a partial seminormalization.

Then π_Y: red$(Y \times_X X') \to Y$ is a partial seminormalization and we have

$$g^* \tau: \text{red}(Y \times_X V) \xrightarrow{\tau'_Y} \text{red}(Y \times_X X') \xrightarrow{\pi_Y} Y.$$

By assumption, π_Y is an isomorphism, so we can compose π_Y^{-1} with the second projection of $Y \times_X X'$ to factor g as $Y \to X' \to X$. This implies that $\pi_* \mathcal{O}_{X'} \subset g_* \mathcal{O}_Y = \mathcal{O}_X$, hence π is an isomorphism. □

Complement 10.15.1 There is a quite useful case of (10.15) when Y is only seminormal but we can conclude that X is normal. This holds if Y is S_2 and every irreducible component of the non-normal locus dominates an irreducible component of X. Indeed, in this case π_Y is an isomorphism over the generic point of every irreducible component of the non-normal locus, hence an isomorphism by (10.14).

10.16 (Seminormalization as a functor) Let $g: X \to Y$ be a morphism of schemes. Let $\tau: V \to Y$ be a morphism and $Y' \to Y$ a partial seminormalization relative to V. Assume that X is seminormal relative to red$(X \times_Y V)$. Then g lifts to $g': X \to Y'$.

To see this, consider the first projection π_1: red$(X \times_Y Y') \to X$. By (10.11.4), it is a partial seminormalization relative to red$(X \times_Y V)$, hence an isomorphism by assumption. Thus we can set $g' := \pi_2 \circ \pi_1^{-1}$.

This implies that seminormalization is a functor: any morphism $g: X \to Y$ lifts uniquely to $g^{\mathrm{sn}}: X^{\mathrm{sn}} \to Y^{\mathrm{sn}}$. In particular, any limit of seminormal schemes is seminormal. We will use this mainly for push-out diagrams (9.30)

$$\begin{array}{ccc} X_1 & \longrightarrow & X_2 \\ \downarrow & & \downarrow \\ X_3 & \longrightarrow & Y. \end{array}$$

We obtain that if the X_i are seminormal then so is Y.

The following is a restatement of Kollár (1996, I.7.2.5).

Lemma 10.17 *Consider a push-out diagram with finite morphisms*

$$\begin{array}{ccc} X_1 & \lhook\joinrel\longrightarrow & X_2 \\ {\scriptstyle u}\downarrow & & \downarrow {\scriptstyle v} \\ X_3 & \lhook\joinrel\longrightarrow & Y \end{array}$$

where X_2 is seminormal and X_3 is reduced. Then Y is seminormal if and only if X_1 is reduced and X_3 is seminormal relative to X_1. $\qquad\square$

10.18 (Construction of seminormal schemes) Let X be a reduced scheme with normalization $\pi: \bar{X} \to X$. Let $Z \subset X$ be the non-normal locus (that is, $X \setminus Z$ is the largest normal open subscheme) and let $\bar{Z} := \mathrm{red}\,\pi^{-1}(Z)$ be the reduced preimage of Z. Let $\sigma_Z: Z^* \to Z$ be the seminormalization of Z in $\bar{Z}$. We have a push-out diagram

$$\begin{array}{ccc} \bar{Z} & \lhook\joinrel\longrightarrow & \bar{X} \\ {\scriptstyle u}\downarrow & & \downarrow {\scriptstyle v} \\ Z^* & \lhook\joinrel\longrightarrow & X^*. \end{array}$$

Note that X^* is seminormal by (10.17) and by the universality of push-out there is a morphism $\sigma_X: X^* \to X$. Let $x \in X$ be a point. If $x \in X \setminus Z$ then σ_X is an isomorphism near x. If $x \in Z$ then

$$k\big(\mathrm{red}\,\sigma_X^{-1}(x)\big) = k\big(\mathrm{red}\,\sigma_Z^{-1}(x)\big) \simeq k(x)$$

since $\sigma_Z: Z^* \to Z$ is a partial seminormalization. Thus $\sigma_X: X^* \to X$ is the seminormalization of X.

Thus, a seminormal scheme X is equivalent to a triple

$$(Z \leftarrow \bar{Z} \hookrightarrow \bar{X})$$

where $\bar{X}$ is normal, $\bar{Z}$ is a reduced subscheme of $\bar{X}$, Z is seminormal in $\bar{Z}$ (thus Z is also reduced) and $Z \leftarrow \bar{Z}$ is not an isomorphism at any of the generic points of Z. From the triple one recovers X by the universal push-out (9.30)

$$
\begin{array}{ccc}
\bar{Z} & \longhookrightarrow & \bar{X} \\
\downarrow & & \downarrow{\scriptstyle\pi} \\
Z & \longhookrightarrow & X.
\end{array}
$$

In particular, $Z \subset X$ and $\bar{Z} \subset \bar{X}$ are the conductors (5.2) and they are reduced.

One can frequently use the above diagram to reduce a problem about a seminormal scheme X to questions involving its normalization $\bar{X}$ and the lower dimensional schemes $Z, \bar{Z}$.

This suggests the following concept. A reduced scheme X is *hereditarily seminormal* if it is seminormal and the conductors $D \subset X$ and $\bar{D} \subset \bar{X}$ are hereditarily seminormal. This approach lies behind the inductive set-up in Section 9.1.

Seminormal reducible schemes

As the examples (10.12) show, the irreducible components of a seminormal scheme need not be seminormal. In principle the following straightforward lemma reduces the computation of seminormalization to the irreducible components.

Lemma 10.19 *Let X be a scheme that is a union of its closed subschemes X_i. Then a regular function ϕ on X^{sn} is determined by giving regular functions ϕ_i on each X_i^{sn} such that $\phi_i(x) = \phi_j(x)$ for every $x \in X_i \cap X_j$.*

Equivalently, fix an ordering of the index set and set $X_{ij} := X_i \cap X_j$. Then there is an exact sequence

$$
0 \to \mathscr{O}_{X^{\mathrm{sn}}} \to \sum_i \mathscr{O}_{X_i^{\mathrm{sn}}} \xrightarrow{r_{ijk}} \sum_{j<k} \mathscr{O}_{X_{jk}^{\mathrm{sn}}} \tag{10.19.1}
$$

where r_{ijk} is the restriction map if $i = j$, minus the restriction map if $i = k$ and zero otherwise.

Proof The maps in (10.19.1) exist since seminormalization is a functor (10.16). To show exactness, let A be the subsheaf that makes the sequence

$$
0 \to A \to \sum_i \mathscr{O}_{X_i^{\mathrm{sn}}} \xrightarrow{r_{ijk}} \sum_{j<k} \mathscr{O}_{X_{jk}^{\mathrm{sn}}} \quad \text{exact.} \tag{10.19.2}
$$

Note that A consists of sequences of regular functions ϕ_i on X_i^{sn} such that $\phi_i(x) = \phi_j(x)$ for every $x \in X_i \cap X_j$. Then A is a sheaf of $\mathscr{O}_X$-algebras containing $\mathscr{O}_{X^{\mathrm{sn}}}$. In order to prove that $A = \mathscr{O}_{X^{\mathrm{sn}}}$, it is enough to show that $g \colon \mathrm{Spec}_X A \to X$ is a partial seminormalization of X. Let $\pi \colon \amalg X_i^{\mathrm{sn}} \to X$ be the natural morphism.

Pick $x \in X$ and assume that it is contained in the X_i for $i \in I(x)$. For each $i \in I(x)$, let $x_i \in X_i^{\mathrm{sn}}$ denote the unique preimage of x. Then

$$k(\operatorname{red} g^{-1}(x)) \subset \sum_{i \in I(x)} k(x_i)$$

consists of those sequences $\{\phi_i(x_i) : i \in I(x)\}$ such that $\phi_i(x_i) = \phi_j(x_j)$ for all $i, j \in I(x)$, where we used the natural identifications $\pi^*: k(x) \simeq k(x_i)$. Thus $k(\operatorname{red} g^{-1}(x))$ is the diagonally embedded $k(x)$. □

Next we study the following question. Given a scheme X with irreducible components X_i, which is the largest partial seminormalization of X whose irreducible components are still the X_i? This is very useful in the study of the moduli of pairs, mostly through (10.21) and (10.22), see Kollár (2011c).

Definition 10.20   Let X be a scheme which is a union of its closed, reduced subschemes X_i. The natural embeddings $X_i \hookrightarrow X$ give a morphism $\amalg_i X_i \to X$ of their disjoint union to X. We study schemes such that X is *seminormal in* $\amalg_i X_i$.

If the X_i are seminormal, then X is seminormal in $\amalg_i X_i$ if and only if it is seminormal. Thus this notion is new and interesting only if the X_i are not seminormal.

The following lemma gives a very useful characterization if there are only two components, but the case of more components is not so well behaved (10.23).

If $X_1 \cup \cdots \cup X_{i+1}$ is seminormal in $(X_1 \cup \cdots \cup X_i) \amalg X_{i+1}$ for $i = 1, \ldots, k-1$ then $X_1 \cup \cdots \cup X_k$ is seminormal in $X_1 \amalg \cdots \amalg X_k$. Thus we will be usually able to use the 2-component case to deal with more components.

Lemma 10.21   *Let X be a reduced scheme and $X_1, X_2 \subset X$ closed, reduced subschemes such that $X = X_1 \cup X_2$. Then X is seminormal in $X_1 \amalg X_2$ if and only if the intersection $X_1 \cap X_2$ is reduced.*

Proof   Let $r_i: \mathscr{O}_{X_i} \to \mathscr{O}_{X_1 \cap X_2}$ and $\bar{r}_i: \mathscr{O}_{X_i} \to \mathscr{O}_{\operatorname{red}(X_1 \cap X_2)}$ denote the restriction maps. Then $\mathscr{O}_X$ sits in an exact sequence

$$0 \longrightarrow \mathscr{O}_X \longrightarrow \mathscr{O}_{X_1} \oplus \mathscr{O}_{X_2} \xrightarrow{(r_1, -r_2)} \mathscr{O}_{X_1 \cap X_2} \longrightarrow 0.$$

The similar sequence

$$0 \longrightarrow A \longrightarrow \mathscr{O}_{X_1} \oplus \mathscr{O}_{X_2} \xrightarrow{(\bar{r}_1, -\bar{r}_2)} \mathscr{O}_{\operatorname{red}(X_1 \cap X_2)} \longrightarrow 0$$

defines a coherent sheaf of $\mathscr{O}_X$-algebras A and $\operatorname{Spec}_X A \to X$ is a partial seminormalization that is dominated by $X_1 \amalg X_2$. Since X is seminormal in $X_1 \amalg X_2$, we see that $A = \mathscr{O}_X$, hence $X_1 \cap X_2 = \operatorname{red}(X_1 \cap X_2)$.

Conversely, let $\mathrm{Spec}_X A \to X$ be a partial seminormalization dominated by $X_1 \amalg X_2$. Then A sits in a diagram

$$0 \longrightarrow A \longrightarrow \mathscr{O}_{X_1} \oplus \mathscr{O}_{X_2} \longrightarrow \mathscr{O}_{X_1 \cap X_2}/J \longrightarrow 0$$

where $J \subset \mathscr{O}_{X_1 \cap X_2}$ is a a nilpotent ideal. If $A \neq \mathscr{O}_X$ then $J \neq 0$, hence $X_1 \cap X_2$ is not reduced. □

Corollary 10.22 *Let X be a scheme that is a union of its closed, reduced subschemes X_i such that X is seminormal in $\amalg_i X_i$. Let Y_1, Y_2 be closed subschemes of X, each a union of some of the X_i and such that $X = Y_1 \cup Y_2$. Then $Y_1 \cap Y_2$ is reduced.*

Proof Since $\amalg_i X_i$ dominates $Y_1 \amalg Y_2$, we see that $Y_1 \cup Y_2$ is seminormal in $Y_1 \amalg Y_2$. Thus $Y_1 \cap Y_2$ is reduced by (10.21). □

The following example shows that even if X and the X_i are seminormal, the intersections $X_i \cap X_j$ can be rather complicated.

Example 10.23 Pick two polynomials $f, g \in k[x, y, z]$ such that fg has no multiple factors. Set

$$X := (u(u - g) = v(v - f) = 0) \subset \mathbb{A}^5(x, y, z, u, v).$$

It is a reduced complete intersection, hence CM. It has four irreducible components, each smooth. The codimension 1 singularities are ordinary double points, thus X is seminormal.

The intersection of the irreducible components $X_1 := (u = v = 0)$ and $X_2 := (u - g = v - f = 0)$ is isomorphic to $(f = g = 0) \subset \mathbb{A}^3(x, y, z)$. Thus the intersection $X_1 \cap X_2$ can be nonreduced.

We have the following special case of (10.15).

Lemma 10.24 *Let $g: Y \to X$ be a proper morphism of reduced schemes such that $g_* \mathscr{O}_Y = \mathscr{O}_X$. Let $X_i \subset X$ be closed, reduced subschemes such that $X = \cup_i X_i$. If Y is seminormal in $\amalg_i$ red $g^{-1}(X_i)$ then X is seminormal in $\amalg_i X_i$.*

Proof This follows from (10.15) once we note that $g^{-1}(X_i) = Y \times_X X_i$. □

There is one useful case when one can conclude that the irreducible components of a seminormal scheme are seminormal.

Lemma 10.25 *Let X be a scheme and $X_1, X_2 \subset X$ closed, reduced subschemes. If $X_1 \cup X_2$ and $X_1 \cap X_2$ are seminormal then X_1, X_2 are both seminormal.*

Proof The question is local, thus we may assume that all these schemes are affine. Let ϕ_1 be a regular function on the seminormalization of X_1. Its restriction is a regular function on the seminormalization of $X_1 \cap X_2$, hence a regular function on $X_1 \cap X_2$. Thus it can be extended to a regular function ϕ_2 on X_2. Thus ϕ_1, ϕ_2 glue to a regular function ϕ on the seminormalization of $X_1 \cup X_2$. The latter is seminormal, hence ϕ is a regular function on $X_1 \cup X_2$ and so $\phi_1 = \phi|_{X_1}$ is a regular function on X_1. $\qquad\square$

Seminormality of subschemes

We study how to descend seminormality conditions of subschemes through finite morphisms.

Lemma 10.26 *Let $g\colon Y \to X$ be a finite morphism of schemes of pure dimension d over a field of characteristic 0. Assume that X is normal. Let $Z \subset X$ be a closed, reduced subscheme and $\tau\colon V \to X$ a morphism. If $\operatorname{red} g^{-1}(Z)$ is normal (resp. seminormal) relative to $\operatorname{red}(Y \times_X V) \to Y$ then Z is normal (resp. seminormal) relative to V.*

Proof As in (10.15), we prove the seminormal case.

We may assume that X, Y are irreducible and affine. Let $\pi\colon Z' \to Z$ be a partial seminormalization relative to V. Pick $\phi \in \mathscr{O}_{Z'}$. Since $\operatorname{red} g^{-1}(Z)$ is seminormal relative to $\operatorname{red}(Y \times_X V)$, the pull back $\phi \circ g$ is a regular function on $\operatorname{red} g^{-1}(Z)$. We can lift it to a regular function Φ_Y on Y. Since X is normal, the normalized trace map (10.27) gives a regular function on X

$$\Phi_X := \tfrac{1}{\deg Y/X} \operatorname{tr}_{Y/X} \Phi_Y \quad \text{and} \quad \Phi_X|_Z = \phi \quad \text{by (10.27.4)}.$$

Thus Z is seminormal relative to V. $\qquad\square$

It may be better to think of this as follows. The normalized trace map splits the injection $\mathscr{O}_X \to g_*\mathscr{O}_Y$ and this induces a splitting of $\mathscr{O}_Z \to g_*\mathscr{O}_{\operatorname{red} g^{-1}(Z)}$. Then the above argument shows the following.

Complement 10.26.1 Let $g\colon Y \to X$ be a finite morphism of schemes such that $\mathscr{O}_X \to g_*\mathscr{O}_Y$ is a split injection. Let $Z \subset X$ be a closed, reduced subscheme and $\tau\colon V \to X$ a morphism. If $\operatorname{red} g^{-1}(Z)$ is normal (resp. seminormal) relative to $\operatorname{red}(Y \times_X V) \to Y$ then Z is normal (resp. seminormal) relative to V. $\qquad\square$

10.27 (Trace maps) Let $\pi\colon Y \to X$ be a finite, flat morphism. Then $\pi_*\mathscr{O}_Y$ is a locally free $\mathscr{O}_X$-sheaf. Multiplication by a section of $\mathscr{O}_Y$ is an $\mathscr{O}_X$-linear endomorphism of $\pi_*\mathscr{O}_Y$, thus it has a trace $\operatorname{tr}_{Y/X}\colon \pi_*\mathscr{O}_Y \to \mathscr{O}_X$.

If π has constant degree (for instance, if X is connected) and $\deg Y/X$ is invertible on X then we get the *normalized trace map*

$$\tfrac{1}{\deg Y/X} \operatorname{tr}_{Y/X}\colon \pi_*\mathscr{O}_Y \to \mathscr{O}_X \quad \text{such that} \quad \left(\tfrac{1}{\deg Y/X} \operatorname{tr}_{Y/X}\right) \circ \pi^* = \mathbf{1}_X.$$
$$(10.27.1)$$

That is, the normalized trace map splits the natural injection $\mathscr{O}_X \to \pi_*\mathscr{O}_Y$.

Let $\pi: Y \to X$ be a finite (not necessarily flat) morphism of schemes of pure dimension d. Assume that X is normal and irreducible. Then π is flat over the smooth locus $X^0 \subset X$, thus we have the trace map $\mathrm{tr}_{Y^0/X^0}: \pi_* \mathscr{O}_{Y^0} \to \mathscr{O}_{X^0}$, where $Y^0 := \pi^{-1}(X^0)$. Since $\mathscr{O}_X$ is S_2 and $X \setminus X^0$ has codimension ≥ 2, tr_{Y^0/X^0} automatically extends to the trace map $\mathrm{tr}_{Y/X}: \pi_* \mathscr{O}_Y \to \mathscr{O}_X$. As before, if $\deg Y/X$ is invertible on X then we get the *normalized trace map*

$$\tfrac{1}{\deg Y/X} \mathrm{tr}_{Y/X}: \pi_* \mathscr{O}_Y \to \mathscr{O}_X \quad \text{such that} \quad \left(\tfrac{1}{\deg Y/X} \mathrm{tr}_{Y/X} \right) \circ \pi^* = \mathbf{1}_X.$$
$$(10.27.2)$$

We also need a pointwise formula for $\mathrm{tr}_{Y/X}$. Pick a point $x \in X$, let its preimages be $y_1, \dots, y_r$ and $e(y_i)$ the local multiplicity of π at y_i, (as defined in Mumford (1976, section 3.A)). We claim that

$$(\mathrm{tr}_{Y/X}(\phi))(x) = \sum_i e(y_i) \cdot \mathrm{tr}_{k(y_i)/k(x)} \phi(y_i). \qquad (10.27.3)$$

In particular, if there is a $c \in k(x)$ such that $\phi(y_i) = c$ for every i then

$$(\mathrm{tr}_{Y/X}(\phi))(x) = \sum_i e(y_i) \cdot \mathrm{tr}_{k(y_i)/k(x)}(c) = \deg(Y/X) \cdot c. \quad (10.27.4)$$

If π is flat over x then (10.27.3) is clear from the definition. In general, choose a 1-dimensional regular local scheme R and a morphism $h: R \to X$ such that h maps the closed point $r \in R$ to x and the generic point to a smooth point of X. Set $Y_R := (Y \times_X R)/(\text{embedded points over } r)$ (but do not normalize). The reduced fiber over x and the local multiplicities are unchanged and (10.27.3) holds for $Y_R \to R$ since $Y_R \to R$ is flat.

If X is only seminormal, then the trace map is only partially defined. There is one important case when a normalized trace map is defined.

Claim 10.27.1 Let $\pi: Y \to X$ be a finite morphism, $Z_X \subset X$ a closed subset and $Z_Y := \mathrm{red}\, \pi^{-1}(Z_X)$. Assume that the characteristic is 0, X is seminormal, $X \setminus Z_X$ is normal, π is generically flat on $Y \setminus Z_Y$ and $\pi: Z_Y \to Z_X$ is a partial seminormalization. Then there is a normalized trace map

$$\tfrac{1}{\text{mult–deg}} \mathrm{tr}_{Y/X}: \pi_* \mathscr{O}_Y \to \mathscr{O}_X \quad \text{such that} \quad \left(\tfrac{1}{\text{mult–deg}} \mathrm{tr}_{Y/X} \right) \circ \pi^* = \mathbf{1}_X.$$

(Note that π can have different degrees over different irreducible components of X, so $\tfrac{1}{\text{mult–deg}} \mathrm{tr}_{Y/X}$ is a symbol, rather than an actual division by an integer.)

Proof To see this, let $\tau: \bar{X} \to X$ be the normalization with irreducible components $\bar{X}_i$. Over each $\bar{X}_i$ we have a normalized trace map which we can think of as $t_i: \mathscr{O}_Y \to \mathscr{O}_{Y \times_X \bar{X}_i} \to \mathscr{O}_{\bar{X}_i}$. Pick any $x \in Z_X$ and $\bar{x}_i \in \tau^{-1}(x) \cap \bar{X}_i$. Then (10.27.4) implies that

$$t_i(\phi)(\bar{x}_i) = \phi(x) \quad \text{for every } \phi \in \mathscr{O}_Y. \qquad (10.27.6)$$

By (10.11.6) this implies that the $t_i(\phi)$ descend to a local section of $\mathscr{O}_X$, giving $\tfrac{1}{\text{mult–deg}} \mathrm{tr}: \pi_* \mathscr{O}_Y \to \mathscr{O}_X$. $\qquad \square$

Example 10.28 The analog of (10.26) does not hold in positive characteristic, not even if g is separable.

For instance, set $Y = \operatorname{Spec} k[x, y, z, t]$ with char $k = 2$. Then $\sigma(x, y, z, t) = (x + y, y, z + t, t)$ determines a $\mathbb{Z}/2$-action; set $X = Y/(\sigma)$. One can compute that $X = \operatorname{Spec} k[x(x + y), y, z(z + t), t, xz + (x + y)(z + t)]$. Set $W := (t = 0) \simeq \operatorname{Spec} k[x, y, z]$. Then W is σ-invariant and $W/(\sigma) \simeq \operatorname{Spec} k[x(x + y), y, z]$. By contrast the image of W in X is the non-normal variety $Z := \operatorname{Spec} k[x(x + y), y, z^2, yz]$.

Note that both $Y \to X$ and $W \to Z$ are generically étale.

The next lemma is a generalization of (10.25).

Lemma 10.29 *Consider a push-out diagram with finite morphisms*

$$
\begin{array}{ccc}
X_1 & \longhookrightarrow & X_2 \\
\downarrow{\scriptstyle u} & & \downarrow{\scriptstyle v} \\
X_3 & \longhookrightarrow & Y.
\end{array}
$$

Let $Z \subset Y$ be a closed, reduced subscheme. Assume that

(1) $\operatorname{red}(X_3 \cap Z)$ *is seminormal and*
(2) $X_1 \cup \operatorname{red}(v^{-1}(Z))$ *is seminormal.*

Then Z is seminormal.

Proof We may assume that Y and hence all the X_i are affine. Let $Z' \to Z$ be a partial seminormalization and pick $\phi \in \mathcal{O}_{Z'}$. By assumption (1), ϕ restricts to a regular function on $\operatorname{red}(X_3 \cap Z)$, hence it can be extended to a regular function ϕ_3 on X_3. Thus $\phi_3 \circ u$ is a regular function on X_1 and $\phi \circ v$ is a function on the seminormalization of $\operatorname{red}(v^{-1}(Z))$, and these agree over the reduced intersection. By (10.19), the pair $(\phi_3 \circ u, \phi \circ v)$ determines a regular function Φ_1 on the seminormalization of $X_1 \cup \operatorname{red}(v^{-1}(Z))$. Since $X_1 \cup \operatorname{red}(v^{-1}(Z))$ is seminormal, we conclude that Φ_1 is a regular function on $X_1 \cup \operatorname{red}(v^{-1}(Z))$. Thus we can extend Φ_1 to a regular function Φ_2 on X_2. Since $\Phi_2|_{X_1} = \phi_3 \circ u$, by the universal property of push-forward, Φ_2 descends to a regular function Φ_Y on Y. Since $\phi = \Phi_Y|_Z$, we conclude that Z is seminormal. $\qquad\square$

Lemma 10.30 *Let $g\colon Y \to X$ be a finite morphism of normal schemes over a field of characteristic 0. Let $Z_i \subset X$ be closed, reduced subschemes. If $\operatorname{red} g^{-1}(\cup_i Z_i)$ is seminormal in $\amalg_i \operatorname{red} g^{-1}(Z_i)$ then $\cup_i Z_i$ is seminormal in $\amalg_i Z_i$.*

Proof This is a special case of (10.26), but here is a direct proof. We may assume that X, Y are irreducible and affine. Pick $\phi_i \in \mathcal{O}_{Z_i}$ that agree on $\operatorname{red}(Z_i \cap Z_j)$. Then $\phi_i \circ g$ is a regular function on $\operatorname{red} g^{-1}(Z_i)$ and any two agree on the reduced intersection. Thus, by assumption, there is a regular function ψ on red $g^{-1}(\cup_i Z_i)$ whose restriction to red $g^{-1}(Z_i)$ is $\phi_i \circ g$. We can lift ψ to a regular function Ψ_Y on Y. Since X is normal,

$$\Psi_X := \tfrac{1}{\deg Y/X} \operatorname{tr}_{Y/X} \Psi_Y$$

is regular on X and $\Psi_X|_{Z_i} = \phi_i$. Thus $\cup_i Z_i$ is seminormal in $\amalg_i Z_i$. □

Lemma 10.31 *Consider a push-out diagram with finite morphisms*

$$
\begin{array}{ccc}
X_1 & \hookrightarrow & X_2 \\
\downarrow{\scriptstyle u} & & \downarrow{\scriptstyle v} \\
X_3 & \hookrightarrow & Y.
\end{array}
$$

Let $Z_i \subset Y$ be closed, reduced subschemes. Assume that

(1) $\cup_i \operatorname{red}(Z_i \cap X_3)$ *is seminormal in* $\amalg_i \operatorname{red}(Z_i \cap X_3)$ *and*
(2) $X_1 \cup \bigcup_i \operatorname{red} v^{-1}(Z_i)$ *is seminormal in* $X_1 \amalg \coprod_i \operatorname{red} v^{-1}(Z_i)$.

Then $Z_1 \cup \cdots \cup Z_r$ is seminormal in $Z_1 \amalg \cdots \amalg Z_r$.

Proof We may assume that Y, and hence all the X_i are affine. Pick $\phi_i \in \mathscr{O}_{Z_i}$ that agree on the reduced intersections. Restricting these to $Z_i \cap X_3$ and using (1) we see that there is a regular function $\psi_3 \in \mathscr{O}_{X_3}$ such that ψ_3 and the ϕ_i agree on the reduced pairwise intersections. Then the $\phi_i \circ v$ and $\psi_3 \circ v$ have the analogous property on the red $v^{-1}(Z_i)$ and X_1. Thus, using (2) they glue to a regular function ψ on red $v^{-1}(Z_1) \cup \cdots \cup \operatorname{red} v^{-1}(Z_r) \cup X_1$, which in turn can be extended to a regular function Ψ_2 on X_2. By construction, $\Psi_2|_{X_1} = \psi_3 \circ u$, thus, by the universal property of push-outs, there is a regular function Ψ_Y on Y such that $\Psi_2 = \Psi_Y \circ v$.

The restriction of Ψ_Y to $Z_1 \cup \cdots \cup Z_r$ is regular, and $\Psi_Y|_{Z_i} = \phi_i$. Thus $Z_1 \cup \cdots \cup Z_r$ is seminormal in $Z_1 \amalg \cdots \amalg Z_r$. □

10.3 Vanishing theorems

Here we summarize the vanishing theorems that we use and derive other variants for which there is no convenient reference.

Assumption In this section we work with schemes (or algebraic spaces) that are of finite type over a base scheme S that is essentially of finite type over a field of characteristic 0.

Most of the results are not known for more general base schemes, for instance for spectra of complete local rings. This is sometimes quite inconvenient. Note, however, that (10.32) is known for morphisms of complex spaces (Nakayama,

1987, section 3) and so are the basic consequences (10.34–10.38). However, the more delicate (10.39–10.41) are known in the algebraic cases only.

The basic result is the following relative version of the Kawamata–Viehweg vanishing theorem. For proofs, see Kawamata *et al.* (1987) or Kollár and Mori (1998).

Theorem 10.32 *Let* $f\colon Y \to X$ *be a proper morphism,* Y *smooth. Let* L *be an* f*-nef* $\mathbb{Q}$*-divisor on* Y *that is big on the generic fiber of* f *and* Δ *an effective simple normal crossing divisor on* Y *such that* $\lfloor \Delta \rfloor = 0$*. Let* M *be a line bundle on* Y *and assume that* $M \sim_{\mathbb{Q},f} K_Y + L + \Delta$*. Then* $R^i f_* M = 0$ *for* $i > 0$. $\square$

If L is not big, Y is not smooth or if $\lfloor \Delta \rfloor \neq 0$, then vanishing fails in general. There are, however, some easy-to-derive consequences if we control the restriction of L to all possible lc centers $Z \subset Y$. To get the sharpest result, we need two definitions.

Definition 10.33 Let Z be a projective variety and D a nef $\mathbb{Q}$-Cartier divisor. The *numerical Kodaira dimension* $\nu(D)$ is the largest integer $m \geq 0$ such that the self-intersection (D^m) is not numerically trivial.

Equivalently, fix an ample divisor H. Then $\nu(D)$ is the largest integer $m \geq 0$ such that $(D^m \cdot H^{\dim Z - m}) > 0$. We see that $0 \leq \nu(D) \leq \dim Z$.

If $H \subset W$ is an ample divisor, then $\nu(D|_H) = \min\{\nu(D), \dim H\}$.

If $\nu(D) = 0$ then D is numerically trivial and if $\nu(D) = \dim Z$ then D is big.

Corollary 10.34 *Let* $(Y, \sum_i D_i)$ *be a semi-snc pair as in* (1.10) *and* $f\colon Y \to X$ *a projective morphism. Let* L *be an* f*-nef* $\mathbb{Q}$*-divisor on* Y *and* M *a line bundle on* Y *such that*

$$M \sim_{\mathbb{Q},f} K_Y + L + \sum_i a_i D_i \quad \text{where } 0 \leq a_i \leq 1.$$

Let $Z \subset Y$ *be an lc center of* $(Y, \sum_i a_i D_i)$ *(that is, a stratum of* $(Y, \lfloor \sum_i a_i D_i \rfloor)$*) and* $F_Z \subset Z$ *the generic fiber of* $f|_Z \colon Z \to f(Z)$*. Set*

$$c := \max\{\dim F_Z - \nu(L|_{F_Z}) : Z \text{ is an lc center}\}.$$

Then $R^i f_* M = 0$ *for* $i > c$.

Proof The proof is through a series of reduction steps, ending up with (10.32).

Step 1 Reduction to the case where Y is irreducible.

We use induction on the number of irreducible components. Let $Y_1 \subset Y$ be any of the irreducible components, Y_1' the union of the other components and $B_1 = Y_1 \cap Y_1'$. We have an exact sequence

$$0 \to M|_{Y_1}(-B_1) \to M \to M|_{Y_1'} \to 0. \tag{10.34.1}$$

Since $K_Y|_{Y_1} \sim K_{Y_1} + B_1$ and $K_Y|_{Y_1'} \sim K_{Y_1'} + B_1$, we see that

$$M|_{Y_1}(-B_1) \sim_{\mathbb{Q},f} K_{Y_1} + L|_{Y_1} + \sum_i a_i D_i|_{Y_1} \quad \text{and}$$
$$M|_{Y_1'} \sim_{\mathbb{Q},f} K_{Y_1'} + L|_{Y_1'} + B_1 + \sum_i a_i D_i|_{Y_1'}.$$

By induction, $R^i f_*(M|_{Y_1}(-B_1))$ and $R^i f_*(M|_{Y_1'})$ vanish for $i > c$. These imply that $R^i f_* M = 0$ for $i > c$.

Step 2 Now Y is irreducible, smooth and we deal with the divisors D_i whose coefficient is $a_i = 1$.

Pick any $D_j \subset \lfloor \sum a_i D_i \rfloor$ and consider the exact sequence

$$0 \to M(-D_j) \to M \to M|_{D_j} \to 0. \tag{10.34.2}$$

Note that $M(-D_j) \sim_{\mathbb{Q},f} K_Y + L + \sum_{i \neq j} a_i D_i$, hence $R^i f_* M(-D_j) = 0$ for $i > c$ by induction. Furthermore,

$$M|_{D_j} \sim_{\mathbb{Q},f} K_{D_j} + L|_{D_j} + \sum_{i \neq j} a_i D_i|_{D_j}$$

and the strata of the restriction are also strata of the original pair. Thus, by induction, $R^i f_*(M|_{D_j}) = 0$ for $i > c$. Putting these into the long exact sequence for the push-forward of (10.34.2), induction on the dimension and on the number of divisors in $\lfloor \sum a_i D_i \rfloor$ completes this step.

Step 3 Now Y is smooth and $(Y, \sum a_i D_i)$ is klt. We reduce to the case where (10.32) applies. The only lc center is Y itself, thus $c = \dim F_Y - \nu(L|_{F_Y})$. If $\nu(L|_{F_Y}) = \dim F_Y$ then (10.32) gives what we need.

Thus assume that $\nu(L|_{F_Y}) < \dim F_Y$ and let H be a smooth sufficiently ample divisor on Y. Consider the exact sequence

$$0 \to M \to M(H) \to M(H)|_H \to 0. \tag{10.34.3}$$

Note that $M(H)|_H \sim_{\mathbb{Q},f} K_H + L_H + \sum_i a_i D_i|_H$.

Since $\nu(L|_{F_Y}) < \dim F_Y$, (10.33) implies that $\nu(L|_{F_H}) = \nu(L|_{F_Y})$. The dimension of the general fiber drops by 1 as we go from Y to H, hence the value of c drops by 1. Therefore, by induction on the dimension, $R^i f_*(M(H)|_H) = 0$ for $i > c - 1$. Since H is sufficiently ample, $R^i f_* M(H) = 0$ for $i > 0$. Putting these into the long exact sequence for the push-forward of (10.34.3) completes the final reduction. □

We get especially simple criteria for the vanishing of $R^i f_* M$ when i is very close to $\dim Y$.

Corollary 10.35 *Notation and assumptions as in* (10.34). *Set* $\Delta = \sum_i a_i D_i$ *and* $n = \dim Y$. *Assume that* f *is generically finite. Then*

(1) $R^n f_* M = 0$.
(2) $R^{n-1} f_* M = 0$ *unless there is a divisor* $B \subset \operatorname{Sing} Y \cup \lfloor \Delta \rfloor$ *such that* $\dim f(B) = 0$.

(3) $R^{n-2} f_* M = 0$ *unless there are divisors* $B_1, B_2 \subset \operatorname{Sing} Y \cup \lfloor \Delta \rfloor$ *such that either* $\dim f(B_1) \leq 1$ *or* $\dim f(B_1 \cap B_2) = 0$. $\square$

The next result shows how to generalize various vanishing theorems from the snc cases considered above to dlt pairs.

Proposition 10.36 *Let* (X, Δ_X) *be a dlt (or semi-dlt) pair* (5.19). *Let* D *be a* $\mathbb{Q}$-*Cartier* $\mathbb{Z}$-*divisor on* X *such that* $D \sim_{\mathbb{Q}} K_X + \Delta_X + M$ *for some* $\mathbb{Q}$-*Cartier* $\mathbb{Q}$-*divisor* M.

Then there is a proper birational morphism $g : Y \to X$, *a Cartier divisor* D_Y *and a* $\mathbb{Q}$-*divisor* $\Delta_{D,Y}$ *on* Y *such that*

(1) $(Y, \Delta_{D,Y})$ *is snc (or semi-snc)*,
(2) $D_Y \sim_{\mathbb{Q}} K_Y + \Delta_{D,Y} + g^* M$,
(3) g *maps every lc center of* $(Y, \Delta_{D,Y})$ *birationally to an lc center of* (X, Δ_X),
(4) $R^i g_* \mathcal{O}_Y(D_Y) = 0$ *for* $i > 0$,
(5) $g_* \mathcal{O}_Y(D_Y) = \mathcal{O}_X(D)$ *and*
(6) *for any proper morphism* $f : X \to S$,

$$R^i f_* \mathcal{O}_X(D) \simeq R^i (f \circ g)_* \mathcal{O}_Y(D_Y) \quad \text{for every } i.$$

Proof By (10.45) and (10.59) there is a thrifty log resolution (resp. semi-log-resolution) $g : Y \to X$ of $(X, D + \Delta_X)$, that is, g maps every lc center of (Y, Δ_Y) birationally to an lc center of (X, Δ_X) where $K_Y + \Delta_Y \sim g^*(K_X + \Delta_X)$.

Note that $g^* D$ is a $\mathbb{Q}$-divisor; we can write it as $\lfloor g^* D \rfloor + \Delta_D$ where $\lfloor \Delta_D \rfloor = 0$ and Δ_D is g-exceptional. We can thus write

$$\Delta_Y - \Delta_D = g_*^{-1} \lfloor \Delta_X \rfloor + \Delta' - B$$

where $\lfloor \Delta' \rfloor = 0$, Δ' has no common irreducible components with $g_*^{-1} \lfloor \Delta_X \rfloor$ and B is an effective, g-exceptional $\mathbb{Z}$-divisor. Finally set

$$\Delta_{D,Y} := g_*^{-1} \lfloor \Delta_X \rfloor + \Delta' \quad \text{and} \quad D_Y := \lfloor g^* D \rfloor + B.$$

Then (2) holds. Since (Y, Δ_Y) and $(Y, \Delta_{D,Y})$ have the same lc centers, (3) also holds and so (10.34) implies (4). (5) follows from (7.30). By (4) the Leray spectral sequence

$$R^i f_* (R^j g_* \mathcal{O}_Y(D_Y)) \Rightarrow R^{i+j}(f \circ g)_* \mathcal{O}_Y(D_Y)$$

gives $R^i f_* (g_* \mathcal{O}_Y(D_Y)) = R^i (f \circ g)_* \mathcal{O}_Y(D_Y)$, and so (5) implies (6). $\square$

As a first consequence, we get the following general version of (10.32).

Theorem 10.37 *Let* (Y, Δ) *be klt*, $f : Y \to X$ *a proper morphism and* L *an* f-*nef* $\mathbb{Q}$-*divisor on* Y *that is big on the generic fiber of* f. *Let* D *be a* $\mathbb{Q}$-*Cartier* $\mathbb{Z}$-*divisor such that* $D \sim_{\mathbb{Q},f} K_Y + L + \Delta$. *Then* $R^i f_* \mathcal{O}_Y(D) = 0$ *for* $i > 0$. $\square$

Using (10.36) and (10.34) together gives the following.

Corollary 10.38 *Let* (Y, Δ) *be a dlt pair or a semi-dlt pair and* $f\colon Y \to S$ *a projective morphism. Let* L *be an* f-*nef* $\mathbb{Q}$-*divisor on* Y *and* D *a* $\mathbb{Q}$-*Cartier* $\mathbb{Z}$-*divisor such that* $D \sim_{\mathbb{Q}, f} K_Y + L + \Delta$. *Then*

(1) $R^i f_* \mathcal{O}_Y(D) = 0$ *for* $i > 0$ *if* $f|_Z\colon Z \to S$ *is generically finite for every lc center* Z *of* (Y, Δ).

(2) $R^i f_* \mathcal{O}_Y(D) = 0$ *for* $i > \max_Z\{\dim F_Z - \nu(L|_{F_Z})\}$ *where* Z *runs through all lc centers of* (Y, Δ) *and* $F_Z \subset Z$ *is the generic fiber of* $f|_Z\colon Z \to S$.

(3) $R^i f_* \mathcal{O}_Y(D) = 0$ *for* $i > 0$ *if* $L|_{F_Z}$ *is big for every lc center* Z *of* (Y, Δ). □

In the next few theorems the difference between semiample and nef divisors (1.4) is crucial.

Theorem 10.39 (Ambro, 2003, 3.2; Fujino, 2009b, 2.39; Fujino, 2009a, 6.3; Fujino, 2012b, 1.1)   *Let* (Y, Δ) *be a dlt pair (or a semi-dlt pair) and* $f\colon Y \to S$ *a proper morphism. Let* L *be an* f-*nef* $\mathbb{Q}$-*divisor on* Y *and* D *a* $\mathbb{Q}$-*Cartier* $\mathbb{Z}$-*divisor such that* $D \sim_{\mathbb{Q}, f} K_Y + L + \Delta$. *Assume that for every lc center* Z *of* (Y, Δ), *the restriction of* L *to the generic fiber of* $Z \to S$ *is semiample.*

Then every associated prime of $R^i f_* \mathcal{O}_Y(D)$ *is the* f-*image of an lc center of* (Y, Δ).[1]

Proof   If (Y, Δ) is simple normal crossing, this is proved in the references. In contrast with (10.34–10.38), this part of the proof is quite difficult and subtle.

The general case can be derived from it using (10.36) and (10.59) or (10.58). □

Two special cases, with easier proofs, have been known earlier.

Corollary 10.40 *Let* (Y, Δ) *be a pair and* $f\colon Y \to S$ *a dominant, projective morphism. Let* L *be an* f-*semiample* $\mathbb{Q}$-*divisor on* Y *and* D *a* $\mathbb{Q}$-*Cartier* $\mathbb{Z}$-*divisor such that* $D \sim_{\mathbb{Q}, f} K_Y + L + \Delta$. *Assume that*

(1) *either* (Y, Δ) *is klt (Kollár, 1986a),*

(2) *or* (Y, Δ) *is dlt and every lc center dominates* S *(Fujino, 2004).*

Then the higher direct images $R^i f_* \mathcal{O}_Y(D)$ *are torsion free.* □

The following theorem is proved in Kollár (1986b, 3.1) for $D = K_Y$. This implies the general case by the usual reduction methods of the general Kodaira vanishing theorems; see Kollár and Kovács (2010, 3.2) and Kollár and Mori (1998, section 2.5) for details.

---

[1] For recent improvements see Ambro (2012) and Fujino (2012a).

Theorem 10.41   *Let $f: Y \to S$ be a projective morphism and D a $\mathbb{Q}$-Cartier $\mathbb{Z}$-divisor on Y. Assume that $D \sim_{\mathbb{Q}, f} K_Y + L + \Delta$ where (Y, Δ) is klt and L is an f-semiample $\mathbb{Q}$-divisor. Then*

$$R f_* \mathcal{O}_Y(D) \simeq_{qis} \sum_i R^i f_* \mathcal{O}_Y(D)[-i].$$

That is, the direct image in the derived category is quasi-isomorphic to the sum of its cohomology sheaves.

In particular, the natural map $\iota\colon f_* \mathcal{O}_Y(D) \to R f_* \mathcal{O}_Y(D)$ admits a left inverse $\iota'\colon R f_* \mathcal{O}_Y(D) \to f_* \mathcal{O}_Y(D)$.   □

The following Du Bois version of Kodaira's vanishing is also useful.

Theorem 10.42 (Kollár, 1993b, 9.12)   *Let X be a projective, pure dimensional scheme that is Du Bois and L an ample line bundle. Then $h^i(X, L^{-m}) = h^i(X, L^{-1})$ for $i < \dim X$ and $m \geq 1$. Thus if X is also CM then $h^i(X, L^{-1}) = 0$ for $i < \dim X$.*   □

The next example shows the limits of the vanishing theorems.

Example 10.43 (Kodaira's vanishing fails on singular varieties)   Let Y be a smooth, projective variety and L a negative line bundle on Y. Set $P := \mathbb{P}_Y(\mathcal{O}_Y^{\oplus n} \oplus L)$. Let $X \subset P$ be the hypersurface defined by a general section of $\mathcal{O}_P(n + 1)$ with projection $\pi: X \to Y$. Thus X has hypersurface singularities, the singular set has codimension $n - 1$ and so X is normal for $n \geq 3$.

Furthermore, $X \to Y$ is a fiber bundle whose fiber is a cone $C_n \subset \mathbb{P}^n$ with trivial canonical class. Note that the singularity of C_n at its vertex is pretty much the simplest non-lc hypersurface singularity since every isolated singularity of a hypersurface of degree $< n + 1$ is lc.

By the adjunction formula $\omega_{X/Y} \simeq \pi^* L$, hence $\pi_* \omega_X \simeq L \otimes \omega_Y$.

Let H be an ample line bundle on Y. Then $H_P := \mathcal{O}_P(1) \otimes \pi^*(H \otimes L^{-1})$ is ample on P. We aim to compute the cohomology of $\omega_X \otimes H_X$ where $H_X := H_P|_X$. The higher direct images are zero, thus

$$
\begin{aligned}
H^i(X, \omega_X \otimes H_P) &\simeq H^i(Y, \pi_*(\omega_X \otimes H_P)) \\
&\simeq H^i(Y, L \otimes \omega_Y \otimes (\mathcal{O}_Y^{\oplus n} \oplus L) \otimes H \otimes L^{-1}) \\
&\simeq H^i(Y, \omega_Y \otimes H)^{\oplus n} \oplus H^i(Y, \omega_Y \otimes H \otimes L) \\
&\simeq H^i(Y, \omega_Y \otimes H \otimes L) \qquad\qquad \text{(for } i \geq 1),
\end{aligned}
$$

where in the last row we used that $H^i(Y, \omega_Y \otimes H) = 0$ by Kodaira's vanishing.

The interesting aspect is that by suitable choice of L, H we can arrange $\omega_Y \otimes H \otimes L$ to be any line bundle M on Y. Indeed, choose H sufficiently ample and set $L := M \otimes \omega_Y^{-1} \otimes H^{-1}$. Thus we have established the following.

Claim 10.43.1   Let Y be a smooth, projective variety and M a line bundle on Y. Then there is a variety X with normal, hypersurface singularities and an ample line bundle H_X on X such that

$$H^i(X, \omega_X \otimes H_X) \simeq H^i(Y, M) \quad \text{for } 1 \leq i \leq \dim Y. \qquad \square$$

The following duality was used in Section 7.3.

Proposition 10.44   *Let $f \colon Y \to X$ be a proper morphism, Y CM. Let M be a vector bundle on Y and $x \in X$ a closed point. Set $n = \dim Y$ and let $W \subset Y$ be a subscheme such that $\operatorname{Supp} W = \operatorname{Supp} f^{-1}(x)$. Then there is a natural bilinear pairing*

$$H_W^i(Y, \omega_Y \otimes M^{-1}) \times (R^{n-i} f_* M)_x \to k(x)$$

which has no left or right kernel, where the subscript denotes the stalk at x.
In particular, if either $H_W^i(Y, \omega_Y \otimes M^{-1})$ or $(R^{n-i} f_* M)_x$ is a finite dimensional $k(x)$-vector space then so is the other and they are dual to each other.

Sketch of proof   Let $mW \subset Y$ be the subscheme defined by the ideal sheaf $\mathcal{O}_Y(-W)^m$. By Grothendieck (1968, II.6)

$$H_W^i(Y, \omega_Y \otimes M^{-1}) = \varinjlim \operatorname{Ext}_Y^i(\mathcal{O}_{mW}, \omega_Y \otimes M^{-1}). \tag{10.44.1}$$

On the other hand, by the theorem on Formal Functions,

$$(R^{n-i} f_* M)^\wedge = \varprojlim H^{n-i}(mW, M|_{mW}) \tag{10.44.2}$$

where $\wedge$ denotes completion at $x \in X$.

It is thus enough to show that, for every m, the groups on the right hand sides of (10.44.1–2) are dual to each other. This gives the required bilinear pairing which has no left or right kernel.

Let us assume first that there is a compactification $\bar{Y} \supset Y$ such that $\bar{Y}$ is CM and M extends to a vector bundle $\bar{M}$ on $\bar{Y}$. Since W is disjoint from $\bar{Y} \setminus Y$,

$$\operatorname{Ext}_Y^i(\mathcal{O}_{mW}, \omega_Y \otimes M^{-1}) = \operatorname{Ext}_{\bar{Y}}^i(\mathcal{O}_{mW}, \omega_{\bar{Y}} \otimes \bar{M}^{-1}) = \operatorname{Ext}_{\bar{Y}}^i(\mathcal{O}_{mW} \otimes \bar{M}, \omega_{\bar{Y}})$$

and, by Serre duality, the latter is dual to

$$H^{n-i}(\bar{Y}, \mathcal{O}_{mW} \otimes \bar{M}) = H^{n-i}(mW, \bar{M}|_{mW}) = H^{n-i}(mW, M|_{mW}).$$

In our applications, Y is smooth or snc and M is a line bundle. Then the above assumptions about $\bar{Y}$ and $\bar{M}$ are satisfied.

Two approaches to the general case are discussed in Kollár (2011a, proposition 19). One can use either an arbitrary compactification $\bar{Y}$ and Grothendieck duality as above, or a more careful comparison of the above Ext groups with cohomology groups on mW. The latter argument also works in the complex analytic case. $\qquad \square$

10.4 Semi-log resolutions

The aim of this section is to discuss resolution theorems that are useful in the study of semi-log canonical varieties.

The basic existence result on resolutions was established by Hironaka (1964). We also need a strengthening of it, due to Szabó (1994), see also Bierstone and Milman (1997, section 12).

Theorem 10.45 (Existence of log resolutions) *Let X be a scheme or an algebraic space of finite type over a field of characteristic 0 and D a Weil divisor on X.*

(1) *Hironaka (1964) (X, D) has a log resolution (1.12).*
(2) *Szabó (1994); Bierstone and Milman (1997) (X, D) has a log resolution $f: X' \to X$ such that f is an isomorphism over the snc locus of (X, D) (1.7).*

Corollary 10.46 (Resolution in families) *Let C be a smooth curve over a field of characteristic 0, $f: X \to C$ a flat morphism and D a divisor on X. Then there is a log resolution $g: Y \to X$ such that $g_*^{-1}D + \mathrm{Ex}(g) + Y_c$ is a snc divisor for every $c \in C$ where Y_c denotes the fiber over a point c.*

Proof Let $p: X' \to X$ be any log resolution of (X, D). There are only finitely many fibers of $f \circ p$ such that $p^{-1}D + \mathrm{Ex}(p) + X'_c$ is not a snc divisor. Let these be $\{X'_{c_i}: i \in I\}$. Let $p': Y \to X'$ be a log resolution of $(X', p^{-1}D + \mathrm{Ex}(p) + \sum_i X'_{c_i})$ that is an isomorphism over $X' \setminus \cup_i X'_{c_i}$ and set $g = p' \circ p$. Then $g_*^{-1}D + \mathrm{Ex}(g) + \sum_i Y_{c_i}$ is a snc divisor. Thus, if $c = c_i$ for some i then $g_*^{-1}D + \mathrm{Ex}(g) + Y_c$ is a snc divisor. For other $c \in C$, the map p' is an isomorphism near Y_c, and $p^{-1}D + \mathrm{Ex}(p) + X'_c$ is an snc divisor, hence so is $g_*^{-1}D + \mathrm{Ex}(g) + Y_c$. $\square$

Next we show how (10.45.2) can be reduced to the Hironaka-type resolution theorems presented in Kollár (2007d). The complication is that the Hironaka method and its variants proceed by induction on the multiplicity. Thus, for instance, the method would normally blow-up every triple point of D before dealing with the non-snc double points. In the present situation, however, we want to keep the snc triple points untouched.

We can start by resolving the singularities of X, thus it is no restriction to assume from the beginning that X is smooth. To facilitate induction, we work with a more general resolution problem.

Definition 10.47 Consider the object $(X, I_1, \dots, I_m, E)$ where X is a smooth variety, the I_j are ideal sheaves of Cartier divisors and E a snc divisor. We say that $(X, I_1, \dots, I_m, E)$ has *simple normal crossing* or *snc* at a point $p \in X$ if

X is smooth at p and there are local coordinates $x_1, \ldots, x_r, x_{r+1}, \ldots, x_n$ and an injection $\sigma : \{1, \ldots, r\} \to \{1, \ldots, m\}$ such that

(1) $I_{\sigma(i)} = (x_i)$ near p for $1 \le i \le r$ and $p \notin \mathrm{cosupp}\, I_j$ for every other I_j;
(2) Supp $E \subset (\prod_{i>r} x_i = 0)$ near p.

Thus $E + \sum_j \mathrm{cosupp}\, I_j$ has snc support near p, but we also assume that no two of E, cosupp $I_1, \ldots,$ cosupp I_m have a common irreducible component near p. Furthermore, the I_j are assumed to vanish with multiplicity 1, but we do not care about the multiplicities in E. The definition is chosen mainly to satisfy the following restriction property:

(3) Assume that X is smooth, I_1 is the ideal sheaf of a smooth divisor $S \subset X$, $E + S$ is a snc divisor and none of the irreducible components of S is contained in E or in cosupp I_j for $j > 1$. Then $(X, I_1, \ldots, I_m, E)$ is snc near S if and only if $(S, I_2|_S, \ldots, I_m|_S, E|_S)$ is snc.

The set of all points where $(X, I_1, \ldots, I_m, E)$ is snc is open. It is denoted by $\mathrm{snc}(X, I_1, \ldots, I_m, E)$.

Definition 10.48 Let $Z \subset X$ be a smooth, irreducible subvariety that has simple normal crossing with E (cf. Kollár (2007d, 3.25)). Let $\pi \colon B_Z X \to X$ denote the blow-up with exceptional divisor $F \subset B_Z X$. Define the *birational transform* of $(X, I_1, \ldots, I_m, E)$ as

$$\left(X' := B_Z X, I'_1, \ldots, I'_m, E' := \pi_{\mathrm{tot}}^{-1} E \right) \qquad (10.48.1)$$

where $I'_j = g^* I_j(-F)$ if $Z \subset \mathrm{cosupp}\, I_j$ and $I'_j = g^* I_j$ if $Z \not\subset \mathrm{cosupp}\, I_j$. Note that if Z has codimension 1, then $X' = X$ but $I'_j = I_j(-Z)$ whenever $Z \subset \mathrm{cosupp}\, I_j$.

By an elementary computation, the birational transform commutes with restriction to a smooth subvariety (cf. Kollár (2007d, 3.62)). As in Kollár (2007d, 3.29) we can define blow-up sequences.

The assertion (10.45.2) will be a special case of the following result.

Proposition 10.49 *Let X be a smooth variety, E an snc divisor on X and I_j ideal sheaves of Cartier divisors. Then there is a smooth blow-up sequence*

$$\Pi : \left(X_r, I_1^{(r)}, \ldots, I_m^{(r)}, E^{(r)} \right) \to \cdots \to \left(X_1, I_1^{(1)}, \ldots, I_m^{(1)}, E^{(1)} \right)$$
$$= (X, I_1, \ldots, I_m, E)$$

such that

(1) $\left(X_r, I_1^{(r)}, \ldots, I_m^{(r)}, E^{(r)} \right)$ *has snc everywhere,*
(2) *for every j, cosupp $I_j^{(r)}$ is the birational transform of (the closure of) cosupp $I_j \cap \mathrm{snc}(X, I_1, \ldots, I_m, E)$ and*
(3) Π *is an isomorphism over $\mathrm{snc}(X, I_1, \ldots, I_m, E)$.*

Proof The proof is by induction on dim X and on m.

Step 1   Reduction to the case where I_1 is the ideal sheaf of a smooth divisor.

Apply order reduction (Kollár, 2007d, 3.107) to I_1. (Technically, to the marked ideal $(I_1, 2)$; see Kollár (2007d, section 3.5).) In this process, we only blow-up a center Z if the (birational transform of) I_1 has order ≥ 2 along Z. These are contained in the non-snc locus. A slight problem is that in Kollár (2007d, 3.107) the transformation rule used is $I_1 \mapsto \pi^* I_1(-2F)$ instead of $I_1 \mapsto \pi^* I_1(-F)$ as in (10.48.1). Thus each blow-up for $(I_1, 2)$ corresponds to two blow-ups in the sequence for Π: first we blow-up $Z \subset X$ and then we blow-up $F \subset B_Z X$.

At the end the maximal order of $I_1^{(r)}$ becomes 1. Since $I_1^{(r)}$ is the ideal sheaf of a Cartier divisor, cosupp $I_1^{(r)}$ is a disjoint union of smooth divisors.

Step 2   Reduction to the case when (X, I_1, E) is snc.

The first part is an easier version of Step (10.49.c), and should be read after it. Let S be an irreducible component of E. Write $E = S + E'$ and consider the restriction $(S, I_1|_S, E'|_S)$. By induction on the dimension, there is a blow-up sequence $\Pi_S \colon S_r \to \cdots \to S_1 = S$ such that $(S_r, (I_1|_S)^{(r)}, (E'|_S)^{(r)})$ is snc and Π_S is an isomorphism over snc$(S, I_1|_S, E'|_S)$. The "same" blow-ups give a blow-up sequence $\Pi \colon X_r \to \cdots \to X_1 = X$ such that $(X_r, I_1^{(r)}, E^{(r)})$ is snc near S_r and Π is an isomorphism over snc(X, I_1, E).

We can repeat the procedure for any other irreducible component of E. Note that as we blow-up, the new exceptional divisors are added to E, thus $E^{(s)}$ has more and more irreducible components as s increases. However, we only add new irreducible components to E that are exceptional divisors obtained by blowing up a smooth center that is contained in (the birational transform of) cosupp I_1. Thus these automatically have snc with I_1. Therefore the procedure needs to be repeated only for the original irreducible components of E.

After finitely many steps, (X, I_1, E) is snc near E and X and cosupp I_1 are smooth. Thus (X, I_1, E) is snc everywhere.

(If we want to resolve just one (X, I_j, E), we can do these steps in any order, but for a functorial resolution one needs an ordering of the index set of E and to proceed systematically.)

Step 3   Reduction to the case when $(X, I_1, \ldots, I_m, E)$ is snc near the cosupport of I_1.

Assume that (X, I_1, E) is snc. Set $S := \mathrm{cosupp}(I_1)$. If an irreducible component $S_i \subset S$ is contained in cosupp I_j for some $j > 1$ then we blow-up S_i. This reduces $\mathrm{mult}_{S_i} I_1$ and $\mathrm{mult}_{S_i} I_j$ by 1. Thus eventually none of the irreducible components of S are contained in cosupp I_j for $j > 1$. Thus we may assume that the $I_j|_S$ are ideal sheaves of Cartier divisors for $j > 1$ and consider the restriction $(S, I_2|_S, \ldots, I_m|_S, E|_S)$.

By induction there is a blow-up sequence $\Pi_S : S_r \to \cdots \to S_1 = S$ such that

$$\left(S_r, (I_2|_S)^{(r)}, \ldots, (I_m|_S)^{(r)}, (E|_S)^{(r)}\right) \quad \text{is snc}$$

and Π_S is an isomorphism over $\text{snc}(S, I_2|_S, \ldots, I_m|_S, E|_S)$. The "same" blow-ups give a blow-up sequence $\Pi : X_r \to \cdots \to X_1 = X$ such that the restriction

$$\left(S_r, I_2^{(r)}|_{S_r}, \ldots, I_m^{(r)}|_{S_r}, E^{(r)}|_{S_r}\right) \quad \text{is snc}$$

and Π is an isomorphism over $\text{snc}(X, I_1, \ldots, I_m, E)$. (Since we use only order 1 blow-ups, this is obvious. For higher orders, one would need the going-up theorem (Kollár, 2007d, 3.84), which holds only for D-balanced ideals. Every ideal of order 1 is D-balanced (Kollár, 2007d, 3.83); that is why we do not need to worry about subtleties here.)

As noted in (10.47.3), this implies that

$$\left(X_r, \mathscr{O}_{X_r}(-S_r), I_2^{(r)}, \ldots, I_m^{(r)}, E^{(r)}\right) \quad \text{is snc near } S_r.$$

Note, furthermore, that $S_r = \text{cosupp}\, I_1^{(r)}$, hence $(X_r, I_1^{(r)}, \ldots, I_m^{(r)}, E^{(r)})$ is snc near $\text{cosupp}\, I_1^{(r)}$.

Step 4   Induction on m.

By Step 3, we can assume that $(X, I_1, \ldots, I_m, E)$ is snc near $\text{cosupp}\, I_1$. Apply (10.49) to $(X, I_2, \ldots, I_m, E)$. The resulting $\Pi : X_r \to X$ is an isomorphism over $\text{snc}(X, I_2, \ldots, I_m, E)$. Since $\text{cosupp}\, I_1$ is contained in $\text{snc}(X, I_2, \ldots, I_m, E)$, all the blow-up centers are disjoint from $\text{cosupp}\, I_1$. Thus $(X_r, I_1^{(r)}, \ldots, I_m^{(r)}, E^{(r)})$ is also snc.

Finally, we may blow-up any irreducible component of $\text{cosupp}\, I_j^{(r)}$ that is not the birational transform of an irreducible component of $\text{cosupp}\, I_j$ which intersects $\text{snc}(X, I_1, \ldots, I_m, E)$. $\square$

Proof of (10.45)   Let D_j be the irreducible components of D. Set $I_j := \mathscr{O}_X(-D_j)$ and $E := \emptyset$. Note that (X, D) is snc at $p \in X$ if and only if $(X, I_1, \ldots, I_m, E)$ is snc at $p \in X$.

If X is a variety, we can apply (10.49) to $(X, I_1, \ldots, I_m, E)$ to get $\Pi : X_r \to X$ and $(X_r, I_1^{(r)}, \ldots, I_m^{(r)}, E^{(r)})$. Note that $E^{(r)}$ contains the whole exceptional set of Π, thus the support of $D' = \Pi_*^{-1} D + \text{Ex}(\Pi)$ is contained in $E^{(r)} + \sum_j \text{cosupp}\, I_j^{(r)}$. Thus D' is snc. By (10.49.3), Π is an isomorphism over the snc locus of (X, D).

The resolution constructed in (10.49) commutes with smooth morphisms and with change of fields (Kollár, 2007d, 3.34.1–2), at least if in (10.48) we allow reducible blow-up centers.

As in Kollár (2007d, 3.42–45), we conclude that (10.45) and (10.49) also hold for algebraic and analytic spaces over a field of characteristic 0.

Starting with (X, D), the above proof depends on an ordering of the irreducible components of D. This is an artificial device, but I don't know how to avoid it. This is very much connected to the difficulties of dealing with general nc divisors.   □

10.50   It should be noted that (10.45.2) fails for nc instead of snc. The simplest example is given by the *pinch point* $D := (x^2 = y^2 z) \subset \mathbb{A}^3 =: X$. Here (X, D) has nc outside the origin. At a point along the z-axis, save at the origin, D has two local analytic branches. As we go around the origin, these two branches are interchanged. We can never get rid of the pinch point without blowing up the z-axis.

Note that (X, D) is not snc along the z-axis, thus in constructing a log resolution as in (10.45.2), we are allowed to blow-up the z-axis.

This leads to the following general problem:

Problem 10.51 (Kollár, 2008a)   For each n, describe the smallest class of singularities $\mathcal{S}_n$ such that for every (X, D) of dimension n there is a proper birational morphism $f \colon X' \to X$ such that

(1) (X', D') has only singularities in $\mathcal{S}_n$, and
(2) f is an isomorphism over the nc locus of (X, D).

It is easy to see that, up to étale equivalence, $\mathcal{S}_2 = \{(xy = 0) \subset \mathbb{A}^2\}$ and Bierstone *et al.* (2011) show that $\mathcal{S}_3 = \{(xy = 0), (xyz = 0), (x^2 = y^2 z) \subset \mathbb{A}^3\}$. The longer list $\mathcal{S}_4$ is also determined in Bierstone *et al.* (2011), but there is not even a clear conjecture on what $\mathcal{S}_n$ should be for $n \geq 5$. The following example illustrates some of the subtleties.

Example 10.52 (Kollár, 2008a)   In $\mathbb{A}^4_{xyzu}$ consider the hypersurface

$$H := ((x + uy + u^2 z)(x + \varepsilon uy + \varepsilon^2 u^2 z)(x + \varepsilon^2 uy + \varepsilon u^2 z) = 0)$$

where ε is a primitive 3rd root of unity. Note that $H \cap (u = c)$ consists of three planes intersecting transversally if $c \neq 0$ while $H \cap (u = 0)$ is a triple plane. The $\mathbb{Z}_3$-action $(x, y, z, u) \mapsto (x, y, z, \varepsilon u)$ permutes the three irreducible components of H and $\mathbb{A}^4_{xyzu}/\mathbb{Z}_3 = \mathbb{A}^4_{xyzt}$ where $t = u^3$. By explicit computation, the image of H in the quotient is the irreducible hypersurface

$$D = (x^3 + ty^3 + t^2 z^3 - 3txyz = 0) \subset \mathbb{A}^4_{xyzt}.$$

The singular locus of D can be parametrized as

$$\operatorname{Sing} D = \operatorname{im}\left[(u, z) \mapsto (\varepsilon u^2 z, -(1 + \varepsilon)uz, z, u^3)\right].$$

Since the quotient map $H \to D$ is étale away from $(t = 0)$, we conclude that $(\mathbb{A}^4_{xyzt}, D)$ is nc outside $(t = 0)$. By explicit computation,

$$(t = 0) \cap \operatorname{Sing} D = (x = y = t = 0).$$

The singularity $D \subset \mathbb{A}^4$ cannot be improved by further smooth blow-ups whose centers are disjoint from the nc locus. Indeed, in our case the complement of the nc locus is the z-axis, hence there are only two choices for such a smooth blow-up center.

(1) Blow-up $(x = y = z = t = 0)$. In the chart $x = x_1 t_1, y = y_1 t_1, z = z_1 t_1, t = t_1$ we get the birational transform

$$D' = \left(x_1^3 + t_1 y_1^3 + t_1^2 z_1^3 - 3 t_1 x_1 y_1 z_1 = 0 \right).$$

(2) Blow-up $(x = y = t = 0)$. In the chart $x = x_1 t_1, y = y_1 t_1, z = z_1, t = t_1$ we get the birational transform

$$D'' = \left(t_1 x_1^3 + t_1^2 y_1^3 + z_1^3 - 3 t_1 x_1 y_1 z_1 = 0 \right).$$

In both cases we get a hypersurface which is, up to a coordinate change, isomorphic to D. (Note, however, that not every birational morphism between smooth 4-folds is a composite of smooth blow-ups, and I do not know if the singularity of D can be improved by some other birational morphism.)

In the coordinates $x_2 = x/z - (y/z)^2$, $y_2 = y/z$, $t_2 = t - \frac{1}{2}(y^3 - 3xy)$ the equation of $D \cap (z = c)$ (where $c \neq 0$) becomes

$$t_2^2 + x_2^2 \left(x_2 + \tfrac{3}{4} y_2^2 \right) = 0.$$

Thus S_4 also contains the 3-variable polynomial $x^2 + y^2(y + z^2)$, which is, however, not in S_3. (Note that the singularity $(x^2 + y^2(y + z^2) = 0) \subset \mathbb{A}^3$ is slc. In the terminology of Kollár and Shepherd-Barron (1988), it is a degenerate cusp of multiplicity 2.)

10.53 (Semi-log resolutions) What is the right notion of resolution or log resolution for non-normal varieties?

The simplest choice is to make no changes and work with resolutions. In particular, if $X = \cup_i X_i$ is a reducible scheme and $f: X' \to X$ is a resolution then $X' = \amalg_i X_i'$ such that each $X_i' \to X_i$ is a resolution. Note that we have not completely forgotten the gluing data determining X since $f^{-1}(X_i \cap X_j)$ is part of the exceptional set, and so we keep track of it.

There are, however, several inconvenient aspects. For instance, $f_* \mathcal{O}_{X'} \neq \mathcal{O}_X$, and this makes it difficult to study the Picard group of X or the cohomology of line bundles on X using X'. Another problem is that although $\mathrm{Ex}(f)$ tells us which part of X_i intersects the other components, it does not tell us anything about what the actual isomorphism is between $(X_i \cap X_j) \subset X_i$ and $(X_i \cap X_j) \subset X_j$.

It is not clear how to remedy these problems for an arbitrary reducible scheme, but we are dealing with schemes that have only double normal crossing in codimension 1. We can thus look for $f: X' \to X$ such that X' has only double normal crossing singularities and f is an isomorphism over codimension 1 points of X.

As in (10.45), this works for simple nc but not in general. We need to allow at least pinch points (1.43).

Theorem 10.54 (Kollár, 2008a) *Let X be a reduced scheme over a field of characteristic 0. Let $X^{\mathrm{ncp}} \subset X$ be an open subset such that X^{ncp} has only smooth points ($x_1 = 0$), double nc points ($x_1^2 - ux_2^2 = 0$) and pinch points ($x_1^2 - x_2^2 x_3 = 0$). Then there is a projective birational morphism $f \colon X' \to X$ such that*

(1) *X' has only smooth points, double nc points and pinch points,*
(2) *f is an isomorphism over X^{ncp},*
(3) *Sing X' maps birationally onto the closure of Sing X^{ncp}.*

If X' has any pinch points then they are on an irreducible component of $B \subset \mathrm{Sing}\, X'$ along which X' is nc but not snc. Then, by (10.54.3), X is nc but not snc along $f(B)$. Thus we obtain the following simple nc version.

Corollary 10.55 *Let X be a reduced scheme over a field of characteristic 0. Let $X^{\mathrm{snc2}} \subset X$ be an open subset which has only smooth points ($x_1 = 0$) and simple nc points of multiplicity ≤ 2 ($x_1 x_2 = 0$). Then there is a projective birational morphism $f \colon X' \to X$ such that*

(1) *X' has only smooth points and simple nc points of multiplicity ≤ 2,*
(2) *f is an isomorphism over X^{snc2},*
(3) *Sing X' maps birationally onto the closure of Sing X^{snc2}.* □

Proof of (10.54) The method of Hironaka (1964) reduces the multiplicity of a scheme starting with the highest multiplicity locus. We can use it to find a proper birational morphism $g_1 \colon X_1 \to X$ such that every point of X_1 has multiplicity ≤ 2 and g_2 is an isomorphism over X^{ncp}. Thus by replacing X by X_1 we may assume to start with that every point of X has multiplicity ≤ 2.

The next steps of the Hironaka method would not distinguish the nc locus (that we want to keep intact) from the other multiplicity 2 points (that we want to eliminate). Thus we proceed somewhat differently.

Let $n \colon \bar{X} \to X$ be the normalization with reduced conductor $\bar{B} \subset \bar{X}$.

It is easy to see (cf. Kollár (2007d, 2.49)) that at any point of X, in local analytic or étale coordinates we can write X as

$$X = (y^2 = g(\mathbf{x}) h(\mathbf{x})^2) \subset \mathbb{A}^{n+1}$$

where $(\mathbf{x}) := (x_1, \ldots, x_n)$ and g has no multiple factors. (We allow g and h to have common factors.) The normalization is then given by

$$\bar{X} = (z^2 = g(\mathbf{x})) \quad \text{where } z = y/h(\mathbf{x}).$$

Here $\bar{B} = (h(\mathbf{x}) = 0)$ and the involution $\tau\colon (z, \mathbf{x}) \mapsto (-z, \mathbf{x})$ is well defined on $\bar{B}$. (By contrast, the τ action on $\bar{X}$ depends on the choice of the local coordinate system.)

Thus we have a pair $(Y_2, B_2) := (\bar{X}, \bar{B})$ plus an involution $\tau_2\colon B_2 \to B_2$ such that for every $b \in B_2$ there is an étale neighborhood U_b of $\{b, \tau_2(b)\}$ such that τ_2 extends (non-uniquely) to an involution τ_{2b} of $(U_b, B_2|_{U_b})$.

Let us apply an étale local resolution procedure (as in Włodarczyk (2005) or Kollár (2007d)) to (Y_2, B_2). Let the first blow-up center be $Z_2 \subset Y_2$. Since the procedure is étale local, we see that $U_b \cap Z_2$ is τ_{2b}-invariant for every $b \in B_2$. Let $Y_3 \to Y_2$ be the blow-up of Z_2 and let $B_3 \subset Y_3$ be the birational transform of B_2. Then τ_2 lifts to an involution τ_3 of B_3 and the τ_{2b} lift to extensions on suitable neighborhoods. Moreover, the exceptional divisor of $Y_3 \to Y_2$ intersected with B_3 is τ_3-invariant. In particular, there is an ample divisor H_3 on Y_3 such that $H_3|_{B_3}$ is τ_3-invariant.

At the end we obtain $g\colon Y_r \to Y_2 = \bar{X}$ such that

(1) Y_r is smooth and $\mathrm{Ex}(g) + B_r$ is an snc divisor,
(2) B_r is smooth and τ lifts to an involution τ_r on B_r and
(3) there is a g-ample divisor H such that $H|_{B_r}$ is τ_r-invariant.

The fixed point set of τ_r is a disjoint union of smooth subvarieties of B_r. By blowing up those components whose dimension is $< \dim B_r - 1$, we also achieve (after replacing $r + 1$ by r) that

(4) the fixed point set of τ_r has pure codimension 1 in B_r.

Let $Z_r := B_r/\tau_r$ and X_r the universal push-out of $Z_r \leftarrow B_r \hookrightarrow Y_r$. As we noted in (1.43), X_r has only nc and pinch points.

Further, let H be a divisor on Y_r such that $H|_{B_r}$ is τ_r-invariant. As noted in (1.43), $2H$ is the pull-back of a Cartier divisor on X_r. In particular, if H is ample then X_r is projective. $\qquad\qquad\square$

We would like not just a semi-resolution of X but a log resolution of the pair (X, D). Thus we need to take into account the singularities of D as well. As we noted in (10.50), this is not obvious even when X is a smooth 3-fold. The following weaker version, which gives the expected result only for the codimension 1 part of the singular set of (X, D), is sufficient for us.

Theorem 10.56 (Kollár, 2008a) *Let X be a reduced scheme over a field of characteristic 0 and D a Weil divisor on X. Let $X^{\mathrm{nc2}} \subset X$ be an open subset which has only nc points of multiplicity ≤ 2 and $D|_{X^{\mathrm{nc2}}}$ is smooth and disjoint* *from* Sing X^{nc2}. *Then there is a projective birational morphism* $f : X' \to X$ *such that*

(1) *the local models for* $(X', D' := f_*^{-1}(D) + \mathrm{Ex}(f))$ *are*
   (a) *(Smooth)* $X' = (x_1 = 0)$ *and* $D' = (\prod_{i \in I} x_i = 0)$ *for some* $I \subset \{2, \ldots, n+1\}$,
   (b) *(Double nc)* $X' = (x_1^2 - u x_2^2 = 0)$ *and* $D' = (\prod_{i \in I} x_i = 0)$ *for some* $I \subset \{3, \ldots, n+1\}$ *or*
   (c) *(Pinched)* $X' = (x_1^2 = x_2^2 x_3)$ *and* $D' = (\prod_{i \in I} x_i = 0) + D_2$ *for some* $I \subset \{4, \ldots, n+1\}$ *where either* $D_2 = 0$ *or* $D_2 = (x_1 = x_3 = 0)$.
(2) f *is an isomorphism over* X^{nc2}.
(3) Sing X' *maps birationally onto the closure of* Sing X^{nc2}.
(4) *Let* $\bar{X} \to X$ *be the normalization,* $\bar{B} \subset \bar{X}$ *the closure of the conductor of* $\bar{X}^{\mathrm{nc2}} \to X^{\mathrm{nc2}}$ *and* $\bar{D} \subset \bar{X}$ *the preimage of* D. *Then* f *is a log resolution of* $(\bar{X}, \bar{B} + \bar{D})$.

As before, (10.56) implies the simple nc version:

Corollary 10.57 *Let* X *be a reduced scheme over a field of characteristic* 0 *and* D *a Weil divisor on* X. *Let* $X^{\mathrm{snc2}} \subset X$ *be an open subset which has only snc points of multiplicity* ≤ 2 *and* $D|_{X^{\mathrm{snc2}}}$ *is smooth and disjoint from* Sing X^{snc2}. *Then there is a projective birational morphism* $f : X' \to X$ *such that*

(1) *the local models for* $(X', D' := f_*^{-1}(D) + \mathrm{Ex}(f))$ *are*
   (a) *(Smooth)* $X' = (x_1 = 0)$ *and* $D' = (\prod_{i \in I} x_i = 0)$ *for some* $I \subset \{2, \ldots, n+1\}$ *or*
   (b) *(Double snc)* $X' = (x_1 x_2 = 0)$ *and* $D' = (\prod_{i \in I} x_i = 0)$ *for some* $I \subset \{3, \ldots, n+1\}$.
(2) f *is an isomorphism over* X^{snc2}.
(3) Sing X' *maps birationally onto the closure of* Sing X^{snc2}.

Proof of (10.56)   First we use (10.54) to reduce to the case when X has only double nc and pinch points. Let $\bar{X} \to X$ be the normalization and $\bar{B} \subset \bar{X}$ the conductor. Here $\bar{X}$ and $\bar{B}$ are both smooth.

Next we want to apply embedded resolution to $(\bar{X}, \bar{B} + \bar{D})$. One has to be a little careful with D since the preimage $\bar{D} \subset \bar{X}$ need not be τ-invariant.

As a first step, we move the support of $\bar{D}$ away from $\bar{B}$. As in Kollár (2007d, 3.102) this is equivalent to multiplicity reduction for a suitable ideal $I_D \subset \mathcal{O}_{\bar{B}}$. Let us now apply multiplicity reduction for the ideal $I_D + \tau^* I_D$. All the steps are now τ-invariant, so at the end we obtain $g : Y_r \to \bar{X}$ such that $B_r + D_r + \mathrm{Ex}(g)$ has only snc along B_r and τ lifts to an involution τ_r.

As in the proof of (10.54), we can also assume that the fixed locus of τ_r has pure codimension 1 in B_r and that there is a g-ample divisor H such that $H|_{B_r}$ is τ_r-invariant.

As in the end of (10.4), let X_r be the universal push-out of $B_r/\tau_r \leftarrow B_r \hookrightarrow Y_r$. Then (X_r, D'_r) has the required normal form along Sing X_r. The remaining singularities of D'_r can now be resolved as in (10.45). $\qquad\Box$

The following analog of (10.45) is proved in Bierstone and Milman (2011) and Bierstone and Pacheco (2011).

Theorem 10.58 *Let X be a reduced scheme over a field of characteristic 0 and D a Weil divisor on X. Let $X^{\mathrm{ssnc}} \subset X$ be the largest open subset such that $(X^{\mathrm{ssnc}}, D|_{X^{\mathrm{ssnc}}})$ is semi-snc. Then there is a projective birational morphism $f: X' \to X$ such that (X', D) is semi-snc and f is an isomorphism over X^{snc}.*

The following weaker version is sufficient for many applications. We do not guarantee that $f: X' \to X$ is an isomorphism over X^{ssnc}, only that f is an isomorphism over an open subset $X^0 \subset X^{\mathrm{ssnc}}$ that intersects every stratum of $(X^{\mathrm{ssnc}}, D|_{X^{\mathrm{ssnc}}})$. This implies that we do not introduce any "unnecessary" f-exceptional divisors with discrepancy -1. The latter is usually the key property that one needs.

Unfortunately, the proof only works in the quasi-projective case.

Proposition 10.59 *Let X be a reduced quasi-projective scheme over a field of characteristic 0 and D a Weil divisor on X. Let $X^0 \subset X$ be an open subset such that $(X^0, D|_{X^0})$ is semi-snc. There is a projective birational morphism $f: X' \to X$ such that*

(1) *(X', D') is an embedded semi-snc pair,*
(2) *f is an isomorphism over the generic point of every lc center of $(X^0, D|_{X^0})$,*
(3) *f is an isomorphism at the generic point of every lc center of (X', D').*

Following (2.79), we call these thrifty *semi-log resolutions.*

Proof In applications it frequently happens that $X + B$ is a divisor on a variety Y and $D = B|_X$. Applying (10.45.2) to $(Y, X + B)$ gives (10.59). In general, not every (X, D) can be obtained this way, but one can achieve something similar at the price of introducing other singularities.

Take an embedding $X \subset \mathbb{P}^N$. Pick a finite set $W \subset X$ such that each log canonical center of $(X^0, D|_{X^0})$ contains a point of W.

Choose $d \gg 1$ such that the scheme theoretic base locus of $\mathcal{O}_{\mathbb{P}^N}(d)(-X)$ is X near every point of W. Taking a complete intersection of $(N - \dim X - 1)$ general members in $|\mathcal{O}_{\mathbb{P}^N}(d)(-X)|$, we obtain $Y \supset X$ such that Y is smooth at every point of W. (Here we use that X has only hypersurface singularities near W.)

For every D_i choose $d_i \gg 1$ such that the scheme theoretic base locus of $\mathcal{O}_{\mathbb{P}^N}(d_i)(-D_i)$ is D_i near every point of W. For each i, let $D_i^Y \in |\mathcal{O}_{\mathbb{P}^N}(d_i)(-D_i)|$ be a general member.

We have thus constructed a pair $(Y, X + \sum D_i^Y)$ such that

(4) $(Y, X + \sum D_i^Y)$ is semi-snc near W and
(5) $(X, \sum D_i^Y|_X)$ is isomorphic to $(X, \sum D_i)$ in a neighborhood of W.

By (10.45.2) there is a semi-log resolution of

$$f : (Y', X' + \sum B_i) \to (Y, X + \sum D_i^Y)$$

such that f is an isomorphism over an open neighborhood of W. Then $f|_{X'}$:
$X' \to X$ is the log resolution we want. $\square$

10.5 Pluricanonical representations

For a smooth projective variety X, the natural action of Bir X on ω_X^m gives the *pluricanonical representations*

$$\varrho_m : \mathrm{Bir}\, X \to \mathrm{GL}\big(H^0(X, \omega_X^m)\big).$$

(As in (3.17.2), the dual group acts naturally.) The finiteness of the image of ϱ_m was proved in Nakamura and Ueno (1973) under some technical assumptions; these were removed in an unpublished note of Deligne, see Ueno (1975, section 14) for details. Essentially without changes the proof works also for klt pairs (X, Δ). The log canonical case, which we do not discuss, requires additional considerations; see Fujino (2000), Gongyo (2010), Fujino and Gongyo (2011) and Hacon and Xu (2011b) for details.

If X is of general type then Bir X itself is finite. If X is uniruled, then frequently Bir X is very large, but $H^0(X, \omega_X^m) = 0$. Thus the above assertion is interesting only for varieties of intermediate Kodaira dimension. There are many examples where Bir X is infinite and the pluricanonical representation is nontrivial. For instance, let E be an elliptic curve. Then $\mathrm{Bir}(E \times E)$ contains the subgroup

$$(x, y) \mapsto (ax + cy, bx + dy) \quad \text{for} \quad \begin{pmatrix} a & b \\ c & d \end{pmatrix} \in \mathrm{GL}(2, \mathbb{Z}).$$

Its action on $H^0(X, \omega_{E \times E})$ is multiplication by the determinant $ad - bc = \pm 1$.

Definition 10.60 Let X be an integral k-variety and denote by $\mathcal{K}(X, \omega_X^{[m]})$ the $k(X)$-vectorspace of rational sections of $\omega_X^{[m]}$. A rational m-form can be pulled-back by a birational map, thus Bir X acts on $\mathcal{K}(X, \omega_X^{[m]})$.

For a finite dimensional k-subspace $V \subset \mathcal{K}(X, \omega_X^{[m]})$, let $\mathrm{Bir}(X, V) \subset \mathrm{Bir}\, X$ denote the stabilizer of V under this action. Our aim is to study the representation

$$\varrho_V: \mathrm{Bir}(X, V) \to \mathrm{GL}(V). \tag{10.60.1}$$

There are two main examples that we are interested in. First, let X be a proper, normal, geometrically integral k-variety and Δ a (not necessarily effective) $\mathbb{Q}$-divisor on X such that $K_X + \Delta \sim_{\mathbb{Q}} 0$. Then $m(K_X + \Delta) \sim 0$ for some $m > 0$ and

$$H^0\big(X, \omega_X^{[m]}(m\Delta)\big) \subset \mathcal{K}\big(X, \omega_X^{[m]}\big) \tag{10.60.2}$$

is a 1-dimensional k-subspace and $\text{Bir}(X, V)$ contains $\text{Bir}^c(X, \Delta)$, the group of crepant-birational maps (2.23). We use this case in Section 5.7.

More generally, let $f \colon (X, \Delta) \to Z$ be a crepant log structure (4.28) and X proper. Then

$$H^0\big(X, \omega_X^{[m]}(\lfloor m\Delta \rfloor)\big) \subset \mathcal{K}\big(X, \omega_X^{[m]}\big) \tag{10.60.3}$$

is a finite dimensional k-subspace and $\text{Bir}(X, V) \supset \text{Bir}^c(X, \Delta)$. We use this case in Section 5.6.

A pair (X, V) is called *lc* (resp. *klt*) if for every $0 \neq \eta \in V$ and for every birational model $g \colon X' \dashrightarrow X$, every pole of $g^*\eta$ has order $\leq m$ (resp. $< m$). As usual (cf. (2.13)), it is sufficient to check this on one model where X' is smooth and the support of $(g^*\eta)$ is an snc divisor.

In the examples (10.60.2–3), (X, V) is klt if (X, Δ) is klt.

Theorem 10.61 *Let X be a smooth projective variety over $\mathbb{C}$ and V a finite dimensional $\mathbb{C}$-subspace of $\mathcal{K}(X, \omega_X^{[m]})$ such that (X, V) is klt. Then the representation*

$$\varrho \colon \text{Bir}(X, V) \to GL(V)$$

defined in (10.60.1) has finite image.

Remark 10.62 The proof relies very much on topological conservations. We do not know what happens in positive characteristic.

The method of Kollár (1992, section 12) shows that (10.61) should also hold if (X, V) is lc, but the proof may need MMP for lc pairs; see also Gongyo (2010), Fujino and Gongyo (2011) and Hacon and Xu (2011b).

On the other hand, (10.61) fails if (X, V) is not lc. For instance, the image of the representation for $(\mathbb{P}^1, \mathbb{C}d(x/y))$ is $\mathbb{C}^*$. Indeed, for $g \colon (x : y) \mapsto (cx : y)$ we have $g^*d(x/y) = d(cx/y) = cd(x/y)$. Here $d(x/y)$ has a double pole at infinity hence $(\mathbb{P}^1, \mathbb{C}d(x/y))$ is not lc.

By far the hardest part of the proof is the special case $\dim V = 1$.

Pick any $0 \neq \eta \in V$ and write our pair as $(X, \mathbb{C}\eta)$. First we reduce to the case when X is smooth, projective and η is a rational section of ω_X. Since the poles are assumed to have order < 1, η has no poles and it is an actual section of ω_X. In the latter case, (10.67) proves that the image of the representation consists of roots of unity whose degree is bounded by the middle Betti number of X. Since there are only finitely many such roots of unity, this will complete the 1-dimensional case of (10.61).

10.63 (Reduction to $\eta \in H^0(X, \omega_X)$) Pick a birational model $(X, \mathbb{C}\eta)$ such that X is smooth and (η) is a snc divisor. Since $(X, \mathbb{C}\eta)$ is klt, we can write $(\eta) = mD - m\Delta$ where D is an effective $\mathbb{Z}$-divisor and $\lfloor \Delta \rfloor = 0$. Then $m(K_X - D) \sim -m\Delta$ and we can view η as an isomorphism

$$\eta : \mathscr{O}_X(-m\Delta) \simeq (\omega_X(-D))^{\otimes m}.$$

Thus, as in (2.44), η defines an algebra structure and a cyclic cover

$$\tilde{X} := X[\sqrt[m]{\eta}] := \mathrm{Spec}_X \sum_{i=0}^{m-1} \omega_X(-D)^{\otimes i}(\lfloor i\Delta \rfloor)$$

with projection $p: \tilde{X} \to X$. Since p ramifies only along the snc divisor Δ, $\tilde{X}$ is klt (2.47.4). Note that $\omega_{\tilde{X}}$ has a section $\tilde{\eta}$, given by the $i = 1$ summand in

$$1 \in H^0(X, \mathscr{O}_X(D)) \subset p_*\omega_{\tilde{X}} = \sum_{i=0}^{m-1} \mathscr{H}om_X(\omega_X(-D)^{\otimes i}(\lfloor i\Delta \rfloor), \omega_X).$$

Assume now that we have $\phi: X \dashrightarrow X$ such that $\phi^*\eta = \lambda\eta$. Fix an mth root $\sqrt[m]{\lambda}$. Then ϕ lifts to a rational algebra map

$$\phi' : \phi^* \sum_{i=0}^{m-1} \omega_X(-D)^{\otimes i}(\lfloor i\Delta \rfloor) \dashrightarrow \sum_{i=0}^{m-1} \omega_X(-D)^{\otimes i}(\lfloor i\Delta \rfloor)$$

which is the natural isomorphism $\phi^*\omega_X \to \omega_X$ multiplied by $(\sqrt[m]{\lambda})^i$ on the ith summand. Thus we get $\tilde{\phi} \in \mathrm{Bir}(\tilde{X}, \mathbb{C}\tilde{\eta})$ with eigenvalue $\sqrt[m]{\lambda}$. Therefore, if $\mathrm{Bir}(\tilde{X}, \mathbb{C}\tilde{\eta}) \to \mathbb{C}^*$ has finite image then so does $\mathrm{Bir}(X, \mathbb{C}\eta) \to \mathbb{C}^*$.

Next we compare the pull-back of holomorphic forms with the pull-back map on integral cohomology. Note that one can pull-back holomorphic forms by rational maps, but one has to be careful when pulling back integral cohomology classes by rational maps.

10.64 Let $f: M \to N$ be a map between oriented, compact manifolds of dimension m. Then one can define a push-forward map

$$f_* : H^i(M, \mathbb{Z})/(\text{torsion}) \to H^i(N, \mathbb{Z})/(\text{torsion})$$

as follows. Cup product with $\alpha \in H^i(M, \mathbb{Z})/(\text{torsion})$ gives

$$\alpha\cup : H^{m-i}(N, \mathbb{Z}) \to H^m(M, \mathbb{Z}) = \mathbb{Z} \quad \text{given by} \quad \beta \mapsto \alpha \cup f^*\beta.$$

Since the cup product $H^i(N, \mathbb{Z}) \times H^{m-i}(N, \mathbb{Z}) \to H^m(N, \mathbb{Z}) = \mathbb{Z}$ is unimodular, there is a unique class $\gamma \in H^i(N, \mathbb{Z})/(\text{torsion})$ such that $\gamma \cup \beta = \alpha \cup f^*\beta$ for every β. Set $f_*\alpha := \gamma$.

Note that if $\alpha = f^*\gamma$ for some $\gamma \in H^i(N, \mathbb{Z})$ then

$$f^*\gamma \cup f^*\beta = f^*(\gamma \cup \beta) = \deg f \cdot (\gamma \cup \beta)$$

shows that $f_*(f^*\gamma) = \deg f \cdot \gamma$. Thus, modulo torsion, $f_* \circ f^* : H^*(N, \mathbb{Z}) \to H^*(N, \mathbb{Z})$ is multiplication by $\deg f$. In particular, if $\deg f = 1$ then $f_* \circ f^*$ is the identity.

Lemma 10.65 *Let X, X', Y be smooth proper varieties over $\mathbb{C}$ and $g: X \dashrightarrow Y$ a map. Let $f: X' \to X$ be a birational morphism such that $(g \circ f): X' \to Y$ is a morphism. Then the following diagram is commutative*

$$\begin{array}{ccc} H^0(Y, \Omega^i_Y) & \hookrightarrow & H^i(Y(\mathbb{C}), \mathbb{C}) \\ \downarrow {\scriptstyle g^*} & & \downarrow {\scriptstyle f_* \circ (g \circ f)^*} \\ H^0(X, \Omega^i_X) & \hookrightarrow & H^i(X(\mathbb{C}), \mathbb{C}). \end{array}$$

Proof For a holomorphic i form ϕ, let $[\phi]$ denote its cohomology class in $H^i(\ , \mathbb{C})$. Pull-back by a morphism commutes with taking cohomology class, thus $[(g \circ f)^*\phi] = (g \circ f)^*[\phi]$. On the other hand, $(g \circ f)^*\phi = f^*(g^*\phi)$. Thus $(g \circ f)^*[\phi] = f^*[g^*\phi]$. As noted in (10.64), $f_* \circ f^*[g^*\phi] = [g^*\phi]$. Thus $f_* \circ (g \circ f)^*[\phi] = f_* \circ f^*[g^*\phi] = [g^*\phi]$. $\qquad\square$

Corollary 10.66 *Let Y be a smooth proper variety over $\mathbb{C}$ and $g: Y \dashrightarrow Y$ a birational map. Then every eigenvalue of $g^*: H^0(Y, \Omega^i_Y) \to H^0(Y, \Omega^i_Y)$ is an algebraic integer of degree $\leq \dim H^i(Y(\mathbb{C}), \mathbb{C})$.*

Proof Let $f: Y' \to Y$ be a birational morphism such that $(g \circ f): Y' \to Y$ is a morphism. By (10.65), every eigenvalue of $g^*: H^0(Y, \Omega^i_Y) \to H^0(Y, \Omega^i_Y)$ is also an eigenvalue of $f_* \circ (g \circ f)^*: H^i(Y(\mathbb{C}), \mathbb{Z}) \to H^i(Y(\mathbb{C}), \mathbb{Z})$. The latter is given by an integral matrix, hence its eigenvalues are algebraic integers. $\qquad\square$

Warning 10.66.1 Although $f_* \circ (g \circ f)^*: H^i(Y(\mathbb{C}), \mathbb{Z}) \to H^i(Y(\mathbb{C}), \mathbb{Z})$ does not depend on the choice of f, the correspondence $g \mapsto f_* \circ (g \circ f)^*$ is not a group homomorphism. In fact, usually $f_* \circ (g \circ f)^*$ is not invertible; it need not even have maximal rank.

Corollary 10.67 *Let Y be a smooth proper variety over $\mathbb{C}$ and $g: Y \dashrightarrow Y$ a birational map. Then every eigenvalue of $g^*: H^0(Y, \omega_Y) \to H^0(Y, \omega_Y)$ is a root of unity of degree $\leq \dim H^{\dim Y}(Y(\mathbb{C}), \mathbb{C})$.*

Proof Assume that η is an eigenform and $g^*\eta = \lambda\eta$. Since $\eta \wedge \bar{\eta}$ is a (singular) volume form,

$$\int_{Y(\mathbb{C})} \eta \wedge \bar{\eta} = \int_{Y(\mathbb{C})} g^*(\eta \wedge \bar{\eta}) = (\lambda\bar{\lambda}) \int_{Y(\mathbb{C})} \eta \wedge \bar{\eta}.$$

Thus $|\lambda| = 1$ and, by (10.66), it is an algebraic integer.

Let $\sigma \in \mathrm{Aut}(\mathbb{C}/\mathbb{Q})$ be any field automorphism. By conjugating everything by σ, we get $g^\sigma \colon Y^\sigma \dashrightarrow Y^\sigma$ such that $(g^\sigma)^* \eta^\sigma = \lambda^\sigma \eta^\sigma$. Thus λ^σ also has absolute value 1. We complete the proof of the 1-dimensional case by noting that an algebraic integer is a root of unity if and only if all of its conjugates have absolute value 1. (This is not hard to prove; see for instance Fröhlich and Taylor (1993, IV.4.5.a).) $\square$

Proof of (10.61)   We prove below that $\varrho(\mathrm{Bir}(X, V))$ consists of elements whose order is uniformly bounded. Then $\varrho(\mathrm{Bir}(X, V))$ is finite by a theorem of Burnside which says that a subgroup $G \subset \mathrm{GL}(n, \mathbb{C})$ is finite if and only if its elements have uniformly bounded order. (To see this, note that if $g^m = 1$ for every $g \in G$ then the same holds for its Zariski closure $\bar{G}$. A connected, positive dimensional, affine algebraic group over $\mathbb{C}$ always contains $\mathbb{C}^*$ or $\mathbb{C}$ as a subgroup. Hence $\dim \bar{G} = 0$ so G is finite. See also Curtis and Reiner (1962, section 36).)

First we prove that all the eigenvalues of elements of $\varrho(\mathrm{Bir}(X, V))$ are roots of unity whose order is uniformly bounded. To see this pick any $g \in \mathrm{Bir}(X, V)$ and let $\eta \in V$ be an eigenvector of $\varrho(g)$. Then $g \in \mathrm{Bir}(X, \mathbb{C}\eta)$, hence, as we have seen, every eigenvalue of $\varrho(g)$ is a root of unity. Moreover, the degree (and hence the order) of this root of unity is uniformly bounded if the middle Betti number of a resolution of the cyclic cover $\tilde{X} = X\big[\sqrt[m]{\eta}\,\big]$ constructed in (10.63) is uniformly bounded.

To see the latter, first pick $\eta \in V$ generic. The construction of $\tilde{X} = X\big[\sqrt[m]{\eta}\,\big]$ and its resolution then extend to all η in a Zariski open subset $V_1 \subset V$. Then work with the generic points of $V \setminus V_1$, and finish by Noetherian induction.

Thus there is an $m = m(X, V)$ such that g^m is unipotent for every $g \in \varrho(\mathrm{Bir}(X, V))$.

If $1 \neq h \in \varrho(\mathrm{Bir}(X, V))$ is unipotent, then there are nonzero $\eta, \zeta \in V$ such that $h^m(\zeta) = \zeta + m\eta$ for every m. Thus $|\zeta| = |\zeta + m\eta|$ for the pseudo-norm defined in (10.68.1). This, however, contradicts (10.68.3).

Thus $\varrho(\mathrm{Bir}(X, V))$ consists of elements whose order is uniformly bounded, hence it is a finite group. $\square$

10.68 (A pseudo-norm)   For $\eta \in \mathcal{K}(X, \omega_X^{[m]})$ define its pseudo-norm as

$$|\eta|^{1/m} := \int_{X(\mathbb{C})} \varepsilon(\dim X)(\eta \wedge \bar{\eta})^{1/m}, \tag{10.68.1}$$

where $\varepsilon(\dim X)$ is a power of $\sqrt{-1}$ (depending on how you orient $X(\mathbb{C})$) so that the integrand is everywhere ≥ 0. This is clearly birationally invariant. We claim that $|\eta| < \infty$ iff η is klt. To see this we may assume that X is smooth and (η) is a snc divisor. Thus locally the integral is of the form

$$\int_{\mathbb{C}^n} u(z_1, \ldots, z_n) \cdot \prod_i (z_i \bar{z}_i)^{c_i} \cdot dz_1 \wedge d\bar{z}_1 \wedge \cdots \wedge dz_n \wedge d\bar{z}_n \tag{10.68.2}$$

where u is nowhere zero and mc_i is the order of vanishing of η along $(z_i = 0)$. By Fubini's theorem and 1-variable integration this is locally finite if and only if $c_i > -1$ for every i. This is exactly the klt condition.

Let $V \subset \mathcal{K}(X, \omega_X^{[m]})$ be a finite dimensional klt subspace. As a very special case of Demailly and Kollár (2001, theorem 0.2), the function $\eta \to |\eta|$ is continuous. Although this needs only the simpler Demailly and Kollár (2001, 3.1), its proof is not elementary calculus.

For our purposes we only need the much weaker assertion that if $\eta \neq 0$ then

$$|\zeta + t\eta| \to \infty \quad \text{as} \quad t \to \infty. \tag{10.68.3}$$

Since the integrand computing $|\zeta + t\eta|$ is everywhere ≥ 0, this can be checked at a smooth point of X where both η, ζ are regular. Now the claim is obvious.

The following corollary is used in Section 5.6. For stronger versions see Fujino (2000), Gongyo (2010), Fujino and Gongyo (2011) and Hacon and Xu (2011b).

Corollary 10.69 *Let Z be a projective variety over $\mathbb{C}$ and $f\colon (X, \Delta) \to Z$ a crepant log structure (4.28) with klt generic fiber. Assume that $K_X + \Delta \sim_{\mathbb{Q}} f^*L$ where L is an ample $\mathbb{Q}$-divisor on Z. Then the image of $\mathrm{Bir}^c(X, \Delta) \to \mathrm{Aut}\, Z$ is finite.*

Proof As in (10.60.3), set

$$V_m := H^0\big(X, \omega_X^{[m]}(\lfloor m\Delta \rfloor)\big) \subset \mathcal{K}(X, \omega_X^{[m]}).$$

Then (X, V_m) is lc and $\mathrm{Bir}^c(X, \Delta)$ acts on V_m. For m sufficiently large and divisible, the linear system $|V_m|$ maps X to Z, hence $\mathrm{Bir}^c(X, \Delta)$ acts by automorphisms on Z. Furthermore, the image of $\mathrm{Bir}^c(X, \Delta) \to \mathrm{Aut}\, Z$ is finite if the image of

$$\varrho_m\colon \mathrm{Bir}^c(X, \Delta) \subset \mathrm{Bir}(X, V_m) \to \mathrm{GL}(V_m)$$

is finite. We are done by (10.61) if (X, V_m) is klt.

In general, let $V_m^0 \subset V_m$ be the subspace of those sections that are klt. Then $\mathrm{Bir}^c(X, \Delta)$ acts on V_m^0 with finite image by (10.61). Furthermore, for any n, m, the natural multiplication map of forms gives $V_n^0 \otimes V_m \to V_{n+m}^0$ which in turn gives a $\mathrm{Bir}^c(X, \Delta)$-equivariant map

$$\phi_{n,m}\colon V_m \to \mathrm{Hom}\big(V_n^0, V_{n+m}^0\big).$$

For every $0 \neq \eta \in V_n^0$, the multiplication $\eta \times V_m \to V_{n+m}^0$ is injective, hence $\phi_{n,m}$ is injective if $V_n^0 \neq 0$.

Now we use that L is ample. Let $W \subset Z$ be the union of all lc centers. Choose $n > 0$ such that nL is Cartier, $n(K_X + \Delta) \sim f^*(nL)$ and $\mathcal{O}_Z(nL)$ has a nonzero section σ that vanishes along W. Then $f^*\sigma$ is a section of $H^0(X, \omega_X^{[n]}(n\Delta))$ that vanishes along all the lc centers, hence it is klt. Thus $V_n^0 \neq 0$. $\square$

Remark 10.70   As a very special case of (10.69), if (X, Δ) is projective, lc and $K_X + \Delta$ is ample then $\mathrm{Bir}^c(X, \Delta)$ is finite. This is, however, easier to prove using the method of Matsumura (1963).

10.6 Cubic hyperresolutions

Here we review the construction of cubic hyperresolutions, as well as several examples. We follow Guillén *et al.* (1988) and mostly use their notation, but some parts and examples are also based upon a set of unpublished notes written by Karl Schwede.

First, fix a small universe to work in. Let Sch denote the category of reduced schemes. Note that the usual fibered product of schemes $X \times_S Y$ need not be reduced, even when X and Y are reduced. On the other hand, it is easy to see that $(X \times_S Y)_{\mathrm{red}}$ is the fibered product in Sch.

We will denote by $\underline{1}$ the category $\{0\}$ and by $\underline{2}$ the category $\{0 \to 1\}$. Let $n \geq -1$ be an integer. We denote by $\square_n^+$ the product of $n + 1$ copies of the category $\underline{2} = \{0 \to 1\}$, compare Guillén *et al.* (1988, I, 1.15). The objects of $\square_n^+$ are identified with the sequences $\alpha = (\alpha_0, \alpha_1, \ldots, \alpha_n)$ such that $\alpha_i \in \{0, 1\}$ for $0 \leq i \leq n$. For an $\alpha = (\alpha_0, \alpha_1, \ldots, \alpha_n) \in \mathrm{Ob}\,\square_n^+$, we will use the notation $|\alpha| = \alpha_0 + \alpha_1 + \cdots + \alpha_n \in \mathbb{N}$. For $n = -1$, we set $\square_{-1}^+ = \{0\}$ and for $n = 0$ we have $\square_0^+ = \{0 \to 1\}$. We denote by $\square_n$ the full subcategory consisting of all objects of $\square_n^+$ except the initial object $(0, \ldots, 0)$. Clearly, the category $\square_n^+$ can be identified with the category of $\square_n$ with an augmentation map from $\underline{1} = \{0\}$.

Definition 10.71   Let C be a category. A *diagram of schemes* is a functor Φ from the category C^{op} to the category of schemes. A *finite diagram of schemes* is a diagram of schemes such that the aforementioned category C has finitely many objects and morphisms; in this case such a functor will be called a C-*scheme*. A morphism of diagrams of schemes $\Phi \colon \mathsf{C}^{\mathrm{op}} \to$ Sch to $\Psi \colon \mathsf{D}^{\mathrm{op}} \to$ Sch is the combined data of a functor $\Gamma \colon \mathsf{C}^{\mathrm{op}} \to \mathsf{D}^{\mathrm{op}}$ together with a natural transformation of functors $\eta \colon \Phi \to \Psi \circ \Gamma$.

With these definitions, the class of (finite) diagrams of schemes can be made into a category. Likewise the set of C-schemes can also be made into a category (where the functor $\Gamma \colon \mathsf{C}^{\mathrm{op}} \to \mathsf{C}^{\mathrm{op}}$ is always chosen to be the identity functor).

Let I be a category. If instead of a functor to the category of reduced schemes, one considers a functor to the category of topological spaces, or the category of categories, one can define I-topological spaces and I-categories in the obvious way.

If $X_\bullet: I^{op} \to$ Sch is an I-scheme, and $i \in \mathrm{Ob}\, I$, then X_i will denote the scheme corresponding to i. Likewise if $\phi \in \mathrm{Mor}\, I$ is a morphism $\phi: j \to i$, then X_ϕ will denote the corresponding morphism $X_\phi: X_i \to X_j$. If $f: Y_\bullet \to X_\bullet$ is a morphism of I-schemes, we denote by f_i the induced morphism $Y_i \to X_i$. If $X_\bullet$ is an I-scheme, a closed sub-I-scheme is a morphism of I-schemes $g: Z_\bullet \to X_\bullet$ such that for each $i \in I$, the map $g_i: Z_i \to X_i$ is a closed immersion. We will often suppress g if no confusion is likely to arise. More generally, any property of a morphism of schemes (projective, proper, separated, closed immersion, etc.) can be generalized to the notion of a morphism of I-schemes by requiring that for each object i of I, g_i has the desired property (projective, proper, separated, closed immersion, etc.)

Now let $\iota_\bullet: \Sigma_\bullet \hookrightarrow X_\bullet$ be a closed I-subscheme and define the I-*pair* $(X_\bullet, \Sigma_\bullet)$ to be the $\square_0^+ \times I$-scheme given by $(X_\bullet, \Sigma_\bullet)_{0\bullet} = \Sigma_\bullet$ and $(X_\bullet, \Sigma_\bullet)_{1\bullet} = X_\bullet$ with the I-morphism $\iota_\bullet: (X_\bullet, \Sigma_\bullet)_{0\bullet} \to (X_\bullet, \Sigma_\bullet)_{1\bullet}$.

Definition 10.72 (Guillén *et al.*, 1988, I.2.2) Given a morphism of I-schemes $f: Y_\bullet \to X_\bullet$, we define the *discriminant of f* to be the smallest closed sub-I-scheme $Z_\bullet$ of $X_\bullet$ such that $f_i: (Y_i - (f_i^{-1}(Z_i))) \to (X_i - Z_i)$ is an isomorphism for all i.

Definition 10.73 (Guillén *et al.*, 1988, I.2.5) Let $S_\bullet$ be an I-scheme, $f: X_\bullet \to S_\bullet$ a proper morphism of I-schemes, and $D_\bullet$ the discriminant of f. We say that f is a *weak resolution* of $S_\bullet$ if $X_\bullet$ is a smooth I-scheme (meaning that each X_i is smooth) and $\dim f_i^{-1}(D_i) < \dim S_i$, for all $i \in \mathrm{Ob}\, I$. If in addition $\dim D_i < \dim S_i$ for all $i \in \mathrm{Ob}\, I$, then f is called a *resolution*. Note that if S_i is irreducible and X_i is equidimensional for all $i \in \mathrm{Ob}\, I$, then a proper morphism of I-schemes, $f: X_\bullet \to S_\bullet$, is a resolution if and only it induces a proper birational morphism $f_i: X_i \to S_i$ for each $i \in \mathrm{Ob}\, I$.

Note that a resolution is a different notion than a (cubic) hyperresolution, compare (10.79).

This definition is slightly different from the one found in Guillén *et al.* (1988). They call a resolution what we call a weak resolution. However, that definition allows the following examples that seem far from what one should call a resolution:

Let $I = \underline{1}$, $X = \mathrm{Spec}\, k[x]$ and $S = \mathrm{Spec}\, k[x, y]/(xy)$. Let $f: X \to S$ be the morphism induced by the map $k[x, y]/(xy) \to k[x]$ that sends y to 0. The discriminant of f is $\mathrm{Spec}\, k[x, y]/(x)$, and the preimage of that is simply the origin on X, which has dimension smaller than 1.

Another example is the following. Choose any variety X of dimension greater than zero and a closed point $z \in X$. Then the discriminant of $g: \{z\} \to X$ is X itself, but the preimage of X is still just z, which has smaller dimension than X itself, by hypothesis.

It turns out that resolutions of I-schemes always exist under reasonable hypotheses.

Let I be a category. The set of objects of I can be given the following pre-order relation, $i \leq j$ if and only if $\mathrm{Hom}_I(i, j)$ is nonempty. We will say that a category I is ordered if this pre-order is a partial order and, for each $i \in \mathrm{Ob}\, I$, the only endomorphism of i is the identity (Guillén *et al.*, 1988, I, 1.9). Note that a category I is ordered if and only if all isomorphisms and endomorphisms of I are the identity.

Theorem 10.74 (Guillén *et al.*, 1988, I.2.6) *Let S be an I-scheme of finite type over a field k. Suppose that k is a field of characteristic 0 and that I is a finite ordered category. Then there exists a resolution of S.*

In order to construct a resolution $Y_\bullet$ of an I-scheme $X_\bullet$, it might be tempting to simply resolve each X_i, set Y_i equal to that resolution, and somehow combine this data together. Unfortunately this does not work, as shown by the example below.

Example 10.75 Consider a pinch point:

$$X = \mathrm{Spec}\, k[x, y, z]/(x^2 y - z^2) = \mathrm{Spec}\, k[s, t^2, st]$$

and let Z be the singular set of X, that is, the closed subscheme defined by the ideal (s, st). Let $I = \underline{2} = \{0 \to 1\}$. Consider the I-scheme defined by $X_0 = X$ and $X_1 = Z$ (with the closed immersion as the map). X_1 is already smooth, and if one resolves X_0 (that is, normalizes it), there is no compatible way to map X_1 (or even another birational model of X_1) to it, since its preimage by normalization will be two-to-one onto $Z \subset X$. The way this problem is resolved is by creating additional components. To construct a resolution $Y_\bullet$ we set $Y_1 = Z = X_1$ (since it was already smooth) and set $Y_0 = \widetilde{X}_0 \coprod Z$ where $\widetilde{X}_0$ is the normalization of X_0. The map $Y_1 \to Y_0$ just sends Y_1 (isomorphically) to the new component and the map $Y_0 \to X_0$ is the disjoint union of the normalization and inclusion maps.

One should note that although the proof of the existence of resolutions of I-schemes in Guillén *et al.* (1988, I, 2.6) is constructive, it is often easier in practice to construct an ad hoc resolution than the one given by the proof.

Now that we have resolutions of I-schemes, we can discuss cubic hyper-resolutions of (diagrams of) schemes. First we will discuss a single iterative step in the process of constructing cubic hyperresolutions. This step is called a 2-resolution.

Definition 10.76 (Guillén *et al.*, 1988, I.2.7) Let S be an I-scheme and $Z_\bullet$ a $\square_1^+ \times I$-scheme. We say that $Z_\bullet$ is a 2-*resolution* of S if $Z_\bullet$ is defined by the Cartesian square (pull-back, or fibered product in the category of (reduced)

I-schemes) of morphisms of *I*-schemes below

$$
\begin{array}{ccc}
Z_{11} & \hookrightarrow & Z_{01} \\
\downarrow & & \downarrow f \\
Z_{10} & \hookrightarrow & Z_{00}
\end{array}
$$

where $Z_{00} = S$, Z_{01} is a smooth *I*-scheme, the horizontal arrows are closed immersions of *I*-schemes, f is a proper *I*-morphism and Z_{10} contains the discriminant of f. In other words, f induces an isomorphism of $(Z_{01})_i - (Z_{11})_i$ over $(Z_{00})_i - (Z_{10})_i$ for all $i \in \mathrm{Ob}\, I$.

Clearly 2-resolutions always exist under the same hypotheses that resolutions of *I*-schemes exist: set Z_{01} to be a resolution, Z_{10} to be the discriminant (or any appropriate proper closed sub-*I*-scheme that contains it) and Z_{11} its (reduced) preimage in Z_{01}.

Consider the following example,

Example 10.77   Let $I = \underline{1} = \{0\}$ and let S be the *I*-scheme $\operatorname{Spec} k[t^2, t^3]$. Let $Z_{01} = \mathbb{A}_k^1 = \operatorname{Spec} k[t]$ and $Z_{01} \to S = Z_{00}$ be the map defined by $k[t^2, t^3] \to k[t]$. The discriminant of this map is the closed subscheme of $S = Z_{00}$ defined by the map $\phi\colon k[t^2, t^3] \to k$ that sends t^2 and t^3 to zero. Finally we need to define Z_{11}. The usual fibered product in the category of schemes is $\operatorname{Spec} k[t]/(t^2)$, but we work in the category of reduced schemes, so instead the fibered product is simply the associated reduced scheme (in this case $\operatorname{Spec} k[t]/(t)$). Thus our 2-resolution is the diagram pictured below.

$$
\begin{array}{ccc}
\operatorname{Spec} k[t]/(t) & \longrightarrow & \operatorname{Spec} k[t] \\
\downarrow & & \downarrow \\
\operatorname{Spec} k[t]/(t) & \longrightarrow & \operatorname{Spec} k[t^2, t^3].
\end{array}
$$

We need one more definition before defining a cubic hyperresolution,

Definition 10.78 (Guillén *et al.*, 1988, I.2.11)   Let n be an integer greater than or equal to 1, and let $X_\bullet^r$ be a $\square_r^+ \times I$-scheme, for $1 \le r \le n$. Suppose that for all r, $1 \le r \le n$, the $\square_{r-1}^+ \times I$-schemes $X_{00\bullet}^{r+1}$ and $X_{1\bullet}^r$ are equal. Then we define, by induction on n, a $\square_n^+ \times I$-scheme

$$
Z_\bullet = \mathrm{red}(X_\bullet^1, X_\bullet^2, \dots, X_\bullet^n)
$$

that we call the *reduction* of $(X_\bullet^1, \dots, X_\bullet^n)$, the following way: if $n = 1$, we define $Z_\bullet = X_\bullet^1$, if $n = 2$ we define $Z_\bullet = \mathrm{red}(X_\bullet^1, X_\bullet^2)$ by

$$
Z_{\alpha\beta} = \begin{cases} X_{0\beta}^1, & \text{if } \alpha = (0,0) \in \mathrm{Ob}\, \square_1^+, \\ X_{\alpha\beta}^2, & \text{if } \alpha \in \mathrm{Ob}\, \square_1 = \mathrm{Ob}\, \square_1^+ \setminus \{(0,0)\} \end{cases}
$$

for all $\beta \in I$, with the obvious morphisms. If $n > 2$, we define $Z_\bullet$ recursively as $\mathrm{red}(\mathrm{red}(X_\bullet^1, \ldots, X_\bullet^{n-1}), X_\bullet^n)$.

The main idea of the above defined reduction is the following: we want to do an iterated process to continuously resolve our schemes more and more precisely. In general a traditional resolution works, a 2-resolution, a $\square_1^+$-scheme is slightly more precise as it records information about the preimage of the discriminant. To get an even better picture we might consider a 2-resolution of the $\underline{2}$-scheme consisting of the discriminant, its preimage and the morphism between them. This 2-resolution is a $\square_2^+$-scheme and the reduction of the original 2-resolution and this one gives a new 2-resolution that combines the two: it gives a $\square_2^+$-resolution of the original scheme; the higher indexed edge of the original 2-resolution is replaced by the truncated 2-resolution of that edge.

Now we are ready to define cubic hyperresolutions.

Definition 10.79 (Guillén *et al.*, 1988, I.2.12)   Let S be an I-scheme. A $\square_n^+ \times I$-scheme $Z_\bullet = \mathrm{red}(X_\bullet^1, \ldots, X_\bullet^n)$ is called an *(augmented) cubic hyperresolution of S* if

(1) $X_\bullet^1$ is a 2-resolution of S,
(2) for $1 \leq r < n$, $X_\bullet^{r+1}$ is a 2-resolution of $X_{1\bullet}^r$ and
(3) Z_α is smooth for all $\alpha \in \square_n$.

Now let $T \subseteq S$ be a closed I-subscheme. Then an *embedded cubic hyperresolution of $T \subseteq S$* is a cubic hyperresolution of the I-pair (S, T).

Note that cubic hyperresolutions exist under reasonable hypotheses:

Theorem 10.80 (Guillén *et al.*, 1988, I.2.15)   *Let S be an I-scheme. Suppose that k is a field of characteristic 0 and that I is a finite (bounded) ordered category. Then there exists $Z_\bullet$, an enhanced cubic hyperresolution of S, such that*

$$\dim Z_\alpha \leq \dim S - |\alpha| + 1, \forall \alpha \in \square_n.$$

Below are some examples of cubic hyperresolutions.

Example 10.81   Let C be a curve and consider the normalization $\pi \colon \widetilde{C} \to C$ and P the set of singular points of C. In particular, P is the discriminant of π. Also let E be the reduced exceptional set of π. We have the following Cartesian square

$$\begin{array}{ccc} E & \longrightarrow & \widetilde{C} \\ \downarrow & & \downarrow{\scriptstyle \pi} \\ P & \longrightarrow & C. \end{array}$$

It is clearly a 2-resolution and thus also a cubic hyperresolution of C.

Example 10.82 Let us now compute a cubic hyperresolution of a scheme X whose singular locus is itself a smooth scheme, and whose reduced exceptional set of a strong resolution $\pi: \widetilde{X} \to X$ is smooth (for example, any cone over a smooth variety). As in the previous example, let Σ be the singular locus of X and E the reduced exceptional set of π, Then the Cartesian square of reduced schemes

$$
\begin{array}{ccc}
E & \longrightarrow & \widetilde{X} \\
\downarrow & & \downarrow{\scriptstyle \pi} \\
\Sigma & \longrightarrow & X
\end{array}
$$

is in fact a 2-resolution of X, just as in the case of curves above.

The general algorithm used to construct cubic hyperresolutions does not necessarily construct hyperresolutions in the most efficient or convenient way possible. For example, it is relatively easy to find a cubic hyperresolution for the three coordinate planes in $\mathbb{A}^3$, but it is not the hyperresolution we would obtain if we follow an algorithm that works in general. See Kovács and Schwede (2011b, 2.16) for the computation of the latter.

Example 10.83 Suppose that S is the union of the three coordinate planes (X, Y and Z) of $\mathbb{A}^3$. Consider the $\square_2$ or $\square_2^+$ scheme defined by the diagram below (where the dotted arrows are those in $\square_2^+$ but not in $\square_2$).

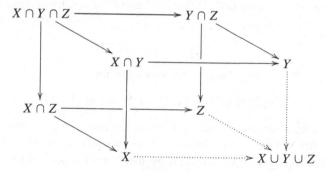

It is easy to verify that this is a cubic hyperresolution of $X \cup Y \cup Z$.

Next we will discuss sheaves on diagrams of schemes as well as the related notions of push-forward and its right derived functors.

Definition 10.84 (Guillén *et al.*, 1988, I.5.3–5.4) Let $X_\bullet$ be an I-scheme (or even an I-topological space). We define a *sheaf (or pre-sheaf) of abelian groups $F^\bullet$ on $X_\bullet$* to be the collection of the following data:

(1) a sheaf (pre-sheaf) F^i of abelian groups over X_i, for all $i \in \mathrm{Ob}\, I$ and
(2) an X_ϕ-morphism of sheaves $F^\phi: F^i \to (X_\phi)_* F^j$ for all morphisms $\phi: i \to j$ of I, required to be compatible in the obvious way.

Notice that by the adjointness of f_* and f^*, the existence of $F^i \to (X_\phi)_* F^j$ is equivalent to the existence of $(X_\phi)^* F^i \to F^j$. For our purposes $F^i \to (X_\phi)_* F_j$ works slightly better.

Given a morphism of diagrams of schemes $f_\bullet\colon X_\bullet \to Y_\bullet$ one can construct a push-forward functor for sheaves on $X_\bullet$.

Definition 10.85 (Guillén *et al.*, 1988, I.5.5)   Let $X_\bullet$ be an I-scheme, $Y_\bullet$ a J-scheme, $F^\bullet$ a sheaf on $X_\bullet$ and $f_\bullet\colon X_\bullet \to Y_\bullet$ a morphism of diagrams of schemes. We define $f_{\bullet *} F^\bullet$ the following way. For each $j \in \mathrm{Ob}\, J$ we set

$$(f_{\bullet *} F^\bullet)^j := \varprojlim (Y_\phi)_* (f_{i *} F^i)$$

where the inverse limit traverses all pairs (i, ϕ) where $\phi\colon f(i) \to j$ is a morphism in J^{op}.

In many applications J will simply be the category $\{0\}$ with one object and one morphism (for example, cubic hyperresolutions of schemes). In that case one can merely think of the limit as traversing I. One can also define a functor f^*, show that it has a right adjoint and that that adjoint is f_* (Guillén *et al.*, 1988, I, 5.5).

Definition 10.86 (Guillén *et al.*, 1988, I, section 5)   Let $X_\bullet$ and $Y_\bullet$ be diagrams of topological spaces over I and J respectively, $\Phi\colon I \to J$ a functor, $f_\bullet\colon X_\bullet \to Y_\bullet$ a Φ-morphism of topological spaces. If $G^\bullet$ is a sheaf over $Y_\bullet$ with values in a complete category C (i.e., all small limits exist in C), one denotes by $f_\bullet^* G^\bullet$ the sheaf over $X_\bullet$ defined by

$$(f_\bullet^* G^\bullet)^i = f_i^* (G^{\Phi(i)}),$$

for all $i \in \mathrm{Ob}\, I$. One obtains in this way a functor

$$f_\bullet^*\colon \mathsf{Sheaves}(Y_\bullet, \mathsf{C}) \to \mathsf{Sheaves}(X_\bullet, \mathsf{C}).$$

Given an I-scheme $X_\bullet$, one can define the category of sheaves of abelian groups $\mathsf{Ab}(X_\bullet)$ on $X_\bullet$ and show that it has enough injectives. Next, define the derived category $D^+(X_\bullet, \mathsf{Ab}(X_\bullet))$ by localizing bounded below complexes of sheaves of abelian groups on $X_\bullet$ by the quasi-isomorphisms (those that are quasi-isomorphisms on each $i \in I$). One can also show that $f_{\bullet *}$ as defined above is left exact so that it has a right derived functor $\mathcal{R} f_{\bullet *}$ (Guillén *et al.*, 1988, I, 5.8–5.9). In the case of a cubic hyperresolution of a scheme $f\colon X_\bullet \to X$,

$$\mathcal{R} f_{\bullet *} F^\bullet = \mathcal{R} \varprojlim (\mathcal{R} f_{i *} F^i)$$

where the limit traverses the category I of $X_\bullet$.

Remark 10.87   We end our excursion into the world of hyperresolutions here. There are many other things to work out, but we will leave them for the interested reader to explore on their own. Many "obvious" statements need to be proved, but most are relatively straightforward once one gets comfortable using the appropriate language. For those and many more statements, including the full details of the construction of the Du Bois complex and many applications, the reader is encouraged to read Guillén *et al.* (1988).

References

Abramovich, D. and Hassett, B. 2011. Stable varieties with a twist. In *Classification of Algebraic Varieties*, C. Faber, G. van Greer, and E. Looijenga (eds), pp. 1–38. Zürich: EMS Ser. Congr. Rep. Eur. Math. Soc. cited on page(s): 267

Alexeev, V. 2008. Limits of stable pairs. *Pure Appl. Math. Q.*, **4**(3, part 2), 767–783. cited on page(s): 242, 243, 244

Alexeev, V. and Hacon, C. D. 2011. Non-rational centers of log canonical singularities. arXiv e-prints, Sept. cited on page(s): 243

Altman, A. and Kleiman, S. 1970. *Introduction to Grothendieck Duality Theory*. Lecture Notes in Mathematics, Vol. 146. Berlin: Springer-Verlag. cited on page(s): 80

Ambro, F. 1999a. The adjunction conjecture and its applications. arXiv.org:math/9903060. cited on page(s): 239

Ambro, F. 1999b. On minimal log discrepancies. *Math. Res. Lett.*, **6**(5–6), 573–580. cited on page(s): 239

Ambro, F. 2003. Quasi-log varieties. *Tr. Mat. Inst. Steklova*, **240**, 220–239. cited on page(s): 151, 167, 171, 173, 196, 275, 321

Ambro, F. 2011. Basic properties of log canonical centers. In *Classification of Algebraic Varieties*, C. Faber, G. van Greer, and E. Looijenga (eds), pp. 39–48. Zürich: EMS Ser. Congr. Rep. Eur. Math. Soc. cited on page(s): 167, 171, 196

Ambro, F. 2012. An Injectivity Theorem. arXiv:1209.6134. cited on page(s): 321

Andreotti, A. and Norguet, F. 1967. La convexité holomorphe dans l'espace analytique des cycles d'une variété algébrique. *Ann. Scuola Norm. Sup. Pisa (3)*, **21**, 31–82. cited on page(s): 306

Andreotti, A. and Bombieri, E. 1969. Sugli omeomorfismi delle varietà algebriche. *Ann. Scuola Norm. Sup Pisa (3)*, **23**, 431–450. cited on page(s): 306

Angehrn, U. and Siu, Y. T. 1995. Effective freeness and point separation for adjoint bundles. *Invent. Math.*, **122**(2), 291–308. cited on page(s): 264

Arapura, D., Bakhtary, P., and Włodarczyk, J. 2011. Weights on cohomology, invariants of singularities, and dual complexes. arXiv e-prints, Feb. cited on page(s): 260

Arbarello, E. and De Concini, C. 1987. Another proof of a conjecture of S. P. Novikov on periods of abelian integrals on Riemann surfaces. *Duke Math. J.*, **54**(1), 163–178. cited on page(s): 264

Arnol'd, V. I., Guseĭn-Zade, S. M., and Varchenko, A. N. 1985. *Singularities of Differentiable Maps. Vols. I–II*. Monographs in Mathematics, Vol. 82. Boston, MA:

Birkhäuser Boston Inc. Translated from the Russian by Ian Porteous and Mark Reynolds. cited on page(s): 251

Artebani, M. and Dolgachev, I. 2009. The Hesse pencil of plane cubic curves. *Enseign. Math. (2)*, **55**(3–4), 235–273. cited on page(s): 169

Artin, M. 1970. Algebraization of formal moduli. II. Existence of modifications. *Ann. of Math. (2)*, **91**, 88–135. cited on page(s): 129, 133, 282

Artin, M. 1975. Wildly ramified $Z/2$ actions in dimension two. *Proc. Amer. Math. Soc.*, **52**, 60–64. cited on page(s): 73, 102, 108

Artin, M. 1977. Coverings of the rational double points in characteristic p. In *Complex Analysis and Algebraic Geometry*, W. L. Baily Jr. and T. Shioda (eds), pp. 11–22. Tokyo: Iwanami Shoten. cited on page(s): 73, 102, 109

Artin, M. 1986. Néron models. In *Arithmetic Geometry (Storrs, Conn., 1984)*, G. Connell and J. H. Silverman (eds), pp. 213–230. New York: Springer. cited on page(s): 50

Atiyah, M. F. 1970. Resolution of singularities and division of distributions. *Comm. Pure Appl. Math.*, **23**, 145–150. cited on page(s): 251

Ax, J. 1968. The elementary theory of finite fields. *Ann. of Math. (2)*, **88**, 239–271. cited on page(s): 265

Barth, W., Peters, C., and Van de Ven, A. 1984. *Compact Complex Surfaces*. Ergebnisse der Mathematik und ihrer Grenzgebiete (3), Vol. 4. Berlin: Springer-Verlag. cited on page(s): 26

Beauville, A. 1983. *Complex Algebraic Surfaces*. London Mathematical Society Lecture Note Series, vol. 68. Cambridge: Cambridge University Press. Translated from the French by R. Barlow, N. I. Shepherd-Barron, and M. Reid. cited on page(s): 26

Bellaccini, B. 1983. Proper morphisms and excellent schemes. *Nagoya Math. J.*, **89**, 109–118. cited on page(s): 282

Benson, D. J. 1993. *Polynomial Invariants of Finite Groups*. London Mathematical Society Lecture Note Series, Vol. 190. Cambridge: Cambridge University Press. cited on page(s): 103

Bernšteĭn, I. N. 1971. Modules over a ring of differential operators. An investigation of the fundamental solutions of equations with constant coefficients. *Funkcional. Anal. i Priložen.*, **5**(2), 1–16. cited on page(s): 251

Białynicki-Birula, A. 2004. Finite equivalence relations on algebraic varieties and hidden symmetries. *Transform. Groups*, **9**(4), 311–326. cited on page(s): 272

Bierstone, E. and Milman, P. D. 1997. Canonical desingularization in characteristic zero by blowing up the maximum strata of a local invariant. *Invent. Math.*, **128**(2), 207–302. cited on page(s): 324

Bierstone, E. and Milman, P. D. 2011. Resolution except for minimal singularities I. arXiv.org:1107.5595. cited on page(s): 333

Bierstone, E. and Pacheco, F. V. 2011. Resolution of singularities of pairs preserving semi-simple normal crossings. arXiv.org:1109.3205. cited on page(s): 333

Bierstone, E., Lairez, P., and Milman, P. D. 2011. Resolution except for minimal singularities II. The case of four variables. arXiv.org:1107.5598. cited on page(s): 328

Birkar, C. 2010. On existence of log minimal models. *Compos. Math.*, **146**(4), 919–928. cited on page(s): 26

Birkar, C. 2011. Existence of log canonical flips and a special LMMP. arXiv:1104.4981. cited on page(s): 27, 261

Birkar, C., Cascini, P., Hacon, C. D., and McKernan, J. 2010. Existence of minimal models for varieties of log general type. *J. Amer. Math. Soc.*, **23**(2), 405–468. cited on page(s): 26, 28, 49, 177

Biswas, I. and dos Santos, J. P. P. 2012. Triviality criteria for vector bundles over separably rationally connected varieties. *To appear.* cited on page(s): 96

Blickle, M. and Schwede, K. 2012. p^{-1}-linear maps in algebra and geometry. arXiv:1205.4577. cited on page(s): 256

Blickle, M., Schwede, K., and Tucker, K. 2011. F-singularities via alterations. arXiv:1107.3807. cited on page(s): 256

Bogomolov, F. and Tschinkel, Y. 2009. Co-fibered products of algebraic curves. arXiv.org:0902.0534. cited on page(s): 209

Boissière, S., Gabber, O., and Serman, O. 2011. Sur le produit de variétés localement factorielles ou Q-factorielles. arXiv e-prints, Apr. cited on page(s): 5

Boyer, C. P., Galicki, K., Kollár, J., and Thomas, E. 2005a. Einstein metrics on exotic spheres in dimensions 7, 11, and 15. *Experiment. Math.*, **14**(1), 59–64. cited on page(s): 265

Boyer, C. P., Galicki, K., and Kollár, J. 2005b. Einstein metrics on spheres. *Ann. of Math. (2)*, **162**(1), 557–580. cited on page(s): 265

Brieskorn, E. 1968. Rationalen Singularitäten komplexer Flächen. *Inv. Math.*, **4**, 336–358. cited on page(s): 73, 103, 123, 135

Brown, G. and Reid, M. 2010. Anyone knows these guys? http://dl.dropbox.com/ u/10909533/anyoneknows.pdf. cited on page(s): 130

Brown, M. V. 2011. Singularities of Cox Rings of Fano Varieties. arXiv e-prints, Sept. cited on page(s): 101

Bruns, W. and Herzog, J. 1993. *Cohen-Macaulay Rings*. Cambridge Studies in Advanced Mathematics, Vol. 39. Cambridge: Cambridge University Press. cited on page(s): 79, 81

Carlson, J. A. 1985. Polyhedral resolutions of algebraic varieties. *Trans. Amer. Math. Soc.*, **292**(2), 595–612. cited on page(s): 215

Cascini, P. and Lazić, V. 2010. New outlook on the Minimal Model Program, I. arXiv: 1009.3188. cited on page(s): 26, 75

Cascini, P. and Lazić, V. 2012. The Minimal Model Program revisited. arXiv:1202,0738. cited on page(s): 26

Chatzistamatiou, A. and Rülling, K. 2009. Higher direct images of the structure sheaf in positive characteristic. arXiv:0911.3599. cited on page(s): 85, 103

Chel'tsov, I. A. and Shramov, K. A. 2009. Extremal metrics on del Pezzo threefolds. *Tr. Mat. Inst. Steklova*, **264**, 37–51. cited on page(s): 265

Chevalley, C. 1955. Invariants of finite groups generated by reflections. *Amer. J. Math.*, **77**, 778–782. cited on page(s): 103

Conrad, B. 2000. *Grothendieck Duality and Base Change.* Lecture Notes in Mathematics, Vol. 1750. Berlin: Springer-Verlag. cited on page(s): 8, 9, 80

Corti, A. 1995. Factoring birational maps of threefolds after Sarkisov. *J. Algebraic Geom.*, **4**(2), 223–254. cited on page(s): 264

Corti, A. (ed.). 2007. *Flips for 3-folds and 4-folds.* Oxford Lecture Series in Mathematics and its Applications, Vol. 35. Oxford: Oxford University Press. cited on page(s): 26

Corti, A., Pukhlikov, A., and Reid, M. 2000. Fano 3-fold hypersurfaces. In *Explicit Birational Geometry of 3-folds*, A. Corti and M. Reid (eds), pp. 175–258. London Mathematical Society Lecture Note Series, Vol. 281. Cambridge: Cambridge University Press. cited on page(s): 264

Cossart, V., Galindo, C., and Piltant, O. 2000. Un exemple effectif de gradué non noethérien associé à une valuation divisorielle. *Ann. Inst. Fourier (Grenoble)*, **50**(1), 105–112. cited on page(s): 31

Curtis, C. W. and Reiner, I. 1962. *Representation Theory of Finite Groups and Associative Algebras.* Pure and Applied Mathematics, Vol. XI. New York–London: Interscience Publishers, a division of John Wiley & Sons. cited on page(s): 338

de Fernex, T., Kollár, J., and Xu, C. 2013. The dual complex of singularities. arXiv:1212.1675. cited on page(s): 261

de Fernex, T. and Mustață, M. 2009. Limits of log canonical thresholds. *Ann. Sci. Éc. Norm. Supér. (4)*, **42**(3), 491–515. cited on page(s): 253

de Fernex, T., Ein, L., and Mustață, M. 2010. Shokurov's ACC conjecture for log canonical thresholds on smooth varieties. *Duke Math. J.*, **152**(1), 93–114. cited on page(s): 253

de Fernex, T., Ein, L., and Mustață, M. 2011. Log canonical thresholds on varieties with bounded singularities. In *Classification of Algebraic Varieties*, C. Faber, G. van der Geer, and E. Looijenge (eds). Zürich: EMS Ser. Congr. Rep. Eur. Math. Soc. cited on page(s): 253

de Jong, A. J. and Starr, J. 2004. Cubic fourfolds and spaces of rational curves. *Illinois J. Math.*, **48**(2), 415–450. cited on page(s): 257

Deligne, P. 1974. Théorie de Hodge. III. *Inst. Hautes Études Sci. Publ. Math.*, 5–77. cited on page(s): 220, 221, 222

Demailly, J.-P. 1993. A numerical criterion for very ample line bundles. *J. Differential Geom.*, **37**(2), 323–374. cited on page(s): 264

Demailly, J.-P. 2001. Multiplier ideal sheaves and analytic methods in algebraic geometry. In *School on Vanishing Theorems and Effective Results in Algebraic Geometry (Trieste, 2000)*. ICTP Lect. Notes, Vol. 6. Trieste: Abdus Salam International Centre Theoretical Physics. cited on page(s): 250

Demailly, J.-P. and Kollár, J. 2001. Semi-continuity of complex singularity exponents and Kähler-Einstein metrics on Fano orbifolds. *Ann. Sci. École Norm. Sup. (4)*, **34**(4), 525–556. cited on page(s): 251, 265, 339

Dolgachev, I. 1975. Automorphic forms, and quasihomogeneous singularities. *Funkcional. Anal. i Priložen.*, **9**(2), 67–68. cited on page(s): 289

Du Bois, P. 1981. Complexe de de Rham filtré d'une variété singulière. *Bull. Soc. Math. France*, **109**(1), 41–81. cited on page(s): 215, 216, 218, 219, 220

Du Val, P. 1934. On isolated singularities of surfaces which do not affect the conditions of adjunction I–II. *Proc. Camb. Phil. Soc.*, **30**(4), 453–465, 483–491. cited on page(s): 48

Durfee, A. 1979. Fifteen characterizations of rational double points and simple critical points. *L'Ens. Math.*, **25**(1), 131–163. cited on page(s): 109

Ein, L. and Lazarsfeld, R. 1993. Global generation of pluricanonical and adjoint linear series on smooth projective threefolds. *J. Amer. Math. Soc.*, **6**(4), 875–903. cited on page(s): 263

Ein, L. and Lazarsfeld, R. 1997. Singularities of theta divisors and the birational geometry of irregular varieties. *J. Amer. Math. Soc.*, **10**(1), 243–258. cited on page(s): 264

Ein, L. and Mustață, M. 2004. Inversion of adjunction for local complete intersection varieties. *Amer. J. Math.*, **126**(6), 1355–1365. cited on page(s): 254

Ein, L. and Mustață, M. 2009. Jet schemes and singularities. In *Algebraic Geometry—Seattle 2005. Part 2*. Proc. Sympos. Pure Math., Vol. 80, pp. 505–546. Providence, RI: American Mathematical Society. cited on page(s): 254

Ein, L., Mustață, M., and Yasuda, T. 2003. Jet schemes, log discrepancies and inversion of adjunction. *Invent. Math.*, **153**(3), 519–535. cited on page(s): 254

Elkik, R. 1981. Rationalité des singularités canoniques. *Inv. Math.*, **64**, 1–6. cited on page(s): 77, 255

Esnault, H. and Viehweg, E. 1982. Revêtements cycliques. In *Algebraic Threefolds (Varenna, 1981)*, A. Conte (ed.), pp. 241–250. Lecture Notes in Mathematics, Vol. 947. Berlin: Springer. cited on page(s): 257

Fedder, R. and Watanabe, K. 1989. A characterization of F-regularity in terms of F-purity. In *Commutative Algebra (Berkeley, CA, 1987)*, M. Hochster, C. Huneke, and J. D. Sally (eds), pp. 227–245. Math. Sci. Res. Inst. Publ., Vol. 15. New York: Springer. cited on page(s): 255

Ferrand, D. 2003. Conducteur, descente et pincement. *Bull. Soc. Math. France*, **131**(4), 553–585. cited on page(s): 281, 282

Flenner, H. 1988. Extendability of differential forms on nonisolated singularities. *Invent. Math.*, **94**(2), 317–326. cited on page(s): 257

Friedman, R. 1983. Global smoothings of varieties with normal crossings. *Ann. of Math. (2)*, **118**(1), 75–114. cited on page(s): 217

Friedman, R. and Morrison, D. R. (eds). 1983. *The Birational Geometry of Degenerations*. Progr. Math., Vol. 29. Boston, MA: Birkhäuser Boston. cited on page(s): 136, 138

Fröhlich, A. and Taylor, M. J. 1993. *Algebraic Number Theory*. Cambridge Studies in Advanced Mathematics, Vol. 27. Cambridge: Cambridge University Press. cited on page(s): 338

Fujino, O. 2000. Abundance theorem for semi log canonical threefolds. *Duke Math. J.*, **102**(3), 513–532. cited on page(s): 26, 172, 261, 334, 339

Fujino, O. 2004. Higher direct images of log canonical divisors. *J. Differential Geom.*, **66**(3), 453–479. cited on page(s): 321

Fujino, O. 2007. What is log terminal? In *Flips for 3-folds and 4-folds*. Oxford Lecture Ser. Math. Appl., Vol. 35. Oxford: Oxford University Press. cited on page(s): 164

Fujino, O. 2009a. Fundamental Theorems for the Log Minimal Model Program. http://www.citebase.org/abstract?id=oai:arXiv.org:0909.4445. cited on page(s): 321

Fujino, O. 2009b. Introduction to the log minimal model program for log canonical pairs. arXiv.org:0907.1506. cited on page(s): 91, 167, 171, 196, 243, 244, 321

Fujino, O. 2010. Semi-stable minimal model program for varieties with trivial canonical divisor. arXiv e-prints. cited on page(s): 26

Fujino, O. 2011. On isolated log canonical singularities with index one. arXiv e-prints, July. cited on page(s): 172, 261

Fujino, O. 2012a. Fundamental theorems for semi log canonical pairs. arXiv:1202.5365. cited on page(s): 321

Fujino, O. 2012b. Vanishing theorems. arXiv:1202.4200. cited on page(s): 321

Fujino, O. and Gongyo, Y. 2011. Log pluricanonical representations and abundance conjecture. arXiv e-prints, Apr. cited on page(s): 334, 335, 339

Fujino, O. and Takagi, S. 2011. On the F-purity of isolated log canonical singularities. arXiv e-prints, Dec. cited on page(s): 101, 261

Fujita, Kento. 2011. Simple normal crossing Fano varieties and log Fano manifolds. arXiv e-print. June. cited on page(s): 140, 143

Fujita, T. 1985. A reative versions of Kawamata-Viehweg's vanishing theorem. Unpublished preprint, University of Tokyo. cited on page(s): 246

Fujita, T. 1987. On polarized manifolds whose adjoint bundles are not semipositive. In *Algebraic Geometry, Sendai, 1985*. Adv. Stud. Pure Math., Vol. 10, pp. 167–178. Amsterdam: North-Holland. cited on page(s): 263

Gel'fand, I. M. and Šilov, G. E. 1958. *Obobshchennye funksii i deistviya iad nimi.* Obobščennye funkcii, Vypusk 1. Generalized Functions, Part 1. Moscow: Gosudarstv. Izdat. Fiz.-Mat. Lit. cited on page(s): 251

Gongyo, Y. 2010. Abundance theorem for numerically trivial log canonical divisors of semi-log canonical pairs. arXiv e-prints, May. cited on page(s): 261, 334, 335, 339

Gongyo, Y., Okawa, S., Sannai, A., and Takagi, S. 2012. Characterization of varieties of Fano type via singularities of Cox rings. arXiv e-prints, Jan. cited on page(s): 101

Goresky, M. and MacPherson, R. 1988. *Stratified Morse Theory.* Ergebnisse der Mathematik und ihrer Grenzgebiete (3), Vol. 14. Berlin: Springer-Verlag. cited on page(s): 136, 259

Graham, R. L., Knuth, D. E., and Patashnik, O. 1989. *Concrete Mathematics.* Reading, MA: Addison-Wesley Publishing Company Advanced Book Program. cited on page(s): 252

Greb, D., Kebekus, S., and Kovács, S. J. 2010. Extension theorems for differential forms, and Bogomolov-Sommese vanishing on log canonical varieties. *Compos. Math.*, **146**(January), 193–219. Published online by Cambridge University Press 11 Dec 2009. cited on page(s): 257, 258

Greb, D., Kebekus, S., Kovács, S. J., and Peternell, T. 2011a. Differential forms on log canonical spaces. *Publ. Math. Inst. Hautes Études Sci.*, **114**, 87–169. cited on page(s): 220, 257, 258

Greb, D., Kebekus, S., and Peternell, T. 2011b. *Reflexive differential forms on singular spaces – geometry and cohomology.* Unpublished manuscript. cited on page(s): 257

Grothendieck, A. 1960. Éléments de géométrie algébrique. I–IV. *Inst. Hautes Études Sci. Publ. Math.* cited on page(s): 36, 282

Grothendieck, A. 1967. *Local Cohomology.* Lecture Notes in Mathematics, Vol. 41. Berlin: Springer-Verlag. cited on page(s): 79

Grothendieck, A. 1968. *Cohomologie Locale des Faisceaux Cohérents et Théorèmes de Lefschetz Locaux et Globaux (SGA 2).* Amsterdam: North-Holland Publishing Co. Augmenté d'un exposé par Michèle Raynaud, Séminaire de Géométrie Algébrique du Bois-Marie, 1962, Advanced Studies in Pure Mathematics, Vol. 2. cited on page(s): 136, 244, 323

Guillén, F., Navarro Aznar, V., Pascual Gainza, P., and Puerta, F. 1988. *Hyperrésolutions Cubiques et Descente Cohomologique.* Lecture Notes in Mathematics, vol. 1335. Berlin: Springer-Verlag. Papers from the Seminar on Hodge-Deligne Theory held in Barcelona, 1982. cited on page(s): 215, 216, 220, 222, 340, 341, 342, 343, 344, 345, 346, 347

Guralnick, R. and Tiep, P. H. 2010. A problem of Kollár and Larsen on finite linear groups and crepant resolutions. arXiv:1009.2535. cited on page(s): 106, 107

Hacon, C. D. 2012. On log canonical inversion of adjunction. arXiv e-prints, Feb. cited on page(s): 161

Hacon, C. D. and McKernan, J. 2012. The Sarkisov program. *To appear in J. Alg. Geom.*, May. cited on page(s): 264

Hacon, C. D., McKernan, J., and Xu, C. 2012. ACC for log canonical thresholds. arXiv:1208.4150. cited on page(s): 253

Hacon, C. D. and Xu, C. 2011a. Existence of log canonical closures. arXiv:1105.1169. cited on page(s): 27, 28

Hacon, C. D. and Xu, C. 2011b. On finiteness of B-representation and semi-log canonical abundance. arXiv:1107.4149. cited on page(s): 171, 261, 334, 335, 339

Hamm, H. 1971. Lokale topologische Eigenschaften komplexer Räume. *Math. Ann.*, **191**(3), 235–252. cited on page(s): 136

Hara, N. 1998a. A characterization of rational singularities in terms of injectivity of Frobenius maps. *Amer. J. Math.*, **120**(5), 981–996. cited on page(s): 256

Hara, N. 1998b. Classification of two-dimensional F-regular and F-pure singularities. *Adv. Math.*, **133**(1), 33–53. cited on page(s): 256

Hara, N. 2005. A characteristic p analog of multiplier ideals and applications. *Comm. Algebra*, **33**(10), 3375–3388. cited on page(s): 256

Hara, N. and Watanabe, K.-I. 2002. F-regular and F-pure rings vs. log terminal and log canonical singularities. *J. Algebraic Geom.*, **11**(2), 363–392. cited on page(s): 256

Hara, N. and Yoshida, K.-I. 2003. A generalization of tight closure and multiplier ideals. *Trans. Amer. Math. Soc.*, **355**(8), 3143–3174 (electronic). cited on page(s): 256

Harris, J. 1995. *Algebraic Geometry*. Graduate Texts in Mathematics, Vol. 133. New York: Springer-Verlag. Corrected reprint of the 1992 original. cited on page(s): 133

Hartshorne, R. 1977. *Algebraic Geometry*. Graduate Texts in Mathematics, Vol. 52. New York: Springer-Verlag. cited on page(s): 4, 8, 9, 44, 64, 76, 77, 97, 98, 149, 292

Hartshorne, R. 1966. *Residues and Duality*. Lecture notes of a seminar on the work of A. Grothendieck, given at Harvard 1963/64. With an appendix by P. Deligne. Lecture Notes in Mathematics, Vol. 20. Berlin: Springer-Verlag. cited on page(s): 8, 9, 80, 81

Hatcher, A. 2002. *Algebraic Topology*. Cambridge: Cambridge University Press. cited on page(s): 138

Hayakawa, T. and Takeuchi, K. 1987. On canonical singularities of dimension three. *Japan. J. Math. (N.S.)*, **13**(1), 1–46. cited on page(s): 76

Heier, G. 2002. Effective freeness of adjoint line bundles. *Doc. Math.*, **7**(1), 31–42 (electronic). cited on page(s): 264

Hernández, D. 2011. F-purity versus log canonicity for polynomials. arXiv:1112.2423. cited on page(s): 256

Hironaka, H. 1964. Resolution of singularities of an algebraic variety over a field of characteristic zero. I, II. *Ann. of Math. (2)*, **79**, 109–203 and 205–326. cited on page(s): 324, 330

Hochster, M. and Huneke, C. 1990. Tight closure, invariant theory, and the Briançon-Skoda theorem. *J. Amer. Math. Soc.*, **3**(1), 31–116. cited on page(s): 256

Hogadi, A. and Xu, C. 2009. Degenerations of rationally connected varieties. *Trans. Amer. Math. Soc.*, **361**(7), 3931–3949. cited on page(s): 261

Holmann, H. 1963. Komplexe Räume mit komplexen Transformations-gruppen. *Math. Ann.*, **150**, 327–360. cited on page(s): 208, 269

Hörmander, L. 1958. On the division of distributions by polynomials. *Ark. Mat.*, **3**, 555–568. cited on page(s): 251

Ito, Y. and Reid, M. 1996. The McKay correspondence for finite subgroups of SL(3, **C**). In *Higher-Dimensional Complex Varieties (Trento, 1994)*, M. Andreatta (ed.), pp. 221–240. Berlin: de Gruyter. cited on page(s): 104

Johnson, J. M. and Kollár, J. 2001a. Fano hypersurfaces in weighted projective 4-spaces. *Experiment. Math.*, **10**(1), 151–158. cited on page(s): 265

Johnson, J. M. and Kollár, J. 2001b. Kähler-Einstein metrics on log del Pezzo surfaces in weighted projective 3-spaces. *Ann. Inst. Fourier (Grenoble)*, **51**(1), 69–79. cited on page(s): 265

Kapovich, M. and Kollár, J. 2011. Fundamental groups of links of isolated singularities. *J. Amer. Math. Soc.* to appear. cited on page(s): 260

Kawakita, M. 2007. Inversion of adjunction on log canonicity. *Invent. Math.*, **167**(1), 129–133. cited on page(s): 151, 158, 161

Kawamata, Y. 1980. On singularities in the classification theory of algebraic varieties. *Math. Ann.*, **251**(1), 51–55. cited on page(s): 73

Kawamata, Y. 1992. Abundance theorem for minimal threefolds. *Invent. Math.*, **108**(2), 229–246. cited on page(s): 261

Kawamata, Y. 1994. Semistable minimal models of threefolds in positive or mixed characteristic. *J. Algebraic Geom.*, **3**(3), 463–491. cited on page(s): 27, 73

Kawamata, Y. 1997. On Fujita's freeness conjecture for 3-folds and 4-folds. *Math. Ann.*, **308**(3), 491–505. cited on page(s): 263

Kawamata, Y. 1998. Subadjunction of log canonical divisors. II. *Amer. J. Math.*, **120**(5), 893–899. cited on page(s): 170

Kawamata, Y. 1999a. Deformations of canonical singularities. *J. Amer. Math. Soc.*, **12**(1), 85–92. cited on page(s): 159

Kawamata, Y. 1999b. Index 1 covers of log terminal surface singularities. *J. Algebraic Geom.*, **8**(3), 519–527. cited on page(s): 27, 73

Kawamata, Y. 2002. *D*-equivalence and *K*-equivalence. *J. Differential Geom.*, **61**(1), 147–171. cited on page(s): 51

Kawamata, Y. and Okawa, S. 2012. Mori dream spaces of Calabi-Yau type and the log canonicity of the Cox rings. arXiv e-prints, Feb. cited on page(s): 101

Kawamata, Y., Matsuda, K., and Matsuki, K. 1987. Introduction to the minimal model problem. In *Algebraic Geometry, Sendai, 1985*, T. Oda (ed.). Adv. Stud. Pure Math., Vol. 10, pp. 283–360. Amsterdam: North-Holland. cited on page(s): 318

Kawasaki, T. 2002. On arithmetic Macaulayfication of Noetherian rings. *Trans. Amer. Math. Soc.*, **354**(1), 123–149 (electronic). cited on page(s): 80, 81

Kebekus, S. and Kovács, S. J. 2010. The structure of surfaces and threefolds mapping to the moduli stack of canonically polarized varieties. *Duke Math. J.*, **155**(1), 1–33. cited on page(s): 257

Keel, S., Matsuki, K., and McKernan, J. 1994. Log abundance theorem for threefolds. *Duke Math. J.*, **75**(1), 99–119. cited on page(s): 172, 261

Kempf, G., Knudsen, F. F., Mumford, D., and Saint-Donat, B. 1973. *Toroidal Embeddings. I*. Lecture Notes in Mathematics, Vol. 339. Berlin: Springer-Verlag. cited on page(s): 85, 262

Kleiman, S. L. 1966. Toward a numerical theory of ampleness. *Ann. of Math. (2)*, **84**, 293–344. cited on page(s): 281

Knutson, D. 1971. *Algebraic Spaces*. Lecture Notes in Mathematics, Vol. 203. Berlin: Springer-Verlag. cited on page(s): 281

Kobayashi, S. 1963. Topology of positively pinched Kaehler manifolds. *Tôhoku Math. J. (2)*, **15**, 121–139. cited on page(s): 265

Kollár, J. 1986a. Higher direct images of dualizing sheaves. I. *Ann. of Math. (2)*, **123**(1), 11–42. cited on page(s): 321

Kollár, J. 1986b. Higher direct images of dualizing sheaves. II. *Ann. of Math. (2)*, **124**(1), 171–202. cited on page(s): 321

Kollár, J. 1991a. Extremal rays on smooth threefolds. *Ann. Sci. École Norm. Sup. (4)*, **24**(3), 339–361. cited on page(s): 27

Kollár, J. 1991b. Flips, flops, minimal models, etc. In *Surveys in Differential Geometry (Cambridge, MA, 1990)*, H. Blaine Lawson and S.-T. Yau (eds), pp. 113–199. Bethlehem, PA: Lehigh University. cited on page(s): 30

Kollár, J. (ed). 1992. *Flips and Abundance for Algebraic Threefolds*. Papers from the Second Summer Seminar on Algebraic Geometry held at the University of Utah, Salt Lake City, Utah, August 1991, Astérisque No. 211 (1992). Société Mathématique de France. cited on page(s): 26, 72, 112, 158, 172, 175, 252, 253, 261, 335

Kollár, J. 1993a. Effective base point freeness. *Math. Ann.*, **296**(4), 595–605. cited on page(s): 264

Kollár, J. 1993b. Shafarevich maps and plurigenera of algebraic varieties. *Invent. Math.*, **113**(1), 177–215. cited on page(s): 130, 259, 260, 322

Kollár, J. 1994. Log surfaces of general type; some conjectures. In *Classification of Algebraic Varieties (L'Aquila, 1992)*, C. Ciliberto, L. Livorni, and A. J. Sommese (eds). Contemp. Math., Vol. 162, pp. 261–275. Providence, RI: Amer. Math. Soc. cited on page(s): 252

Kollár, J. 1995. *Shafarevich Maps and Automorphic Forms*. M. B. Porter Lectures. Princeton, NJ: Princeton University Press. cited on page(s): 225, 264

Kollár, J. 1996. *Rational Curves on Algebraic Varieties*. Ergebnisse der Mathematik und ihrer Grenzgebiete. 3. Folge., Vol. 32. Berlin: Springer-Verlag. cited on page(s): 96, 255, 307, 310

Kollár, J. 1997. Singularities of pairs. In *Algebraic Geometry—Santa Cruz 1995*, J. Kollár, R. Lazarsfeld, and D. R. Morrison (eds). Proc. Sympos. Pure Math., Vol. 62, pp. 221–287. Providence, RI: American Mathematical Society. cited on page(s): 34, 43, 80, 251, 252, 263

Kollár, J. 2005. Einstein metrics on five-dimensional Seifert bundles. *J. Geom. Anal.*, **15**(3), 445–476. cited on page(s): 265

Kollár, J. 2007a. A conjecture of Ax and degenerations of Fano varieties. *Israel J. Math.*, **162**, 235–251. cited on page(s): 261, 265

Kollár, J. 2007b. Einstein metrics on connected sums of $S^2 \times S^3$. *J. Differential Geom.*, **75**(2), 259–272. cited on page(s): 265

Kollár, J. 2007c. Kodaira's canonical bundle formula and adjunction. In *Flips for 3-folds and 4-folds*. Oxford Lecture Ser. Math. Appl., Vol. 35, pp. 134–162. Oxford: Oxford University Press. cited on page(s): 170

Kollár, J. 2007d. *Lectures on Resolution of Singularities*. Annals of Mathematics Studies, Vol. 166. Princeton, NJ: Princeton University Press. cited on page(s): 324, 325, 326, 327, 330, 331, 332

Kollár, J. 2008a. Semi log resolutions. arXiv:0812.3592. cited on page(s): 328, 330, 331

Kollár, J. 2008b. Which powers of holomorphic functions are integrable? arXiv.org: 0805.0756. cited on page(s): 253

Kollár, J. 2009. Positive Sasakian structures on 5-manifolds. In *Riemannian Topology and Geometric Structures on Manifolds*. Progr. Math., Vol. 271, pp. 93–117. Boston, MA: Birkhäuser Boston. cited on page(s): 265

Kollár, J. 2010a. Continuous closure of sheaves. arXiv e-prints, Oct. cited on page(s): 308

Kollár, J. 2010b. Exercises in the birational geometry of algebraic varieties. In *Analytic and Algebraic Geometry*, J. McNeal and M. Mustaţă (eds), IAS/Park City Math. Ser., Vol. 17, pp. 495–524. Providence, RI: American Mathematical Society. cited on page(s): 49

Kollár, J. 2011a. A local version of the Kawamata-Viehweg vanishing theorem. Pure Appl. Math. Q., **7**(4, Special Issue: In memory of Eckart Viehweg), 1477–1494. cited on page(s): 242, 247, 323

Kollár, J. 2011b. New examples of terminal and log canonical singularities. arXiv:1107.2864. cited on page(s): 93, 129, 131

Kollár, J. 2011c. Seminormal log centers and deformations of pairs. arXiv.org:1103.0528. cited on page(s): 196, 234, 286, 312

Kollár, J. 2011d. Sources of log canonical centers. arXiv:1107.2863. cited on page(s): 151, 176, 179, 182, 195, 261

Kollár, J. 2011e. Two examples of surfaces with normal crossing singularities. *Sci. China Math.*, **54**(8), 1707–1712. cited on page(s): 32, 262, 292

Kollár, J. 2012a. Dual graphs of exceptional divisors. arXiv e-prints, Mar. cited on page(s): 132, 261

Kollár, J. 2012b. Moduli of varieties of general type. In *Handbook of Moduli. Advanced Lectures in Mathematics*, G. Farkas and I. Morrison (eds). Somerville MA: International Press. to appear. cited on page(s): 262, 263

Kollár, J. 2012c. Quotients by finite equivalence relations. Pages 227–256 of: In *Current Developments in Algebraic Geometry*. Math. Sci. Res. Inst. Publ., Vol. 59, pp. 227–256. Cambridge: Cambridge University Press. With an appendix by Claudiu Raicu. cited on page(s): 190, 208, 269, 271, 273, 278, 281, 282

Kollár, J. 2013. *Moduli of Varieties of General Type*. (book in preparation). cited on page(s): 262

Kollár, J. and Kovács, S. J. 2010. Log canonical singularities are Du Bois. *J. Amer. Math. Soc.*, **23**(3), 791–813. cited on page(s): 3, 151, 171, 172, 179, 196, 214, 224, 226, 228, 229, 255, 261, 262, 321

Kollár, J. and Larsen, M. 2009. Quotients of Calabi-Yau varieties. In *Algebra, Arithmetic, and Geometry: In Honor of Yu. I. Manin. Vol. II*, Y. Tschinkel and Y. Zarhin (eds). Progr. Math., Vol. 270, pp. 179–211. Boston, MA: Birkhäuser Boston Inc. cited on page(s): 106

Kollár, J. and Mori, S. 1992. Classification of three-dimensional flips. *J. Amer. Math. Soc.*, **5**(3), 533–703. cited on page(s): 305

Kollár, J. and Mori, S. 1998. *Birational Geometry of Algebraic Varieties*. Cambridge Tracts in Mathematics, Vol. 134. Cambridge: Cambridge University Press. With the collaboration of C. H. Clemens and A. Corti, Translated from the 1998 Japanese original. cited on page(s): ix, 4, 8, 9, 13, 16, 21, 31, 41, 42, 45, 50, 56, 61, 68, 76, 78, 82, 85, 91, 92, 96, 109, 143, 144, 161, 175, 225, 243, 244, 255, 318, 321

Kollár, J. and Shepherd-Barron, N. I. 1988. Threefolds and deformations of surface singularities. *Invent. Math.*, **91**(2), 299–338. cited on page(s): 87, 187, 262, 329

Kollár, J., Smith, K. E., and Corti, A. 2004. *Rational and Nearly Rational Varieties*. Cambridge Studies in Advanced Mathematics, Vol. 92. Cambridge: Cambridge University Press. cited on page(s): 106, 125, 129, 158, 175, 264

Kondō, S. 1992. Automorphisms of algebraic $K3$ surfaces which act trivially on Picard groups. *J. Math. Soc. Japan*, **44**(1), 75–98. cited on page(s): 97

Kovács, S. J. 1999. Rational, log canonical, Du Bois singularities: on the conjectures of Kollár and Steenbrink. *Compositio Math.*, **118**(2), 123–133. cited on page(s): 227, 228, 255

Kovács, S. J. 2000a. A characterization of rational singularities. *Duke Math. J.*, **102**(2), 187–191. cited on page(s): 83, 255

Kovács, S. J. 2000b. Rational, log canonical, Du Bois singularities. II. Kodaira vanishing and small deformations. *Compositio Math.*, **121**(3), 297–304. cited on page(s): 226, 255

Kovács, S. J. 2011a. Du Bois pairs and vanishing theorems. *Kyoto J. Math.*, **51**(1), 47–69. cited on page(s): 218, 222, 223, 227, 228

Kovács, S. J. 2011b. Irrational centers. Pure Appl. Math. Q., **7**(4, Special Issue: In memory of Eckart Viehweg), 1495–1515. cited on page(s): 80, 87

Kovács, S. J. 2012a. The intuitive definition of Du Bois singularities. In *Geometry and Arithmetics*, C. Faber, G. Farkas, and R. de Jong (eds), pp. 257–266. EMS Series on Congress Reports. European Mathematical Society. cited on page(s): 215

Kovács, S. J. 2012b. Singularities of stable varieties. In *Handbook of Moduli*, G. Farkas and I. Morrison (eds). Advanced Lectures in Mathematics. Somerville MA: International Press. to appear. cited on page(s): 8

Kovács, S. J. 2012c. The splitting principle and singularities. In *Compact Moduli Spaces and Vector Bundles*. Contemp. Math., Vol. 564, pp. 195–204. Providence, RI: American Mathematical Society. cited on page(s): 226, 227, 228

Kovács, S. J. and Schwede, K. 2011a. Du Bois singularities deform. arXiv:1107.2349. preprint. cited on page(s): 229

Kovács, S. J. and Schwede, K. E. 2011b. Hodge theory meets the minimal model program: a survey of log canonical and Du Bois singularities. In *Topology of Stratified Spaces*, G. Friedman, E. Hunsicker, A. Libgdeer, and L. Maxim (eds). Math. Sci. Res. Inst. Publ., Vol. 58, pp. 51–94. Cambridge: Cambridge University Press. cited on page(s): 215, 345

Kovács, S. J., Schwede, K., and Smith, K. E. 2010. The canonical sheaf of Du Bois singularities. *Adv. Math.*, **224**(4), 1618–1640. cited on page(s): 220

Kurke, H., Pfister, G., and Roczen, M. 1975. *Henselsche Ringe und algebraische Geometrie*. Mathematische Monographien, Band II. Berlin: VEB Deutscher Verlag der Wissenschaften. cited on page(s): 178

Küronya, A. 2003. A divisorial valuation with irrational volume. *J. Algebra*, **262**(2), 413–423. cited on page(s): 31

Laufer, H. B. 1977. On minimally elliptic singularities. *Amer. J. Math.*, **99**(6), 1257–1295. cited on page(s): 75

Lazarsfeld, R. 2004. *Positivity in Algebraic Geometry. I-II*. Ergebnisse der Mathematik und ihrer Grenzgebiete. 3. Folge., Vol. 48–49. Berlin: Springer-Verlag. cited on page(s): 23, 250, 256

Lipman, J. 1969. Rational singularities, with applications to algebraic surfaces and unique factorization. *Inst. Hautes Études Sci. Publ. Math.*, **36**, 195–279. cited on page(s): 26, 53, 109, 300, 301, 303

Łojasiewicz, S. 1958. Division d'une distribution par une fonction analytique de variables réelles. *C. R. Acad. Sci. Paris*, **246**, 683–686. cited on page(s): 251

Maeda, H. 1986. Classification of logarithmic Fano threefolds. *Compositio Math.*, **57**(1), 81–125. cited on page(s): 140

Matsumura, H. 1963. On algebraic groups of birational transformations. *Atti Accad. Naz. Lincei Rend. Cl. Sci. Fis. Mat. Natur. (8)*, **34**, 151–155. cited on page(s): 340

Matsumura, H. 1986. *Commutative Ring Theory*. Cambridge Studies in Advanced Mathematics, Vol. 8. Cambridge: Cambridge University Press. Translated from the Japanese by M. Reid. cited on page(s): 33, 77, 79, 282

McNeal, J. and Mustață, M. (eds). 2010. *Analytic and Algebraic Geometry*. IAS/Park City Mathematics Series, Vol. 17. Providence, RI: American Mathematical Society. cited on page(s): 250

Mehta, V. B. and Srinivas, V. 1991. Normal *F*-pure surface singularities. *J. Algebra*, **143**(1), 130–143. cited on page(s): 255

Mehta, V. B. and Srinivas, V. 1997. A characterization of rational singularities. *Asian J. Math.*, **1**(2), 249–271. cited on page(s): 256

Milne, J. S. 1980. *Étale Cohomology*. Princeton Mathematical Series, Vol. 33. Princeton, NJ: Princeton University Press. cited on page(s): 178

Miyaoka, Y. 1988. Abundance conjecture for 3-folds: case $\nu = 1$. *Compositio Math.*, **68**(2), 203–220. cited on page(s): 261

Mori, S. 1985. On 3-dimensional terminal singularities. *Nagoya Math. J.*, **98**, 43–66. cited on page(s): 76

Mori, S. 1987. Classification of higher-dimensional varieties. In *Algebraic Geometry, Bowdoin, 1985 (Brunswick, Maine, 1985)*. Proc. Sympos. Pure Math., Vol. 46, pp. 269–331. Providence, RI: American Mathematical Society. cited on page(s): 169

Mori, S. 1988. Flip theorem and the existence of minimal models for 3-folds. *J. Amer. Math. Soc.*, **1**(1), 117–253. cited on page(s): 26

Mori, S., Morrison, D. R., and Morrison, I. 1988. On four-dimensional terminal quotient singularities. *Math. Comp.*, **51**(184), 769–786. cited on page(s): 106

Morrison, D. R. 1985. Canonical quotient singularities in dimension three. *Proc. Amer. Math. Soc.*, **93**(3), 393–396. cited on page(s): 106

Morrison, D. R. and Stevens, G. 1984. Terminal quotient singularities in dimensions three and four. *Proc. Amer. Math. Soc.*, **90**(1), 15–20. cited on page(s): 106

Mumford, D. 1961. The topology of normal singularities of an algebraic surface and a criterion for simplicity. *Inst. Hautes Études Sci. Publ. Math.*, **9**, 5–22. cited on page(s): 61, 135, 260

Mumford, D. 1970. *Abelian Varieties*. Tata Institute of Fundamental Research Studies in Mathematics, No. 5. Bombay: Published for the Tata Institute of Fundamental Research. cited on page(s): 281

Mumford, D. 1976. *Algebraic Geometry. I.* Complex projective varieties, Grundlehren der Mathematischen Wissenschaften, Vol. 221. Berlin: Springer-Verlag. cited on page(s): 315

Mustaţă, M. 2002. Singularities of pairs via jet schemes. *J. Amer. Math. Soc.*, **15**(3), 599–615 (electronic). cited on page(s): 254

Mustaţă, M. 2010. Ordinary varieties and the comparison between multiplier ideals and test ideals II. arXiv:1012.2915. cited on page(s): 256

Mustaţă, M. and Srinivas, V. 2010. Ordinary varieties and the comparison between multiplier ideals and test ideals. arXiv:1012.2818. cited on page(s): 256

Nadel, A. M. 1990. Multiplier ideal sheaves and Kähler-Einstein metrics of positive scalar curvature. *Ann. of Math. (2)*, **132**(3), 549–596. cited on page(s): 264, 265

Nagata, M. 1965. *Lectures on the Fourteenth Problem of Hilbert*. Bombay: Tata Institute of Fundamental Research. cited on page(s): 148

Nakamura, I. and Ueno, K. 1973. An addition formula for Kodaira dimensions of analytic fibre bundles whose fibre are Moišezon manifolds. *J. Math. Soc. Japan*, **25**, 363–371. cited on page(s): 334

Nakayama, N. 1986. Invariance of the plurigenera of algebraic varieties under minimal model conjectures. *Topology*, **25**(2), 237–251. cited on page(s): 159

Nakayama, N. 1987. The lower semicontinuity of the plurigenera of complex varieties. In *Algebraic Geometry, Sendai, 1985*, T. Oda (ed.). Adv. Stud. Pure Math., Vol. 10, pp. 551–590. Amsterdam: North-Holland. cited on page(s): 317

Namikawa, Y. 2001. Extension of 2-forms and symplectic varieties. *J. Reine Angew. Math.*, **539**, 123–147. cited on page(s): 257

Odaka, Y. and Xu, C. 2012. Log-canonical models of singular pairs and its applications. *Math. Res. Lett.*, **19**(2), 325–334. cited on page(s): 28, 33

Ogoma, T. 1983. Descent of P-property by proper surjective morphisms. *Nagoya Math. J.*, **92**, 175–177. cited on page(s): 282

Orlik, P. and Wagreich, P. 1975. Seifert *n*-manifolds. *Invent. Math.*, **28**, 137–159. cited on page(s): 289

Patakfalvi, Z. 2010. Base change for the relative canonical sheaf in families of normal varieties. arXiv e-prints:1005.5207. cited on page(s): 243

Păun, M. 2010. Quantitative extensions of twisted pluricanonical forms and non-vanishing. In *Proceedings of the International Congress of Mathematicians. Volume II*, R. Bhaita (ed.), pp. 540–557. New Delhi: Hindustan Book Agency. cited on page(s): 26

Peters, C. A. M. and Steenbrink, J. H. M. 2008. *Mixed Hodge Structures*. Ergebnisse der Mathematik und ihrer Grenzgebiete. 3. Folge., Vol. 52. Berlin: Springer-Verlag. cited on page(s): 215

Pinkham, H. 1977. Normal surface singularities with C^* action. *Math. Ann.*, **227**(2), 183–193. cited on page(s): 289

Pukhlikov, A. V. 2007. Birationally rigid varieties. I. Fano varieties. *Uspekhi Mat. Nauk*, **62**(5(377)), 15–106. cited on page(s): 264

Pukhlikov, A. V. 2010. Birationally rigid varieties. II. Fano fibrations. *Uspekhi Mat. Nauk*, **65**(6(396)), 87–180. cited on page(s): 264

Raoult, J.-C. 1974. Compactification des espaces algébriques. *C. R. Acad. Sci. Paris Sér. A*, **278**, 867–869. cited on page(s): 282

Reid, M. 1976. Elliptic Gorenstein singularities of surfaces. Unpublished preprint. cited on page(s): 75

Reid, M. 1980. Canonical 3-folds. In *Journées de Géometrie Algébrique d'Angers, Juillet 1979/Algebraic Geometry, Angers, 1979*. Alphen aan den Rijn: Sijthoff & Noordhoff. cited on page(s): 14, 37, 65, 73, 74, 75, 105, 130

Reid, M. 1983. Minimal models of canonical 3-folds. In *Algebraic Varieties and Analytic Varieties (Tokyo, 1981)*, S. Iitaka (ed.). Adv. Stud. Pure Math., Vol. 1, pp. 131–180. Amsterdam: North-Holland. cited on page(s): 74

Reid, M. 1987. Young person's guide to canonical singularities. In *Algebraic Geometry, Bowdoin, 1985 (Brunswick, Maine, 1985)*, C. H. Clemens and S. Bloch (eds). Proc. Sympos. Pure Math., Vol. 46, pp. 345–414. Providence, RI: American Mathematical Society. cited on page(s): 37, 76, 102, 106, 130, 257

Reid, M. 2002. La correspondance de McKay. *Astérisque*, **276**, 53–72. Séminaire Bourbaki, Vol. 1999/2000. cited on page(s): 104

Saito, M. 2000. Mixed Hodge complexes on algebraic varieties. *Math. Ann.*, **316**(2), 283–331. cited on page(s): 255

Schlessinger, M. 1971. Rigidity of quotient singularities. *Invent. Math.*, **14**, 17–26. cited on page(s): 102

Schwartz, L. 1950. *Théorie des Distributions. Tome I*. Actualités Sci. Ind., no. 1091 = Publ. Inst. Math. Univ. Strasbourg 9. Paris: Hermann & Cie. cited on page(s): 251

Schwede, K. 2007. A simple characterization of Du Bois singularities. *Compos. Math.*, **143**(4), 813–828. cited on page(s): 215

Schwede, K. 2009. *F*-injective singularities are Du Bois. *Amer. J. Math.*, **131**(2), 445–473. cited on page(s): 256

Schwede, K. and Takagi, S. 2008. Rational singularities associated to pairs. *Michigan Math. J.*, **57**, 625–658. Special volume in honor of Melvin Hochster. cited on page(s): 87

Schwede, K. and Tucker, K. 2012. A survey of test ideals. In *Progress in Commutative Algebra 2. Closures, Finiteness and Factorization*, C. Francisco, L. C. Klinger, S. M. Sather-Wagstaff, and J. C. Vassiler (eds), pp. 39–99. Berlin: Walter de Gruyter GmbH & Co. KG. cited on page(s): 256

SGA3. 1970. *Schémas en Groupes. I: Propriétés Générales des Schémas en Groupes.* Séminaire de Géométrie Algébrique du Bois Marie 1962/64 (SGA 3). Dirigé par M. Demazure et A. Grothendieck. Lecture Notes in Mathematics, Vol. 151. Berlin: Springer-Verlag. cited on page(s): 289

Shafarevich, I. R. 1966. *Lectures on Minimal Models and Birational Transformations of Two Dimensional Schemes*. Notes by C. P. Ramanujam. Tata Institute of Fundamental Research Lectures on Mathematics and Physics, Vol. 37. Bombay: Tata Institute of Fundamental Research. cited on page(s): 26, 53

Shafarevich, I. R. 1974. *Basic Algebraic Geometry*. Translated from the Russian by K. A. Hirsch, Die Grundlehren der mathematischen Wissenschaften, Band 213. New York: Springer-Verlag. cited on page(s): 281

Shephard, G. C. and Todd, J. A. 1954. Finite unitary reflection groups. *Canadian J. Math.*, **6**, 274–304. cited on page(s): 103, 107

Shibuta, T. and Takagi, S. 2009. Log canonical thresholds of binomial ideals. *Manuscripta Math.*, **130**(1), 45–61. cited on page(s): 256

Shokurov, V. V. 1988. Problems about Fano varieties. *Birational Geometry of Algebraic Varieties: Open Problems*. Taniguchi Kogyo Shoreikai. cited on page(s): 252, 253

Shokurov, V. V. 1992. Three-dimensional log perestroikas. *Izv. Ross. Akad. Nauk Ser. Mat.*, **56**(1), 105–203. cited on page(s): 26, 151, 158, 172

Shokurov, V. V. 2006. *Letters of a Bi-Rationalist: VII. Ordered termination.* arXiv.org:math/0607822. cited on page(s): 26

Singh, A. K. 2003. Cyclic covers of rings with rational singularities. *Trans. Amer. Math. Soc.*, **355**(3), 1009–1024 (electronic). cited on page(s): 96

Siu, Y.-T. 2008. Finite generation of canonical ring by analytic method. *Sci. China Ser. A*, **51**(4), 481–502. cited on page(s): 26, 75

Sloane, N. J. A. 2003. *The On-Line Encyclopedia of Integer Sequences*. AT&T Research. www.research.att.com/ njas/sequences/. cited on page(s): 252

Smith, K. E. 1997. *F*-rational rings have rational singularities. *Amer. J. Math.*, **119**(1), 159–180. cited on page(s): 255

Smith, K. E. 2000. The multiplier ideal is a universal test ideal. *Comm. Algebra*, **28**(12), 5915–5929. Special issue in honor of Robin Hartshorne. cited on page(s): 256

Smith, L. 1995. *Polynomial Invariants of Finite Groups*. Research Notes in Mathematics, vol. 6. Wellesley, MA: A K Peters Ltd. cited on page(s): 103

Smith, R. and Varley, R. 1996. Multiplicity g points on theta divisors. *Duke Math. J.*, **82**(2), 319–326. cited on page(s): 264

Steenbrink, J. H. M. 1985. Vanishing theorems on singular spaces. *Astérisque*, **130**, 330–341. Differential systems and singularities (Luminy, 1983). cited on page(s): 215, 217

Stepanov, D. A. 2008. A note on resolution of rational and hypersurface singularities. *Proc. Amer. Math. Soc.*, **136**(8), 2647–2654. cited on page(s): 260

Stevens, J. 1998. Degenerations of elliptic curves and equations for cusp singularities. *Math. Ann.*, **311**(2), 199–222. cited on page(s): 76

Szabó, E. 1994. Divisorial log terminal singularities. *J. Math. Sci. Univ. Tokyo*, **1**(3), 631–639. cited on page(s): 324

Takagi, S. 2004. An interpretation of multiplier ideals via tight closure. *J. Algebraic Geom.*, **13**(2), 393–415. cited on page(s): 256

Takagi, S. 2008. A characteristic p analogue of plt singularities and adjoint ideals. *Math. Z.*, **259**(2), 321–341. cited on page(s): 256

Takagi, S. 2011. Adjoint ideals and a correspondence between log canonicity and F-purity. arXiv:1105.0072. cited on page(s): 256

Takagi, S. and Watanabe, K. 2004. On F-pure thresholds. *J. Algebra*, **282**(1), 278–297. cited on page(s): 256

Takayama, S. 2003. Local simple connectedness of resolutions of log-terminal singularities. *Internat. J. Math.*, **14**(8), 825–836. cited on page(s): 130, 259, 260

Thuillier, A. 2007. Géométrie toroïdale et géométrie analytique non archimédienne. Application au type d'homotopie de certains schémas formels. *Manuscripta Math.*, **123**(4), 381–451. cited on page(s): 260

Totaro, B. 2010. The ACC conjecture for log canonical thresholds, (after de Fernex, Ein, Mustaţă, Kollár). *Asterisque*, **1025**, Sem. Bourbaki, 2009–2010. cited on page(s): 253

Traverso, C. 1970. Seminormality and Picard group. *Ann. Scuola Norm. Sup. Pisa (3)*, **24**, 585–595. cited on page(s): 306

Tsuji, H. 1994. Global generation of adjoint bundles. Unpublished preprint. cited on page(s): 264

Ueno, K. 1975. *Classification Theory of Algebraic Varieties and Compact Complex Spaces*. Lecture Notes in Mathematics, Vol. 439. Berlin: Springer-Verlag. Notes written in collaboration with P. Cherenack. cited on page(s): 169, 334

van Straten, D. and Steenbrink, J. 1985. Extendability of holomorphic differential forms near isolated hypersurface singularities. *Abh. Math. Sem. Univ. Hamburg*, **55**, 97–110. cited on page(s): 257

Varčenko, A. N. 1976. Newton polyhedra and estimates of oscillatory integrals. *Funkcional. Anal. i Priložen.*, **10**(3), 13–38. cited on page(s): 251

Włodarczyk, J. 2005. Simple Hironaka resolution in characteristic zero. *J. Amer. Math. Soc.*, **18**(4), 779–822 (electronic). cited on page(s): 331

Wolf, J. A. 1967. *Spaces of Constant Curvature*. New York: McGraw-Hill Book Co. cited on page(s): 107

Xu, C. 2012. Finiteness of algebraic fundamental groups. arXiv:1210.5564. cited on page(s): 260

Index

Printed in the United States
By Bookmasters